T0318746

Human Papillomavirus

Proving and Using a Viral Cause for Cancer

Human Papillomavirus

Proving and Using a Viral Cause for Cancer

Edited by

David Jenkins

Emeritus Professor of Pathology, University of Nottingham, United Kingdom
Consultant in Pathology to DDL Diagnostic laboratories, Rijswijk, The Netherlands

F. Xavier Bosch

Cancer Epidemiology Research Program, Catalan Institute of Oncology, Barcelona, Spain

Academic Press is an imprint of Elsevier
125 London Wall, London EC2Y 5AS, United Kingdom
525 B Street, Suite 1650, San Diego, CA 92101, United States
50 Hampshire Street, 5th Floor, Cambridge, MA 02139, United States
The Boulevard, Langford Lane, Kidlington, Oxford OX5 1GB, United Kingdom

British Library Cataloguing-in-Publication Data
A catalogue record for this book is available from the British Library

Library of Congress Cataloging-in-Publication Data
A catalog record for this book is available from the Library of Congress

ISBN: 978-0-12-814457-2

For Information on all Academic Press publications
visit our website at https://www.elsevier.com/books-and-journals

Publisher: Andre Wolff
Acquisition Editor: Kattie Washington
Editorial Project Manager: Leticia M. Lima
Production Project Manager: Maria Bernard
Cover Designer: Miles Hitchen

Typeset by MPS Limited, Chennai, India

Contents

SECTION 2 HUMAN PAPILLOMAVIRUS BEYOND CERVICAL CANCER

List of Contributors

Laia Alemany
Cancer Epidemiology Research Program, Catalan Institute of Oncology (ICO) - Bellvitge Biomedical Research Institute (IDIBELL), Barcelona, Spain; CIBER en epidemiología y salud pública CIBERESP, Barcelona, Spain

Marc Arbyn
Unit of Cancer Epidemiology, Belgian Cancer Centre, Sciensano, Brussels, Belgium

Harshita Beeravolu
Department of Developmental, Molecular and Chemical Biology, Tufts University School of Medicine, Boston, MA, United States

Christine Bergeron
Department of Pathology, Laboratoire Cerba, Cergy Pontoise, France; Laboratoire Cerba, Parc d'activité "les Béthunes" 95310 Saint Ouen L'aumone, Paris, France

Johannes Berkhof
Department of Epidemiology and Biostatistics, Amsterdam UMC, Vrije Universiteit, Amsterdam, The Netherlands

Neerja Bhatla
Department of Obstetrics and Gynaecology, All India Institute of Medical Sciences, New Delhi, India

F. Xavier Bosch
Catalan institute of Oncology (ICO), Barcelona, Spain; Bellvitge Research Institute (IDIBELL), Barcelona, Spain; Open University of Catalonia (UOC), Barcelona, Spain

Thomas R. Broker
Biochemistry & Molecular Genetics, University of Alabama, AL, United States

Laia Bruni
Cancer Epidemiology Research Program, Catalan Institute of Oncology (ICO) - Bellvitge Biomedical Research Institute (IDIBELL), Barcelona, Spain; CIBER en oncología CIBERONC, Madrid, Spain

Veronica Canarte
Department of Developmental, Molecular and Chemical Biology, Tufts University School of Medicine, Boston, MA, United States; Post-Baccalaureate Research and Education Program, Tufts University School of Medicine, Boston, MA, United States

Karen Canfell
Cancer Research Division, Cancer Council NSW, Sydney, NSW, Australia; Prince of Wales Clinical School, The University of New South Wales, Sydney, NSW, Australia; School of Public Health, University of Sydney, Sydney, NSW, Australia

Brenda Corcoran
National Immunisation Office (formerly), Dublin, Ireland

Heather Cubie
Global Health Academy, University of Edinburgh, Edinburgh, United Kingdom

Kate Cuschieri
Scottish HPV Reference Laboratory, Royal Infirmary of Edinburgh, United Kingdom

J. Cuzick
Wolfson Institute of Preventive Medicine, Queen Mary University of London, London, United Kingdom

Lynette Denny
Department of Obstetrics and Gynaecology, University of Cape Town/Groote Schuur Hospital, Cape Town, South Africa; South African Medical Research Council Gynaecology Cancer Research Centre, University of Cape Town, Cape Town, South Africa

John Doorbar
Department of Pathology, University of Cambridge, Cambridge, United Kingdom

Tapati Dutta
Department of Applied Health Science, Prevention Insights, Rural Center for AIDS/STD Prevention, Indiana University Bloomington School of Public Health, Bloomington, IN, United States

Carole Fakhry
Department of Otolaryngology-Head and Neck Surgery, Johns Hopkins University School of Medicine, Baltimore, MD, United States; Department of Epidemiology, Johns Hopkins Bloomberg School of Public Health, Baltimore, MD, United States

Farhoud Faraji
Division of Otolaryngology-Head and Neck Surgery, University of California San Diego School of Medicine, San Diego, CA, United States

Alice Forster
Research Department of Behavioural Science and Health, University College London Institute of Epidemiology and Health Care, London, United Kingdom

Silva Franceschi
Aviano Cancer Center (CRO) IRCCS, Aviano, Italy

Eduardo L. Franco
McGill University, Montreal, QC, Canada

Suzanne M. Garland
Department of Obstetrics and Gynaecology, University of Melbourne, VIC, Australia; Director Centre Women's Infectious Diseases Research, VIC, Australia; Honorary Research Fellow, Infection & Immunity, Murdoch Children's Research Institute, VIC, Australia

Daan Geraets
DDL Diagnostic Laboratory, Rijswijk, The Netherlands

Anna R. Giuliano
Center for Immunization and Infection Research in Cancer (CIIRC), Moffitt Cancer Center and Research Institute, Tampa, FL, United States

Miranda Grace
Department of Developmental, Molecular and Chemical Biology, Tufts University School of Medicine, Boston, MA, United States

Nuria Guimera
DDL Diagnostic Laboratory, Rijswijk, The Netherlands

Sharon Hanley
Department of Reproductive and Developmental Medicine, Hokkaido University Graduate School of Medicine, Sapporo, Japan

D.A.M. Heideman
Amsterdam UMC, Vrije Universiteit Amsterdam, Pathology, Cancer Center Amsterdam, Amsterdam, Netherlands

David Jenkins
Emeritus Professor of Pathology, University of Nottingham, United Kingdom; Consultant in Pathology to DDL Diagnostic laboratories, Rijswijk, The Netherlands

Elmar Joura
Department of Gynecology and Obstetrics, Medical University of Vienna-AKHComprehensive Cancer Center Vienna, Vienna, Austria

W.W. Kremer
Amsterdam UMC, Vrije Universiteit Amsterdam, Pathology, Cancer Center Amsterdam, Amsterdam, Netherlands

Charles J.N. Lacey
York Biomedical Research Institute, University of York, York, United Kingdom

Annemiek Leeman
DDL laboratories, Rijswijk, the Netherlands

Attila Lorincz
Wolfson Institute of Preventive Medicine, Queen Mary University of London, London, United Kingdom

Lauri E. Markowitz
Associate Director for Science, HPV National Center for Immunizations and Respiratory Diseases, Centers for Disease Control and Prevention, Atlanta, GA, United States

Chris J.L.M. Meijer
Amsterdam UMC, Vrije Universiteit Amsterdam, Pathology, Cancer Center Amsterdam, Amsterdam, The Netherlands

Beth E. Meyerson
Department of Applied Health Science, Rural Center for AIDS/STD Prevention, Indiana University Bloomington School of Public Health, Bloomington, IN, United States

Anna-Barbara Moscicki
Division of Adolescent and Young Adult Medicine, Department of Pediatrics, University of California, Los Angeles, Los Angeles, CA, United States

Karl Munger
Department of Developmental, Molecular and Chemical Biology, Tufts University School of Medicine, Boston, MA, United States; Molecular Microbiology Ph.D. Program, Tufts University School of Medicine, Boston, MA, United States; Post-Baccalaureate Research and Education Program, Tufts University School of Medicine, Boston, MA, United States; Biochemistry Ph.D. Program, Tufts University School of Medicine, Boston, MA, United States

Nubia Muñoz
Cancer Institute of Colombia, Bogotá, Republic of Colombia

Jorma Paavonen
University of Helsinki, Helsinki, Finland

Joel M. Palefsky
Professor of Medicine, University of California, San Francisco, CA, United States

Kevin G. Pollock
School of Health and Life Sciences, Glasgow Caledonian University, Glasgow, Scotland

Wim Quint
DDL Diagnostic Laboratory, Rijswijk, The Netherlands

G. Ronco
International Agency for Research on Cancer, Lyon, France

Mark Schiffman
Division of Cancer Epidemiology and Genetics, National Cancer Institute, Bethesda, MD, United States

Surendra Sharma
Department of Developmental, Molecular and Chemical Biology, Tufts University School of Medicine, Boston, MA, United States; Biochemistry Ph.D. Program, Tufts University School of Medicine, Boston, MA, United States

Albert Singer
Gynecological Research, University College, London, United Kingdom

Margaret Stanley
Department of Pathology, University of Cambridge, Tennis Court Road, Cambridge, United Kingdom

Mark H. Stoler
Professor (Emeritus) of Pathology and Clinical Gynecology, Associate Director of Surgical and Cytopathology, Department of Pathology, University of Virginia Health Health System, Charlottesville, VA, United States

Magnus von Knebel Doeberitz
Department of Applied Tumor Biology, Institute of Pathology, Heidelberg University Hospital, Heidelberg, Germany

Cosette Marie Wheeler
University of New Mexico Comprehensive Cancer Center, Albuquerque, NM, United States

Sharon C. Wu
Department of Developmental, Molecular and Chemical Biology, Tufts University School of Medicine, Boston, MA, United States; Molecular Microbiology Ph.D. Program, Tufts University School of Medicine, Boston, MA, United States

Gregory D. Zimet
Department of Pediatrics, Indiana University School of Medicine, Indianapolis, IN, United States

Preface

Science, including medical science, works to very strict rules of experiment and evidence to ensure as much as possible that the findings can be applicable across the world. But, it is also a creative, very human activity depending on individual insight and inspiration, and a social activity in which ideas bounce from one researcher to another, developing or withering in the process. The development of science is largely about collaboration, increasingly on a very international scale, but there is also argument and dispute as any scientific meeting will show. Inspired insights are important but there is often a history behind these changes of view. The ideal is that the results should be reliable and reproducible, and improve practice through better accuracy of diagnosis, selection for treatment or new prevention, or treatment. This ideal is not easy or always achievable. Medical science of imperfect reliability, depending on personal skill and interpretation, like morphological pathology can remain a key medical tool for over 100 years. The use of microscopy like clinical diagnosis or colposcopy of the cervix requires expert skill to achieve some success. Skill is also important as medical science has to be fitted to the demands and person of the individual patient, and to that of society, competing with other priorities in politics. Prevention whether through screening or vaccination requires large scale organization. It can be very costly and the economics of medicine and research and development cannot escape the political background of different societies.

This book "The HPV story: finding and using the viral cause of important cancers" is the story of the scientific, clinical, and the epidemiological and other public health research behind the demonstration of the importance of the human papillomavirus as a cause of human cancer. It also tells how the development of tests and vaccines has changed the understanding of cervical and a few other cancers and provided new means of prevention more effective than those previously available. The story is told by the researchers involved and as these are all scientists of different specialisms the story is carefully referenced, but the stories are told in their own different ways. A few vignettes provide a very personal insight into the science at different stages. The book, however, also aims to provide evidence-based information on the history of HPV-related cancers which is necessary for those currently working in the field. The evidence of the role of HPV and the benefits of new testing and vaccine tools in cancer prevention is overwhelming to convince any sceptics open to evidence of the importance of spreading these new tools to countries beyond the privileged few in the Western world.

Much of the early work was driven by and centered on cervical cancer, finding and proving the infectious cause, and improving its prevention by taking advantage of the key role of HPV in adding HPV testing to screening. HPV's cause other disease, not all cancer, and also other cancers, particularly anal cancer, vulval cancer, penile, and scrotal cancer but also some head and neck cancers. These have become more important in relation to potential cancer prevention with the development of prophylactic HPV vaccines.

This is not a textbook or a review of every latest development, it is the science story told by some of the key researchers involved, many of whom have worked in HPV research for many years. This is not to say that there are many important researchers who have not been able to contribute. Many are still very busy with current projects. Over the years the number of researchers has grown from a few tens in Europe and the USA to many thousands globally. There are a number of important international societies that have contributed to collaborative research and the spread of knowledge. We recognize their important contributions and those of many other who have worked for a while in this field of research, or who have died in the 40 years since research on HPV started to take off following the key research of Harald zur Hausen, Lutz Gissmann, Matthias Durst, and others in Heidelberg, for which zur Hausen was awarded a Nobel prize.

All the opinions expressed in the chapters and in the other sections are the authors' or editors' own views, recognizing that we cannot be totally inclusive or comprehensive. The book is dedicated to all who have been and are now involved in HPV research and to those who will carry on the progression to global prevention of cervical cancer and other HPV-related cancers. It is important to remind the world that cervical cancer is still as important in the developing world as it has been historically in Europe. We would also thank Lionel Crawford for his contribution to the account of early studies of HPV and other viruses, and who provided access to the vignette of the story of Jian Zhou in the development of virus-like particles used in the HPV vaccines.

Without the enthusiastic support of all the authors and the contribution of the editorial staff at Elsevier, this project for telling the story of HPV science would have been unachievable. David Jenkins has received particular support and help from Wim Quint, Anco Molijn, Nuria Guimera and Annemiek Leeman at DDL in the Netherlands. He is also very grateful for the consistent unflinching support from his wife, Pauline, and tolerance by his family in the UK. Albert Singer kindly provided a vignette of the impact of the discovery of HPV in cervical cancer on gynaecology research in the UK in the 1980's and was responsible for involving David Jenkins in this area of research from an early stage.

General Introduction—The Background to Human Papillomavirus and Cancer Research

David Jenkins[1,2] and F. Xavier Bosch[3]

[1]Emeritus Professor of Pathology, University of Nottingham, United Kingdom [2]Consultant in Pathology to DDL Diagnostic laboratories, Rijswijk, The Netherlands [3]Catalan institute of Oncology (ICO), Barcelona, Spain

WHY WORRY ABOUT HUMAN PAPILLOMAVIRUS?

In 2018 the World Health Organization announced its aim to eliminate cervical cancer and made a global call for action [1]. Success with this would result in the disappearance of a cancer with a global incidence of around 530,000 cases per year, approximately half of whom will die of the disease. Cervical cancer is almost entirely caused by a group of closely related small DNA viruses with a tightly organized little genome of only 8 kb, which have been around for millions of years and found a very special ecological niche in the human anogenital region, exploiting the power of sex. Prevention of all human papillomavirus (HPV)-related cancers would also remove another 80,000 cancers of anogenital and oropharyngeal origin globally, accounting for about 5% of all cancers [2]. This is a small but important contribution to eliminating cancer, especially for the low- and middle-income countries where there is no effective screening for cervical cancer.

Australia was one of the first countries to introduce a national HPV vaccination program in 2007 and moved from cytology to primary HPV testing for cervical screening in 2017. In 2018 Australian mathematical modelers, using an extensively validated dynamic model of the natural history of sexually transmitted HPV, and the subsequent development of cervical cancer and the impact of cervical screening and prevention of HPV infection by vaccination, predicted that Australia could reduce cervical cancer below a threshold for considering a cancer as "rare" of 6 new cases per 100,000 women annually and by 2034 below one death per 100,000 women [3]. The WHO and Australian statements show the extent of knowledge about HPV, cancer, and cancer prevention built up over the 40 years since the start of discoveries about HPV in cervical cancer.

THE BOOK

This book aims to tell the story of how the science and medical use of HPV developed through the accounts of the scientists, laboratory-based, clinical researchers, and epidemiologists, developed this knowledge. The book is part history of the research into HPV and cancer but also a review of the knowledge built up over the years intended for all workers from different backgrounds who are involved in putting the knowledge and the new tools into global use. The book follows each scientific discipline involved and its development: each chapter is written by experts who have made major contributions to research into HPV.

This introduction provides an overview of some major events. The story of HPV is an intriguing example of how modern medical science can work extremely well under the right circumstances, how a discovery can be developed through the efforts of a small number of enthusiasts and then a growing number of scientific and clinical researchers, and epidemiologists, and also the role of those seeking to exploit business opportunities with diagnostics, instruments, and vaccines once the basic evidence of the importance of the role of HPV had been discovered. None of the research has happened in isolation, and the focus of research has moved around between basic science, pathology, epidemiology, clinical research, mathematical modeling, and health economics at different times. Importantly, the numbers of scientists involved shows how research expanded from the interests of a small number of enthusiasts coming together from different scientific and medical disciplines in the 1980s, and often seen as eccentric and misdirected by their colleagues to huge clinical and epidemiological studies in the 1990s and the first 10 years of the second millennium, involving tens of thousands of subjects, thousands of clinicians, and hundreds of local investigators.

In many ways the development of HPV science follows the pattern set out in the development of scientific medicine through the earliest study of bacterial and viral infections and their prevention in the 19th century by Pasteur, Koch, and others. It has its heroes in those who first disclosed the role of HPV in Harald zur Hausen's laboratory, but this not only owed its success to the Pasteurian maxim "Fortune favors the prepared mind" but followed a long chain of plausible but ultimately irrelevant potential infectious causes of cervical cancer from parasites to Herpes virus. The finding of HPV depended on what was then new technology of Southern blotting and remained of interest and fascination to a small number of enthusiasts until the almost universal role of HPV in cervical cancer was established. Later the story was repeated in the development of vaccines based on virus-like particles.

As the French sociologist of science Bruno Latour demonstrated in "The Pasteurization of France," it is the network of forces set up by the many different clinicians, public health workers, and scientists responding to the excitement of a particular finding that translates a piece of science into a useful social change [4]. This is what makes the important medical intervention different from the 99% of biomedical science that remains of limited value and interest scientifically and to health providers.

FIRST CLASS MEDICAL SCIENCE IS NOT THE ONLY ISSUE

The scientific achievement has to be tempered by the recognition that successful prevention of HPV-related cancers is only within immediate grasp in a few high-income countries with organized and well-funded cervical screening and HPV vaccination programs. In particular it has been very, or at least partially and usefully successful in countries in Northern Europe, such as the Netherlands, Scandinavia, and the United Kingdom, and in the United States, Canada, and

Australia. But even in wealthy countries, there have been problems around changing technology for screening and around public perceptions of vaccine safety, with reasonable concerns, but also sometimes anger driven by deliberate misinformation.

The big problem for middle- and low-income countries remains finance for preventive programs. This is complicated by the cost and availability of current vaccines and optimal screening technology. Designing programs of prevention that are adapted to the different countries needs and resources is an important issue. There are also the difficulties of obtaining and maintaining good data on which to base such programs, and ensuring that governments, public health experts, clinicians, and people in general, are informed and motivated to take part in the prevention of HPV-related cancer.

The issue of ensuring participation not only of health professionals but of women around the world and of political leaders has become increasingly dominant as the principles of prevention of HPV-related disease through screening and vaccination have been established. The importance of maintaining enthusiasm has been seen in the declining participation among young women in screening in the United Kingdom despite its success, related possibly to concerns about body image and exposure to others during the screening procedures as well as decreasing awareness of cervical cancer as a result of the effectiveness of screening and vaccination. Some of these problems may be solved by new technology such as self-sampling. In the United Kingdom and the Netherlands, important organizational problems have come up about managing and coordinating the change in the technology of primary screening from cervical cytology to HPV testing and the partial redundancy of cytological skills.

The naming by WHO in 2019 of vaccine hesitancy as one of the top 10 threats to global health reflects the importance of the antivaccine movement, and the personal and emotional conflict and uncertainty it has generated, despite the absence of any balanced scientific base for concerns. The success of HPV-based prevention in some countries mixed with the issues limiting its application is a part of the long history of how medicine and science have interacted with society. These issues have often been important in a disease that is fundamentally linked to sexual relationships and the fashions, customs, and taboos that surround these relationships.

CERVICAL CANCER WAS THE DRIVER OF HUMAN PAPILLOMAVIRUS RESEARCH

Although it is now recognized that HPV causes important disease beyond the female cervix, vagina, and vulva, producing anal neoplasia in men and women and also a large fraction of penile and scrotal cancer in men and oropharyngeal cancer, it is cervical cancer prevention that has driven most research. This book begins by telling the story of cervical cancer in Chapter 1, A Brief History of Cervical Cancer, prior to the first recognition of the potential importance of HPV. It was only with the development of microscopic pathology in the mid-19th century that cancer was defined as a specific process of abnormal cell growth. Before this, the effects of a cancer with an obvious and unpleasant nature producing pain, unpleasant vaginal discharge, bleeding, debility, and ultimately early death and its effects on relationships between male and female have been recorded since the time of classical Greece and Rome.

The development of cervical cancer prevention through early detection, then screening and treatment of precancer from the 1940s provided the basis on which clinicopathological and epidemiological understanding of the relation between HPV and cervical cancer could grow.

FINDING THE ROLE OF HUMAN PAPILLOMAVIRUS IN CERVICAL CANCER

The story of HPV is taken up in Chapter 2, Linking Human Papillomavirus to Human Cancer and Understanding the Carcinogenetic Mechanisms, by Magnus von Knebel Doeberitz from the Heidelberg group. The finding of HPV16 in Heidelberg in 1983 using Southern blot hybridization with HPV11 DNA [5] was the key opening HPV science of cancer. This followed the hypothesis put forward by Harald zur Hausen [6], initially in 1972, based on anecdotal reports linking genital warts to squamous carcinomas and supported by finding koilocytes as cytological evidence of HPV infection in cervical dysplasia [7]. All this had been preceded by a long history of a possible link between sexual behavior and cervical cancer, and by study of papillomavirus-related cancers in animals, and failure to demonstrate the role of other transmissible factors including sperm DNA and herpes virus. Importantly, some of the enthusiasts for other factors recognized the importance of HPV at an early stage and formed important collaborations that played a part in establishing in different countries and settings the role of HPV in cervical and other cancers and precancers.

This research in Heidelberg is described in Chapter 2, Linking Human Papillomavirus to Human Cancer and Understanding Its Carcinogenic Mechanisms, along with important preceding work on the nature of papovaviruses (polyoma, papilloma, and vacuolating agents) by Lionel Crawford in the 1960s in the United Kingdom and his studies of the transforming power of viral proteins that led to the discovery of the significance of p53 protein in 1979 [8]. The work in Heidelberg earned a Nobel prize for Harald zur Hausen and was the key initiator of a developing pattern of research that sought initially to understand how the presence of the recognizable HPV types 6, 11, 16, and then 18 drove the development of cervical dysplasia with either resolution of infection or an established precancer which, if left untreated, could lead to cancer. This was followed by detailed study in many laboratories and clinics of how certain HPV types led to progression through the stages leading to cervical cancer, at the molecular, cellular, microscopic tissue, and clinical levels described further in Chapters 2–6. An example of how small groups of enthusiasts set up research projects in different countries during the early 1980's confirming the role of HPV is given in the vignette by Albert Singer at the end of Chapter 2. Chapter 9, The Natural History of Human Papillomavirus Infection in Relation to Cervical Cancer, describes how the natural history of HPV infection, and the ease of transmission of HPV by skin contact in sexual activity was studied.

PROVING THE CAUSAL ROLE OF HUMAN PAPILLOMAVIRUS IN CERVICAL CANCER

General scientific acceptance of the key role of HPV16 and a small number of other HPV types in cervical and other cancers was not always made easily as it went against a view that had been established for many years, that "solid" cancers were one of the "lifestyle" diseases associated with modern, industrialized society. Chapter 8, Proving the Causal Role of Human Papillomavirus in

Cervical Cancer: A Tale of Multidisciplinary Science, shows how the role of some HPVs as class 1 carcinogens was accepted by the World Health Organization cancer research arm, IARC. The 1999 Walboomers paper supported the role of HPV as an (almost) necessary cause of cervical cancer. It became clear eventually with extensive molecular epidemiological studies globally that a specific group of "high-risk" HPVs were the key, causal, risk factor, "necessary but not sufficient" in over 95%–97% of cervical cancer. Subsequent work has shown a very small contributions from some "low-risk HPVs" that were either infrequent or of low carcinogenic activity, and a small group of cervical adenocarcinomas driven by germline or other somatic mutations, once misdiagnoses, were excluded as discussed in Chapter 5, Biology of the Human Papillomavirus Life Cycle: The Basis for Understanding the Pathology of PreCancer and Cancer, Chapter 6, The Pathology of Cervical Precancer and Cancer, and Chapter 10, Low-Risk Human Papillomavirus—Genital Warts, Cancer, and Laryngeal Papillomatosis.

THE IMPORTANCE OF HUMAN PAPILLOMAVIRUS DNA TESTING FOR CLINICAL RESEARCH INTO HUMAN PAPILLOMAVIRUS AND CANCER AND ITS PREVENTION

The development of clinical research into HPV and cancer has depended on continuing improvements in the detection and study of HPV in human tissue. Chapter 7, Developing and Standardizing Human Papillomavirus Tests, describes the development of HPV testing. This has depended almost entirely on developments in HPV DNA detection technology that has provided the basis for developing knowledge of the importance of HPV types in cancers and precancers and the understanding of the complexities of HPV infections, including the existence of infections with multiple HPV types and understanding the importance of these. The quality of the new methods as well as their molecular sensitivity and sensitivity for detecting cancer and precancer has been an important issue. HPV DNA testing has also been one of the earlier biotechnologies to find an important use in clinical research and then in routine cervical screening practice. The earliest clinical research into HPV depended on Southern blotting (developed by Edwin Southern in the United Kingdom in 1975) to detect HPV DNA sequences in whole biopsies of fresh tissue, extended with the development of in situ hybridization of histological sections, but these depended on radioactive isotopes to obtain sufficient sensitivity. However, it is the development of tests that detect more sensitively a range of HPV types that have provided the impetus to expand clinical and epidemiological knowledge.

Further developments in HPV testing occurred in Europe and the United States, within a changing political, legal, and economic framework surrounding biotechnology that differed widely between countries. In the United Kingdom and much of Europe, there was a reluctance to patent medical data on gene sequences, regarding these as a common good. In the United States, there was emphasis on patenting and commercialization of methods and DNA sequences, alongside the development of new very sensitive methods for DNA diagnostics, particularly the polymerase chain reaction (PCR) invented by Kary Mullis in 1983 at Cetus laboratories. There was a rapid and extensive development of HPV diagnostics from the mid-1980s. The complexity of test development also caused some confusion as investigators at different centers generated conflicting results, often as a consequence of the extreme sensitivity of PCR in detecting a few molecules of HPV DNA. This created the potential for false-positive results from contamination in the laboratory or in the sample, as well as variation in results because some tests were much more sensitive at detecting unexpressed latent infections, whereas others only detected very productive lesions or high-grade precancers with HPV in every cell.

In the United States, Michelle Manos at Cetus developed the MY09-011 primer set for HPV detection by PCR, and this was patented in 1989 [9]. In 1988 Bethesda Research Laboratories—Life Technologies obtained FDA approval for the ViraPap HPV test kit, but this had only limited clinical uptake with widespread skepticism among clinicians and cytologists about HPV testing. The test and associated intellectual property was sold to Digene in 1990 for $3.6 million. This was developed into the hybrid capture 2 (HC2) test that became the de facto standard of HPV testing in the United States following the ALTS trial, and in 2007 Qiagen bought Digene for US$1.6 billion. In contrast, in Europe the GP5/6 test and its successor were developed by the academic group in the pathology department at the Free University of Amsterdam based on PCR technology but not produced initially as a commercially available standardized clinical test. GP5 + / 6 + along with HC2 nonetheless became the standard by which other clinical HPV testing was to be judged. While most new tests were aimed at a performance for the detection of high-grade precancers in cytological samples, testing was also developed in the Netherlands by Wim Quint. This SPF10 test aimed at optimizing HPV detection in archival samples held for years in pathology departments, allowing extension of early studies showing that HPV had been a cause of cervical cancer for many years and was not a "new" virus, and also providing more precise information on the role of HPV in cervical and other cancers on a large scale across the globe. The development of HPV testing is discussed in detail in Chapter 7, Developing and Standardizing Human Papillomavirus Tests, and the ALTS Trial in Chapter 15, Triage of Women With ASCUS and LSIL Abnormal Cytology: The ALTS Experience and Beyond.

THE COMPLEX INTERACTIONS OF ACADEMIC SCIENCE, CLINICAL MEDICINE, AND THE CORPORATE WORLD

The close interaction between the corporate, the medical expert, and academic HPV research community, particularly in the United States has been an important driver in the development and application of new technology in HPV-related cancer prevention. The regulatory authorities, especially in the USA (FDA-Federal Drugs Agency) and also in Europe (EMEA – European Medicines and Equipment Agency) and other government agencies involved in public health and disease prevention have played an increasingly important part in modulating this interaction and ensuring that a high level of evidence is required before major changes in clinical and preventive practice are made and in mitigating risk from such changes. The development of HC2 continued through ALTS and beyond, and many more HPV tests were developed by different companies in the United States and elsewhere as companies fought and continued to fight for market share in the developing use of HPV testing for triage of women with abnormal cytology globally and then primary screening by HPV testing.

It is difficult to underestimate the importance of the commercial and academic links in developing HPV testing and later vaccination, and the tensions and differences this has created as well as leading to rapid development of the technology. These links that involved many researchers played an important part in moving from the small-scale research of individual groups of enthusiasts from diverse backgrounds to the large-scale technical developments and clinical trials needed to implement global cancer prevention. The very active, strong, marketing strategy of Digene, targeting directly physicians and also women as well as laboratory directors contributed to the demand for and progress of clinical HPV testing but also contributed to creating tensions with the cytology and gynecological communities and with academics, particularly in Europe, who considered that the rapid involvement of companies in this research prejudiced the independence of investigators and introduced bias into research design and interpretation. One key issue was the participation in and financial and practical support for clinical trials and other clinical studies, leading the company to claim that from 2002 it had been involved in clinical studies of over 90,000 women on four continents [10].

HUMAN PAPILLOMAVIRUS BEYOND CERVICAL CANCER

Section 2 addresses the role of HPV beyond cervical cancer. Since the discovery of HPV16, there has been steadily growing evidence of the wider role of this and certain other HPV types in cancer outside the cervix, mainly other male and female anogenital cancers but also oropharyngeal cancer, and study of the importance of genital warts and some rare types of cancer, associated with HPV6,11 and other "low-risk" HPV types. This has become increasingly important in discussions about vaccination policy, especially whether boys should be included as well as girls. In vaccine development the role of HPV6,11 in genital warts has been important in the Merck promotion of quadrivalent vaccination of boys as well as girls to reduce the burden of genital warts and also low-grade cervical and other squamous intraepithelial lesions, laryngeal papillomas, and occasionally cancers.

HPV infection in men is discussed in Chapter 11, Human Papillomavirus and Related Diseases Among Men. An important issue for both men and women is anal cancer. There is the possibility of screening, similar to cervical screening by cytology and HPV testing, to prevent anal cancer in high-risk groups such as men who have sex with men (MSM), particularly HIV-positive men. This is an attractive proposition but such MSM often have complex multiple HPV infections, and the nature and outcome of anal lesions histologically similar to cervical precancers are not necessarily the same.

In both vulval and penile carcinogenesis, there are HPV-related and non–HPV-driven pathways. These are morphologically and clinically distinctive. Although the pathway of HPV carcinogenesis are less clearly defined in the oropharynx, HPV-related cancer is very specific to certain sites such as tonsillar-type epithelium, and the outcome of HPV-related and unrelated oropharyngeal cancer is very different. The specific issues around HPV in these cancers and HPV infection and cancer in men are discussed in Chapters 10–14.

THE KEY ROLE OF HUMAN PAPILLOMAVIRUS IN PREVENTING CERVICAL CANCER

Nonetheless, most research activity has been driven by the role of HPV in cervical cancer. The interaction of laboratory and clinical research translating the science into practical developments in clinical and preventive practice is the key topic of Section 3. This covers the study of HPV testing in triage of women with low-grade smear abnormalities found on cervical cytology, the studies of primary cervical screening by HPV testing, the development of biomarkers to improve the triage of women who are HPV-positive or have a low-grade smear abnormality, and the development of understanding of immune responses to HPV and the development of vaccines based on artificial noninfectious viral particles, and finally, the clinical trials that showed these vaccines worked well and safely.

The work on prevention of cervical cancer using HPV ran in parallel with the work establishing the key causal role of certain HPV types that came to be designated as high-risk HPV. In the 1980s and early 1990s there was an increasing amount of information about the different HPV types involved in cervical and other neoplasia. There was also a wide exploration of the natural history of HPV infection, particularly the relation between the presence and persistence of different HPV types and cervical dysplasia, precancer, and cancer, a recognition that HPV DNA could be present in apparently normal cervical epithelium, and also that its presence in cervical smears that were morphologically negative could indicate the presence of precancer in the cervix itself.

Such studies were mostly initially conducted within existing cervical screening systems, in women attending for screening or having further investigation of abnormal smears, especially those undergoing colposcopy and biopsy. All this fed an interest in the application of HPV testing to cervical screening both within the existing cytological screening systems to improve the management of women with common low-grade abnormalities that did not indicate the definite presence of an important precancer and as a possible replacement for cytological screening.

CERVICAL SCREENING BEFORE HUMAN PAPILLOMAVIRUS

As is discussed in Chapter 1, A Brief History of Cervical Cancer, prior to the 1980s, cervical screening by cytology had, in most countries, grown up on a very ad hoc basis depending on the enthusiasm of individual clinicians and pathologists, only partly driven by national and local political initiatives. With the exception of some Scandinavian countries and British Columbia screening was not an organized and carefully monitored integrated program. There was marked variation in screening and follow-up practice between countries and within some countries. In the places with highly organized screening systems and high levels of coverage, there was a marked impact on cervical cancer incidence and mortality. In other countries the overall impact was small, even nonexistent and was very variable for women depending on access to screening and quality of the process. In the United Kingdom, screening was carried out every 5 years in women from 25 to 60 years within the National Health Service; in the United States, annual gynecological examination including cervical cytology from 18 years onward was standard private practice for paying patients and an important part of gynecologists' incomes.

There was also considerable variation in the management of women with abnormal smears. In the United Kingdom, this was the responsibility mainly of the pathology laboratory that examined the cervical smears with gynecologists only involved when a definite persistent or high-grade abnormality was present. Performing colposcopy was restricted to a limited, but growing, number of enthusiasts. In the United States, screening was managed mostly within the private practice of individual gynecologists who were more inclined to resort directly to colposcopy as a tool for immediate examination of the cervix in the presence of any cytological abnormality and follow this by biopsy of any possible abnormality.

REDEFINING THE PATHOLOGY OF PRECANCER IN THE LIGHT OF HUMAN PAPILLOMAVIRUS—THE BETHESDA SYSTEM AND ITS SUCCESSORS

One important early response to the finding of different HPV types in cervical cancers and their precursors in the latter part of the 1980s was for pathologists to start looking more closely at the way cervical cytology was reported and how cervical and other anogenital biopsies should be examined and classified in the light of knowledge about high and low-risk HPV—those HPVs associated strongly with cervical cancer and those that were not. Working in combination with molecular virologists and with developing the science of molecular pathology in their own laboratories, pathologists in Europe and the United States reexamined how cervical pathology was reported and began to train pathologists reporting cervical biopsies and cytology about HPV during the 1980s. This led to an update of the previous well-established concepts of a progressive dysplasia leading to a carcinoma in situ and developed into the cervical intraepithelial neoplasia classification. It was rapidly established that high-grade lesions were mostly due to high-risk HPV, and the higher the grade, the more likely it was to be due to high-risk HPV; low and high-risk HPV could not be distinguished morphologically in low-grade lesions. This and the presence of a particular pattern of nuclear abnormality and cell vacuolation (koilocytosis) and positivity for bovine papillomavirus L1 capsid antigen began to be studied as a possible HPV test [11]. This was expressed only in superficial layers of low-grade lesions and supported the concept that low-grade lesions were essentially productive infections that were often self-limiting, while high-grade lesions were variously transformed by the HPV E6 and E7 genes whose nature and actions on cells were being investigated at the molecular and biochemical level by virologists. Histopathology did not reliably distinguish, however, the type of HPV present [12].

The most influential early outcome of this study of HPV in relation to cytological screening was the Introduction of the Bethesda system of classification in the United States, named after the location in Bethesda, Maryland, the United States where NIH has its center and the key conference was held [13–15]. This classification was based on the concept of low-grade productive lesions and high-grade transforming lesions with features that could be recognized on cytology and histology and linked to clinical practice in which it implied all high-grade lesions needed excision or destructive treatment and did not solve the difficult problem that women with low-grade cytology abnormalities not infrequently had underlying high-grade disease on further examination at direct examination of the cervix by colposcopy and biopsy. The Bethesda classification was first developed in 1988, with revisions in 1991, 2001, and 2014. It was the direct ancestor of the contemporary US lower anogenital tract (LAST) system that extends these concepts to other anogenital sites and attempts to define more precisely how a distinction between high and low-grade histological lesions should be made [16].

The Bethesda and LAST systems have been widely adopted in the United States and internationally but have important limitations to their value, in particular the absence of a direct relationship between the cytology grade and the similarly termed histology grade, their unreliability and poor reproducibility across different pathologists, and especially their failure to distinguish progressive precancer from regressing lesions leading to overtreatment of lesions likely to regress. These issues have fueled the search for better biomarkers of HPV regression and transformation and understanding of the "soil" (i.e., the particular characteristics of the very susceptible epithelium of the cervical transformation zone), in which the carcinogenetic "seed" of HPV lands. These issues are discussed in Chapters 5, 6, 12, 14, 15, and 17.

THE MOVE TO CHANGE CERVICAL SCREENING BASED ON HUMAN PAPILLOMAVIRUS SCIENCE

A number of studies during the 1980s in different countries showed that within the population of women with minor morphological abnormalities on their cytology screening smears, a substantial proportion had more severe precancerous changes on biopsy after colposcopy than the smear disclosed, while many had nonprogressive lesions. Progression had been linked to the presence of high-risk HPV and its persistence from early studies such as that of [17] for HPV16. This kind of study, the finding that a limited range of HPV types were responsible for almost all cervical cancer and precancer and the developments of reliable tests for these HPV types (as described previously), led to the investigation of the possible use of HPV testing in cervical screening. Early mathematical models of cervical screening supported the potential for HPV testing in triage and particularly for improving primary screening. The varied clinical and screening practices in different countries led to important differences in the clinical research response.

It is not possible to overestimate the importance of the ALTS trial in the United States, as discussed by Atilla Lorincz, who was the principal scientist in Digene at the time, in Chapter 15, Triage of Women With ASCUS and LSIL Abnormal Cytology: The ALTS Experience and Beyond, involving both the scientific and commercial interests and activities of Digene and the developing global research interest in HPV testing for cervical screening. ALTS began in 1996 and focused on the investigation of women with low-grade smear abnormalities. This was a common problem with cytological cervical screening, affecting 5%–8% of women screened: in the United States about 55 million Pap smears were performed each year and about 3 million women had low-grade smear abnormalities. In the United States, screening at that time started soon after sexual activity, and women with low-grade abnormalities were often managed by colposcopy and biopsy, whereas in many other countries, they were followed by cytology for a year or more to determine if there was regression of cytological abnormality. There was considerable controversy about which approach was best, with arguments about cancer development during follow-up versus overtreatment of regressing lesions associated with risks to future pregnancies. There was also uncertainty about the negative psychological impact of both approaches.

The motivation for the ALTS study was to establish a definitive triage scheme through a randomized clinical trial of the use of HPV status against colposcopy and repeat cytology for women with abnormal smears. The establishment of a national trial in the United States using a commercial test followed the War on Cancer policy for NCI introduced in 1971 encouraged NCI collaboration with private industry such as Digene.

ALTS began in November 1996 and concluded in 2000. It recruited about 5000 women at four centers across the United States, comparing immediate colposcopy with cytological follow-up and combined HPV testing with cytology. It generated around 130 publications on many aspects of natural history of HPV and cervical precancer as well as changing practice globally with the widespread implementation of HPV testing in triage, and drawing attention to the variation, even among experts of interpretation of cervical smears and biopsies and the value of standardized HPV testing.

The background to existing cervical screening and the response to the growing information on HPV in cervical cancer and its precursors in Northern European countries were very different. Many Scandinavian countries had organized and successful cytological screening programs that provided high coverage of the female population, not dependent on ability to pay. In the United Kingdom, cytological screening was introduced in the 1950s but was not well-organized. However, evidence from the growing use of colposcopy and from audit of performance of the program led to major reorganization of the established program to improve coverage and the quality of cytology in an attempt to improve its sensitivity and reliability for detection of precancer 1988. This produced an accelerated decline in mortality, but there was still evidence that the sensitivity of cytology was inadequate [18]. The ambiguity of the UK response was reflected in the inconclusive Health Technology Assessment concerning HPV testing (1999), and the decision of the medical research council (MRC) that the main focus of the UK trial of the management of low-grade smear abnormalities involving over 4000 women started in October 1999 was on cytology versus colposcopy and on the side effects, psychological and economic aspects of these approaches. Nonetheless, this study produced important evidence of the probable regression of much CIN2 without treatment under management by cytological follow-up [19].

In the Netherlands, as described in Chapter 16, Primary Screening by Human Papillomavirus Testing: Development, Implementation, and Perspectives, by Chris Meijer, thanks to the major interest in HPV testing at the Free University of Amsterdam. The effective modeling of existing data (as described in Chapter 20) suggested the potential value of primary HPV screening and the organized national screening program allowed a population-based randomized controlled trial of high-risk HPV testing to be established (POBASCAM) recruiting between 1999 and 2002, 44,102 women 29–61 years [20]. This was pivotal for the development of HPV-based primary screening and has been followed by other studies in Europe, and elsewhere, with the adoption in the Netherlands of a tightly managed, national primary HPV screening system using a standardized HPV testing system with cytology only for reflex follow-up of HPV-positive women in 2015, despite strong and fierce opposition from some cytologists.

Implementation of primary HPV screening was a huge change in laboratory and public health practice, vastly diminishing the work of cytology and seen as threatening the profession and careers of cytopathologists and cytoscreeners. In Europe, meta-analyses of several large studies in different countries have been important in promoting the change [21]. In the United States, there has been a steady change in cervical screening guidelines since 2002 with the HC2 test receiving FDA approval in 2003 for both reflex testing of ASCUS smears in women over 21 years and primary HPV screening in women over 30 years. Different guidelines from 2012, 2015, and 2016 progressively favored primary HPV screening or cotesting for women older than 25 or 30 years every 3 years, with cytology every 3 years from 21 to 29 years. In the United States additional evidence from the ATHENA trial [22] secured approval from FDA for the Cobas HPV test in primary screening over the age of 25 years. A study of over 1 million women in the Kaiser Permanente healthcare organization at Oakland, California, confirmed the value of primary HPV screening. This led to the US Preventive Services Task Force guidelines [23] recommending cervical screening of women from 21 to 29 years every 3 years by cytology, and then every 5 years using HPV testing or cotesting up to 65 years. Nonetheless, this approach in the United States still leaves important racial and regional disparities in cervical cancer incidence and mortality [24].

In Britain the HART study [25] and the ARTISTIC trial [26] were important steps in developing the application of primary HPV screening within the national screening program. These led to the decision in 2016 to implement primary HPV screening without co-testing nationally in 2019.

APPLYING HUMAN PAPILLOMAVIRUS SCREENING GLOBALLY

While these developments in using HPV screening were in progress in Europe, North America, and other Western countries, the main burden of invasive cervical cancer remained in low and middle-income countries where there was little chance of introducing cytological screening effectively. Extending knowledge and application of the new testing methods and approaches to cervical cancer prevention was important. This developed initially through individual clinicians and investigators in developing countries and through professional societies. The enthusiastic attendance of the United States and European researchers in both clinical and basic science, at national and local medical and scientific meetings in Latin America, Asia, and Africa, reporting the newest studies, meant that local experts and practicing clinicians and pathologists were learning about the latest developments and changing practice. Meetings of the International Papillomavirus Society Conference and Eurogin outside Europe and North America provided a development of this approach, particularly since the vaccine manufacturers became involved.

WHO recognized the importance of global cervical cancer prevention and produced guidelines for comprehensive cervical cancer control in 2005. The book returns to a detailed discussion of the issues involved in extending prevention of cervical cancer and HPV-related disease globally in Chapter 23, Expanding Prevention of Cervical Cancer in Low- and Middle-Income Countries, and in the Epilogue: Looking Forward to Cervical Cancer Elimination.

DEVELOPMENT OF BIOMARKERS

This is the subject of Chapter 17, Infection to Cancer—Finding Useful Biomarkers for Predicting Risk of Progression to Cancer. While the clinical research into HPV testing in cervical screening was evolving, there was continuing basic science research in a number of directions to understand more of the mechanism by which HPV acts on epithelial cells in viral replication and in transformation and of the cellular events associated with transformation. Much detailed knowledge of the actions of HPV types has accumulated and shown the complexity of the process from infection to cancer. However, colposcopic interpretation by a gynecologist and the morphological pathology of cervical biopsies remained the basis of the decision on whether there is a lesion needing treatment. This has been largely unchanged in practice since the 1980s, despite increasing awareness that both techniques require expertise and are poorly reproducible.

Although the triage process of selection for colposcopy may be supplemented by HPV genotyping, only one other marker has been widely used at the biopsy stage and in cytology, the immunohistochemical (IHC) marker p16INK4a, which acts as a surrogate marker for hrHPV E7 gene activity. Unfortunately, while extensive diffuse expression is a reproducible feature of high-grade cervical precancer, the lower cutoff for expression and expression in different patterns in lrHPV lesions and physiologically in metaplasia limits the value of this marker on its own [16]. Other new molecular markers including those based on methylation of viral genes and of somatic tumor suppressor genes during progress of precancer to cancer may supplement or replace this and are needed for full molecular screening, particularly if self-sampling of the vagina is likely to replace current doctor or nurse taken liquid-based cytology, as this has replaced previous cytological sampling of the cervix.

NATURAL AND ARTIFICIAL IMMUNE RESPONSES TO HUMAN PAPILLOMAVIRUS AND HUMAN PAPILLOMAVIRUS VACCINES

In the further development of understanding how hrHPV generates precancer and cancer, the study of the natural humoral and cellular immune responses to HPV infection and also the generation of an artificial immune response to HPV capsid as the basis for prophylactic vaccination have been important stages in the development of HPV science. This is discussed in Chapter 18, Immune Responses to Human Papillomavirus and the Development of Human Papillomavirus Vaccines, along with the effects of HPV on the infected epithelial cells in Chapter 4, Finding How Human Papillomaviruses Alter the Biochemistry and Identity of Infected Epithelial Cells. By infecting the basal layer of the cervical epithelium and having no viremic phase, HPVs are isolated from circulating immune cells until the infection is established. The production and release of virions in only the superficial layers of maturing epithelium limits innate immune responses that normally occur in relation to cell death. HPVs also appear to generate tolerance by action on downregulating local antigen-presenting cells and CD4 and CD8 T cells as established in early IHC studies of cervical lesions [27]. CD8 cytotoxic lymphocytes are important in the regression of warts and experimental HPV infections. The absence of such cells may be important in the progression of persistent infection with transformation, as is a relatively weak B cell and antibody response. Altered cytokine patterns such as marked expression of the immunosuppressive IL-10 may be also important in maintaining persistence [28]. Whether after resolution of lesions how much HPV can persist as a latent infection and the clinical importance of this is still under investigation.

The complex mechanisms that can permit HPV to escape an effective immune response have allowed HPV to achieve its niche way of life and become one of the most common sexually transmitted infections worldwide.

However, while the role of the immune system in natural infection is of uncertain importance in clearing infection and exploiting this to treat lesions by therapeutic vaccination has proved difficult, prophylactic vaccines, circumventing the problems of natural antigen presentation by direct, have proved an enormous success. Direct injection into the arm of HPVL1 capsid antigen in the form of virus-like particles which present the L1 antigen in the correct conformational shape to induce high titers of effective neutralizing antibodies leads to a strong and effective immune response avoids the unique protective features that HPV has evolved in natural infection.

DEVELOPING HUMAN PAPILLOMAVIRUS VACCINES

The clinical trials of the HPV vaccines developed by Merck and GSK are reported in Chapter 19, Clinical Trials of Human Papillomavirus Vaccines. The early history is described in Chapter 18, Immune Responses to Human Papillomavirus and the Development of Human Papillomavirus Vaccines, and in the vignette on Jian Zhou. The early history is also described in Muller and Gissmann [29]. In 1930 Findlay reported that during a series of self-inoculation experiments with wart extracts, he became immune to wart induction. However, effective and safe presentation of the L1 antigen only became possible with virus-like particles. Success with mouse polyomavirus led to the production of papillomavirus-like particles in the early 1990s by different researchers. Zhou et al. working initially in Lionel Crawford's ICRF Tumour Virus Group at the Department of Pathology in Cambridge, England, where Ian Frazer was also working, and then completing the work in Brisbane, Australia, with Ian Frazer in 1991 [30]. Jian Zhou moved from the United Kingdom with an Australian passport as the British authorities were not prepared to be flexible about his immigration, even for a brilliant scientist who had survived the Chinese cultural revolution. He used a *Vaccinia virus* expression system to produce HPV16 L1 VLPs; however, the yield was very low and did not permit full characterization. This was followed in 1992 by HPVL1 expression in mammalian cells that showed the self-assembly of L1 into VLPs and produce neutralizing antibodies, although no VLPs were actually produced. After working on bovine papillomavirus (BPV) and HPV11 at NCI and Rochester University, Kirnbauer et al. reported that HPV16 from early virus-producing lesions self-assembled more efficiently than the mutated HPV prototype genome integrated in a cancer, making structural and immunological studies possible and providing the basis for the clinical development of HPV vaccines by Merck and GSK. This also established the scientific importance of the VLP production system and the HPV sequences used. The incremental process of initial scientific invention of the L1 vaccine led to a complex patent situation with a prolonged adjudication on overlapping claims, including a judgment in favor of the priority of Jian Zhou and Ian Frazer. The patent situation was even more complex than this, and all participants held some patents. Importantly Merck and GSK had extensive cross-licensure of the patents of all parties. The cross-licensing and other patent issues later proved one important factor for commercial decisions about the future of the two vaccines that went into clinical development and for manufacture of low-cost HPV vaccines needed for public health use in low- and middle-income countries.

The clinical trials of the prophylactic HPV vaccines are described in Chapter 20, The Key Role of Mathematical Modeling and Health Economics in the Public Health Transitions in Cervical Cancer Prevention. The first clinical trial with VLPs involving human subjects was initiated in 1996. With improved procedures for production of VLPs in yeast or insect cells with recombinant Baculovirus, phase 1 safety and immunogenicity trials were successful. Phase 2, proof of principle, study of an HPV16 vaccine began in 1998, followed by phase 3 studies of the Merck HPV6,11,16,18 vaccine in 2002, aiming at genital warts and recurrent laryngeal papillomatosis as well as female genital cancers. At this time a phase 2 trial of the GSK HPV16/18 vaccine adjuvanted for extra immunogenicity with ASO4 that mimics a Toll-like receptor 4 agonist providing direct stimulation to antigen-presenting cells was running and then followed rapidly by two large phase 3 studies of this as an anticancer vaccine.

The large and extensive clinical trials in young women, older women, girls, and later boys are reported in the chapter on vaccine development. These demonstrated the efficacy of both vaccines in preventing cervical precancer associated with HPV16/18, with the quadrivalent vaccine also providing protection against genital warts associated with HPV6/11, and the bivalent vaccine offering better cross-protection against cervical precancer associated with other hrHPV types. This led to the introduction of a nine-valent vaccine by Merck against HPVs 6/11/16/18/31/33/45/52/58.

Both vaccine companies conducted large, expensive, international multicenter phase 3 clinical efficacy trials. The two powerful new tools of HPV screening and vaccination raised the prospect of global prevention of HPV-related disease, especially, but not exclusively cervical cancer. The involvement of two large, wealthy, and at a certain level, highly competitive pharmaceutical companies in HPV as well as the potential for primary screening using a selected, standardized commercial HPV test meant a complete change of gear within the scale of clinical research activity and the nature of the professional community involved in HPV research and development. Both drug companies were actively promoting their vaccines and had intensive marketing campaigns to drive financial success for their new products. This was important as the development of highly profitable, universally, clinically important "blockbuster" drugs was getting more difficult, and the nature of conventional drug development especially in the cancer field is changing. However, marketing expensive new vaccines, which were directed at a cancer-causing sexually transmitted virus, was seen as difficult by both companies and involved huge investment by the vaccine manufacturers in research into how vaccines could be promoted in different countries. The activities of both the vaccine manufacturers, especially their departments of "Medical Affairs" included "mapping" of key players in countries such as politicians, officials, key scientists and doctors, women's, and other groups, looking for potential lobbying support for these products. The companies invited large numbers of experts, clinicians, and lobbyists to the International HPV Conferences and the Eurogin conferences in the first decade of the second millennium. The cost of the initial three-dose regime, however, of the Merck and GSK vaccines rendered these inaccessible to low and middle-income countries where most cervical cancer deaths occur and raised issues of intellectual property, technology transfer, and manufacture outside Europe and the United States [31].

As well as those likely to support vaccination, the companies sought to know about those opposed to vaccination, the "antivaxxers," and groups promoting sexual abstinence, and seeing HPV prevention being achieved through this approach. The companies undertook reviews of adolescent sexual behavior and women's concerns around cervical cancer, screening, and how to measure this for health economic analyses, as well as monitoring favorable and unfavorable news media responses. The huge potential of HPV vaccines meant that both companies saw their vaccines as "blockbuster" products, an unusual situation for a vaccine. One senior executive of a medical background in charge of his company's HPV vaccine program, although not responsible for the science, said, when he made an announcement about the success of a clinical trial that pushed up the company's share price, that this was the proudest day of his life.

There was, however, considerable concern within the companies about the successful marketing of a vaccine aimed at preventing sexually transmitted viruses. While this was an issue, it proved to be generally acceptable to religious leaders in focus groups and less important than initially thought in most countries and among most girls and their parents. The general strategy of each company was different: GSK was based in the conservative Belgian region of Wallonia and favored a cancer focus, fitting with its available HPV16/18 vaccine and minimizing the sexual element; Merck was based in the United States and favored a more confrontational approach to sexually transmitted disease based on the additional protection from genital warts in its quadrivalent vaccine, even if this contributed very little to cancer protection and other evidence at the time favored the bivalent vaccine for long-term cancer prevention.

Merck very actively promoted Gardasil, the most expensive vaccine in history, in the United States from before licensure in 2006 through sales representatives and contributions to political campaigns and women's groups and TV adverts. This seemed to work well initially, but the issue of mandating vaccination of preteen girls led in 2007 to a backlash from religious moral crusaders and an increasingly vocal group of families and consumer advocates who felt that the only mechanism for US government to promote universal vaccination was to require vaccination for school or college entry attacked the right of parents to make decisions for their children [32]. This did not happen in Britain where early introduction of the bivalent vaccine in 2008 was done mainly through school vaccine programs which had been previously developed as a public health tool, with an active opt-out required for parents not supporting vaccination. However, the argument for prevention of genital warts led to a decision to move to Gardasil in 2012.

The advantage of genital wart protection possessed by Gardasil, and subsequently, the more extensive protection against other types by Gardasil 9, led to a substantial drop in sales of Cervarix and a commercial decision to stop promoting the vaccine, despite potential advantages, particularly in low-income countries. The battle between the companies about whether Gardasil vaccines could potentially deliver such good and lasting protective immune responses with a more limited does schedule as the different expression system and adjuvant of Cervarix continued unresolved in the absence of any totally accepted correlate of protection. The effectiveness of this vaccine was shown through routine cervical screening at age 20 years in Scotland as described in Chapter 21, Twelve Years of Vaccine Registration and the Consequences [33].

THE IMPACT OF CORPORATE HUMAN PAPILLOMAVIRUS VACCINE TRIALS AND MARKETING

On the scientific side the two drug companies, although both had a mixed history of drug development and especially recording and recognition of side effects, did conduct successful large phase 3 trials that brought attention to HPV as a cause of cervical and other anogenital cancers to many who had not been previously aware as well as demonstrating the impressive efficacy of HPV prophylactic vaccination. The whole exercise opened up the importance of HPV to national and international bodies involved in cancer, including WHO, PAHO, and NGOs such as PATH and the global vaccine alliance initiative for the poorest countries—GAVI, as well as to the large numbers of doctors and scientists who were paid for by Merck and GSK to attend the major international conferences on HPV organized by the International Papillomavirus Society and Eurogin. It provided a new tool that no national or international authority was prepared to develop and opened up some difficult social questions about sexuality. All this was at a price. Drug companies are about share price, and this means profits in a complex and difficult situation of developing new drugs and vaccines, many of which fail for businesses that have been used to big profits and matching salaries and bonuses for senior executives who deliver on share price. The cost of the new HPV vaccines was very high ($360 for three doses of Gardasil in 2007). The cost limited their application in many countries outside the high-income parts of the globe and the wealthier classes elsewhere: a world in which HPV vaccination is an added protection to existing cervical screening. With intellectual property rights being the key to a company's success, issues of technology transfer, and manufacture dominate the making of low-cost vaccines in countries such as China and India.

As well as its major public health possibilities and limitations, the vaccine development also had an impact on requiring further studies of HPV-type distribution globally to ensure that the right mix of HPV types was included for all countries, and on issues such as multiple HPV infections and whether all cervical cancer was due to HPV and therefore theoretically preventable by vaccination and HPV screening. Vaccination not only reduced the value of screening but also affected negatively the performance of cytology in vaccinated women, mainly by reducing the frequency of abnormal smears indicating serious underlying lesions and increasing the proportion of mildly abnormal smears due to lrHPVs not included in the vaccines or other non-HPV causes, creating another push toward primary HPV screening.

There were also studies of alternative vaccination schedules, vaccination of boys, and very importantly of the effects of HPV vaccines in real life on precancers, cancer, and screening activity. Trying to predict effectiveness and cost-effectiveness through mathematical modeling became of renewed importance and complexity as models had to cope with the acquisition and spread of HPV infection in adolescence and the natural history of HPV infection rather than just the progression of precancer modeled in early studies. This is described in detail along with the development and current use of mathematical modeling in Chapter 20, The Key Role of mathematical Modeling and Health Economics in the Public Health Transitions in Cervical Cancer Prevention.

HUMAN PAPILLOMAVIRUS VACCINATION IN PRACTICE

HPV vaccination has now been put into practice over the last 12 years enabling study of the different real-life aspects of its effectiveness through monitoring the introduction and first years of HPV vaccination. This is reviewed in Chapter 21, Twelve Years of Vaccine Registration and the Consequences. One important issue that has come up regularly is the failure or delay to HPV vaccination programs from antivaccine beliefs and actions and the exploitation of concerns about vaccine safety through social media. This is the subject of Chapter 21, Twelve Years of Vaccine Registration and the Consequences. There is a long and complex history behind public concerns about the existence of any potential serious risks of vaccination to a small number of individuals, despite the obvious overall population benefits in preventing infectious disease. Concerns about the potential of vaccines for damage to vaccinees as well as protection have existed since the 18th century days of the dangerous practice of variolation, attempting prevention of serious smallpox by infecting a child with a mild strain of smallpox virus [34]. This practice was succeeded by vaccination using the *V. virus* from cowpox to reduce the complications. Ever since then there have been concerns around vaccine safety, related to many issues including contamination such as that of polio vaccine with the SV40 virus, whole cell pertussis vaccine as a trigger of rare neurological disease in children, preservatives such as mercury and adjuvants that aim to stimulate a strong, but not specific immune response. The issues of producing uncontaminated vaccines have been addressed through careful regulation of the vaccine industry.

There are, however, particularly difficult scientific, medical, emotional, and sometimes political issues concerning rare events, including sudden death and neurological disease that occur spontaneously in a young population such as girls or boys receiving the HPV vaccines. Distinguishing a causal relationship to vaccination for such events from a casual temporal association is not always easy and can be difficult for public health officials in the context of distressed families, media clamor, and misinformation from militant antivaxxers. Clinical trials of each HPV vaccine studied in phase 3 included 100,000 women over the many years of different trials under strictly controlled conditions gave clear data about many safety issues. Importantly, some of the concerns about safety involved neurological disease that is very rare and can only be addressed in careful monitoring of the introduction of a vaccine in practice.

Addressing vaccine safety also requires sensitivity to the bitterness that can be generated by a system that appears to be uncaring and unwilling to address the problems of the individual child or adolescent. Vaccine mistrust is recognized as an important issue by WHO, and weighing up the benefits and risks of vaccination can be a real, problematic concern for many parents who are not well informed about immunology and vaccines and faced with much misinformation but concerned about avoiding harm to their children. This kind of anxiety and attitude amongst mothers has been discussed from the nonexpert point of view by Eula Bliss in "On Immunity" [35]. It has been intensified and manipulated by militant antivaccination campaigners, supported by some opportunist doctors and others, especially the poor-quality gastroenterological studies of Andy Wakefield who linked autism to measles vaccination. Although this science has been criticized since its first presentation, the key paper was withdrawn by the Lancet, and the British General Medical Council (GMC) withdrew his medical license; he has become an important figure in the United States among antivaccine activists who gather around their own versions of the truth and has been taken up by populist politics [36].

Chapter 22, Political and Public Responses to Human Papillomavirus Vaccination, points out the importance of preparing the ground, engaging support nationally and locally before any vaccination is introduced; the need to involve local community groups as well as national experts and politicians in ensuring there is proper support for the introduction of vaccination. This can also build strong advocacy to address any misinformation from antivaccine pressure groups. A further lesson is that it is important to respond to any serious adverse events occurring in those being vaccinated very rapidly and effectively to avoid unnecessary suspension of a program because of misplaced concerns.

HUMAN PAPILLOMAVIRUS AND MODERN MEDICAL SCIENCE

The chapters in this book explore the issues outlined here in detail and reflect the experience as well as the science of scientists and doctors who have developed this field over about 40 years. Once there is a key into the causal mechanism of a disease, modern science and the academic and commercial structure it supports can provide some effective solutions to the problem. Cervical cancer and other HPV-driven cancers turned out, on the whole, and fortunately for the science and the patients, to be relatively simple compared with many other cancers that seem to be driven by a complex variety of human genetic and environmental factors. The history of HPV shows how far science has gone and can go in providing solutions to a small but important group of cancers affect women and men worldwide. The new screening technology and the vaccines can contribute to the ambition of preventing these cancers globally and rapidly.

The development of the effective preventive tools for cervical cancer and other HPV-related cancer reflects very strongly the nature of modern medical science with the international base for the laboratory, clinical and epidemiological research, the involvement of public health authorities, and the interactions of large and small private companies with academic scientific and clinical research. One very important tool is the ability to model mathematically the possible use and impact in public health of different options through computing. This is extremely valuable in assessing likely good options and clarifying the discussions. Chapter 21, Twelve Years of Vaccine Registration and the Consequences, looks at how mathematical modeling can support health economic analysis of proposed changes in prevention introduction of screening and treatment and of vaccination into low and middle-income countries can speed up global prevention of cervical cancer.

Chapter 23, Expanding Prevention of Cervical Cancer in Low- and Middle-Income Countries, looks at some examples of how effective cervical cancer prevention is being developed for lower and middle-income countries where the main burden of HPV-related cancers is currently to be found. The problem remains, however, of the political issues and engaging the population of the world: creating the will to extend cancer prevention to low- and middle-income countries.

THE FUTURE

HPV science has produced excellent tools for cancer prevention and will continue to refine and develop ways of testing for cancer risk and managing people already infected with HPV as well as improving coverage by screening and vaccination. It is an inevitable problem of all medical interventions that none are perfect, even if as good as HPV vaccination and screening. These bring some inconvenience and at least anxiety about hazard as well as benefits. Every approach has its failures, complications, and side effects, although fortunately for cervical cancer prevention these are minimal compared to cancer treatments and the inconveniences of screening for other cancers. Nonetheless, how these are treated publicly and honestly with respect for the quality, accuracy, and also limitations of evidence and the balance between individual and population-based public health concerns are important. It is to the people who have to ponder all the different issues and those that will be dealing with these issues in future, as well as those who have developed the science and application of HPV so far, that this book is dedicated. The story of HPV and cervical cancer prevention is a mirror to the success of modern technological medicine. Once there is a key from basic science, such as the likely possibility of a viral cause of cervical cancer, this can be exploited very effectively. The science and technology of both cervical screening and HPV vaccination have been developed and can be developed further. Great successes have already been delivered. These, however, are limited to some parts of the world and to countries or individuals who can pay for access. The inevitable economic, political and consequent human issues remain. Defining some of these and beginning to address global prevention of cervical cancer and HPV related disease is the subject of the epilogue to this book.

REFERENCES

[1] Baussano J, Bray F. Modelling cervical cancer elimination. Lancet Public Health 2019;4:PE2–3.
[2] Forman D, de Martel C, Lacey CJ. Global burden of human papillomavirus and related diseases. Vaccine 2012;30S:F12–23.
[3] Hall MT, Simms K, Lew J-B, et al. The projected timeframe until cervical cancer elimination in Australia: a modelling study. Lancet Public Health 2018.
[4] Latour B. The Pasteurization of France [Sheridan A, Law J, Trans]. Cambridge, MA: Harvard University Press; 1988.
[5] Gissmann L, Wolnik L, Ikenberg H, et al. Human papillomavirus type 6 and 11 sequences in genital and laryngeal papillomas and some cervical cancers. Proc Natl Acad Sci USA 1983;80:560–3.
[6] Zur Hausen H. Papillomaviruses in the causation of human cancers—a brief historical account. Virology 2009;384:260–5.
[7] Meisels A, Fortin R. Condylomatous lesions of the cervix and vagina.1. Cytologic patterns. Acta Cytol 1976;20:505–9.
[8] Crawford LV. A study of human papillomavirus DNA. J Mol Biol 1965;8:489–95.
[9] Manos MM, Ting Y, Wright DK, et al. Use of polymerase chain reaction amplification for the detection of genital human papillomaviruses. Cancer Cells 1989;7:209–14.
[10] Hogarth S, Hopkins M, Rotolo D. Chapter 5. Technological accretion in diagnostics: HPV testing and cytology in cervical cancer screening (https://www.ncbi.nim.nih.gov/books/NBK379795/) In: Consoli D, Mina A, Nelson RR, et al., editors. Medical innovation: science, technology and practice. New York: Routledge; 2015.
[11] Jenkins D, Tay SK, Maddox P. Routine papillomavirus antigen staining of cervical punch biopsy specimens. J Clin Pathol 1987;40:1212–16.
[12] Jenkins D, Tay SK, McCance DJ, Campion MJ, Clarkson PK, Singer A. Histological and immunocytochemical study of cervical intraepithelial neoplasia (CIN) with associated HPV 6 and HPV16 infections. J Clin Pathol 1986;39:1177–80.
[13] Solomon D. The 1988 Bethesda System for reporting cervical/vaginal cytologic diagnoses: developed and approved at the National Cancer Institute workshop in Bethesda, Maryland, MD, December 12-13 1988. Diagn Cytopathol 1989;5:31–4.
[14] Solomon D, Davey D, Kurman R, et al. The 2001 Bethesda system: terminology for reporting results of cervical cytology. JAMA 2002;287:2114–19.
[15] Nayar R, Wilbur D. The Bethesda system for reporting cervical cytology: definitions, criteria and explanatory notes. Springer; 2015.
[16] Darragh TM, Colgan TJ, Cox JT, et al. The lower anogenital squamous terminology standardization project for HPV-associated lesions. Arch Pathol Lab Med 2012;136:1266–97.
[17] Campion MJ, McCance DJ, Cuzick J, et al. Progressive potential of mild cervical atypia, prospective cytological, colposcopic and virological study. Lancet 1986;ii:237–40.
[18] Reynolds LA, Tansey EM. History of cervical cancer and the role of human papillomavirus 1960-2000. In: Wellcome witnesses to twentieth century medicine; 2009. p. 38.
[19] TOMBOLA group. Cytological surveillance compared with immediate referral for colposcopy in management of women with low-grade cervical abnormalities: multicentre randomised controlled trial. BMJ 2009;339:b2546.
[20] Bulkmans NW, Rozendaal L, Snijders PJ, et al. POBASCAM, a population-based randomized controlled trial for implementation of high-risk HPV testing in cervical screening: design, methods and baseline data of 44,102 women. Int J Cancer 2004;110:94–101.

[21] Cuzick J, Clavel C, Petry K-U, et al. Overview of the European and North American studies on HPV testing in primary cervical cancer screening. Int J Cancer 2006;109:1–7.

[22] Wright TC, Stoler MH, Behrens CM, et al. Primary cervical cancer screening with human papillomavirus: end of study results from the ATHENA study using HPV as the first-line screening test. Gynecol Oncol. 2015;136:189–97.

[23] US Preventive Services Task Force. Screening for cervical cancer: US preventive services task force recommendation statement. JAMA, 320. 2018. p. 674–86.

[24] Yoo W, Kim S, Huh WK, et al. Recent trends in racial and regional disparities in cervical cancer incidence and mortality in United States. PLoS One 2017;12(2):e0172548. Available from: https://doi.org/10.1371/journal.pone.0172548.

[25] Mesher D, Szarewski A, Cadman L, et al. Long-term follow-up of cervical disease in women screened by cytology and HPV testing: results from the HART study. Br J Cancer 2010;102:1405–10.

[26] Kitchener HC, Almorite M, Gilham C, et al. ARTISTIC: a randomised trial of human papillomavirus (HPV) testing in primary cervical screening. Health Technol Assess 2009;13:1–150.

[27] Barton SE, Jenkins D, Cuzick J, et al. Effect of cigarette smoking on cervical epithelial immunity: a mechanism for neoplastic change? Lancet 1988;ii:652–6.

[28] Einstein MH, Schiller JT, Viscidi RP, et al. Clinician's guide to human papillomavirus immunology: knowns and unknowns. Lancet Infect Dis 2009;9:347–56.

[29] Muller M, Gissmann L. A long way: history of the prophylactic papillomavirus vaccine. Dis Markers 2007;23:331–6.

[30] McNeil C. Who invented the VLP cervical cancer vaccines? JNCI 2006;98:433.

[31] Padmanabhan S, Amin T, Sampat B, et al. Intellectual property, technology transfer and manufacture of low-cost HPV vaccines in India. Nat Biotechnol 2010;28:671–8.

[32] Editorial. Flogging Gardasil. Nat Biotechnol 2007;25:261.

[33] Palmer T, Wallace L, Pollock KG, et al. Prevalence of cervical disease at age 20 after immunisation with bivalent HPV vaccine at age 12-13 in Scotland: retrospective population study. BMJ 2019;365:l1161.

[34] Allen A. Vaccine: the controversial story of medicine's greatest lifesaver. New York: WW Norton and Company; 2007.

[35] Bliss E. On immunity: an inoculation. Graywolf Press; 2014.

[36] Mnookin S. The panic virus. New York: Simon and Schuster; 2011.

FURTHER READING

Walboomers JM, Jacobs MV, Manos MM, et al. Human papillomavirus is a necessary cause of invasive cancer worldwide. J Pathol. 1999;189:12–19.

National Cancer Institute, Division of Cancer Prevention. The ASCUS/LSIL triage study for cervical cancer (ALTS), <https://prevention.cancer.gov/c;inical-trials/major-trials/ascuslsil-traige-study> [accessed 22.01.19].

Cuzick J, Sasieni P, Davies P, et al. A systematic review of the role of human papillomavirus testing within a cervical screening programme. Health Technol Assess 1999;3:14.

CHAPTER

A BRIEF HISTORY OF CERVICAL CANCER

1

David Jenkins

Emeritus Professor of Pathology, University of Nottingham, United Kingdom; Consultant in Pathology to DDL Diagnostic laboratories, Rijswijk, The Netherlands

Cervical cancer is not a new disease. The history of how it was experienced, viewed, and managed in the past has been studied as a part of women's history and the history of medicine and gynecology. The story is important for understanding how current knowledge developed and provides a link to how it is still seen in some societies outside the range of the modern Western medical tradition. This becomes very relevant to guiding attempts at global eradication.

The way in which diseases are defined and understood depends on the whole framework of ideas and customs that a society has developed, not just the doctors' view or that of the patient. Cervical cancer has been for at least two millenia the archetypal "woman's cancer" because of its frequency and obvious unpleasnat symptoms." The knowledge of the nature of cancer has, of course, changed hugely: clear pathological definition of cancer itself did not develop until the second half of the 19th century in a way that would be recognizable to modern medicine. Instead, the dominance of cervical cancer was a result of its frequent and dramatic, unpleasant, external manifestations of uncontrolled vaginal bleeding, and offensive discharge with severe pain and an often isolated death. Sometimes examination of the vagina would reveal ulcers and swelling, "tumor," of the cervix to provide support for this diagnosis.

This led to cancer as a diagnosis, being seen as a "woman's disease" before scientific autopsy provided better recognition of other cancers. The strict limitation of cancer to women also meant that the relationship of an almost entirely male medical profession to women's encounter with cancer was complicated. The involvement of doctors was often limited to the very late stage of the disease when management of the illness within the home became impossible. Furthermore, midwives dealt with many of women's reproductive problems and medical intervention in cervical tumors and ulcers was ineffective and sometimes fatal if a cure was attempted. Otherwise, it was limited to trying to relieve symptoms such as offensive discharge or pain.

It was not until pathology was extended from the 1830s by the application of the microscope to normal and abnormal tissues and cells that cervical cancer began to be defined more precisely. Right from the start microscopy was criticized for its solipsism and problems of reproducibility with arguments about its clinical utility, but nonetheless with Virchow's *Cellular Pathology* published in 1858, there developed the foundations of modern understanding of the process of neoplasia and a definition of cancer based on tissue of origin and the behavior of cells. By the end of the century, pathologists were giving advice on whether a tumor should be removed, but even Virchow

Human Papillomavirus. DOI: https://doi.org/10.1016/B978-0-12-814457-2.00001-5

himself made misdiagnoses, very unfortunately in the case of the laryngeal cancer of Frederick III, Emperor of Germany. However, by the early 20th century, the definition of cervical cancer was now firmly based on microscopic pathology and not on symptoms, limited clinical examination, and ultimately the death of the unfortunate woman.

UNDERSTANDING CERVICAL CANCER—FROM PALEOPATHOLOGY TO THE 19TH CENTURY

Papillomaviruses have been found to be the key to cervical cancer and some other important cancers within the last 40 years. However, modern phylogenetic research into the almost unlimited range of different papillomavirus genomes, the host specificity to many different animals and slow mutation rate has led to awareness that this virus family can be traced back for several hundred million years. It is therefore likely that cervical cancer, unlike cancers that are related to risk factors of the modern lifestyle, has been an important disease of women from prehistory. This gives a perspective that some types of Human Papillomavirus (HPV) have taken advantage of ordinary human sexuality as a part of their evolutionary approach to maintaining their existence, with unfortunate side effects for some humans, as is discussed in Section 1 of this book.

The problem with determining how long cervical cancer has existed with any certainty is that cancer of the cervix is unlikely to be reflected in purely skeletal remains, so most archeology is not that helpful, and retrospective diagnosis on mummies is often uncertain. There is a description of HPV in a genital warty lesion in a medieval mummy of an Italian Renaissance noblewoman who died in 1568, which was shown to contain HPV18 (*Lancet* 2003), but otherwise the early history relies entirely on the surviving descriptions of symptoms and signs, death as an outcome, and naked-eye appearances of cervical lesions by ancient medical authors. Cervical cancer, as we understand it today, was not really defined until the development of microscopic cellular pathology in the second half of the 19th century and early 20th century.

Nonetheless there are descriptions of probable uterine tumors in an Egyptian papyrus of 1700 BCE, and in the Hippocratic writings of the 4th century BCE. One of the most learned, critical and lucid, surviving, ancient texts is that of the Greek physician Soranus who was born in Ephesus, studied in Alexandria and eventually practiced in Rome in the early 2nd century CE. His *Gynecology* includes a description of a kind of vaginal speculum and also of cervical scirrhous and ulcerating lesions that almost certainly included cervical cancers (Figs. 1.1 and 1.2). He was a "Methodist" eschewing the theoretical study of etiology and humoral ideas, but looking for patterns of disease rather than unfiltered experience. Other classical medical authors also discussed "malignant thymus" of the womb and Aetius of Amida (CE 502–575) summarized available information on "uterine chancres" and divided them into ulcerative and nonulcerative types. The Dutch surgeon Nikolaas Tulp (1593–1674) is credited with the first successful surgical removal of the cervix and Jean Astruc (1684–1766) described both pelvic inflammatory disease and "tumors of the cervix"—the distinction between benign and malignant growths being made using the patient's ultimate fate. In the 17th and 18th centuries many drugs, including belladonna, strychnine and lead were applied to uterine tumors, but other doctors such as Herman Boerhaave treated these with milder substances.

FIGURE 1.1

Pictures of uterus and cervix from a manuscript of about CE 900. There are no surviving illustrations from the works of Soranus, but this is from a treatise on gynecology by Muscio around CE 500 who borrowed heavily from Soranus.

From Ilberg. Die Uberlieferung der Gynakologie des Soranus von Ephesos. Leipzig: Teubner; 1910. Courtesy Wellcome collection.

TRANSFORMING WESTERN MEDICINE AND THE RISE OF MEDICAL SCIENCE

This picture of cervical cancer began to change with the development of morbid anatomy in the 18th century. The first drawing of a cancerous uterus appeared in Matthew Baillie's *The Morbid Anatomy of Some of the Most Important Parts of the Human Body* in 1793, but whether this was cervical cancer is unclear. Doctors were interested in this new morphological approach but could not see the relevance to practice based on symptoms and trying to relieve them. In postrevolutionary France in the 1790s, the Paris school formed the basis of Western medicine in the first half of the 19th century, based on detailed observation of patients, routine autopsy, and the use of statistics. The French surgeon and gynecologist Joseph Recamier was credited as reviving the use of the vaginal speculum, although its main use was to detect venereal disease in prostitutes. The speculum was not permitted for midwives and its use linked to deviant female behavior, which made it an unwelcome symbol of male power for women (Fig. 1.3). Nonetheless, it provided the basis for the scientific development of gynecology, and the use of local treatments, although separation of scirrhous and ulcerating lesions from "true" cancer remained difficult.

Ideas on the origin of cancer were also developing. In 1806 the French doctor Auguste Rossignol explained that "the majority of experts agree that the cancerous virus is produced at the site of developments of this disease and infects the whole system only when it is carried through the circulation." This was connected to the idea that cancer could be prevented or cured through elimination of lesions that might become cancerous. Opportunity for this was, however, limited by difficulties in diagnosing an early cancer from other "scirrhous" lesions.

FIGURE 1.2

Use of vaginal speculum c.1657. This shows a three pronged speculum similar to the kind in use since classical times.

Courtesy Wellcome collection.

Although some "heroic," painful, and often fatal surgery, cautery, and application of destructive chemicals took place in the treatment of suspect cancers, it was not until the development of anesthesia and antiseptic, then aseptic surgery that surgical treatment began to become more than occasional. Fig. 1.4 shows the surgical specimen of one excision of a suspect cervical cancer by the Victorian gynecologist, Sir James Simpson, a few years before he introduced anesthesia into obstetrics. Unfortunately, this might well be a giant condyloma not a cancer. The woman survived. The 1870s and 1880s were a period of rapid expansion of surgery. The German gynecologist Karl Schroeder (1838–87) recommended high amputation of the cervix; in 1878 Wilhelm Alexander Freund successfully removed a cancerous uterus by abdominal hysterectomy; and in 1891 Frederich Schauta proposed vaginal hysterectomy. This only carried a 25%–30% operative mortality compared to over 50% for more radical surgery. By the end of the 19th century hysterectomy had become the standard for tumors confined to the uterus, although there were few long-term survivals. In 1898 the Austrian gynecologist Ernst Wertheim introduced his radical abdominal hysterectomy, and within 10 years, he had reduced mortality from 20%–30% to 10%. Long term survival was still only 10%–20% at 5 years but this was better than vaginal hysterectomy.

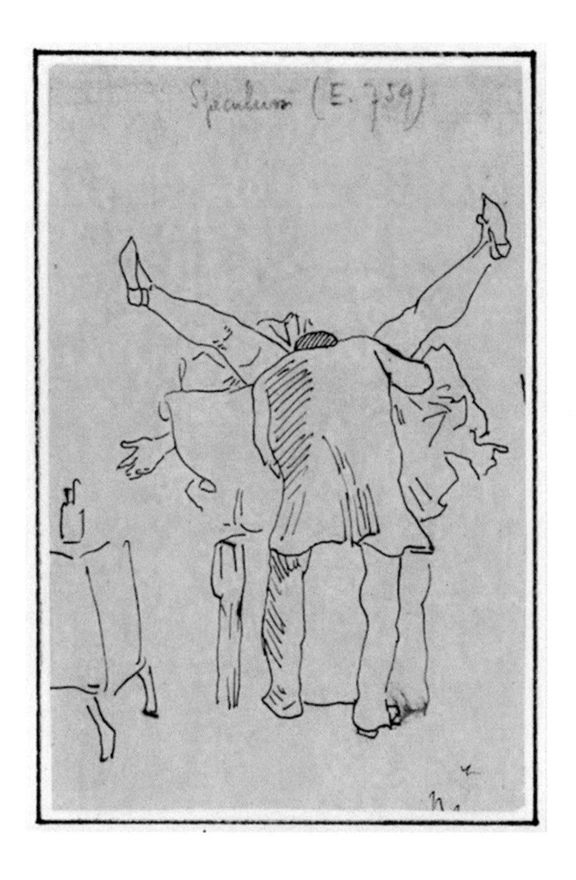

FIGURE 1.3

Felicien Rops (1833–98) was a Belgian artist who moved to Paris and whose work mingled sex, death, and irreligious images. He drew this satirical interpretation of a 19th century speculum examination.

Courtesy Wellcome collection.

In the second half of the 19th century, cancer was becoming increasingly more precisely defined by microscopy. Following the approach of Rudolf Virchow, two doctors from the gynecological clinics of the University of Berlin, Carl Ruge and Johann Veit in 1880 analyzed tissues from women operated on by Karl Schroeder, found only half had confirmed malignancy and recommended biopsy before surgery. As well as defining the differences between endometrial and cervical cancer and showing the much higher frequency than of cervical cancer, the first description of carcinoma-in-situ emerged in 1886 by John Williams, and by the early 20th century, other such cases were being reported from time to time. There was interest in possible infectious causes of cancer, including protozoa (Fig. 1.5).

Soon after the discoveries of X-rays and of radium, in the early 1900s, they were being employed to treat gynecological malignancies as well as skin cancers. This introduced a new treatment with lower mortality than radical surgery, but with new and sometimes severe complications. It did not provide a complete solution.

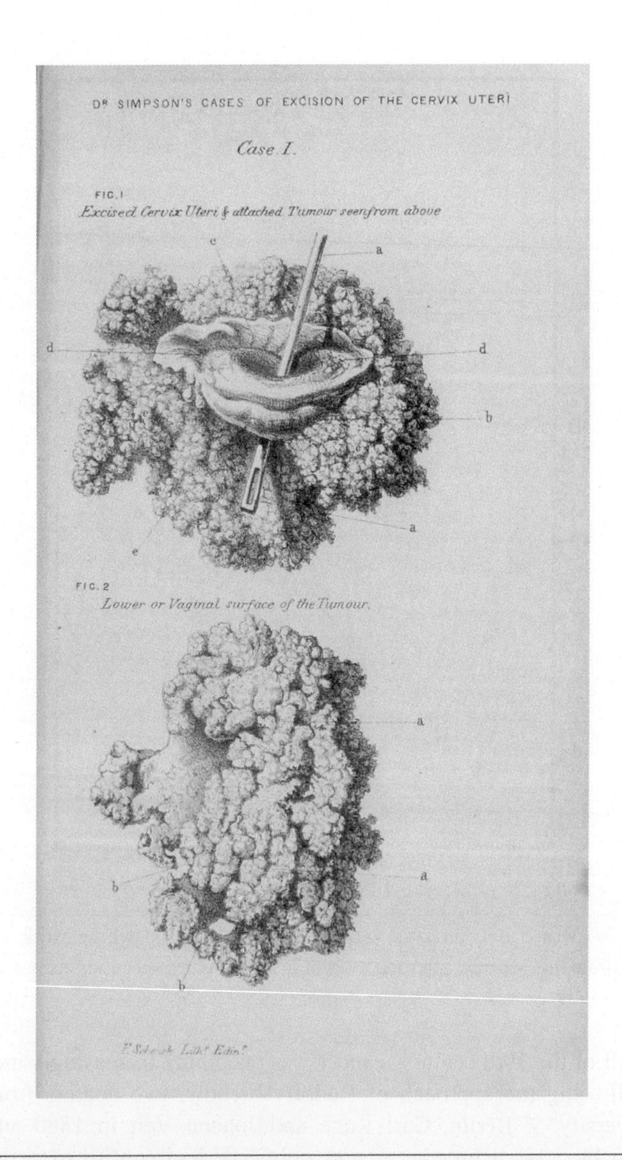

FIGURE 1.4

A specimen of excised cervix from Sir James Simpson and reported in "Cases of excision of the cervix uteri for carcinomatous disease" in 1846. This was probably preanesthetic (1840) and the patient had a subsequent pregnancy and was alive 6 years after surgery. Was this really a giant condyloma?

Courtesy Wellcome collection.

THE IDEA OF PREVENTION (FIG. 1.6)

In the 1920s and 1930s, several factors led to the idea that prevention was better than cure, and might be achievable. In Britain, at the request of a committee set up by the Minister for Health, Janet

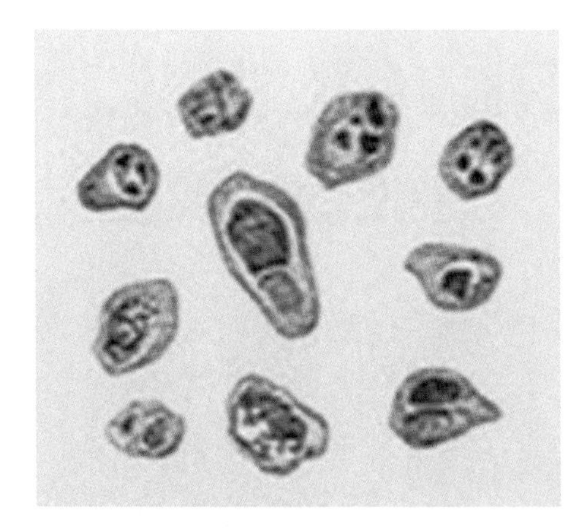

FIGURE 1.5

Cytological drawing of carcinoma cells showing so-called protozoal contents, probably karrhyorectic fragments, showing the continual search for an infectious cause.

From Carl Abel, Gynecological Pathology, 1901. Courtesy Wellcome collection.

Lane-Claypon studied the statistics on the effectiveness of cervical cancer therapies. She identified that many cancers were only detected late and were nearly always fatal. Cancers confined to the cervix had a reasonably good prognosis but many were locally advanced tumors with variable, unpredictable outcomes. Overall rates of cure were low, around 18%. The frequency of presenting with late-stage tumors (75% at the Mayo clinic) was confirmed when standard staging was introduced and adopted widely in the 1930s. All this led to pressure to detect cervical cancer early by some kind of regular examination.

In 1924 the German gynecologist Hans Hinselman constructed the colposcope to allow direct low-power microscopic examination of a woman's cervix, using acetic acid to disclose abnormal epithelium. Unfortunately, his prominent role in the National Socialist Party affected the spread of the use of colposcopy, particularly in the United Kingdom. In 1928 Walter Schiller added iodine as a marker of abnormal epithelium. The extent of surgical treatment of any carcinoma-in-situ found was controversial, until a Philadelphia gynecologist Catherine Macfarlane started a pilot program in 1938 for the early detection of cervical cancer with removal of the entire affected area by circular biopsy or trachelectomy, introducing local treatment.

In the same period, the use of vaginal smears to detect cervical cancer by microscopic examination was recognized by Babes in Romania and Georges Papanicolaou in the United States in 1928. Papanicolaou did not pursue this until a publication with Herbert Traut in 1941. In 1943 they showed that medical technologists could be trained to read cervical smears. Direct sampling was rediscovered by Ayres in 1949, who recognized "halo" cells, later called koilocytes and formed part of the linking of cervical carcinogenesis to HPV.

The goal of this exercise was part of a general attempt to detect invasive cancer at an early stage when treatment should be possible, thus reducing cancer mortality, but it soon became evident that a range of lesions apart from and less severe than early invasive cancer or carcinoma-in-situ was being found very commonly. These often required only simple local treatment or a biopsy to treat, as shown by Lionel Koss. In the 1960s, cervical screening by cytology and

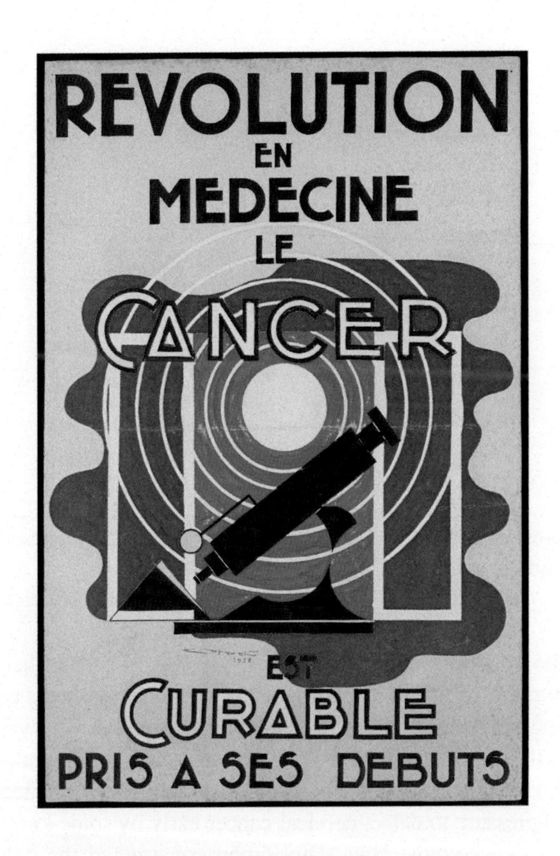

FIGURE 1.6

A French poster from the early 20th century with the message of early detection of cancer.

Courtesy Wellcome.

treatment by cone biopsy introduced the age of prevention. However, as early as 1956, a study of 25 US pathologists showed that interpretation of cytology was not very reproducible, indicating one of the major problems that was to come as cervical screening became more widespread.

CERVICAL CANCER PREVENTION BY CERVICAL CYTOLOGY SCREENING

The development of Pap screening from the 1950s to the 1970s was somewhat haphazard and varied from country to country and even area to area, with the variable introduction of cervical screening depending on a mixture of local enthusiasm and political will, both within the medical profession and in national and local politics. In British Columbia, Arthur Boyes showed that screening could be done using a central laboratory. Based on this pilot study, a government program was established in 1949. In some countries, notably Finland, Sweden, and Denmark and also Japan, highly organized screening was introduced in the 1960s and produced a reduction in incidence of cervical cancer of 50%–75%. In the United States, the American Cancer Society (ACS) sponsored a network of 125 cancer detection clinics dedicated mainly to cervical malignancies, but while the ACS endorsed more research and standards for diagnosis, the maintenance of cervical screening as predominantly the activity of gynecologists in private practice limited the overall success of screening with marked differences in cervical cancer rates racially and between states.

In Britain, cervical screening initially grew in a way determined by individual interest and enthusiasm. Support from pathologists and gynecologists was very variable, and this was not seen as important for the NHS. Much of the patchy cytology service in the early 1960s was provided by a small number of married women doctors who saw a professional opportunity. The Ministry of Health developed a national cervical cytology service from 1964 to 1966, supported and driven by the National Cervical Cancer Prevention Group representing an alliance, among others, of cytologists, medical women, left-wing politicians, and the Family Planning Association. One of the arguments they put forward was that the introduction of the contraceptive pill might increase cancer risk, but no one mentioned the changing position of women and how the long sexual revolution was progressing.

The NHS initially decided to establish regional screening centers, but although the numbers of cervical smears and technicians increased dramatically, the screening was not successful and finally screening was switched to being a responsibility of general practitioners. The important UK epidemiologist Richard Doll concluded, mistakenly, in 1968 from a study in Canada that cervical cytology was not working, and this delayed interest and improvements in the United Kingdom until the Walton report in 1976 in Canada showed the effect of screening and separated this from the reduced mortality as a result of developments in treatment. In the 1970s and 1980s, the formation of the British Society for Colposcopy and Cervical Pathology led to a more detailed understanding of the relationship between cytology interpretations and the underlying pathology, and also began to focus attention on the role of the male in cervical cancer, and the search for a transmissible agent, especially once the probes for HPV DNA were made available from Heidelberg.

An important issue in the United Kingdom was that cervical screening, overall, was not working well, and in 1988, a national program was organized including, a defined age range and screening interval, payment to GP's for achieving high coverage of eligible women, attention to better sampling, quality control of cytology, and guidelines for different aspects of screening, colposcopy, and pathology practice, with audit of practice and failures to prevent cervical cancer.

The result of this was to establish, by the 1990s, a functioning, effective cervical screening program based on conventional cytology and cytology follow-up of women with low-grade abnormal smears, supplemented by growing use of colposcopy, but not one favoring the use of HPV-based technology. In the early 1980s, it became evident to those who were actively working in the field in the United Kingdom that HPV was important and by 1985–86, clearly to those who were interested that this was the cause of cervical cancer, although it took more than another 10 years to provide the absolute evidence to convince the rest of the United Kingdom establishment and other skeptics that HPV provided a key way forward.

ENCOURAGING PARTICIPATION IN CERVICAL CANCER SCREENING

Successful cervical screening needs active and positive involvement of women who should be screened to ensure their regular participation in screening, necessary follow-up, and treatment.

In the 1960s, the Women's National Cancer Control Campaign in the United Kingdom produced a series of upbeat short films aimed at working-class women suggesting that any treatment was unproblematic and was their duty toward their family. In the United States, similar films emphasized personal happiness and freedom from cancer threat. None of these solved the problems of participation in cervical screening programs. Between 1950 and 1960, the oral contraceptive pill was developed and marketed. This pushed the sexual revolution that had been developing since 1800 into a new phase in which sexual intercourse for women was separated from economic and social consequences, changing the position of women in western society dramatically. In the 1980s, women's experience of cervical screening and its outcomes became much more of a concern among those running screening programs or managing women who had abnormal results. Activists were willing to express that the screening process generated a lot of anxiety. Studies from the 1980s onward showed screening distress and an impact of treatment on sexual attitudes, and provided a basis for future studies on the psychology of screening.

In parallel with this, there was increasing concern about the costs of scientific medicine and implementing new treatments and preventive programs. This led to the involvement from the 1980s of mathematical modelers and health economists to examine the cost-effectiveness of different screening intervals, diagnostic approaches (particularly HPV-testing vs cytology and colposcopy) and evaluate new potential and actual approaches to cervical cancer prevention.

EVOLUTION OF IDEAS ON THE CAUSES OF CERVICAL CANCER, MALE ROLE, AND TRANSMISSION OF A CAUSATIVE AGENT

In the early 1800s, there were many different suggestions about the cause of cancer of the womb. Cervical cancer was not yet separated as a disease from endometrial cancer. Cancer of the womb was linked by some to sexual excess. The Canadian doctor Guilliame Vallee argued in 1826 that the frequency in lower-class urban women could only be explained by "their greater moral laxity." In 1842 an Italian doctor, Domenico Rigoni-Stern, presented a paper on "statistical facts relative to the disease of cancer," pointing out that uterine cancer was infrequent but breast cancer was common in nuns. A British physician JCW Lever had observed in 1839 that cancer of the womb was rare in single women. By 1908 a study by Luisa Garrett Anderson and Kate Platt in London showed that cancer of the cervix occurred in younger women and those with many children, whereas endometrial cancer was a disease of older women and frequent in the childless. At the turn of the 20th century also, evidence was added that cervical cancer was infrequent in Jewish women, but common in African Americans. Although circumcision seemed to have a protective role, laceration in childbirth, and chronic inflammation/irritation were seen as the likely causes of cervical cancer. This was the doctrine presented in Boyd's standard textbook of *Surgical Pathology* up to 1942.

Some kind of transmissible factor seemed plausible. Its nature was elusive and sperm was considered a possibility. Although a viral cause of cancer in animals was demonstrated from early in the 20th century with the Rous Sarcoma Virus and the carcinogenetic potential of rabbit papillomaviruses in cotton-tail and domestic rabbits, it was not until the 1960s that the Epstein–Barr virus was shown to be associated with the rare Burkitt's lymphoma in African children. The sexual nature of genital warts was recognized by the Greeks and Romans, and the cell-free transmission of warts established in 1907 by Ciuffo. Electron microscopic analysis of particle agglutination studies in 1969 showed that genital and skin warts were different.

This background set the scene for the key studies that established the role of HPV in cervical and then other anogenital and oropharyngeal cancers. Harald zur Hausen recounts that he and his colleagues in Heidelberg began studies trying to establish a relationship between papillomavirus infection and cancer in 1972, based on anecdotal reports of malignant conversion of genital warts into carcinomas. This was boosted in his group by the failure to find herpes simplex type 2 virus in cervical cancer biopsies, although it led to skepticism elsewhere when HPV followed as a potential causal agent. The isolation of novel HPV types HPV 6,11 from genital warts and laryngeal papillomas and of HPV from an invasive giant condyloma in 1980–82 was followed by the finding of HPV16 in cervical cancer by Southern blot hybridization in 1983 and then HPV18 in 1984 by the same group, including Lutz Gissmann and Matthias Durst.

Other evidence for the role of papillomaviruses in neoplasia was also accumulating from cervical cytology and from studies of bovine papillomavirus (BPV). The finding by Ayres of "halo cells" in cervical smears, renamed "koilocytotic atypia" by Koss and Durfee in 1956 and noted by Koss in 1961 to be similar to warty atypia, was reevaluated in 1976–77 by Meisels, Fortin, and Roy and by Eva Savia and E. Purola in Finland as representing the cytopathic effect of HPV infection and a link between cervical cancer and condyloma. From being a casual finding, disregarded in routine pathology practice, it became an important observation in determining whether HPV might be involved in a woman's cervical lesion. The antibody against the BPV virus particle LI coat protein became the first routine and straightforward, immunohistochemical test available in routine pathology to demonstrate the presence of a productive HPV infection.

This chapter shows how cervical cancer historically was an unpleasant, painful, slowly fatal disease of women for millennia, as it still is in many parts of the world despite being susceptible to control. It also shows how ideas of its pathology, treatment, prevention, and cause evolved gradually for two centuries since the enlightenment. It sets the medical, scientific, and social background for the rapid growth in knowledge and understanding that happened from the 1970s, especially since the first International Papillomavirus Meeting in Lyons, France, in 1975. The knowledge acquired since then forms the detailed explorations of HPV science in the following chapters of this book.

FURTHER READING

Boyd W. Surgical pathology. 5th ed. Philadelphia: Saunders; 1942.

Bynum WF, Hardy A, Jacyna S, et al. The Western medical tradition 1800-2000. Cambridge: Cambridge University Press; 2006.

Davis D. The secret history of the war on cancer. New York: Basic Books; 2007.

Duffin J. History of medicine. Toronto: University of Toronto Press; 1999.

Fornaciari G, Zavagkia K, Giusti L, et al. Human papillomavirus in a 16th century mummy. Lancet 2003;362:1160.

Gasparini R, Panatto D. Cervical cancer: from Hippocrates through Rigoni-Stern to zur Hausen. Vaccine 2008;27:A4–5.

Lowy I. Preventive strikes: women, precancer and prophylactic surgery. Baltimore: The Johns Hopkins University Press; 2010.

Lowy I. A woman's disease: the history of cervical cancer. Oxford: Oxford University Press; 2011.

Moscucci O. The science of woman. Cambridge: Cambridge University Press; 1990.

Temkin O. Soranus gynecology. Baltimore: Johns Hopkins University Press; 1956.

Virchow R. Cellular pathology - as based upon physiological and pathological histology trans: Frank chance from 2nd German Edition (1863). New York: Dover; 1971.

Wellcome Witnesses to Twentieth Century Medicine. History of cervical cancer and the role of the human papillomavirus 1960-2000. Wellcome Trust Centre for the History of Medicine, UCL, vol. 38; 2009.

Zur Hausen H. Papillomaviruses in the causation of human cancers-a brief historical account. Virology 2009;384:260−5.

SECTION 1

PROVING THE ROLE OF HUMAN PAPILLOMAVIRUS IN CERVICAL CANCER: STUDIES IN THE LABORATORY, CLINIC, AND COMMUNITY

INTRODUCTION

Following the identification of HPV6 and HPV16 in a sample of cervical cancers by Harald zur Hausen, Lutz Gissmann and Matthias Durst, the immediate questions were how did Human Papillomavirus (HPV) act and how important was it likely to be in causing cervical cancer and precancer? Some investigators were excited by the discovery, but others, from the experience with Herpes virus, were sceptical. In the early 1980s HPV16 and HPV18 appeared to be present in up to 50% of the limited numbers of cancers and precancers studied by Southern blotting and then by radioactive such methods as in situ hybridization, but the nature of the remaining fraction of cervical neoplasia, apart from those low-grade and intermediate lesions found by cytological screening and clearly associated with HPV6 and HPV11, was uncertain.

The discovery of the association between HPV and cervical cancer and precancer, and the mechanism of action of HPV as an oncogene is described by Magnus vonKnebel Doeberitz and coauthors in Chapter 2, Linking Human Papillomavirus to Human Cancer and Understanding Its Carcinogenic Mechanisms, together with an account of how the key interactions between HPV genes E6 and E7 in HPV16 and the tumor suppressor genes p53 and retinoblastoma gene (Rb) respectively were identified. The chapter also shows the importance to defining the role of HPV in cancer of the work done in many laboratories in different countries on papillomaviruses and other oncogenic animal viruses. These findings opened up a more detailed exploration of the range and genetic relationships of HPV's and subsequent studies continuing over many years and still continuing to distinguish the presence of different HPV types in cervical and other cancers and precancers.

The studies leading to the recognition of the substantial range of HPV types involved in different degrees in neoplasia, and their relationships and characteristics are detailed in Chapter 3, Demonstrating the Importance of Different HPVs in Cervical Cancer and Other HPV-Related Cancers. Chapter 4, Finding How Human Papillomavirus Alter the Biochemistry and Identity of Infected Epithelial Cells, discusses the development of knowledge of how high-risk HPV affects the biochemistry of the infected epithelial cell, its growth control, and relations to other epithelial cells and to the local specific and non-specific defence systems. This chapter shows in detail how a small DNA virus can produce extremely complex effects on the chemistry of infected epithelial cells, blocking the innate immune system, triggering, and subverting cell defence mechanisms to survive and reproduce in a replicating, moving epithelium that is normally programmed to terminal differentiation, by ensuring continued cell replication of cells moving up the epithelium and disrupting cellular genes, with carcinogenesis as a by-product of the virus' replicative needs. In Chapter 5, Biology of the Human Papillomavirus Life Cycle: The Basis for Understanding the Pathology of Precancer and Cancer, this discussion is extended to consider specifically the way in which HPV interacts with the structural differentiation of cervical epithelium as seen microscopically. The chapter explores how the interaction of the mechanisms of HPV replication and carcinogenesis with the complexities of cervical microanatomy influences the outcome of HPV infection and the patterns of abnormality seen microscopically. It explores how this is affected by different HPV types and also the role of HPV infection in the specific issue of cervical adenocarcinoma.

Understanding how HPV led to the development of intraepithelial neoplasia and ultimately invasive cancer has been pursued in different ways. Attempts to relate HPV to the morphological pathways defined by cytology and histology prior to its discovery have been practically important since the start of clinical HPV research. The clinical management of screen-detected lesions demands biopsy of any visible lesion at colposcopy and then classification of lesions by histopathology into categories that can then be used to decide and guide treatment. Using knowledge of HPV to develop better pathological classification systems has proved an important and complex issue, but one essential to good clinical practice.

Chapter 6, The Pathology of Cervical Precancer and Cancer and its Importance in Clinical Practice, explores some of the issues involved in attempts to develop a practical, reproducible clinicopathological classification for use in routine practice that recognizes the role of productive and transforming HPV infection and the contribution of different HPV types have been made since the 1980s. Of the several attempts, the Bethesda classification developed in the USA has proved the most long-lasting, powerful, and useful system for both histology and cytology. It still has, however, a number of problems especially its limited reproducibility between pathologists and the simple division into low-grade lesions which are considered to be primarily self-limiting infections producing new virus particles and high-grade lesions, all of which are considered as precancerous and requiring treatment. This approach has the virtue of simplicity and minimizes undertreatment, but the distinction between low-grade and high-grade on histology is poorly reproducible and the classification while avoiding the likely failure to treat a progressive lesion does lead to overtreatment of nonprogressive and regressive lesions and the negative as well as positive consequences of the necessary surgical excision or destructive treatment. Attempts over many years to identify biomarkers that can provide better standardization of pathological classification and selection for treatment are discussed later in Section 3 (Chapter 17: Infection to Cancer — Finding Useful Biomarkers for Predicting Risk of Progression to Cancer).

The ability to determine accurately the presence of different HPV types in cancers and precancers across the world and hence measure the importance of HPV have depended on the development of a range of tests for different types and combinations of types of HPV, mostly based on the detection of specific DNA sequences. The development of this important technology is discussed in detail in Chapter 7, Developing and Standardizing Human Papillomavirus Tests, by Attila Lorincz and colleagues. The development of HPV detection has been strongly influenced by the discovery and development of the polymerase chain reaction although one very important early clinical HPV test, hybrid capture 2, used a different approach to detecting HPV DNA. The different tests had very different performance characteristics in terms of range of HPV types detected, sensitivity of HPV detection. Tests suitable for epidemiological and investigative pathology needed a high level of molecular sensitivity of HPV DNA detection in very small routinely prepared clinical samples. This is important for understanding the full extent of the role of HPV in global disease and the significance of the frequent multiple HPV infections seen in real life. In those tests for clinical use in triage or primary cervical screening, however, extreme molecular sensitivity leads to lack of diagnostic specificity in clinical detection of relevant precancer or cancer. Over the years test requirements and performance have become better defined and standardized, but in the early stages of research this was a major source of confusion and disagreement among experts.

In the 1980s and the 1990s the combination of studying the molecular and pathological epidemiology of HPV infection and neoplasia proved the key to developing firm knowledge, convincing to the most sceptical of the role and importance of HPV in cervical cancer on which the whole development of a new HPV-based technology depended. In Chapter 8, Proving the

Causal Role of Human Papillomavirus in Cervical Cancer: A Tale of Multidisciplinary Science, Eduardo Franco and Xavier Bosch discuss how the different views on causality between laboratory scientists with their concentration on disease mechanisms and epidemiologists looking at the importance of a cause of disease in whole human populations were reconciled. The criteria for demonstrating the causal role and importance of HPV in cervical cancer and precancer were defined, and this finally led to the formal recognition by IARC on behalf of WHO of a group of HPV types as carcinogens while others were considered as potentially having some role in carcinogenesis.

One key aspect of proving the causal role of HPV in cervical neoplasia was to demonstrate that HPV infection precedes the development of neoplasia and is not a consequence of pre-existing neoplasia. The sexual transmission of HPV raised a number of issues in relation to the nature of transmission and its relation to sexual activity and sexual behavior. The finding that HPV infection was ubiquitous and with very ordinary sexual behavior and not related specifically to "promiscuity" or "high-risk" behavior was of great importance. Study of the acquisition of cervical HPV and the role of persistent infection with an HPV type in progression of neoplasia is described in Chapter 9, The Natural History of Human Papillomavirus Infection in Relation to Cervical Cancer, along with discussion of the progression and regression of HPV-associated morphological lesions. Subsequently, the development of good HPV tests, understanding of the pathogenesis, and pathology of associated lesions and the natural history of HPV infection and lesions proved important. All was necessary in developing understanding of the possibilities of using HPV testing in triage of women with abnormal cervical smears and especially for primary screening by HPV testing and developing approaches to prophylactic vaccination. These are discussed in Sections 3 and 4.

CHAPTER

2 LINKING HUMAN PAPILLOMAVIRUS TO HUMAN CANCER AND UNDERSTANDING ITS CARCINOGENIC MECHANISMS

Magnus von Knebel Doeberitz[1], Heather Cubie[2], Thomas R. Broker[3] and David Jenkins[4,5]

[1]Department of Applied Tumor Biology, Institute of Pathology, Heidelberg University Hospital, Heidelberg, Germany [2]Global Health Academy, University of Edinburgh, Edinburgh, United Kingdom [3]Biochemistry & Molecular Genetics, University of Alabama, AL, United States [4]Emeritus Professor of Pathology, University of Nottingham, United Kingdom [5]Consultant in Pathology to DDL Diagnostic laboratories, Rijswijk, The Netherlands

Cancer develops over many years without initially causing any clinical symptoms or signs. A few cells within a tissue might gain the capacity to modify their genome, then acquire additional genomic alterations with subsequent rounds of cell renewal. These may in turn enable escape from immune recognition, promote replication, and enable cell survival. Repeated cycles of tissue wounding and healing contribute to the possibility of genetic damage. The long delay, sometimes several decades, between the initial stages and the clinical disease often obscures the causes that trigger the development of cancer.

Although "life-style" risk factors such as diet, obesity, alcohol, and smoking may variably contribute, cancers are more closely associated with genetic predisposition as well as accumulation of random mutations that lead to selective growth advantages [1]. However, some coincidences observed in the 19th century suggested a causal relationship between exposure to a suspected agent and certain distinct types of cancer. These observations were the key to developing scientific methods that led to the establishment of *cancer epidemiology*. Systematic surveys provide data that allow formulation of a scientific hypothesis concerning interaction with a possible causative agent and the later development of disease. This approach ultimately requires experimental proof using laboratory based mechanistic experiments to either verify or refute the hypothesis.

The long history of papillomavirus discovery and association with epithelial lesions and clinical diseases was well summarised through 1960s–1980s [2–8]. Bernardino Ramazzini [9], an Italian doctor, reported on the virtual absence of cancer of the uterus but a relatively high incidence of breast cancer in nuns and wondered whether the bias was related to their celibate lifestyle. The observation was an important step toward identifying sexually-transmitted infectious agents and evaluating their possible role in development of cervical cancer. This indication was substantially corroborated by the Italian pathologist Domenico Rigoni-Stern [10], who reviewed 79 years of death records in Verona to demonstrate that married women and widows were at a substantially higher risk of cancer of the uterine cervix in comparison to women who were virgins or nuns.

Human Papillomavirus. DOI: https://doi.org/10.1016/B978-0-12-814457-2.00002-7

The striking epidemiological link to sexual activity uncovered by the two physicians clearly suggested an infectious etiology and triggered an ambitious chase to identify the responsible cancer-causing agent over the course of the entire 20th century. Guiseppe Ciuffo [11] narrowed the search when he disrupted a hand wart (verrucous vulgaris), passed the extract through a Berkefeld Normal diatomaceous earth ultra-filter able to trap bacteria, and self-innoculated the filtrate into the skin of his finger. Some weeks later warts developed, thus demonstrating the infectious agent to be among the recently discovered class of microbes called viruses.

Two seminal reports in 1928, one by George Papanicolaou and the other by the Romanian pathologist Aurel Babeş [12], independently described a rapid and nonsurgical approach to diagnosis of cervical precancer and cancer (for historical evaluation, see [13]). Papanicolaou at Cornell Medical Center and gynecological pathologist Herbert Traut at New York Hospital further developed and validated the "Pap-Test" in which cytological changes in cervical keratinocytes reflect the stepwise progression of precancerous lesions to invasive cancers [14]. With widespread public health implementation by the 1950s, cervical cytology became a transformative clinical screening tool for diagnosing and preventing cervical cancer [15], as outlined in Chapter 1. Richart [16] described abnormal induction of host cell DNA replication in suprabasal cells of cervical dysplasias. Additional definitive cytological and pathological connections linking genital tract warts and condylomata with the risk of progression to dysplasias and carcinomas were made by Koss and Durfee [17], Oriel and Whimster [18], Meisels, Fortin and Roy [19] and Laverty et al. [20].

ANIMAL PAPILLOMAVIRUSES: FROM PAPILLOMAS TO CANCERS

Certain animal papillomaviruses were recognized early in the 20th century as potential carcinogens. The pioneering studies of cottontail rabbit papillomavirus (CRPV) by Richard Shope and his colleagues [21] established a crucial animal model of papillomavirus carcinogenesis that remains in use today for mechanistic studies and for evaluating antiviral strategies. CRPV research was fundamental to the emerging concepts of viral latency and reactivation, of neoplastic changes over time from papilloma to carcinoma [22–24], and of the co-carcinogenic synergy of radiation damage [25] or chemical insults [26]. The rabbit—CRPV model also enabled the earliest biophysical studies of papillomaviral proteins, nucleic acids, and virions. For example, paracrystalline arrays of virus particles were imaged by electron microscopy and correlated with the histopathologies of lesions [27]. Tumors could be induced in rabbits by transfecting Shope papillomavirus DNA [28]. Transplantable tumors that could be passaged provided a consistent and ongoing source of experimental material [29], and immunological suppression of such tumors by vaccines [30,31] were all achieved with the CRPV system. Many of those concepts and capabilities were subsequently reproduced with the multiple types of bovine papillomavirus, notably the association of sub-genomic portions of the bovine papillomavirus DNA with naturally arising equine sarcoid tumors [32] as well as the promotion of BPV type 4 lesions to esophageal carcinomas in cows that grazed on bracken ferns containing geno-toxins [33–36]. Patient-derived xenografts of HPV lesions in immunodeficient mice resulted in the first laboratory production of human papillomaviruses [37,38]. The newly discovered and characterized laboratory mouse papillomavirus [39–42] has finally resulted in an efficient and effective small animal model for papillomavirus—host research.

THE SEARCH FOR INFECTIOUS AGENTS ASSOCIATED WITH CERVICAL CANCER

Microbes such as *Treponema pallidum* and *Trichomonas vaginalis* were considered as possible pathogens responsible for cervical cancer [43,44]. However, neither of these agents had the epidemiological characteristics or biological properties consistent with an oncogenic role in cervical cancer. Herpesvirus genitalis (HSV 2) [45] is sexually transmitted and can cause persisting infections. HSV 2 was isolated from exfoliated cells of a cervical cancer [46], consistent with complement-fixing antibodies in cervical cancer patients and HSV 2 antigens in atypical exfoliated cervical cells [47]. In addition, extensive serological studies provided a positive correlation of neutralizing antibodies in women with cervical (pre-) cancer [45]. Large epidemiological studies seemed to point to genital herpesvirus 2 as a cause of cervical cancers [48]. This notion was bolstered by the observation of the Hungarian veterinarian Joszef Marek that a highly contagious T-cell lymphoma in chicken was caused by a close relative of the human Herpes simplex viruses [49].

THE FAILURE OF THE HERPES HYPOTHESIS

A young German virologist, Harald zur Hausen, working in the laboratory of Gertrude and Walter Henle at the University of Pennsylvania in Philadelphia, identified another recently discovered herpes virus, namely the Epstein–Barr Virus (EBV), in cultured cells derived from a Burkitt lymphoma [50] and also in nasopharyngeal carcinoma cells [51]. While the initial identification was based on finding viral proteins in patient sera or observing virions by electron microscopy, zur Hausen and his colleagues used new DNA hybridization techniques in which radioactively labeled cRNA probes complimentary to viral nucleic acids enabled detection of genomic sequences of viruses in DNA from tumor cells. However, all attempts to find nucleic acid fragments of HSV 2 in cervical carcinoma cells consistently failed [52,53]. In an editorial comment on a review in Cancer Research [54], zur Hausen [2] suggested that the "condyloma causing papillomavirus" might account for the neoplastic transformation observed with some genital tract infections over time. This became the focus of the zur Hausen Lab over the coming decades.

MOLECULAR TOOLS

The first biophysical analysis of any duplex DNA having a short, essentially homogeneous length was conducted with Shope CRPV. Using analytical ultracentrifugation, the study by Watson and Littlefield [55] accurately calculated the molecular weight of the genome at 5,300,000 Da, corresponding to 7852 base pairs, remarkably close to the 7868 determined a quarter century later following CRPV genome cloning and sequencing [56].

From the early 1960s, Lionel Crawford and E.M. Crawford at the MRC Institute of Virology in Glasgow explored the physical characteristics of ''Papova'' viruses (Papilloma, Polyoma, and Vacuolating agents) [57], not only their proteins but also the nature of their DNA, resulting in seminal publications on their size, icosahedral capsid structure, and protein mass [58–61]. Notably, they demonstrated that papilloma virions isolated from rabbits, cows, dogs, and highly productive human plantar warts were very similar in size, but much larger than mouse polyomavirus. The morphological issue of how viral capsid proteins could assemble spontaneously into complex icosahedral structures was resolved [62,63] and this became central to the design and manufacture of the HPV vaccines decades later.

The five papillomaviruses known at the time had significantly distinguishable A + T / G + C base composition ratios [61]. HPV, CRPV, and their host cell DNAs share similar nearest neighbor frequencies for the 16 possible dinucleotides, with the sole exception of substantial deficiency of the CpG motif in the viral DNA relative to statistical expectation [64]. Only much later were the small DNA viruses shown to undergo epigenetic methylation of the cytosine in CpG for regulation of RNA transcription (cf. [65–67], reviewed in von Knebel Doeberitz and Prigge [68]).

The initial development of ribo-probes for in situ hybridization to papilloma and carcinoma tissues associated with Shope papillomavirus were combined with immunofluorescence detection of the viral capsid proteins [69]. Seminal observations determined that viral infection induces host cell replication in noncycling, differentiated keratinocytes, that vegetative replication of CRPV DNA is differentiation-dependent, that late capsid protein synthesis follows viral DNA amplification, and that carcinomas do not support productive viral DNA replication. Additional physical studies of HPV DNA and capsid protein followed [70], along with the discovery that the virion DNA is condensed into chromatin by the typical cellular histones [71].

THE HUMAN PAPILLOMAVIRUS HYPOTHESIS

Harald zur Hausen's arguments to foster his new hypothesis included the notion that cases of condylomata acuminata share similar risk factors, in particular sexual promiscuity and the observation that some warts may (rarely) convert into invasive carcinomas. He therefore proposed an expanded search for a "condyloma agent" In the 1970s it was not known whether the anatomic site was responsible for morphological and pathological differences in patient epithelial lesions, or possibly that there were genetically distinguishable HPVs [72]. Thus early investigations of the roles of human antigenically or papillomaviruses in disease necessitated primary isolation of HPV DNA (and later the RNA) from patient lesions followed by recombinant nucleic acid cloning to create reproducible reagents.

To generate essential reagents to be used in the search for the "condyloma agent," zur Hausen's group placed plantar warts in culture medium and froze this tissue at −20°C. The frozen material was thawed, cut into small fragments, and ground with sterilized sea sand. These extracts were subsequently centrifuged to yield an aqueous supernatant which was then further purified by high speed ultracentrifugation in an equilibrium density gradient. Electron microscopy of the appropriate fraction revealed virus particles (Fig. 2.1). DNA was then extracted from these and transcribed into radioactively-labeled cRNA that was used as probes for "Southern blots" of patient lesions.

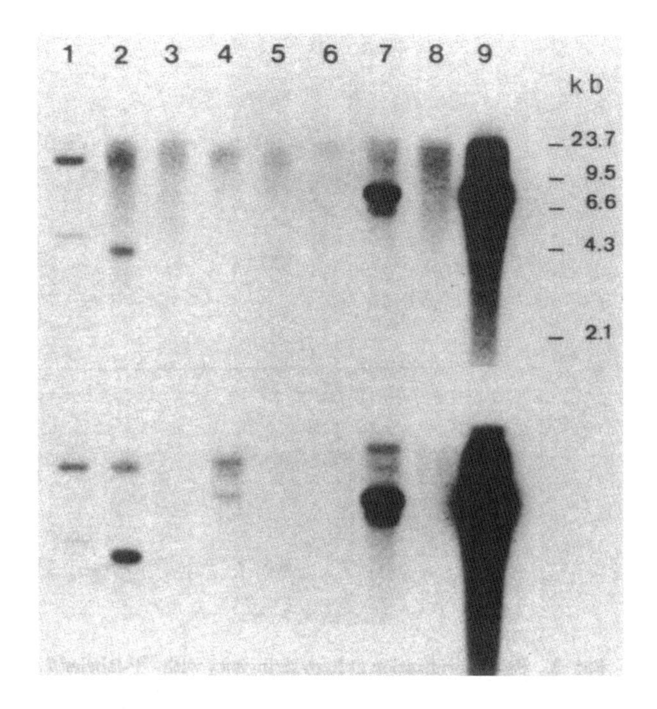

FIGURE 2.1

The identification of HPV 16 in cervical carcinoma biopsy samples: The original blot published by Dürst et al., in PNAS 1983 (PMID: 6304740). As Fig. 2.4 a Southern blot hybridization with 32P-labeled 7.2-kb (HPV 16) DNA was shown. Cellular DNAs from invasively growing cervical carcinomas (lanes 2, 4, 5, 7, and 9), one dysplasia (lane6), two carcinomas in situ of the cervix (lanes 1 and 5), and one vulval carcinoma (lane 3) were cleaved by BamHl. Hybridization and washing at low stringency (upper) and rewashing at high stringency (lower) are shown. Samples 1, 2, 4, 5, 7, and 9 were positive under both conditions. Positive bands in 2, 4, and 7 are seen more clearly after removal of unspecific background at high temperature.

SOUTHERN BLOTTING

"Southern blotting" was described for the first time in 1975 by Edwin M. Southern [73] to detect specific DNA sequences in a complex mixture of DNA. Specifically, DNA isolated from patient lesion samples was digested by sequence-specific restriction endonucleases and separated by agarose gel electrophoresis according to the size of the generated DNA fragments. The DNA within the gels was then denatured by alkaline treatment and transferred by capillary action onto nitrocellulose, or later nylon, filters. The denatured DNA was then exposed to radioactively labeled single-stranded RNA probes. Any complementary nucleic acid duplexes that formed remained on the filters after washing and were detected after exposure to x-ray films. This procedure allowed for identification of both closely matched and more distantly related sequences, depending on the solvent and temperature conditions.

CLONING HUMAN PAPILLOMAVIRUS DNAS

The zur Hausen Lab set out to recover molecularly cloned and characterized human papillomaviruses, starting with density gradient centrifugation fractionation techniques. Three of the isolates from plantar warts were closely related while a fourth was significantly different (Gissmann and zur Hausen, 1976, 1978). Viral particles had been repeatedly recovered from genital warts in 1968–1970 and identified by electron microscopy [2]. The next objective was to isolate the nucleic acids of the "condyloma related agent," as zur Hausen called it. Viral particles had been repeatedly recovered from genital warts in 1968–1970 and identified by electron microscopy [2]. Multiple genital wart specimens yielded a new virus type, which was cloned in bacterial plasmids and designed as HPV 6 [74]. This in turn was labeled with ^{32}P-ATP as a hybridization probe for DNA isolated from a large set of patient condylomata. In 41 of 44 specimens, the hybridization signals demonstrated the presence of identical or closely related nucleic acids, establishing that HPV 6 was the "condyloma associated agent" [75].

EPIDERMODYSPLASIA VERRUCIFORMIS

The rare skin disorder known as epidemodysplasia verruciformis, with susceptibility now recognized as due to a recessive genetic deficiency impacting the immune system (de Jong et al., 2018) has long been known to pose high risk of neoplastic conversion in skin exposed to solar radiation. The lesions progress to invasive carcinomas with high mortality [76]. Parallel to research in zur Hausen's lab in Erlangen, the laboratory team of Gerard Orth at CNRS Institute Gustav-Roussy working with colleagues at the Pasteur Institute in Paris and in long-standing collaborations with the dermatology clinic and lab of Stephania Jablonska in Warsaw, Poland investigated papillomavirus sequences in cutaneous skin lesions, notably those from plantar warts [70], and from epidermodysplasia verruciformis patients [77–79]. These efforts resulted in the isolation and characterization of HPV types 1, 5, 7, 8, and 9 and eventually many others (also see [80]). The isolation of multiple papillomavirus types in different laboratories necessitated a plan to compare the clones of HPV DNAs as well as animal papillomaviruses and establish a standardized procedure and reference lab for nomenclature, as adopted by a meeting in 1978 [81].

IDENTIFYING HPV 11, HPV 16, AND HPV 18

Using the Southern blot technique, Lutz Gissmann identified HPV 11 in DNA extracted from laryngeal papillomas by using HPV 6 as a hybridization probe [82]. Then in 1982 Matthias Dürst and Lutz Gissmann recovered a novel HPV sequence in one DNA sample isolated from a cervical carcinoma and molecularly cloned it using phage lambda vectors. This DNA hybridized only under nonstringent conditions to the previously identified HPV sequences and was thus regarded as a new type and tentatively assigned as "HPV 16" [83] (Fig. 2.2). Under high stringency hybridization conditions, probes demonstrated HPV 16 infections in 11 of 18 DNA samples isolated from cervical carcinoma biopsies from patients in Germany and 8 of 23 biopsies received from cervical cancer patients from Kenya and Brazil. Some vulvar and penile cancer biopsy samples also hybridized to

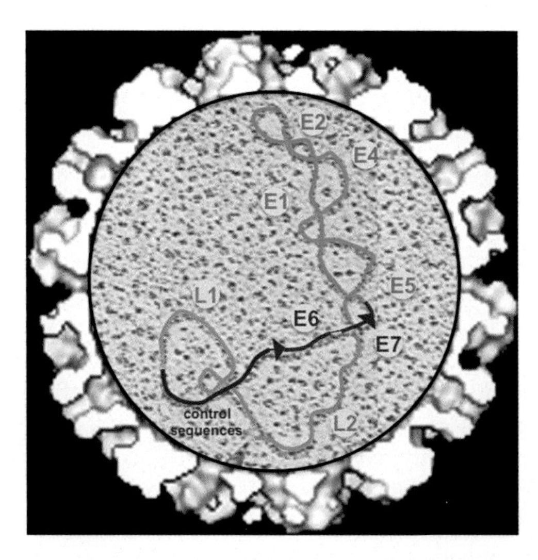

FIGURE 2.2 Transmission electron micrograph of human papillomavirus (HPV) particles.

HPV particles, colored transmission electron micrograph (TEM). HPV particles (virions) consist of a protein capsid (orange) enclosing DNA (deoxyribonucleic acid), the genetic material of the virus. HPV causes warts, which mostly occur on the hands and feet. Certain strains also infect the genitals. Although most warts are nonmalignant (not cancerous), some strains of HPV have been associated with cancers, especially cervical cancer.

this new probe, while benign condylomata accuminata apparently contained substantially less or no HPV 16 DNA. In subsequent experiments, Hans Ikenberg, a MD student in the zur Hausen lab, used ^{32}P-labeled DNA of HPV 16 to screen cellular DNAs obtained from different biopsies of Bowenoid papulosis lesions [84]. Continuing the same strategy, HPV 18 was identified in cervical carcinoma biopsies by Michael Boshart, another student working on his MD thesis in the zur Hausen lab [85]. DNA that hybridized with HPV 18 probes under stringent conditions was detected in several cervical carcinomas and in a penile carcinoma.

EXTENDING THE RANGE OF HPV TYPES

By the mid-1980s, the identification of distinct human papillomaviruses led to the investigative and clinical imperative to distinguish which HPV types are associated with various epithelial pathologies. At the time, serology using patient-derived antibodies was not able to make reproducible identifications since the isolates were immunologically cross-reactive. However, the DNA restriction enzyme digestion patterns and DNA:DNA or DNA:RNA hybridization methods were definitive. With the initial cloning of human papillomaviruses and bovine papillomavirus 1 (cf. [86]), transmission electron microscopic analysis of heteroduplexes formed between the first clonal isolates of HPV and BPV 1 DNAs and examined at different levels of base pairing stringency (taking advantage of duplex melting temperatures) demonstrated (1) that the various genomes were collinear and (2) that some genetic regions were quite highly conserved while others were fairly different, ranging from $>90\%$ matched to $<60\%$ similar [87–89]. Several of the cloned human and animal papillomavirus DNAs were sequenced between 1982 and 1985, revealing their similar genomic organization and protein coding potentials (cf. [86,90,91]). Each PV genome was approximately 7900 (+/− 300) base pairs long, with eight open reading frames along only one of the two DNA strands. Work from several laboratories demonstrated that only one strand was transcribed into messenger RNAs, unlike the polyoma viruses and the adenoviruses, in which both DNA strands gave rise to messenger RNAs. However, as with the other DNA tumor viruses, the papillomavirus genomes contained an early transcribed E region and a late transcribed (L) region, as well as what turned out to be an untranscribed region containing the DNA replication origin and transcriptional regulatory elements (enhancers and early region promoter). Putting the EM heteroduplex mapping and the sequence analysis together indicated that the two most conserved open reading frames (ORFs) were L1 and E1, while the noncoding region was least conserved.

While building a molecular biological foundation toward defining causality was underway, additional clinical pathological correlates across the disease spectrum were also characterized in which benign papillomas and condylomas were linked to neoplastic progression to cancers based on proximity and morphological transitions [92–95] (for a review, see [96]). Notably, John Kreider, Mary Kay Howett and colleagues at Penn State-Hershey Medical Center were able to transform human cervical epithelial cells with papillomavirus recovered from a cervical condyloma and experimentally generate dysplastic changes comparable to those defining clinical lesions [97].

THE ADVENT OF PCR

The basic technique of polymerase chain reaction (PCR) amplification of DNA was described [98,99]. Tom Broker, Steven Wolinsky and Louise Chow at the University of Rochester, NY suggested a strategy to Cetus Corporation by which human patient lesions could be screened for known and unknown papillomavirus genotypes using generic PCR oligonucleotide primers matched to specific sites in the conserved L1 open reading frame, with allowance for some mismatched target sequences using the nucleoside inosine as a generic base pairing partner for thymidine, cytosine, and adenosine. They designed universal primers for HPVs based on the only six HPV DNA sequences known at that time and presented them to Cetus. Subsequently, the company hired a new team to reduce the concept to practice. The primers termed MY09 and MY11 worked as predicted and remain in use to this day for HPV screening [100]. Other labs also developed PCR based strategies for screening (cf. [101–103]). Notably, application of PCR technologies to screening patient lesions from around the world led to the isolation and ultimately sequencing of large numbers of new HPV genotypes [104]. Much more recently, next-generation shotgun sequencing of genomic material for cutaneous as well as mucosal lesions has resulted in an explosive increase the numbers of HPV genotypes, now numbering in the many hundreds (cf. [105,106]).

STUDYING HPV GENE EXPRESSION IN LESIONS

Research in the electron microscopy lab of Tom Broker and Louise Chow at Cold Spring Harbor Laboratory, New York and then at University of Rochester, NY demonstrated that primary transcripts of papillomaviruses undergo alternative RNA splicing to generate diverse messenger RNAs, comparable to that found with other DNA tumor viruses and eukaryotic genes. Single molecule analyses using electron microscopic visualization of RNA: genomic DNA heteroduplexes revealed a very complex array of early and late mRNAs, arising from at least four promoters located in the early region. Broker [107–109]. Over the years, detailed transcription and splicing maps were established for many different HPV genotypes based upon direct analyses or upon prediction according to highly conserved processing sequences evident in the genomic DNA (cf. [110]). Regularly updated papillomaviral genome sequence information, transcription maps, comprehensive protein analyses, and numerous bioinformatics tools are now available through the NIH website www.pave.niaid.nih.gov.

With the knowledge of the initial PV genomic sequences and their open reading frames as well as the structures of the spliced RNAs, probes could be designed that would hybridize specifically to presumed early (E) and likely late (L) expressed coding regions. Such probes for in situ hybridization to biopsies of epithelial HPV lesions from patients demonstrated distinct spatial and temporal patterns, wherein early transcripts were expressed in a differentiation-dependent manner, primarily in the lower- to mid-spinous keratinocytes of squamous epithelia, while viral DNA amplification occurred in mid-to upper-spinous cells, thereafter followed by late region transcripts encoding the L2 minor capsid protein and the L1 major capsid protein and finally capsid assembly and maturation in the most superficial keratinocytes [111–113]. The patterns of HPV gene expression changed dramatically over the transition from low grade dysplasias to invasive cancers, with an increase in E6-E7-E1 expression in cycling cells that were no longer restricted to basal and parabasal strata, and an absence of the other mRNAs, from E2 through the late capsid genes [114]. Dependency of the stages of HPV gene expression on epithelial differentiation provided a clear explanation for why it had not been possible to grow papillomaviruses in monolayer cell cultures.

HPV 16 AND 18 IN NEOPLASTIC TRANSFORMATION: E6 AND E7 GENES

HPV 18-related DNA was found in the HeLa, KB, C4-1, and 756 cell lines. Stimulated by these findings, Elisabeth Schwarz in the zur Hausen lab started to explore the genomic structure and transcriptional activity in HPV 18 positive cervical cancer cell lines. She found that the HPV 18 genome had become integrated into the cellular genome [115]. Interestingly, the viral genome was disrupted at the end of the "early region" into two portions in each integrated copy. In HeLa and C4-1 cells, a segment of HPV 18-specific sequences from the E2 to L2 coding regions was missing, but the HPV genome was specifically transcribed from the E6-E7-E1 region into poly(A)-positive RNAs that retained at their 3'ends cellular sequences derived from the individual integration locus of the HPV 18 genome. Similar genomic arrangements were also found for HPV 16 in the long established cervical cancer cell lines SiHa and CaSki and in biopsies derived from HPV positive cervical cancers. The consistent genomic integration and transcription patterns of HPV 16 and 18 genomes in the cervical cancer cells clearly suggested a major functional role of the E6 and E7 genes with regard to initiation and possibly maintenance of the neoplastic growth of cervical and other HPV-related cancer cells, and they came to be seen as major drivers of carcinogenesis.

To establish these concepts, the zur Hausen lab used two experimental settings. First, they transfected full length HPV genomes into primary human keratinocytes and observed that after weeks and months of further culture, transformed "immortal" cell clones emerged that showed features of preneoplastic cell clones [116], with similar integration and transcription patterns to cervical cancer cells. The second key experiment was to block the activity of the HPV E6 and E7 gene functions specifically in HPV associated cervical cancer cell lines in vitro, using complimentary "antisense"-RNA molecules to the HPV E6-E7 transcripts under control of promoter elements that could be switched on and off by steroid hormones in HPV 18 positive C4-1 cells. Treatment of the respective cells with glucocorticoids blocked the translation of E6-E7 mRNA molecules and resulted in immediate degradation [117,118]. This method was known as the antisense RNA technique [119]. Importantly, expression of these complementary mRNA molecules resulted in growth arrest of the respective cells, finally proving that indeed the HPV E6-E7 gene transcripts are responsible for accelerated growth of HPV-associated cancer cells.

These findings stimulated intense research activities as to how the viral genes, E6 and E7, might provoke such devastating consequences for the infected host. In particular the laboratories of Peter Howley and Douglas Lowy at the US National Institutes of Health and of Lionel Crawford in Cambridge, UK investigated the transforming functions of papillomavirus genes in rodent and human cells using mutant forms of these two viral genes. The experiments soon revealed critical genetic functions that were linked with the transforming properties of the HPV E6 and E7 genes. Distinct sequence stretches of the viral E6 and E7 genes of oncogenic HPV types showed close homologies to other known transforming viral genes as, for example, in the simian virus 40 large T-antigen (SV40-LT) and the adenovirus E1A and E1B gene products. These proteins were known to form complexes with certain cellular tumor suppressor proteins including the retinoblastoma tumor suppressor pRB and a protein referred to as p53 that became later known as the "guardian of the genome" [120–123]. Not surprisingly, the E6 and E7 proteins of the oncogenic papillomaviruses similarly bound to and triggered premature degradation of these cellular tumor suppressor proteins, whereas the E6 and E7 proteins of benign HPV types that were not linked to malignant lesions did not show these properties [124,125]. These findings offered a rational explanation as to why only certain papillomaviruses can trigger the pathogenesis of human cancer. Notably, these biochemical studies were facilitated by earlier work of Lionel Crawford's laboratory undertaken in the London labs of the Imperial Cancer Research Foundation. The research culminated in the discovery of the significance of the p53 protein with his post-doctoral researcher David Lane [126] and further work on tumorigenesis, oncogenes, and cancer control.

ADDITIONAL PIONEERING STUDIES OF IMMORTALIZATION AND TRANSFORMATION

Once it had been established that HPV associated cancers, primarily cervical tumors, contained integrated high-risk HPV-16, HPV-18, or their close relatives, several laboratories set out to determine more about the nature of papillomavirus oncogenes [127–129]. The contributing genes of high-risk HPVs include E5 during initiation (leading up to viral DNA integration), but E5 becomes expendable during long term maintenance of the immortalized and then transformed cells which depend only on high level E6 and E7 gene expression. Suppressing either or both of these results in the death of the cells; thus cell survival is addicted to HPV E6 and E7 expression [130]. However, the animal papillomaviruses were not always congruent with the HPVs. The transforming genes of bovine papillomavirus are E7 and E5, while the cancer drivers of certain rabbit papillomaviruses are E7 plus an alternative reading frame that overlaps the E6 gene.

In contrast to the mucosotropic human papillomaviruses, current models for human cutaneous squamous cell carcinomas and epidermodysplasia verruciformis include initiating events driven by high-risk beta-family papillomaviruses, but without persistent integration of the HPV DNA. Rather the viral sequences are lost and the tumor is presumably a result of a "hit and run" mechanism in which the initiating events included viral caused mutagenesis of host cell regulatory genes (Hasche et al., 2018; [131]). Thus there is no single pathway or process by which different animal or human papillomaviruses trigger carcinogenesis. Over the years, comprehensive understanding of the interaction partners of E6 and E7 and the mechanisms of oncogenic action have been achieved (cf. [132–136]). In addition, the E6-E7 driven reactivation of DNA synthesis in noncycling differentiated keratinocytes necessary to enable vegetative HPV DNA replication, the resulting DNA damage responses, and factors leading to inadvertent DNA breakage and integration of the high-risk HPV genome into cellular chromosomes all provide complementary perspectives bearing on the stages of oncogenesis [137].

RECAPITULATION OF THREE-DIMENSIONAL EPITHELIAL TISSUES IN RAFT CULTURES AS AN EXPERIMENTAL MODEL FOR INVESTIGATIONS OF HPV-HOST TISSUE INTERACTIONS AND MECHANISMS OF CARCINOGENESIS

To establish an environment wherein papillomaviruses could carry out their reproductive program, a tissue culture system capable of forming terminally differentiating squamous epithelium would be essential. The possible means to promote keratinocyte differentiation was suggested by the pioneering work of Eugene Bell and colleagues at MIT who created "raft" cultures of skin tissue biopsies grown at the liquid medium/air interface for the purpose of expanding patient-derived tissues for re-engraftment to cover the donor's severe wounds and burns. McCance et al. [138] demonstrated that primary human keratinocytes or an immortalized cell line transfected with the HPV-16 genome could be developed into three-dimensional squamous epithelial raft cultures but that viral gene expression delays and alters normal keratinocyte differentiation, comparable in pathology to clinical dysplasias. To recapitulate productive infection, Dollard et al. [139] placed HPV-11 infected genital tract tissue on a dermal equivalent consisting of a collagen support harboring feeder fibroblasts. After 2 weeks of culturing in which new tissue grew out from the original biopsy, stratified and completely differentiated, HPV virions were generated and matured into structures identical to authentic particles, Meyers et al. [140] used a patient-derived cell line harboring HPV-31 plasmid DNA to achieve similar results. Thus the parameters necessary to investigate virus–host interactions in squamous epithelia were validated. The laboratories of Denise Galloway and James McDougall at the Fred Hutchison Cancer Center in Seattle [141–143] as well as the Broker-Chow lab in Rochester and then at the University of Alabama at Birmingham (cf. [144–146]) focused on expressing oncogenic as well as nononcogenic E6 protein alone, E7 alone or both together in raft cultures developed from primary human keratinocytes or from cell lines to characterize their interactions with host cell proteins and pathways. To assess global transcriptional changes driven by oncogenic HPVs, raft cultures comprised of HPV-18 E6-E7 expressing primary human keratinocytes [147] as well as early passage

HPV-16 immortalized keratinocytes competent for differentiation and late passage lines that are unable to differentiate have been compared to normal primary human keratinocytes and to patient derived cervical lesions ranging from CIN to cancer [148]. Without understanding the normal functions of the E6 and E7 proteins in the viral reproductive cycle, it would be difficult to explain their aberrant actions during neoplastic progression and cancer (for review, see [149]).

Subsequently, it finally became possible to recapitulate the entire infection cycle of whole genomic HPV 18 in raft cultures developed from primary human keratinocytes, with robust yields of infectious virions [150]. This system enables detailed genetic studies of virus–host cell interactions [151,152] as well as investigations of virus neutralization by sera following HPV vaccination [153] and also antiviral drug discovery and validation [154,155]. In addition, raft cultures developed from HPV 16 immortalized cervical keratinocytes that retain the ability to produce virions after stratification and tissue differentiation have enabled assessments of changes in global host gene expression compared to 3D cultures from normal cervical epithelia [156]. Together, such investigations are identifying the host cell proteins and pathways modulated by oncogenic HPVs across the disease spectrum and enabling informed consideration of targets for antiviral and anti-cancer therapeutic drug discovery.

CONCLUSION

Taken together, the early days between the discovery of HPV genomes in cervical cancers and the experimental validation of their oncogenic activity and key driver functions in 1990s and beyond revealed that HPV is the causative carcinogenic agent in uterine cervix and in malignancies of epithelia at other anatomic sites in males as well as females. Papillomavirus infections can introduce foreign viral genes into epithelial cells that may become integrated and become part of the genome of the emerging cancer cells. These genes clearly have oncogenic activities and largely explain the complex biology of these cancers caused by a virus that is not able to reproduce and multiply once integrated and partially deleted. Seen from the virus' perspective, these are clearly "bad luck" chance aberrations of interactions between viral and human genes. For humans, the "bad luck" can be prevented by blocking primary infection through vaccination against HPV. Potentially, HPV lesions can be treated, if we find ways to inactivate the expression or function of the viral E6 and E7 oncogenes in cells on their way to cancer or, even earlier, by selectively killing keratinocytes that harbor primary viral infections before they become dysplastic or experience HPV DNA integration, an irreversible event in neoplastic progression.

A VIGNETTE: EARLY CLINICAL EXPERIENCE WITH HUMAN PAPILLOMAVIRUS

Albert Singer
Gynecological Research, University College, London, United Kingdom

This is my account of early clinical experiences with human papillomavirus (HPV) between 1965 and 1990. During this time basic clinical investigations occurred which eventually led to the acceptance that HPV was the major cause of cervical and lower genital tract neoplasia.

In 1965, I joined the eminent colposcopist, the late Professor Malcolm Coppleson in Sydney Australia and Bevan Reid a vastly experienced cellular pathologist who had proposed that human sperm DNA could be the coital mutagen for the development of cervical neoplasia [157].

In 1967 Coppleson undertook a unique clinical trial which had a profound effect on our knowledge of the natural history of cervical HPV infections. In his textbook [158] he wrote that, "in one state in Australia detained promiscuous girls are under legal compulsion to submit to bacteriological examination of the vagina and opportunity was taken with permission to undertake colposcopy/biopsy."

These young women, usually with a sexually abused background, started intercourse between 10 and 13 and colposcopy revealed specific, unusual features in the epithelium which later were shown associated with a HPV infection. Coppleson took biopsies and commented that he was observing "the initial moment of the complicated sequence leading to the development of squamous cancer."

Feeling that the causative agent for neoplastic change gained access via mucus an electron microscopic we examined the ultrastructure of cervical mucus (Singer and Reid, 1970). Results showed that in the early post menarcheal years mucus had "open channels" that possibly facilitated entry of a coital mutagen to the epithelium. We proposed in a letter to the British Medical Journal in 1969 [159] that this mutagen might be spermatozoa smegma or viral DNA *after examining the biopsies from the adolescent girls which showed changes described by Koss [160] as "warty atypia."*

By early 1970s it became obvious that the male played a significant role in the origins of cervical cancer [161]. Epidemiologic, social, and behavioral data pointed to our hypothesis [162] of the "high risk male" whose partners were at an increased risk of developing cervical cancer. This led to a number of studies that showed first the various morphological patterns of both clinical

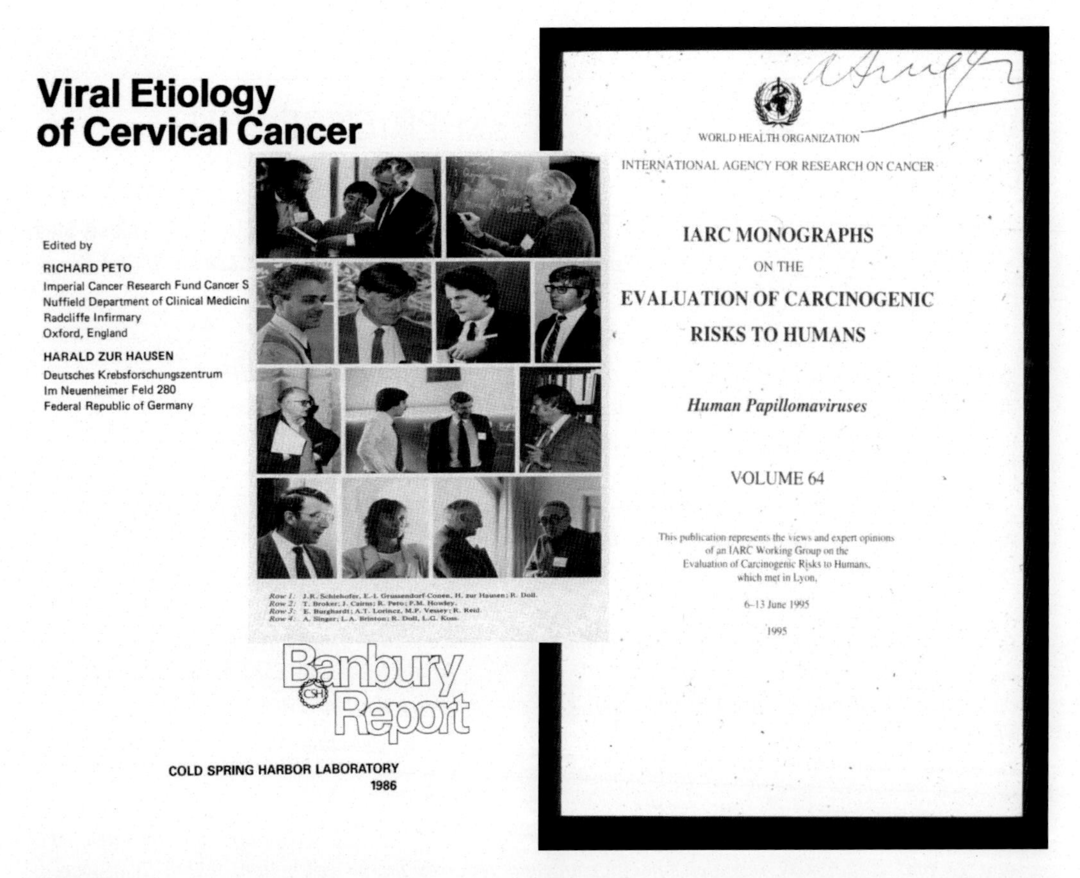

FIGURE V2.1

Front covers and inset of two of the most important meetings in the early days of clinical HPV studies. To the left is the cover of a meeting held in the Cold Spring Harbor Laboratory in 1986 and the inset shows some of the attendees. The right hand cover is that of one of the IARC monographs of 1995 evaluating carcinogenic risks to humans with volume 64 devoted to human papilloma virus. The working group of IARC who commissioned this monograph concluded that there was a carcinogenic risk to humans with HPV. This document was the start of the acceptance of the malignant potential of HPV infections.

and subclinical forms of penile HPV, and secondly that the presence of HPV in the male lead to an increase risk of cervical neoplasia in his female partners [163].

The difficult clinical question in the early 1980s was how many women with minor cytological smear changes (ASCUS/LSIL) regressed or progressed and what were the factors influencing these changes? We followed a number of these women for 3 years without biopsy and showed that 26 of the 100 with mild changes (CIN 1) turned into severe disease (CIN2/3) and 85% of these had HPV 16, as shown by hybridization studies using probes for HPV 6, 11, 16, and 18 cloned by H zur Hausen and colleagues in Heidelberg [164].

In the early 1980s there was still widespread skepticism as to the role of HPV as a potential mutagen, addressed by several editorials. It was hard to convince people about the role of HPV.

In 1986 a meeting entitled "Viral aetiology of cervical cancer" [165]. at Cold Spring Harbor Laboratory of multidisciplinary HPV experts gave the impetus for subsequent clinical and basic research that culminated in the IARC monograph in 1995 (Fig. V2.1) with the title "Evaluation of carcinogenic risks to humans; human papilloma virus" [166]. It was only then that clinicians started to take note of the possible causal relationship between HPV and cervical and lower genital tract neoplasia.

REFERENCES

[1] Tomasetti C, Vogelstein B. Cancer etiology: variation in cancer risk among tissues can be explained by the number of stem cell divisions. Science 2015;347:78–81.

[2] zur Hausen H. Condylomata acuminata and human genital cancer. Cancer Res 1976;36:794.

[3] Broker TR, Botchan M. Papillomaviruses: retrospectives and prospectives. In: Botchan M, Grodzocker T, Sharp PA, editors. DNA Tumor Viruses: Control of Gene Expression and Replication (1984 Meeting). Cancer Cells 4. NY: Cold Spring Harbor Laboratory; 1985. p. 17–36 (no PMID).

[4] Karamanou M, Agapitos E, Kousoulis A, Androutsos G. From the humble wart to HPV: a fascinating story throughout the centuries. Oncol Rev 2010;4:133–5 (Springer) (no PMID).

[5] Orth G, Breitburd F, Favre M, Croissant O. Papillomaviruses: possible role in human cancer. In: Hiatt HH, Watson JD, Winsten JA, editors, Origins of Human Cancer. Cold Spring Harbor Conferences on Cell Proliferation, vol. 4 (Book B); 1977. pp. 1043–68 (no PMID).

[6] Rowson KE, Mahy BW. Human papova (wart) virus. Bacteriol Rev 1967;31:110–31 PMID: 5342713; PMCID: PMC378272.

[7] Syrjänen S, Syrjänen K. The history of papillomavirus research. Cent. Eur J Publ Health 2008;16 (Suppl):S7–13 PMID: 18681303.

[8] zur Hausen H. Human papillomaviruses and their possible role in squamous cell carcinomas. Curr Topics Microbiol Immunol 1977;78:1–30. Available from: https://doi.org/10.1007/978-3-642-66800-5_1.

[9] Ramazzini, B. De Morbis Artificum. Padua, Italy (no PMID); 1713.

[10] Rigoni-Stern D. Fatti statistici relativi alle malattie cancerose che servivono di base alle poche cose dette dal dottore. Giornale Servire Progresso Patologia i Terapia (Series 2) 1842;2:507–17 (no PMID).

[11] Ciuffo G. Imnesto positivo con filtrato di verruca vulgare. Giorn Ital Mal Venereol. 1907;48:12–17 (no PMID).

[12] Babeş A. Diagnosis of cancer of the uterine cervix by smears. Le Press Médicale 36: 451-454 (translated andreprinted in 1967). Acta Cytol. 1928;11:217–24 PMID: 5232023.

[13] Wright Jr. JR. Cytopathology: why did it take so long to thrive? Diagn Cytopathol 2015;43:257–63. Available from: https://doi.org/10.1002/dc.23246 Epub 2015 Jan 21.

[14] Papanicolaou GN, Traut HF. The diagnostic value of vaginal smears in carcinoma of the uterus. Arch Pathol Lab Med 1941;121:211–24 PMID: 9111103.

[15] Papanicolaou GN. The cancer-diagnostic potential of uterine exfoliative cytology. CA Cancer J Clin 1957;7:124–35 PMID: 13446725.

[16] Richart RM. A radioautographic analysis of cellular proliferation in dysplasia and carcinoma in situ of the uterine cervix. Am J Obstet Gynecol 1963;86:925–30. Available from: https://doi.org/10.1016/s0002-9378(16)35248-6 PMID: 14047515.
[17] Koss LG, Durfee GR. Cytological changes preceding the appearance of in situ carcinoma of the uterine cervix. Cancer 1955;8:295–301 PMID: 14352168.
[18] Oriel JD, Whimster IM. Carcinoma in situ associated with virus-containing anal warts. Br J Dermatol 1971;84:71–3 PMID: 4324719.
[19] Meisels AR, Fortin, Roy M. Condylomatous lesions of the cervix II. Cytological, colposcopic and histopathological study. Acta Cytologica 1977;21:379–90 PMID: 268117.
[20] Laverty CR, Russell P, Hills E, Booth N. The significance of noncondylomatous wart virus infection of the cervical transformation zone: a review with discussion of two illustrative cases. Acta Cytol 1978;22:195–201 PMID: 364903.
[21] Shope RE, Hurst EW. Infectious papillomatosis of rabbits: with a note on the histopathology. J Exp Med 1933;58:607–24 PMID: 19870219; PMCID: PMC2132321.
[22] Ito Y. Relationship of components of papilloma virus to papilloma and carcinoma cells. Cold Spring Harb Symp Quant Biol 1962;27:387–94 PMID: 1395.
[23] Kidd JG, Rous P. A transplantable rabbit carcinoma originating in a virus-induced papilloma and containing the virus in masked or altered form. J Exp Med 1940;71:813–38 PMID: 19871000.
[24] Syverton JT, Dascomb HE, Wells EB, Koomen Jr. J, Berry GP. The virus-induced rabbit papilloma-to-carcinoma sequence. II. Carcinomas in the natural host, the cottontail rabbit. Cancer Res 1950;10:440–4 PMID: 15427083.
[25] Syverton JT, Harvey RA, Berry GP, Warren SL. The Roentgen radiation of papilloma virus (Shope): I. the effect of X-rays upon papillomas on domestic rabbits. J Exp Med 1941;73:243–8 PMID: 19871075.
[26] Rous P, Kidd JG. Conditional neoplasms and subthreshold neoplastic states: a study of the tar tumors of rabbits. J Exp Med 1941;73:365–90 PMID: 19871084.
[27] Strauss MJ, Bunting H, Melnick JL. Virus-like particles and inclusion bodies in skin papillomas. J Invest Dermatol 1950;15:433–44 PMID: 14794997.
[28] Ito Y, Evans CA. Induction of tumors in domestic rabbits with nucleic acid preparations from partially purified Shope papilloma virus and from extracts of the papillomas of domestic and cottontail rabbits. J Exp Med 1961;114:485–500 PMID: 19867197.
[29] Ito Y, Evans CA. Tumorigenic nucleic acid extracts from tissues of a transplantable carcinoma, VX7. J Natl Cancer Inst 1965;34:431–7 PMID: 14342590.
[30] Evans CA, Gorman LR, Ito Y, Weiser RS. Antitumor immunity in the Shope papilloma-carcinoma complex of rabbits. II. Suppression of a transplanted carcinoma, Vx7, by homologous papilloma vaccine. J Natl Cancer Inst 1962;29:287–92 PMID: 13890.
[31] Evans CA, Gorman LR, Ito Y, Weiser RS. A vaccination procedure which increases the frequency of regressions of Shope papillomas of rabbits. Nature 1962;193:288–9 PMID: 13890963.
[32] Lancaster WD, Olson C, Meinke W. Bovine papilloma virus: presence of virus-specific DNA sequences in naturally occurring equine tumors. Proc Natl Acad Sci USA 1977;74:524–8. Available from: https://doi.org/10.1073/pnas.74.2.524 PMID: 191813; PMCID: PMC392322.
[33] Jarrett WF. Transformation of warts to malignancy in alimentary carcinoma in cattle. Bull Cancer 1978;65:191–4 PMID: 212146.
[34] Jarrett WF. Bovine papillomaviruses. Clin Dermatol 1985;3:8–19 PMID: 2850862.
[35] Olson C. Animal papillomas: historical perspectives. In: Salzman NP, Howley PM, editors. The Papovaviridae, Vol. 2: The Papillomaviruses. New York, London: Plenum Press; 1987. p. 39–62 (no PMID).
[36] Sundberg JP. Papillomavirus infections in animals. In: Syrjänen K, Gissmann L, Koss LG, editors. Papillomaviruses and Human Disease. Berlin, New York: Springer-Verlag; 1987. p. 40–103 (no PMID).

[37] Bonnez W, DaRin C, Borkhuis C, Jensen KdM, Reichman RC, Rose RC. Isolation and propagation of human papillomavirus type 16 in human xenografts implanted in the severe combined immunodeficiency mouse. J Virol 1998;72:5256–61 PMID: 9573300; PMCID: PMC110112.

[38] Kreider JW, Howett MK, Leure-Dupree AE, Zaino RJ, Weber JA. Laboratory production in vivo of infectious human papillomavirus type 11. J Virol 1987;61:590–3 PMID: 3027386.

[39] Hu J, Cladel NM, Budgeon LR, Balogh KK, Christensen ND. The mouse papillomavirus infection model. Viruses 2017;9. Available from: https://doi.org/10.3390/v9090246 pii: E246. PMID: 28867783.

[40] Joh J, Jenson AB, King W, Proctor M, Ingle A, Sundberg JP, et al. Genomic analysis of the first laboratory-mouse papillomavirus. J Gen Virol. 2011;92(Pt 3):692–8. Available from: https://doi.org/10.1099/vir.0.026138-0 Epub 2010 Nov 17. PMID: 21084500.

[41] Schulz E, Gottschling M, Ulrich RG, Richter D, Stockfleth E, Nindll I. Isolation of three novel rat and mouse papillomaviruses and their genomic characterization. PLoS One 2012;7:e47164. Available from: https://doi.org/10.1371/journal.pone.0047164 PMID: 23077564; PMCID: PMC3471917;.

[42] Uberoi A, Lambert PF. Rodent papillomaviruses. Viruses 2017;9. Available from: https://doi.org/10.3390/v9120362 pii: E362. PMID: 29186900; PMCID: PMC5744137.

[43] Patten Jr. SF, Hughes CP, Reagan JW. An experimental study of the relationship between Trichomonas vaginalis and dysplasia in the uterine cervix. Acta Cytol 1963;7:187–90.

[44] Rojel J. The interrelation between uterine cancer and syphilis; a pathodemographic study. Acta Pathol Microbiol Scand Suppl 1953;97:1–82.

[45] Kessler II. Perspectives on the epidemiology of cervical cancer with special reference to the herpesvirus hypothesis. Cancer Res 1974;34:1091–110.

[46] Aurelian L, Strandberg JD, Melendez LV, Johnson LA. Herpesvirus type 2 isolated from cervical tumor cells grown in tissue culture. Science 1971;174:704–7.

[47] Aurelian L, Royston I, Davis HJ. Antibody to genital herpes simplex virus: association with cervical atypia and carcinoma in situ. J Natl Cancer Inst 1970;45:455–64.

[48] Rotkin ID. A comparison review of key epidemiological studies in cervical cancer related to current searches for transmissible agents. Cancer Res 1973;33:1353–67 PMID: 4352371.

[49] Boodhoo N, Gurung A, Sharif S, Behboudi S. Marek's disease in chickens: a review with focus on immunology. Vet Res 2016;47:119.

[50] zur Hausen H, Henle W, Hummeler K, Diehl V, Henle G. Comparative study of cultured Burkitt tumor cells by immunofluorescence, autoradiography, and electron microscopy. J Virol 1967;1:830–7.

[51] zur Hausen H, Schulte-Holthausen H, Klein G, Henle W, Henle G, Clifford P, et al. EBV DNA in biopsies of Burkitt tumours and anaplastic carcinomas of the nasopharynx. Nature 1970;228:1056–8.

[52] zur Hausen H, Schulte-Holthausen H, Wolf H, Dorries K, Egger H. Attempts to detect virus-specific DNA in human tumors. II. Nucleic acid hybridizations with complementary RNA of human herpes group viruses. Int J Cancer 1974;13:657–64.

[53] zur Hausen H, Meinhof W, Scheiber W, Bornkamm GW. Attempts to detect virus-specific DNA in human tumors. I. Nucleic acid hybridizations with complementary RNA of human wart virus. Int J Cancer 1974;13:650–6 PMID: 4367340.

[54] Goldberg RJ, Gravell M. A search for herpes simplex virus type 2 markers in cervical carcinoma. Cancer Res 1976;36:795–9.

[55] Watson JD, Littlefield JW. Some properties of DNA from Shope papilloma virus. J Mol Biol 1960;2:161–5 (no PMID).

[56] Giri I, Danos O, Yaniv M. Genomic structure of the cottontail rabbit (Shope) papillomavirus. Proc Natl Acad Sci USA 1985;82:1580–4. Available from: https://doi.org/10.1073/pnas.82.6.1580 PMID: 2984661; PMCID: PMC397315.

[57] Melnick JL. Papova virus group. Science 1962;135:1128–30 PMID: 14472429.

[58] Follett EA, Crawford LV. Electron microscope study of the denaturation of human papilloma virus DNA. I. Loss and reversal of supercoiling turns. J Mol Biol 1967;28:455–9.
[59] Crawford LV. A study of human papilloma virus DNA. J Mol Biol 1965;13:362–72.
[60] Crawford LV. Nucleic acids of tumor viruses. Adv Virus Res 1969;14:89–152.
[61] Crawford LV, Crawford EM. A Comparative Study of Polyoma and Papilloma Viruses. Virology 1963;21:258–63.
[62] Caspar DL, Klug A. Physical principles in the construction of regular viruses. Cold Spring Harb Symp Quant Biol 1962;27:1–24 PMID: 14019094.
[63] Klug A, Finch JT. Structure of viruses of the papilloma-polyoma type. I. Human wart virus. J Mol Biol 1965;11:403–23 PMID: 14290353.
[64] Morrison JM, Keir HM, Subak-Sharpe H, Crawford LV. Nearest neighbour base sequence analysis of the deoxyribonucleic acids of a further three mammalian viruses: Simian virus 40, human papilloma virus and adenovirus type 2. J Gen Virol 1967;1:101–8.
[65] Johannsen E, Lambert PF. Epigenetics of human papillomaviruses. Virology 2013;445:205–12. Available from: https://doi.org/10.1016/j.virol.2013.07.016 Epub 2013 Aug 13. PMID: 23953230.
[66] Van Tine BA, Kappes JC, Banerjee NS, Knops J, Lai L, Steenbergen RDM, et al. Clonal selection for transcriptionally active viral oncogenes during progression to cancer by DNA methylation-mediated silencing. J Virol 2004;78:11172–86 PMID: 15452237; PMCID: 521852.
[67] Wettstein FO, Stevens JG. Shope papilloma virus DNA is extensively methylated in non-virus-producing neoplasms. Virology 1983;126:493–504 PMID: 6305000.
[68] von Knebel Doeberitz M, Prigge ES. Role of DNA methylation in HPV associated lesions. Papillomavirus Res. 2019;7:180–183. Available from: https://doi.org/10.1016/j.pvr.2019.03.005. Epub 2019 Apr 10. Review. PMID: 30978415.
[69] Orth G, Jeanteur P, Croissant O. Evidence for and localization of vegetative viral DNA replication by autoradiographic detection of RNA-DNA hybrids in sections of tumors induced by Shope papilloma virus. Proc Natl Acad Sci U S A 1971;68:1876–80 PMID: 4331563.
[70] Favre M, Orth G, Croissant O, Yaniv M. Human papillomavirus DNA: physical map. Proc Natl Acad Sci U S A 1975;72:4810–14 PMID: 174077.
[71] Favre M, Breitburd F, Croissant O, Orth G. Chromatin-like structures obtained after alkaline disruption of bovine and human papillomaviruses. J Virol 1977;21:1205–9 PMID: 191643; PMCID: PMC515661.
[72] Almeida JD, Oriel JD, Stannard L. Characterisation of the virus found in genital warts. Microbios 1969;3:225–32.
[73] Southern EM. Detection of specific sequences among DNA fragments separated by gel electrophoresis. J Mol Biol 1975;98:503–17.
[74] de Villiers EM, Gissmann L, zur Hausen H. Molecular cloning of viral DNA from human genital warts. J Virol 1981;40:932–5.
[75] Gissmann L, deVilliers EM, zur Hausen H. Analysis of human genital warts (condylomata acuminata) and other genital tumors for human papillomavirus type 6 DNA. Int J Cancer 1982;29:143–6.
[76] Jablonska S, Dabrowski J, Jakubowicz K. Epidermodysplasia verruciformis as a model in studies on the role of papovaviruses in oncogenesis. Cancer Res 1972;32:583–9 PMID: 5061309.
[77] Kremsdorf D, Jablonska S, Favre M, Orth G. Human papillomaviruses associated with epidermodysplasia verruciformis. II. Molecular cloning and biochemical characterization of human papillomavirus 3a, 8, 10, and 12 genomes. J Virol 1983;48:340–51.
[78] Orth G, Jablonska S, Breitburd F, Favre M, Croissant O. The human papillomaviruses. Bull Cancer 1978;65:151–64.

[79] Orth G, Jablonska S, Favre M, Croissant O, Jarzabek-Chorzelska M, Rzesa G. Characterization of two types of human papillomaviruses in lesions of epidermodysplasia verruciformis. Proc Natl Acad Sci U S A 1978;75:1537–41.

[80] Ostrow RS, Bender M, Niimura M, Seki T, Kawashima M, Pass F, et al. Human papillomavirus DNA in cutaneous primary and metastasized squamous cell carcinomas from patients with epidermodysplasia verruciformis. Proc Natl Acad Sci U S A 1982;79:1634–8. Available from: https://doi.org/10.1073/pnas.79.5.1634 PMID: 6280194; PMCID: PMC346030.

[81] Coggins Jr. JR, zur Hausen H. Workshop on papillomaviruses and cancer. Cancer Res 1979;39:545–6 (no PMID).

[82] Gissmann L, Diehl V, Schultz-Coulon HJ, zur Hausen H. Molecular cloning and characterization of human papilloma virus DNA derived from a laryngeal papilloma. J Virol 1982;44:393–400.

[83] Durst M, Gissmann L, Ikenberg H, zur Hausen H. A papillomavirus DNA from a cervical carcinoma and its prevalence in cancer biopsy samples from different geographic regions. Proc Natl Acad Sci U S A 1983;80:3812–15 PMCID: PMC394142.

[84] Ikenberg H, Gissmann L, Gross G, Grussendorf-Conen EI, zur Hausen H. Human papillomavirus type-16-related DNA in genital Bowen's disease and in Bowenoid papulosis. Int J Cancer 1983;32:563–5.

[85] Boshart M, Gissmann L, Ikenberg H, Kleinheinz A, Scheurlen W, zur Hausen H. A new type of papillomavirus DNA, its presence in genital cancer biopsies and in cell lines derived from cervical cancer. EMBO J 1984;3:1151–7.

[86] Chen EY, Howley PM, Levinson AD, Seeburg PH. The primary structure and genetic organization of the bovine papillomavirus type 1 genome. Nature 1982;299:529–34. Available from: https://doi.org/10.1038/299529a0 PMID: 6289124.

[87] Broker TR, Chow LT. Electron microscopic analyses of DNA heteroduplexes formed with HPV types, 6, 11 and 18. In: Botchan M, Grodzocker T, Sharp PA, editors. DNA Tumor Viruses: Control of Gene Expression and Replication (1984 Meeting). Cancer Cells 4. NY: Cold Spring Harbor Laboratory; 1985. p. 589–94 (no PMID).

[88] Chow L, Broker T. Homology relationships of human and bovine papilloma viruses. Cold Spring Harbor Annual Report. New York: Cold Spring Harbor; 1981. p. 45–50 (no PMID).

[89] Croissant O, Testanière V, Orth G. Detection and mapping of conserved nucleotidic sequences between the genomes of human papillomavirus 1a and bovine papillomavirus 1 by electron microscope heteroduplex analysis. C R Seances Acad Sci III 1982;294:581–6 PMID: 6286053.

[90] Danos O, Katinka M, Yaniv M. Human papillomavirus 1a complete DNA sequence: a novel type of genome organization among papovaviridae. EMBO J 1982;1:231–6 PMID: 6325156.

[91] Danos O, Giri I, Thierry F, Yaniv M. Papillomavirus genomes: sequences and consequences. J Invest Dermatol 1984;83(1 Suppl):7s–11s.

[92] Kurman RJ, Jenson AB, Lancaster WD. Papillomavirus infection of the cervix. II. Relationship to intraepithelial neoplasia based on the presence of specific viral structural proteins. Am J Surg Pathol 1983;7:39–52 PMID: 6299124.

[93] Okagaki T, Twiggs LB, Zachow KR, Clark BA, Ostrow RS, Faras AJ. Identification of human papillomavirus DNA in cervical and vaginal intraepithelial neoplasia with molecularly cloned virus-specific DNA probes. Int J Gynecol Pathol 1983;2:153–9 PMID: 6313534.

[94] Reid R, Stanhope CR, Herschman BR, Booth E, Phibbs GD, Smith JP. Genital warts and cervical cancer. I. Evidence of an association between subclinical papillomavirus infection and cervical malignancy. Cancer 1982;50:377–87 PMID: 6282442.

[95] Syrjänen K, Syrjänen S, Lamberg M, Pyrhönen S, Nuutinen J. Morphological and immunohistochemical evidence suggesting human papillomavirus (HPV) involvement in oral squamous cell carcinogenesis. Int J Oral Surg 1983;12:418–24 PMID: 6325356.

[96] Syrjänen KJ. Papillomavirus infections and cancer. In: Syrjänen K, Gissmann L, Koss LG, editors. Papillomaviruses and Human Disease. Berlin, New York: Springer-Verlag; 1987. p. 467–503 (no PMID).

[97] Kreider JW, Howett MK, Wolfe SA, Bartlett GL, Zaino RJ, Sedlacek TV, et al. Morphological transformation in vivo of human uterine cervix in papillomavirus from condylomata acuminata. Nature 1985;317:639–41 PMID 2997616.

[98] Mullis K, Faloona F, Scharf S, Saiki R, Horn G, Erlich H. Specific enzymatic amplification of DNA in vitro: the polymerase chain reaction. Cold Spring Harbor Symp Quant Biol 1986;51(Pt 1):263–73 PMID: 3472723.

[99] Saiki R, Scharf S, Faloona F, Mullis K, Horn G, Erlich H, et al. Enzymatic amplification of beta-globin genomic sequences and restriction site analysis for diagnosis of sickle cell anemia. Science 1985;230:1350–4. Available from: https://doi.org/10.1126/science.2999980 PMID 2999980.

[100] Manos MM, Ting Y, Wright DK, Lewis AJ, Broker TR, Wolinsky SM. Use of polymerase chain reaction amplification for the detection of genital human papillomaviruses. Cancer Cells 1989;7:209–14 (no PMID).

[101] Shibata DK, Arnheim N, Martin WJ. Detection of human papilloma virus in paraffin-embedded tissue using the polymerase chain reaction. J Exp Med 1988;167:225–30. Available from: https://doi.org/10.1084/jem.167.1.225 PMID: 2826637; PMCID: PMC2188813.

[102] Snijders PJF, van den Brule AJC, Schrijnemakers HFJ, Snow G, Meijer CJLM, Walboomers JMM. The use of general primers in the polymerase chain reaction permits the detection of a broad spectrum of human papillomavirus genotypes. J Gen Virol 1990;71:173–81. Available from: https://doi.org/10.1099/0022-1317-71-1-173 PMID: 2154534.

[103] van den Brule AJC, Snijders PJF, Meijer CJLM, Walboomers JMM. PCR based detection of genital HPV genotypes: An update and future perspectives. In: Lacey C, editor. Papillomavirus Reviews: Current research on papillomaviruses. Leeds University Press; 1996. p. 181–8 (no PMID).

[104] de Villiers E-M, Fauquet C, Broker TR, Bernard H-U, zur Hausen H. Classification of papillomaviruses. Virology 2004;324:17–27 PMID: 15183049.

[105] Van Doorslaer K, Chen Z, Bernard H-U, Chan PKS, DeSalle R, Dillner J, et al. ICTV Virus Taxonomy Profile: Papillomaviridae. J Gen Virol 2018;99:989–90. Available from: https://doi.org/10.1099/jgv.0.001105 Epub 2018 Jun 21. PMID: 29927370; PMCID: PMC6171710.

[106] Willemsen A, Bravo IG. Origin and evolution of papillomavirus (onco)genes and genomes. Philos Trans R Soc Lond B Biol Sci 2019;374:20180303. Available from: https://doi.org/10.1098/rstb.2018.0303 PMID: 30955499.

[107] Broker TR. Animal virus RNA processing. In: Apirion D, editor. Processing of RNA. Boca Raton, FL: CRC Press; 1984. p. 181–212 (no PMID).

[108] Chow LT, Hirochika H, Nasseri M, Stoler MH, Wolinsky SM, Chin MT, et al. Human papillomavirus gene expression. Cancer Cells 1987;5:55–72 (no PMID).

[109] Chow LT, Pelletier AJ, Galli R, Brinckmann U, Chin M, Arvan D, et al. Transcription of human papillomavirus types 1 and 6. In: Botchan M, Grodzocker T, Sharp PA, editors. DNA Tumor Viruses: Control of Gene Expression and Replication (1984 Meeting). Cancer Cells 4. NY: Cold Spring Harbor Laboratory; 1985. p. 603–14.

[110] Rotenberg MO, Chow LT, Broker TR. Characterization of rare human papillomavirus type 11 mRNAs coding for regulatory and structural proteins using the polymerase chain reaction. Virology 1989;172:489–97 PMID: 2552659.

[111] Broker TR, Chow LT, Chin MT, Rhodes CR, Wolinsky SM, Whitbeck A, et al. A molecular portrait of human papillomavirus carcinogenesis. Cancer Cells 1989;7:197–208 (no PMID).

[112] Crum CP, Nagai N, Levine RU, Silverstein S. In situ hybridization analysis of HPV 16 DNA sequences in early cervical neoplasia. Am J Pathol 1986;123:174–82 PMID: 2421580; PMCID: PMC1888145.

[113] Stoler MH, Broker TR. In situ hybridization detection of human papilloma virus DNAs and messenger RNAs in genital condylomas and a cervical carcinoma. Human Pathol 1986;17:1250–7 PMID: 3025074.

[114] Stoler MH, Rhodes CR, Whitbeck A, Wolinsky S, Chow LT, Broker TR. Gene expression of human papillomavirus type 16 and 18 in cervical neoplasias. Human Pathol 1992;23:117–28 PMID: 1310950.

[115] Schwarz E, Freese UK, Gissmann L, Mayer W, Roggenbuck B, Stremlau A, et al. Structure and transcription of human papillomavirus sequences in cervical carcinoma cells. Nature 1985;314:111–14.

[116] Durst M, Dzarlieva-Petrusevska RT, Boukamp P, Fusenig NE, Gissmann L. Molecular and cytogenetic analysis of immortalized human primary keratinocytes obtained after transfection with human papillomavirus type 16 DNA. Oncogene 1987;1:251–6.

[117] von Knebel Doeberitz M, Oltersdorf T, Schwarz E, Gissmann L. Correlation of modified human papilloma virus early gene expression with altered growth properties in C4-1 cervical carcinoma cells. Cancer Res 1988;48:3780–6.

[118] von Knebel Doeberitz M, Rittmuller C, zur Hausen H, Durst M. Inhibition of tumorigenicity of cervical cancer cells in nude mice by HPV E6-E7 anti-sense RNA. Int J Cancer 1992;51:831–4.

[119] Izant JG, Weintraub H. Constitutive and conditional suppression of exogenous and endogenous genes by anti-sense RNA. Science 1985;229:345–52.

[120] Androphy EJ, Hubbert NL, Schiller JT, Lowy DR. Identification of the HPV-16 E6 protein from transformed mouse cells and human cervical carcinoma cell lines. EMBO J 1987;6:989–92 PMID: 3036495.

[121] Dyson N, Howley PM, Münger K, Harlow E. The human papilloma virus-16 E7 oncoprotein is able to bind to the retinoblastoma gene product. Science 1989;243:934–7 PMID: 2537532.

[122] Scheffner M, Werness BA, Huibregtse JM, Levine AJ, Howley PM. The E6 oncoprotein encoded by human papillomavirus types 16 and 18 promotes the degradation of p53. Cell 1990;63:1129–36 PMID: 2175676.

[123] Vousden KH, Doniger J, DiPaolo JA, Lowy DR. The E7 open reading frame of human papillomavirus type 16 encodes a transforming gene. Oncogene Res 1988;3:167–75 PMID: 2852339.

[124] Howley PM, Munger K, Romanczuk H, Scheffner M, Huibregtse JM. Cellular targets of the oncoproteins encoded by the cancer associated human papillomaviruses. Princess Takamatsu Symp 1991;22:239–48.

[125] Barbosa MS, Vass WC, Lowy DR, Schiller JT. In vitro biological activities of the E6 and E7 genes vary among human papillomaviruses of different oncogenic potential. J Virol 1991;65:292–8 PMID: 1845889.

[126] Lane DP, Crawford LV. T antigen is bound to a host protein in SV40-transformed cells. Nature 1979;278:261–3.

[127] Pater MM, Dunne J, Hogan G, Ghatage P, Pater A. Human papillomavirus types 16 and 18 sequences in early cervical neoplasia. Virology 1986;155:13–18 PMID: 3022464.

[128] Pirisi L, Creek KE, Doniger J, DiPaolo JA. Continuous cell lines with altered growth and differentiation properties originate after transfection of human keratinocytes with human papillomavirus type 16 DNA. Carcinogenesis 1988;9:1573–9. Available from: https://doi.org/10.1093/carcin/9.9.1573 PMID: 2457456.

[129] Yasumoto S, Burkhardt AL, Doniger J, DiPaolo JA. Human papillomavirus type 16 DNA-induced malignant transformation of NIH 3T3 cells. J Virol 1986;57:572–7 PMID: 3003388.

[130] Hwang ES, Riese 2nd DJ, Settleman J, Nilson LA, Honig J, Flynn S, et al. Inhibition of cervical carcinoma cell line proliferation by the introduction of a bovine papillomavirus regulatory gene. J Virol 1993;67:3720–9 PMID: 8389903; PMCID PMC237735.

[131] Rollison DE, Viarisio D, Amorrortu RP, Gheit T, Tommasino M. An emerging issue in oncogenic virology: the role of beta human papillomavirus types in the development of cutaneous squamous cell carcinoma. J Virol 2019;93. Available from: https://doi.org/10.1128/JVI.01003-18 pii: e01003-18. Print 2019 Apr 1. PMID: 30700603; PMCID: PMC6430537.

[132] Gaglia MM, Münger K. More than just oncogenes: mechanisms of tumorigenesis by human viruses. Curr Opin Virol 2018;32:48–59. Available from: https://doi.org/10.1016/j.coviro.2018.09.003 Epub 2018 Sep 27. PMID: 30268926; PMCID: PMC6405337.

[133] Lambert PF. Transgenic mouse models of tumor virus action. Annu Rev Virol 2016;3:473–89 PMID: 27741405.

[134] Meyers JM, Grace M, Uberoi A, Lambert PF, Münger K. Inhibition of TGF-β and NOTCH signaling by cutaneous papillomaviruses. Frontiers Microbiol 2018;9:389. Available from: https://doi.org/10.3389/fmicb.2018.00389 eCollection 2018. PMID: 29568286; PMCID: PMC5852067.

[135] Thomas M, Narayan N, Pim D, Tomaić V, Massimi P, Nagasaka K, et al. Human papillomaviruses, cervical cancer and cell polarity. Oncogene 2008;27:7018–30. Available from: https://doi.org/10.1038/onc.2008.351 PMID: 19029942.

[136] White EA, Münger K, Howley PM. High-risk human papillomavirus E7 proteins target PTPN14 for degradation. mBio 2016;7(5). Available from: https://doi.org/10.1128/mBio.01530-16 pii: e01530-16. PMID: 27651363; PMCID: PMC5030362.

[137] Chow LT, Broker TR. Human papillomaviruses infections: warts or cancer? In: Bell SD, Méchali M, DePamphilis ML, editors. DNA Replication and Human Disease. Cold Spring Harbor, NY: Cold Spring Harbor Laboratory Press; 2013. p. 469–85. PMID: 23685995; PMCID: 3685896.

[138] McCance DJ, Kopan R, Fuchs E, Laimins LA. Human papillomavirus type 16 alters human epithelial cell differentiation in vitro. Proc Natl Acad Sci U S A 1988;85:7169–73. Available from: https://doi.org/10.1073/pnas.85.19.7169 PMID: 2459699; PMCID: PMC282145.

[139] Dollard SC, Wilson JL, Demeter LM, Bonnez W, Reichman RC, Broker TR, et al. Production of human papillomavirus and modulation of the infectious program in epithelial raft cultures. Genes Dev 1992;6:1131–42 PMID: 8382318.

[140] Meyers C, Frattini MG, Hudson JB, Laimins LA. Biosynthesis of human papillomavirus from a continuous cell line upon epithelial differentiation. Science 1992;257:971–3 PMID: 1323879.

[141] Galloway DA, McDougall JK. Human papillomaviruses and carcinomas. Adv Virus Res 1989;37:125–71 PMID: 2557758.

[142] Halbert CL, Demers GW, Galloway DA. The E7 gene of human papillomavirus type 16 is sufficient for immortalization of human epithelial cells. J Virol 1991;65:473–8 PMID: 1845902.

[143] Kaur P, McDougall JK. Characterization of primary human keratinocytes transformed by human papillomavirus type 18. J Virol 1988;62:1917–24 PMID: 2452896.

[144] Cheng S, Schmidt-Grimminger D-C, Murant T, Broker TR, Chow LT. Differentiation-dependent upregulation of the human papillomavirus E7 gene reactivates cellular DNA replication in suprabasal differentiated keratinocytes. Genes Dev 1995;9:2335–49 PMID: 7557386.

[145] Chien W-M, Noya F, Benedict-Hamilton HM, Broker TR, Chow LT. Alternative fates of keratinocytes transduced by human papillomavirus type 18 E7 during squamous differentiation. J Virol 2002;76:2964–72 PMID: 11861862; PMCID: 114328.

[146] Banerjee NS, Genovese NJ, Noya F, Chien W-M, Broker TR, Chow LT. Conditionally activated E7 proteins of high-risk and low-risk human papillomaviruses induce S-phase in post-mitotic, differentiated human keratinocytes. J Virol 2006;80:6517–24 PMID: 16775338.

[147] Garner-Hamrick PA, Fostel JM, Chien W-M, Banerjee NS, Chow LT, Broker TR, et al. Global effects of human papillomavirus 18 (HPV-18) E6/E7 in an organotypic culture system. J Virol 2004;78:9041–50 PMID: 15308700.

[148] Wan F, Miao X, Quraishi I, Kennedy V, Creek KE, Pirisi L. Gene expression changes during HPV-mediated carcinogenesis: a comparison between an in vitro cell model and cervical cancer. Int J Cancer 2008;123:32–40. Available from: https://doi.org/10.1002/ijc.23463 PMID: 18398830; PMCID: PMC2872618.

[149] Chow LT, Broker TR, Steinberg BM. The natural history of human papillomavirus infections of the mucosal epithelia. APMIS - Acta Pathologica, Microbiologica et Immunologica Scandinavica 2010;118:422–49 PMID: 20553526.

[150] Wang H-K, Duffy AA, Broker TR, Chow LT. Robust production and passaging of infectious HPV in squamous epithelium of primary human keratinocytes. Genes Dev 2009;23:181–94 PMID: 19131434; PMCID: 2648537.

[151] Banerjee NS, Wang H-K, Broker TR, Chow LT. Human papillomavirus (HPV) E7 induces prolonged G2 following S-phase reentry in differentiated human keratinocytes. J Biol Chem 2011;286:15473–82 PMID: 21321122; PMC3083224.

[152] Kho EY, Wang H-K, Banerjee NS, Broker TR, Chow LT. HPV-18 E6 mutants reveal p53 modulation of viral DNA amplification in organotypic cultures. Proc Natl Acad Sci U S A 2013;110:7542–9 PMID: 23572574; PMCID: 3651465.

[153] Wang H-K, Wei Q, Moldoveanu Z, Huh WK, Vu HL, Broker TR, et al. Characterization of serum antibodies from women immunized with Gardasil: a study of HPV-18 infection of primary human keratinocytes. Vaccine 2016;34:3171–7 PMID: 27113165; PMCID: 4987144.

[154] Banerjee NS, Wang H-K, Beadle JR, Hostetler KY, Chow LT. Evaluation of ODE-Bn-PMEG, an acyclic nucleoside phosphonate prodrug, on HPV DNA amplification in organotypic epithelial cultures. Antiviral Res 2018;150:164–73 PMID 29287913; PMCID: PMC5800947.

[155] Banerjee NS, Moore DW, Broker TR, Chow LT. Vorinostat, a pan-HDAC inhibitor, abrogates productive HPV-18 DNA amplification. November 20, 2018. Proc Nat Acad Sci U S A Plus 2018;115(47): E11138–47. Available from: https://doi.org/10.1073/pnas.1801156115 PMID: 30385631; PMCID: PMC6255162.

[156] Kang SD, Chatterjee S, Alam S, Salzberg AC, Milici J, van der Burg SH, et al. Effect of productive human papillomavirus 16 infection on global gene expression in cervical epithelium. J Virol 2018;92. Available from: https://doi.org/10.1128/JVI.01261-18 pii: e01261-18. PMID: 30045992; PMCID: PMC6158420.

[157] Reid BL. The behaviour of human sperm toward cultured fragments of human cervix uteri. Lancet 1964;1(7323):21–3.

[158] Coppleson M, Pixley E, Reid B. Colposcopy: a scientific and practical approach to the cervix in health and disease. Springfield, IL: Charles C Thomas; 1971. p. p82–5.

[159] Singer A, Shearman R. Letter: oral contraceptives and cervical carcinoma. Br Med J 1969;4:108.

[160] Koss LG. Diagnostic cytology and it's histopathologic bases. Association With Grace R Durfee. 1st ed. London: Pitman Medical Publishing Company; 1961. Philadelphia, PA: J B Lippincott Co.

[161] Beral V. Cancer of the cervix: a sexually transmitted infection? Lancet 1974;1(7865):1037–40.

[162] Singer A, Reid BL, Coppleson M. A hypothesis: the role of a high-risk male in the etiology of cervical carcinoma: a correlation of epidemiology and molecular biology. Am J Obstet Gynecol 1976;126 (1):110–15.

[163] Campion MJ, McCance DJ, Mitchell HS, Jenkins D, et al. Subclinical penile human papillomavirus infection and dysplasia in consorts of women with cervical neoplasia. Genitourin Med 1988;64:90–9.

[164] Campion MJ, McCance DJ, Cuzick J, Singer A. Progressive potential of mild cervical atypia: prospective cytological, colposcopic, and virological study. Lancet 1986;2(8501):237–40.

[165] Peto R, zur Hausen H, editors. Viral aetiology of cervical cancer. In: Based on a conference held on 14-17 April 1985 at the Banbury Center of Cold Springs Harbor Laboratory. Cold Spring Harbor, Long Island, NY: Cold Spring Harbor Laboratory; 1986.

[166] IARC. Monographs on the evaluation of the carcinogenic risks to humans, Human Papillomavirus, vol. 64; 1995.

CHAPTER

DEMONSTRATING THE IMPORTANCE OF DIFFERENT HPVS IN CERVICAL CANCER AND OTHER HPV-RELATED CANCERS*

3

Nuria Guimera[1], Laia Alemany[2,3], Laia Bruni[2,4] and Nubia Muñoz[5]

[1]DDL Diagnostic Laboratory, Rijswijk, The Netherlands [2]Cancer Epidemiology Research Program, Catalan Institute of Oncology (ICO) - Bellvitge Biomedical Research Institute (IDIBELL), Barcelona, Spain [3]CIBER en epidemiología y salud pública CIBERESP, Barcelona, Spain [4]CIBER en oncología CIBERONC, Madrid, Spain [5]Cancer Institute of Colombia, Bogotá, Republic of Colombia

INTRODUCTION

The search for the cause of cervical cancer has been long: in 1842 when Rigoni Stern suggested that a sexually transmitted agent could be linked to cervical cancer, various biological agents have been considered. In the 1960s and 1970s, Herpes simplex virus type 2 (HSV-2) was the leading candidate, and although in the 1970s several investigators suspected that human papillomavirus (HPV) could also be involved in the etiological pathway of cervical cancer, solid epidemiological evidence for this association was reported in 1992 [1] (see Chapter 8: Proving the Causal Role of Human Papillomavirus in Cervical Cancer: A Tale of Multidisciplinary Science).

The demonstration that infection with certain types of HPV is not only the main cause but also a necessary cause of cervical cancer has led to great advances in the disease prevention by the introduction of prophylactic HPV vaccines and increase of cervical cancer screening accuracy and management. After the discovery of HPV as the main cause of cervical cancer, multiple studies have been conducted in other anogenital and head and neck anatomical sites as they share the same route of transmission. These studies also proved the carcinogenic role of HPVs in cancers originated in these other noncervical sites. In this chapter, we will summarize the epidemiological studies involved in the discovery of HPV association to different cancers and the importance of different HPV types as causative agents.

*This chapter is dedicated to the memory of our dear colleague Keerti Shah, who made possible the majority of the molecular epidemiologic studies described here.

Human Papillomavirus. DOI: https://doi.org/10.1016/B978-0-12-814457-2.00003-9

HUMAN PAPILLOMAVIRUS AS ETIOLOGICAL AGENT OF CERVICAL CANCER

A major discovery in human cancer etiology was the recognition that cervical cancer is a rare consequence of an infection by some mucosa-tropic types of HPV. In public health terms, the importance of this finding is comparable to the association between chronic infections of hepatitis B and C viruses and the risk of liver cancer, or between cigarette smoking and lung cancer.

The early steps in the discovery of HPV as a cause of cervical cancer go back to 1974–76 when Nubia Muñoz as an epidemiologist in the unit of biological carcinogenesis of IARC collected a large number of cervical cancer specimens from Colombia, Uganda, Brazil, and Iran. These surveys aimed to detect markers of three viruses that suspected to be associated with cervical cancer: HPV, HSV-2, and cytomegalovirus (CMV) [2,3]. Hybridization and serological assays to look for DNA of HSV-2 and CMV ended up with negative results. The association between HPV and cervical cancer was explored in collaboration with Adonis de Carvalho, a local pathologist from Recife, who noted the high incidence of cervical cancer and high frequency of giant condylomas of the cervix, vulva, vagina, anus, and penis and with Gerard Orth from the Pasteur Institute in Paris who at that time was the expert on HPV. He used electron microscopy, the only technique available at that time to look for HPV. A few HPV particles were observed in biopsies from giant genital condylomas, but none in biopsies from cervical cancer and other genital cancers. Nowadays, it is known that in genital cancers, HPV might be integrated into the human genome during the carcinogenic process; therefore, HPV viral particles are no longer detected in malignant cells by electron microscopy.

In parallel, anecdotal reports of rare malignant conversion of genital warts into squamous cell carcinomas were reported [4]. Molecular characterization and cloning of the first HPV types in the early 1980s made possible the development of hybridization assays to look for HPV gene fragments in human tissue. In 1983, by using HPV11 as a probe, it was possible to isolate HPV16 from cervical cancer biopsies and in a subset of vulva and penile cancers [5]. Such hybridization assays that are able to detect different HPV types could then be used to assess exposure to HPV in molecular epidemiologic studies and lead to the link of HPV6 and HPV11 to benign lesions (condylomas) and HPV16 and 18 to cervical cancer. At that time, other HPV types, such as 31, 33, or 35, were rarely detected.

However, formal epidemiological evidence of an association between HPV and cervical cancer was still lacking in the late 1980s [6]. In 1985 the unit of field and intervention studies at the IARC, led by Nubia Munoz, initiated a series of case-control studies of cervical cancer described later in this chapter. The first difficulty faced was the selection of an accurate test to detect HPV DNA, since a pilot study during the case-control study in Spain and Colombia comparing the two assays available at that time [dot blot and filter in-situ hybridization (FISH)] gave conflicting results. The dot blot assay performed at the laboratory of Gerard Orth was positive for HPV DNA in almost all cases of cervical cancer and in very few control women without cancer. Contrarily, the FISH assay performed at the laboratory of Harald zur Hausen showed no differences in HPV DNA positivity between cases and controls. In view of the conflicting results obtained with the dot blot and the FISH assays, collaboration was established with Keerti Shah at the Johns Hopkins University in Baltimore, who agreed to test over 2000 specimens derived from the pioneer case-control studies on cervical cancer in Spain and Colombia using southern blot assays, the gold standard at that time, even though this assay was very laborious and expensive. After the testing with southern blot was completed, all specimens were also tested with Virapap, the first commercially developed assay and the first PCR assay developed [7].

In 1988 and 1992 the Unit of Field and Intervention Studies at IARC convened two meetings with epidemiologists, virologists, pathologists, molecular biologists, and clinicians, to critically review the available epidemiological evidence linking HPV to cervical cancer and to stimulate the initiation of the collaborative molecular epidemiological studies needed to clarify the role of HPV in cervical cancer (see Chapter 8: Proving the Causal Role of Human Papillomavirus in Cervical Cancer: A Tale of Multidisciplinary Science).

HUMAN PAPILLOMAVIRUS ASSOCIATION WITH CERVICAL CANCER: CASE-CONTROL STUDIES

In 1992 and 1995 results from two population-based case-control studies initiated in Colombia and Spain in 1985 were published, showing a very strong association between HPV16, 18, 31, 33, and 35 with invasive cervical cancer and cervical intraepithelial neoplasia grade 3 (CIN3). These studies were carried out in two countries with contrasting rates of cervical cancer, Colombia having an incidence rate of about eight times higher than Spain. In the first case control, the presence of HPV DNA was studied in 436 cases of invasive cervical cancer and 387 randomly selected population controls. In the second case-control study, 525 CIN3 and 512 matched controls were also investigated. For invasive cancer, the adjusted odd ratios and 95% confidence intervals were 46.2 (18.5–115.1) in Spain and 15.6 (6.9–34.7) in Colombia and for CIN III, they were 56.9 (24.8–130.6) in Spain and 15.5 (8.2–29.4) in Colombia. In this second case-control study, in addition to HPV16, 18, 31, 33, and 35, other HPV types not yet characterized were also detected [1,8]. As a result from these studies, HPV16 and 18 were classified as human carcinogens by the IARC monograph 1995 (see Chapter 8: Proving the Causal Role of Human Papillomavirus in Cervical Cancer: A Tale of Multidisciplinary Science).

The recognition of HPV as a human carcinogen stimulated the biotechnology and pharmaceutical industry to develop tests to improve screening programs and to develop vaccines to target genital HPVs.

In 2003 the epidemiologic risk classification for HPV was published [9]. This study pooled data from 11 IARC case-control studies involving 1918 women with cervical cancer and 1928 women as controls. As a result, the number of HPV genotypes studied was extended and showed that some HPVs have more cervical carcinogenic risk than others. Thus the relative HPV distribution between women with normal cytology and women suffering cervical cancer permitted the identification of 15 HPV genotypes 16, 18, 31, 33, 35, 39, 45, 51, 52, 56, 58, 59, 68, 73, and 82 as high risk (HR)-HPVs, and HPV types 26, 53, and 66 as probably HR-HPVs. Twelve HPV types (6, 11, 40, 42, 43, 44, 54, 61, 70, 72, 81, and 89) were categorized as low risk (LR)-HPVs, and three HPV types (34, 57, and 83) were considered undetermined risk because they were not found in any tumor specimens.

HUMAN PAPILLOMAVIRUS PREVALENCE AND DISTRIBUTION IN CERVICAL CANCER AND OTHER HUMAN PAPILLOMAVIRUS-ASSOCIATED CARCINOMAS

In parallel with the Spain–Colombia case-control studies, an international biological study on cervical cancer (IBSCC) was conducted by the Unit of Field and Intervention studies at the IARC [10]. In this study, more than 1000 frozen cervical cancer specimens were collected from 22 countries and reported a worldwide HPV prevalence of 93% based on the MY09/11 PCR assay which targets a 450 bp fragment within the HPV L1 ORF capable of detecting more than 25 different HPV types. HPV16 was present in 50% of the specimens, HPV18 in 14%, HPV45 in 8%, and HPV31 in 5%. No significant variation of HPV positivity was detected among countries and HPV16 was the predominant type in all countries except Indonesia, where HPV18 was more common. This study demonstrated that more than 20 different genital HPV types are associated with cervical cancer.

In 1999, in collaboration with J. Walboomers, the HPV-negative cases from the IARC IBSCC study were reanalyzed by three different HPV PCR assays targeting different HPV open reading frames. It was considered that the HPV prevalence of 93% could be an underestimation due to sample inadequacy or integration events affecting the HPV L1 gene, which was the target of the PCR assay used. Combining the original results with the ones obtained with the three new PCR assays, and excluding the inadequate specimens, the worldwide HPV prevalence in cervical carcinoma was 99.7%. The presence of HPV in virtually all cervical cancers implied the highest worldwide attributable fraction so far reported for a specific cause of any major human cancer. HPV was not only the main cause of cervical cancer but also a necessary cause [11].

Subsequently, several studies have been published on the distribution of types in cervical lesions. The IARC has performed systematic reviews and pooled analyses with the purpose of estimating HPV-type distribution in invasive cervical cancer to predict the potential impact of HPV type-specific vaccines and screening tests, and to understand the carcinogenicity of HPV types. In 2011 they published an update of these pooled analyses, data published from 1990 to 2010, including a total of 243 studies and 30,848 invasive cervical cancer (ICC). The proportion of ICC associated with HPV16 and/or 18 (HPV16/18) was between 68% and 82% in all world regions except Asia. The 12 most common HPV types identified, in order of decreasing prevalence, were HPV16 (57%), 18 (16%), 58, 33, 45, 31, 52, 35, 59, 39, 51, and 56 [12].

Furthermore, at the Catalan Institute of Oncology (ICO), international surveys aiming at giving robust estimates on the HPV prevalence and type-specific relative contribution (types distribution among HPV-positive cases) in HPV-related cancers and time trend analyses were conducted from 2003 to 2014. Original data from the ICO international studies on HPV in HPV-related cancers included 10,575 cervical cancer cases, 496 anal, 1709 vulvar, 408 vagina, 1010 penile, and 3685 head and neck cancers [13–18]. In these studies, formalin fixed biopsies were processed and tested with the same methodology (SPF_{10} PCR-DEIA-LiPA25) to provide accurate HPV DNA prevalence and distribution estimates, as HPV positivity rate is influenced by the material and methodology used to detect HPV. In other anatomical sites different than cervix, HPV positivity was based on at least two biomarkers related to HPV (HPV DNA positivity and at least p16 or mRNA E6*I positivity). Additional HPV-related biomarkers are usually used to ensure the etiological relation of HPV with these cancers as HPV might be a passenger and not related to these noncervical carcinomas. In cervical cancer cases the worldwide HPV DNA positivity was 85% and ranged from 88% in anal cancer to around 6%–7% in laryngeal and oral cavity cancers, respectively, other percentages of HPV positivity were 74% vagina, 33% penis, 29% vulvar, and 25% oropharyngeal cancers. These estimates were similar to the most updated systematic review reported by de Martel and colleagues [19]. For noncervical cancers when considering positivity for at least mRNA or p16 in addition to the HPV DNA detection, the HPV-related proportions were 83% in anal cancer, 71% vagina, 28% penis, 25% vulvar, and 22% oropharyngeal cancers, and around 4% in laryngeal and oral cavity cancers [20] (Fig. 3.1). Regarding HPV detection in oropharyngeal cancer, it is worth noting that a high geographical variation in the attributable fraction of HPV was described.

For type-specific contributions, we consistently estimated that HPVs 16/18 account for 70% of cervical cancers worldwide. The additional 5 HR types included in the nine-valent vaccine accounted together with 16/18 for about 90% of the entire cervical cancer burden in the world. For

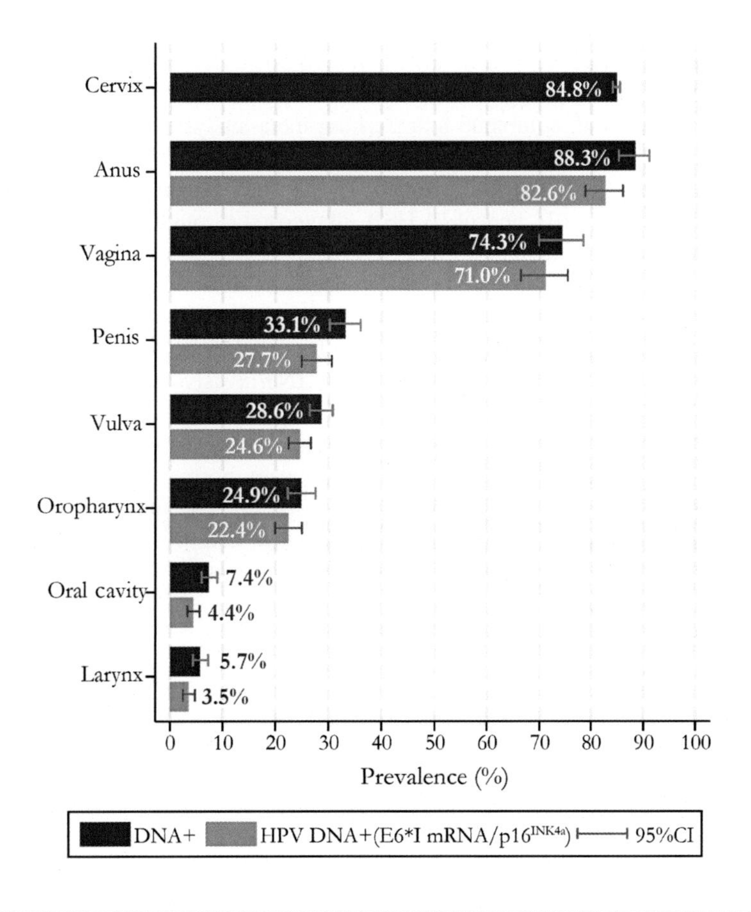

FIGURE 3.1

HPV prevalence by anatomical sites. *HPV*, Human papillomavirus.

de Sanjose et al. [20].

other anogenital and head and neck cancers, HPV16 was by far the most frequently detected type across all anatomical sites. HPVs 16/18/31/33/45/52/58/6/11 accounted for more than 80% of the HPV-positive cases surpassing 90% of HPV-related anal and oropharyngeal cancers, in which HPV16 was extremely common (Figs. 3.2 and 3.3). It is expected that HPV vaccines will have an important impact in reducing not only cervical cancer burden but also that of other HPV-related cancers; our estimations suggest a potential impact of the nine-valent HPV vaccine in reducing around 90% of cervical cancer cases and a global reduction of 50% of all the cases at HPV-related cancer sites [20].

HUMAN PAPILLOMAVIRUS-TYPE DISTRIBUTION OF ONCOGENIC HUMAN PAPILLOMAVIRUS TYPES ACROSS THE FULL SPECTRUM OF CERVICAL DISEASE

In 2012 Guan et al. confirmed the steady rise in HPV16 positivity through the classification of cervical disease from normal cytology to invasive cervical cancer, known as "enrichment," in all world regions [21]. Guan et al. combined four existing meta-analysis on HPV prevalence in women with normal cytology, low-grade precancerous lesions, high-grade precancerous lesions, and invasive cervical cancer. The final analysis included 33,154 HPV-positive women with normal cytology, 6810 ASC-US, 13,480 LSIL, and 6616 HSIL diagnosed through cytology, and as histological diagnosis 8106 CIN1, 4068 CIN2, and 10,753 CIN3 and 36,374 invasive cervical cancers from 423 PCR-based studies worldwide. No strong differences in HPV-type distribution were apparent between normal cytology, ASC-US, LSIL, or CIN1. A steady rise of HPV-type contribution through cervical carcinogenesis stages was observed for HPV16 from normal/ASC-US/LSIL/CIN1 (20%–28%), through CIN2/HSIL (40%/47%) to CIN3/invasive cervical cancer (58%/63%). HPV16 and 18 were, respectively, 3.07 and 1.87 times more frequent in invasive cervical cancers than in normal cytology. HPV45 was equally distributed. However, other oncogenic types, HPVs 31, 33, 35, 52, and 58, presented strong enrichment between normal cytology and CIN3, but their relative importance dropped between CIN3 and invasive cervical cancer. This suggests that their relative carcinogenic potential in comparison to HPV16, 18, and 45 might be overestimated based on CIN3. Rest of the oncogenic types accounted for important proportions of HPV-positive precancerous lesions (both low and high grade), but their contribution also dropped dramatically in invasive cervical cancer.

HUMAN PAPILLOMAVIRUS CLASSIFICATION AND ITS ASSOCIATION WITH CARCINOGENICITY

Nowadays, more than 200 HPV types have been identified and characterized. There is a strong concordance of HPV phylogeny, viral natural history, and carcinogenicity. For humans the clinically most relevant genus is the alpha [22]. HPVs are classified as HR-HPV and LR-HPV based on the strength of the association of each type with cervical cancer, where risk implies the potential for malignant transformation. The World Health Organization and the International Agency of

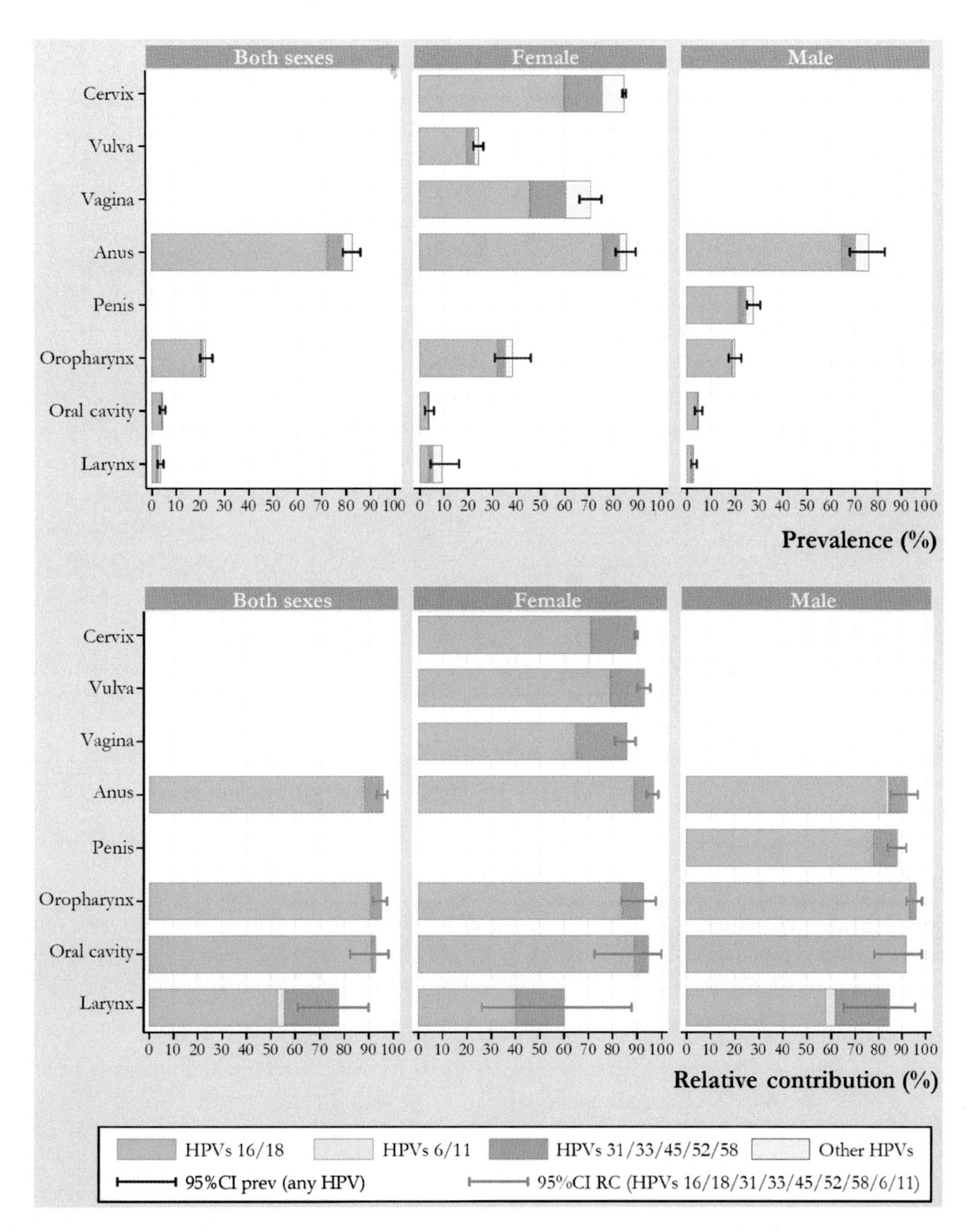

FIGURE 3.2

Relative contribution (%) of nine vaccine types among HPV-positive* anogenital and head and neck cancers. *HPV*, Human papillomavirus.

*de Sanjose et al. [20]. *Cervix based on HPV DNA positivity; others based on HPV DNA positivity and mRNA or p16 positivity.*

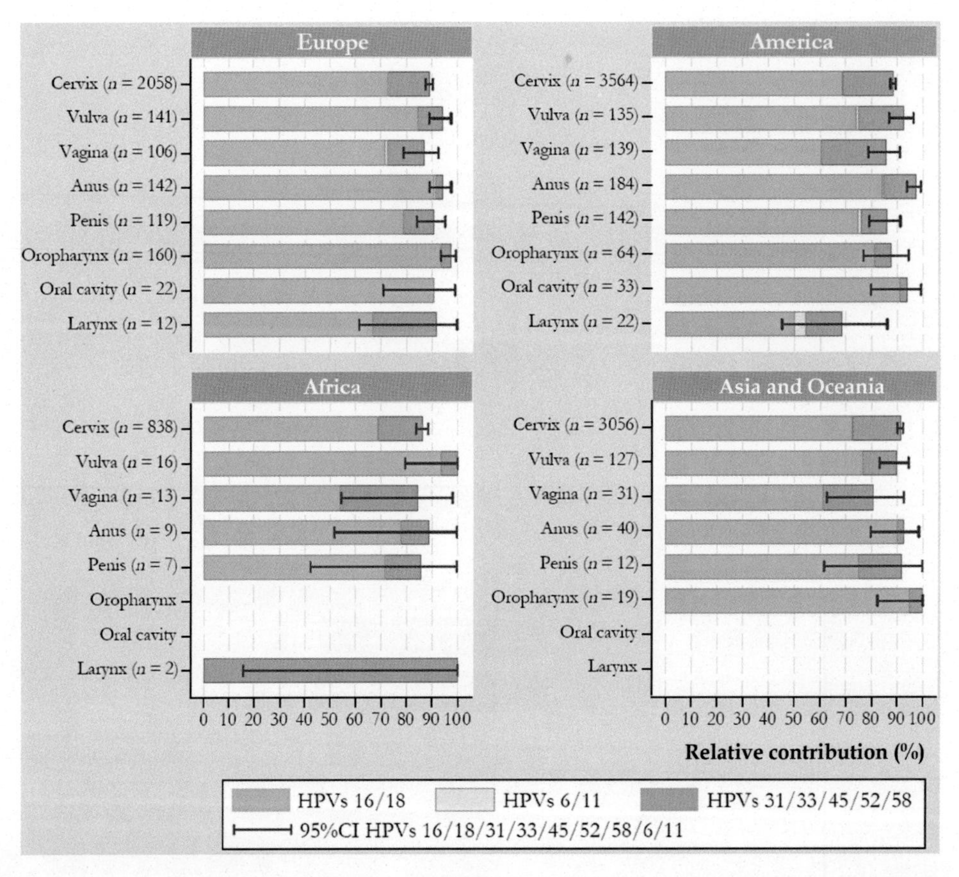

FIGURE 3.3

Relative contribution (%) of nine vaccine types among HPV-positive* anogenital and head and neck cancers by regions. *HPV*, Human papillomavirus.

*de Sanjose et al. [20]. *Cervix based on HPV DNA positivity; others based on HPV DNA positivity and mRNA or p16 positivity.*

Research on Cancer (IARC) consider some HR-HPVs as formal, carcinogenic bioagents in humans. IARC classifies carcinogens as carcinogenic (Group 1), probably carcinogenic (Group 2A), possibly carcinogenic (Group 2B), and not classifiable (Group 3) [23,24]. The categorization of different HPV genotypes in this carcinogenic risk classification is based on epidemiologic evidence according to an algorithm that included comparison of HPV type-specific prevalence in cervical cancer with that among women with normal cytology. The main studies that provided these epidemiological evidence have been summarized in this chapter. Other HPV types were classified as possible carcinogens based solely on phylogenetic relatedness. Of the ~60 alpha-HPV types, 12 have been classified as carcinogenic to humans (Group 1: HPV16, 18, 31, 33, 35, 39, 45, 51, 52, 56, 58, and 59), 1 HPV as probably carcinogenic (Group 2A: HPV68), and 12 HPV types as possibly carcinogenic (Group 2B: HPV 26, 30, 34, 53, 66, 67, 69, 70, 73, 82, 85, and 97). HPV30, 34, 69, 85, and

97 had inadequate carcinogenic evidence and were classified as possibly carcinogenic based on phylogenetic analogy. LR-HPVs 6 and 11 were considered not classifiable by IARC. HPV16 is the most carcinogenic HPV type and together with HPV18, it is associated with ~70% of all cervical cancer (Fig. 3.2). HPV carcinogenicity has only been established for cervical cancer, as HPV carcinogenic studies in other sites different than cervix are limited to epidemiological evidence. But in the majority of other HPV-related anogenital cancers, HPV16 is the most prevalent HPV type and thus the most carcinogenic.

After the IARC classification in 2009, some investigators discussed and acknowledged the limitations of epidemiological studies to inform public health policy [25,26]. The main carcinogenic HPV types for public health purposes and to be considered in screening tests and vaccines have already been identified. However, only a small fraction of all HPV types identified have any known carcinogenic potential and weakest carcinogenic HPV types are very challenging to classify. A type-by-type evaluation proves to be very difficult and stretches epidemiology to its limits and carcinogenicity is solely based on phylogenetic relatedness. Some carcinogenic HPV types might be uncommon, very difficult to detect in epidemiological studies and masked by the presence of the most powerful carcinogenic types if present as multiple infections. The definitive proof of carcinogenicity of an HPV type is finding transcriptionally active specifically in tumoral cells, but this evidence is very limited for the vast majority of carcinogenic HPV types, as all epidemiological studies are mainly based on HPV DNA testing in tissue biopsies. Such studies would be very laborious and expensive. The carcinogenic potential of less prevalent and less carcinogenic HPV types would be of interest for the refinement of future prevention measures (see Chapter 2: Linking Human Papillomavirus to Human Cancer and Understanding Its Carcinogenic Mechanisms, and Chapter 6: The Pathology of Cervical Precancer and Cancer and its Importance in Clinical Practice).

IMPLICATIONS

The research performed during the last decades on the carcinogenicity of certain types of HPV has been translated into preventive strategies against HPV-related disease. Vaccination programs against HPV have been developed to reduce cervical cancer incidence and mortality (primary prevention) with early detection strategies using HPV testing to detect and remove precancerous lesions through screening programs (secondary prevention).

Since 2006, three efficacious prophylactic vaccines against HPV have been licensed in more than 100 countries worldwide and introduced in the National Immunization Programs of 86 countries. The quadrivalent and nine-valent vaccines (Gardasil) contain virus-like particles (VLPs) of HPVs 16 and 18, or HPV16, 18, 31, 33, 45, 52, and 58 the most carcinogenic HPVs responsible of around 70%–90% of cervical cancers, and VLPs of HPVs 6 and 11, which cause about 90% of genital warts. The bivalent vaccine (Cervarix) contains only VLPs of HPVs 16 and 18. These vaccines have been shown to have high efficacy for the prevention of high-grade precancerous lesions of the cervix. In addition, these vaccines can also prevent high-grade precancerous lesions of the vulva and vagina caused by HPVs 16 and 18 and of genital warts caused by HPV6 and 11 (see Chapter 22: Political and Public Responses to Human Papillomavirus Vaccination).

Nowadays, several studies have shown that HPV DNA detection assays including the most carcinogenic HPV types have greater test sensitivity than Pap cytology in identifying cervical high-grade precancerous lesions and cervical cancer and suggest that they should be used as the primary screening test followed by triage with cytology, other HPV markers, or visual inspection according to the facilities existing in the various regions. A few countries have already adopted HPV DNA assays as primary screening test (see Chapter 17: Infection to Cancer—Finding Useful Biomarkers for Predicting Risk of Progression to Cancer).

The history of research on HPV, cervical cancer, and other HPV-related carcinomas is an excellent example of the close collaboration and interaction needed between epidemiologists and virologists to move cancer research forward from causation to prevention. Public heath action can only be advocated after rigorous evaluation of the scientific evidence based both on laboratory and epidemiological studies. Public health strategies based on HPV vaccination and HPV assays for primary screening have been proposed for the elimination of cervical cancer.

REFERENCES

[1] Muñoz N, Bosch FX, de Sanjosé S, Tafur L, Izarzugaza I, Gili M, et al. The causal link between human papillomavirus and invasive cervical cancer: a population-based case-control study in Colombia and Spain. Int J Cancer 1992;52:743–9.

[2] Muñoz N. Model systems for cervical cancer. Cancer Res 1976;36:792–3.

[3] Muñoz N. From causality to prevention - the example of cervical cancer: my personal contribution to this fascinating history. Public Health Genomics 2009;12(5–6):368–71.

[4] zur Hausen H. Condylomata acuminata and human genital cancer. Cancer Res 1976;36:794.

[5] Dürst M, Gissmann L, Ikenberg H, zur Hausen H. A papillomavirus DNA from a cervical carcinoma and its prevalence in cancer biopsy samples from different geographic regions. PNAS 1983;80: 3812–15. Available from: https://doi.org/10.1073/pnas.80.12.3812.

[6] Muñoz N, Bosch X, Kaldor JM. Does human papillomavirus cause cervical cancer? The state of the epidemiological evidence. Br J Cancer 1988;57:1–5.

[7] Guerrero E, Daniel RW, Bosch FX, Castellsagué X, Muñoz N, Gili M, et al. Comparison of ViraPap, Southern hybridization, and polymerase chain reaction methods for human papillomavirus identification in an epidemiological investigation of cervical cancer. J Clin Microbiol 1992;30(11):2951–9.

[8] Bosch FX, Muñoz N, de Sanjosé S, Izarzugaza I, Gili M, Viladiu P, et al. Risk factors for cervical cancer in Colombia and Spain. Int J Cancer 1992;52:750–8.

[9] Muñoz N, Bosch FX, de Sanjosé S, Herrero R, Castellsagué X, Shah KV, et al. Epidemiologic classification of human papillomavirus types associated with cervical cancer. N Engl J Med 2003;348(6):518–27.

[10] Bosch FX, Manos MM, Muñoz N, Sherman M, Jansen AM, Peto J, et al. Prevalence of human papillomavirus in cervical cancer: a worldwide perspective. J Natl Cancer Inst 1995;87:796–802. Available from: https://doi.org/10.1093/jnci/87.11.796.

[11] Walboomers JM, Jacobs MV, Manos MM, Bosch FX, Kummer JA, Shah KV, et al. Human papillomavirus is a necessary cause of invasive cervical cancer worldwide. J Pathol 1999;189(1):12–19.

[12] Li N, Franceschi S, Howell-Jones R, Snijders PJ, Clifford GM. Human papillomavirus type distribution in 30,848 invasive cervical cancers worldwide: variation by geographical region, histological type and year of publication. Int J Cancer 2011;128(4):927–35. Available from: https://doi.org/10.1002/ijc.25396.

[13] de Sanjosé S, Quint WG, Alemany L, et al. Human papillomavirus genotype attribution in invasive cervical cancer: a retrospective cross-sectional worldwide study. Lancet Oncol 2010;11:1048–56.

[14] Alemany L, Saunier M, Alvarado-Cabrero I, Quirós B, Salmeron J, Shin H-R, et al. Human papillomavirus DNA prevalence and type distribution in anal carcinomas worldwide. Int J Cancer 2015;136 (1):98–107.

[15] de Sanjosé S, Alemany L, Ordi J, Tous S, Alejo M, Bigby SM, et al. Worldwide human papillomavirus genotype attribution in over 2000 cases of intraepithelial and invasive lesions of the vulva. Eur J Cancer 2013;49(16):3450–61.

[16] Alemany L, Saunier M, Tinoco L, Quirós B, Alvarado-Cabrero I, Alejo M, et al. Large contribution of human papillomavirus in vaginal neoplastic lesions: a worldwide study in 597 samples. Eur J Cancer 2014;50(16):2846–54.

[17] Alemany L, Cubilla A, Halec G, Kasamatsu E, Quirós B, Masferrer E, et al. Role of human papillomavirus in penile carcinomas worldwide. Eur Urol 2016;69:953–61.

[18] Castellsagué X, Alemany L, Quer M, Halec G, Quirós B, Tous S, et al. HPV involvement in head and neck cancers: comprehensive assessment of biomarkers in 3680 patients. J Natl Cancer Inst 2016;108(6): djv403.

[19] de Martel C, Plummer M, Vignat J, Franceschi S. Worldwide burden of cancer attributable to HPV by site, country and HPV type. Int J Cancer 2017;141(4):664–70. Available from: https://doi.org/10.1002/ijc.30716.

[20] de Sanjosé S, Serrano B, Tous S, Alejo M, Lloveras B, Quirós B, et al. Burden of human papillomavirus (HPV)-related cancers attributable to HPVs 6/11/16/18/31/33/45/52 and 58. JNCI Cancer Spectr 2019;2 (4):pky045. Available from: https://doi.org/10.1093/jncics/pky045.

[21] Guan P, Howell-Jones R, Li N, Bruni L, de Sanjosé S, Franceschi S, et al. Human papillomavirus types in 115,789 HPV-positive women: a meta-analysis from cervical infection to cancer. Int J Cancer 2012;131(10):2349–59. Available from: https://doi.org/10.1002/ijc.27485.

[22] Bernard HU, Burk RD, Chen Z, van Doorslaer K, zur Hausen H, de Villiers EM. Classification of papillomaviruses (PVs) based on 189 PV types and proposal of taxonomic amendments. Virology 2010;401 (1):70–9. Available from: https://doi.org/10.1016/j.virol.2010.02.002.

[23] Bouvard V, Baan R, Straif K, Grosse Y, Secretan B, El Ghissassi F, et al. WHO International Agency for Research on Cancer Monograph Working Group A review of human carcinogens – Part B: biological agents. Lancet Oncol 2009;10:321–2. Available from: https://doi.org/10.1016/S1470-2045 (09)70096-8.

[24] IARC Monographs on the Evaluation of Carcinogenic Risks to Humans. Volume 100B. Human Papillomaviruses. Lyon, France: IARC; 2012.

[25] Schiffman M, Clifford G, Buonaguro FM. Classification of weakly carcinogenic human papillomavirus types: addressing the limits of epidemiology at the borderline. Infect Agents Cancer 2009;4:8. Available from: https://doi.org/10.1186/1750-9378-4-8.

[26] Combes J-D, Guan P, Franceschi S, Clifford GM. Judging the carcinogenicity of rare human papillomavirus types. Int J Cancer 2015;136:740–2. Available from: https://doi.org/10.1002/ijc.29019.

CHAPTER

FINDING HOW HUMAN PAPILLOMAVIRUSES ALTER THE BIOCHEMISTRY AND IDENTITY OF INFECTED EPITHELIAL CELLS

4

Sharon C. Wu[1,2], Veronica Canarte[1,3], Harshita Beeravolu[1], Miranda Grace[1], Surendra Sharma[1,4] and Karl Munger[1,2,3,4]

[1]*Department of Developmental, Molecular and Chemical Biology, Tufts University School of Medicine, Boston, MA, United States* [2]*Molecular Microbiology Ph.D. Program, Tufts University School of Medicine, Boston, MA, United States* [3]*Post-Baccalaureate Research and Education Program, Tufts University School of Medicine, Boston, MA, United States* [4]*Biochemistry Ph.D. Program, Tufts University School of Medicine, Boston, MA, United States*

PARASITIC LIFESTYLES

Viruses are obligate intracellular parasites and their replication cycles are acutely dependent on their host cells. In general, one can envision two strategies that would ensure successful persistence of these biological entities (Fig. 4.1). The first is based on establishing long-term persistence and production of viral progeny only when the survival of the host organism is jeopardized. This is epitomized by bacteriophages, such as the lambda phage, which inserts its genome into the bacterial chromosome during its lysogenic, dormant phase. The activation of the lytic, virulent phase of the life cycle is engaged under conditions of cellular stress and involves excision of the phage genome and synthesis of infectious phage particles. The infected cell undergoes lysis and the newly synthesized phage progeny can infect a new, healthy host cell [1]. Many herpesviruses have adopted such a lifestyle. After the initial infection, they establish life-long latency in immune privileged, postmitotic cells with occasional bouts of reactivation and virus production, when the host organism encounters various forms of stress [2]. A second strategy is to establish a productive infection with ample production and release of infectious viral progeny. This requires vigorous suppression of host immune responses but allows for infection of new hosts before the original infective lesion is eliminated by the immune system or the demise of the host. Many papillomaviruses cause large, productive lesions, warts, which, at least in an immunocompetent host organism, spontaneously regress at a high rate. High-risk human papillomaviruses (HPVs), however, often establish long-term, persistent and productive infections [3,4] (Fig. 4.1).

Human Papillomavirus. DOI: https://doi.org/10.1016/B978-0-12-814457-2.00004-0

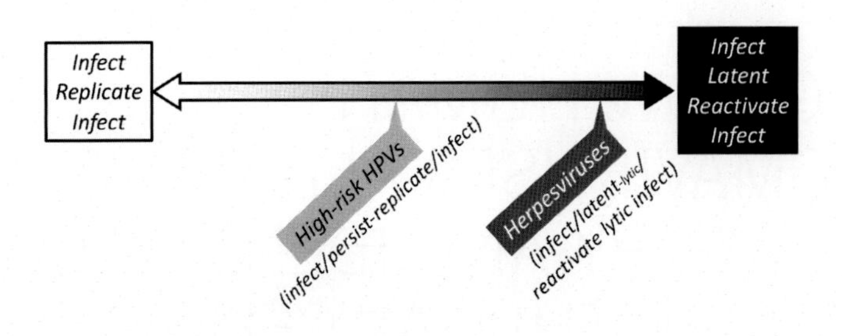

FIGURE 4.1

The spectrum of viral lifestyles.

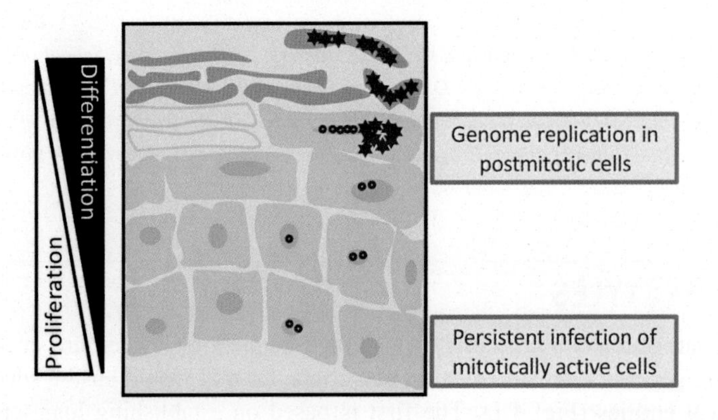

FIGURE 4.2

The enigmatic life cycle of high-risk HPVs. Viral genomes are shown as circles and infectious viral progeny is represented by stars.

THE ENIGMATIC, ILLOGICAL LIFE CYCLE OF HIGH-RISK HUMAN PAPILLOMAVIRUSES

Unlike herpesviruses that establish life-long latent infections in postmitotic cells and replicate in mitotically active cells, the molecular architecture of squamous epithelia necessitates that high-risk HPVs establish and maintain persistent infections in mitotically active, basal epithelial cells, while synthesis of progeny virus has to take place in the outermost layers of the skin that consist of terminally differentiated, postmitotic cells (Fig. 4.2). Hence, high-risk HPVs evolved to successfully execute two overtly opposing, paradoxical feats during different phases of the viral life cycle. In order to limit the risk of viral genome loss during mitosis, they constrain division of infected basal epithelial cells and viral proteins which may trigger innate and adaptive immune responses that cause the elimination of persistently infected cells, are expressed at low levels. On the other hand, during the productive phase of the viral life cycle, HPVs need to retain differentiated, normally postmitotic epithelial cells in a replication competent state to enable viral genome synthesis and production of infectious progeny [3,4] (Fig. 4.2).

The necessary HPV-mediated rewiring of host cells triggers a variety of innate, cell autonomous cytostatic or cytotoxic defense responses that also need to be muted in order to successfully establish and sustain a persistent infection [5]. Despite the fact that all papillomaviruses face similar challenges for completing their replicative life cycles in squamous epithelia and that their genomic organization is remarkably conserved, they evolved strikingly different replication strategies that all appear to be evolutionarily successful. The different replication strategies may reflect whether their life cycles evolved to favor long-term persistent infections as opposed to generating large infectious lesions that facilitate efficient infection of new hosts. Because an in-depth discussion of these different strategies exceeds what can be covered here, we will focus in the following on how mucosal, high-risk HPVs have evolved to balance the need for establishing and maintaining long-term, persistent infections of basal epithelial cells while establishing a replication competent cellular milieu in growth arrested, terminally differentiated epithelial cells. We will also discuss how this strategy creates cellular vulnerabilities that provide unique opportunities for developing therapeutic approaches.

INHIBITION OF INNATE IMMUNE SIGNALING

Cells have evolved multiple cell autonomous defense mechanisms to provide protection from invading pathogens, including viruses. Innate immune defense pathways are triggered in response to sensing pathogen-associated molecular patterns (PAMPs) and culminate in the activation of various inflammatory, cytostatic, and cytotoxic responses, most prominently type-I interferon signaling. In addition, expression of cytokines and chemokines is activated, which can trigger recruitment of antigen presenting cells that elicit induction of effective adaptive immune responses [6,7]. We will not discuss non-cell autonomous, adaptive immune responses in this short chapter but refer the reader to excellent review articles [8–10]. Instead we will focus on cell-intrinsic, innate defense responses and mechanisms.

The major cellular PAMP sensors are toll-like receptors (TLRs), RIG-I/MAVS and cGAS/STING. TLRs are not only expressed on the plasma- and intracellular membranes of immune cells but also in epithelial cells, where they can sense a variety of PAMPs. The RIG-I/MAVS pathway senses RNA-based PAMPs, whereas cGAS/STING senses cytoplasmic DNA [6,7]. By triggering any of these PAMP sensors, an invading pathogen causes activation of proinflammatory type-I interferon signaling.

After cell attachment and binding to an entry receptor, HPVs enter and are trafficked inside cells enveloped within a transport vesicle. Nuclear entry of the virus-containing transport vesicles is during mitosis after breakdown of the nuclear matrix and involves vesicle transport on microtubules [11]. This effectively shields the invading virus from recognition by intracellular innate immune surveillance mechanisms. Despite the fact that HPVs have evolved such a stealthy mechanism for cellular and nuclear invasion, these viruses have developed multiple mechanisms to inhibit cellular innate immune signaling both at the level of initial PAMP sensing and downstream signaling.

Only very few studies have comprehensively investigated the role of TLR signaling during the initial phases of HPV infection, but expression levels of some TLRs positively correlate with regression of premalignant lesions [12,13]. This suggests that TLRs importantly contribute to surveying and eliminating HPV-infected cells. Whether this is mediated by recruitment of antigen presenting cells, or represents a purely innate effect of TLR signaling within the infected epithelial cell, or whether this is executed by a combination of these two mechanisms, remains unclear.

Moreover, it has been reported that HPV gene expression can affect expression levels of TLRs as well as of signaling molecules involved in TLR signaling, and E6 has been reported to interact with TLR signaling components including the master regulator Myeloid differentiation primary response 88 (MYD88) [14,15].

The HPV18 E7 protein was shown to bind and inhibit the DNA sensor Stimulator of interferon genes (STING). The E7 sequences necessary for STING binding overlap with the LXCXE (L, leucine; C, cysteine; E, glutamic acid; X, any amino acid) motif that is also necessary for binding the retinoblastoma tumor suppressor family of proteins [16]. STING inactivation may be important for sustaining the viral life cycle given that expression of HPV proteins, particularly E7, causes an increased incidence of double-strand DNA breaks, which serves the viral life cycle by providing preferred sites for HPV genome replication [17–20]. However, persistent double-strand DNA breaks also generates DNA fragmentation [18,20]. This results in the generation of cytoplasmic DNA fragments that can trigger the cGAS/STING sensor. It is noted that the cytoplasmic DNA sensor STING is key to signaling oncogene-induced senescence (OIS) [21,22]. It will be interesting to determine whether accumulation of cytoplasmic DNA fragments represents the oncogenic stimulus generated by high-risk HPV E7 expression that triggers an OIS response through induction of the KDM6B lysine demethylase, epigenetic de-repression of $p16^{INK4A}$ and activation of the retinoblastoma tumor suppressor (RB1) [23]. According to such a model, STING inhibition, in concert with RB1 destabilization, may at least in part be related to the necessity of thwarting OIS as is described in more detail in a later section.

Retinoic acid-Inducible Gene I (RIG-I) sensing and signaling of RNA-based PAMPs requires interaction with the mitochondrial antiviral signaling protein (MAVS). This interaction is stabilized by lysine (K)63-linked RIG-I ubiquitination by the tripartite motif Containing 25 (TRIM25) ubiquitin ligase [24,25]. The ubiquitin specific protease 15 (USP15) deubiquitinating enzyme protects TRIM25 from rapid proteasomal degradation by removing K48 linked ubiquitin chains. The resulting increase in TRIM25 steady-state levels increases the pool of K63-linked ubiquitinated RIG-I that can efficiently interact with MAVS [26]. HPV E6 proteins can form complexes with TRIM25 and USP15 [27,28]. As a consequence, TRIM25 is destabilized, which results in decreased K63-linked RIG-I ubiquitination and, in turn, reduces RIG-I/MAVS complex formation. Hence, HPV E6 expression inhibits RIG-I/MAVS interaction, which dampens activation of type-I interferon signaling in response to viral infection [28]. One might wonder why DNA viruses such as HPVs evolved to inhibit RNA sensing. One possibility is that some of the HPV mRNAs that are produced during infection are sensed as PAMPs. Alternatively, HPVs may dysregulate expression of certain RNA binding proteins that function to shield structures on endogenous RNAs that are recognized as PAMPs, as was recently reported for herpes simplex virus 1 infections [29]. In addition to directly impeding the functionality of innate PAMP sensors, HPVs also inhibit interferon signaling by binding to interferon regulatory factors (IRFs) and other downstream mediators [30–34].

MECHANISMS THAT MINIMIZE HUMAN PAPILLOMAVIRUS GENOME LOSS DURING CELL DIVISION

The viral DNA genome is released from intranuclear transport vesicles only after the nuclear envelope has re-formed at the completion of mitosis [11]. The genome then binds to PML nuclear bodies and early viral proteins including the replication proteins E1 and E2 are synthesized to allow for an initial burst of viral genome synthesis. This is followed by occasional viral genome replication to avoid dilution of HPV genomes in daughter cells after each cell division [35,36]. Since the nuclear envelope breaks down during mitosis, every round of cell division poses a danger of viral genome loss.

To minimize the possibility that, after the nuclear envelope has re-formed, viral genomes end up in the cytoplasm where they may be sensed and trigger innate immune responses, the viral E2 protein tethers viral genomes to mitotic chromosomes [37,38]. This allows for genome partitioning to the daughter cells' nuclei after cell division. Basal epithelial cells can divide asymmetrically, giving rise to one basal and one differentiating suprabasal daughter cell. HPV genome retention in the basal daughter cell is the key to maintaining a persistent infection, whereas HPV genomes that are partitioned into the suprabasal daughter cell are destined to produce infectious viral progeny. Hence, it will be interesting to determine whether there are mechanisms that regulate equal or unequal genome partitioning as a means to controlling the persistent versus infectious phases of the viral life cycle. One intriguing possibility is that the unique ability of high-risk HPV E6 proteins to target cellular proteins that contain postsynaptic density protein, disc large tumor suppressor, zonula occludens-1 (PDZ) domains [39,40] may be related to controlling the persistent versus infectious phases of the life cycle by directly dysregulating asymmetric cell division. PDZ proteins and cellular PDZ binding proteins have been shown to control the geometry of basal epithelial cell division. E6-mediated targeting of cellular PDZ proteins may thus distort this process and generate unequal numbers of basal and suprabasal daughter cells [41,42]. This process may be regulated since the phosphorylation status of the E6 PDZ binding motif, which affects binding efficiency to cellular PDZ proteins, is altered in response to cellular stress such as DNA damage [43].

A second strategy may be for HPVs to reside in cells that do not frequently divide. Epithelial stem cells, for example, only divide infrequently and when there is a need to regenerate the basal layers of the epithelium in response to obliterative trauma. Indeed, cutaneous HPV genomes can be detected in hair follicles, which are known to harbor stem cells [44]. The location of stem cells in the mucosal epithelium is unknown, but the cell type at the cervical squamo-columnar transformation zone that is preferentially infected by HPVs has stem cell properties [45]. Hence, mucosal high-risk HPVs may preferentially infect cells with stem-like properties. Moreover, the unique ability of high-risk HPV E7 proteins to target the protein tyrosine phosphatase, nonreceptor type 14 (PTPN14) for proteasomal degradation inhibits epithelial differentiation and may also serve to retain high-risk HPV-infected cells in an undifferentiated state, thereby supporting persistent infection [46–48].

Interestingly, E6 and E7 have each been reported to modulate gene expression and/or steady-state levels of proteins that control aspects of cellular stemness [49,50]. HPV E7 causes increased expression of (sex determining region Y)-box 2 (SOX2), octamer-binding transcription factor 4 (OCT4), homeobox protein NANOG, and Kruppel like factor 4 (KLF4) and dysregulated homeobox (HOX) gene expression [51–53]. HPV E6 enhances NANOG and homolog of enhancer of split 1 (HES1) expression and has also been reported to enhance KLF4 levels [52,54]. Moreover, E6 stimulates expression and activity of telomerase (TERT), an enzyme that thwarts telomere erosion [55]. Telomerase activation in conjunction with inactivation of RB1-mediated OIS signaling has been shown to immortalize keratinocytes and retain them in a stem-like state [56,57]. Hence, it is tempting to speculate that expression of the HPV E6 and E7 proteins may lock infected basal epithelial cells in a stem-like state and/or endow infected cells with stem-like properties [49].

HIGH-RISK HPV E6 AND E7 TRIGGER AND SUBVERT CELL-INTRINSIC DEFENSE MECHANISMS

Despite their small size, high-risk HPV E6 and E7 proteins are multifunctional and cause abnormalities in almost all cellular processes that have been referred to as the "hallmarks of cancer" [58]. Most HPV-associated cancers represent nonproductive infections that terminate the viral life cycle and, eventually, the life of the host, as well. Hence, oncogenic transformation is a rare and evolutionarily undesirable outcome of viral rewiring of the host cell to establish and sustain a persistent HPV infection and to support the productive viral life cycle.

High-risk HPV E7 expression is sensed by the cell as an oncogenic insult and threat. The molecular nature of the signal and the cellular sensor are both unknown, but the cellular defense program that is triggered has been termed OIS [59]. Senescent cells remain metabolically active but have terminally withdrawn from the cell division cycle [60]. OIS was first reported as a consequence of ectopic RAS oncogene expression in normal human cells [61]. Evidence of OIS has also been observed in vivo in RAS-driven premalignant lesions. OIS provides an effective barrier to cancer progression, and additional mutations need to be acquired for malignant progression to occur [62,63]. RAS-driven OIS has been linked to RAF/MEK/ERK kinase-driven AP-1 transcription factor activation. This leads to increased expression of the KDM6B lysine demethylase, which causes demethylation of repressive trimethylation marks on lysine 27 of histone H3 (H3K27me3) and epigenetic de-repression of specific target genes, including the cyclin-dependent kinases 4 and 6 (CDK4/CDK6) inhibitor $p16^{INK4A}$ [64,65]. CDK4/CDK6 inhibition leads to accumulation of the hypophosphorylated, active form of the retinoblastoma tumor suppressor, RB1, which, in addition to a G1 cell cycle arrest, imposes cellular senescence [66,67]. Similar to RAS, HPV E7 expression signals OIS through KDM6B-mediated epigenetic de-repression of the $p16^{INK4A}$ tumor suppressor and RB1 tumor suppressor activation, although the molecular steps are not clear [53]. Since an OIS response would quell the viral life cycle, high-risk HPVs evolved to abrogate OIS. Consistent with this model, HPV16 E7 preferentially and efficiently binds hypophosphorylated RB1 and targets it for proteasomal degradation [68–70]. The ability of HPV16 E7 to target RB1 for degradation is necessary to defeat RB1-induced senescence [71]. As a consequence, despite expressing the $p16^{INK4A}$ tumor suppressor at high levels, HPV-associated cancers remain proliferatively active as assessed by expression of the Ki67 proliferation marker [72]. The detection of $p16^{INKA}$/Ki67 double positive cells is a hallmark of high-risk HPV-associated premalignant lesions and cancers that has been exploited for diagnosis [73–76].

HPV E7-mediated RB1 inactivation causes dysregulation of E2F transcription factor activity, which leads to persistent expression of S-phase specific genes that control DNA synthesis and replication. The deterioration of G1/S checkpoint control prompts uncontrolled, aberrant proliferation, and replication stress that can lead to single- and double-strand DNA breaks. Aberrant S-phase entry guarantees the availability of cellular proteins and metabolites that are necessary for sustaining viral genome replication in intrinsically growth-arrested cells and the presence of double-strand DNA breaks provides sites for viral genome replication [77]. In, otherwise normal cells, the presence of double-strand DNA breaks also triggers activation of a cellular suicide pathway that is controlled by the TP53 tumor suppressor. Consistent with this model, high-risk HPV E7 expression in normal human cells causes TP53 stabilization and activation and this correlates with the ability of E7 to destabilize RB1. HPV16 E7 expressing cells are sensitized to undergo TP53-mediated cell death particularly when they are deprived of extrinsic growth factors. Growth factor deprivation creates a situation of conflicting growth signals; a pro-proliferative stimulus generated by oncogene expression conflicts with an antiproliferative signal due to lack of extrinsic growth factors. This potentially dangerous condition is resolved by engaging a TP53-mediated cell death program [78]. To avoid the risk of elimination, high-risk HPVs evolved a strategy to eliminate TP53 [79]. The HPV E6 protein acts as a platform for formation of a tripartite complex with the UBE3A (E6AP) ubiquitin ligase and TP53, which results in TP53 ubiquitination and proteasomal degradation [80,81]. Hence, the RB1 and TP53 tumor suppressor pathways each are muted by the viral E7 and E6 proteins, respectively, whereas components of these two tumor suppressor pathways are frequently mutated in cancers not caused by HPV infections [82,83].

Reactivation of the dormant RB1 and/or TP53 tumor suppressors provides an attractive therapeutic opportunity for HPV-associated lesions and cancers. In addition, the rewiring of cellular signaling circuits by HPV E6 and E7 causes collateral sensitivities to inhibition of signaling pathways that are nonessential in normal cells. Over the years, a number of such vulnerabilities have been identified. Loss-of-function screens revealed that signaling pathways controlled by cellular kinases were subverted; E7 expression was shown to complement loss the CDK6, ERBB3, FYN, AAK1, and TSSK2 kinases that are necessary for survival of normal cells [84]. In contrast, HPV16 E6-expressing cells were sensitive to loss of PAK3 and SGK2 expression, and that this was shown to be a consequence of E6-mediated TP53 inactivation [85]. By stimulating KDM6A and KDM6B lysine demethylase expression, HPV16 E7 expressing cells and clinical lesions show a marked global decrease of the repressive H3K27me3 mark [53]. Surprisingly, high-risk HPV E7 expressing cells acquire an acute dependence on KDM6A and KDM6B and their downstream transcriptional targets, $p21^{CIP1}$ and $p16^{INK4A}$, respectively [23,86]. In addition, HPV-associated cancers have been reported to be dependent on the USP46 deubiquitinase, which is recruited by the E6 protein to stabilize the $CRL4^{CDT2}$ ubiquitin ligase, which stimulates cell cycle progression [87].

HUMAN PAPILLOMAVIRUS INFECTIONS RENDER HOST CELLS GENOMICALLY UNSTABLE

Expression of HPV E7 causes an increase in double-strand DNA breaks [18], which are preferred sites for HPV genome replication [36]. Double-strand DNA breaks arise as sequelae of E7 induced replication stress due to RB1 tumor suppressor inactivation as well as through other pathways [77]. Double-strand DNA breaks can give rise to chromosomal fusions that can generate chromosomal translocations through breakage–fusion–bridge cycles [88]. E7 also induces synthesis of supernumerary centrosomes, which can give rise to multipolar mitoses and aneuploidy. Moreover, E6-mediated TP53 inactivation generates polyploidy and aneuploidy tolerance, which allows cells with abnormal chromosome numbers to remain in the proliferative pool [89]. STING depletion has also been shown to favor survival of cells with chromosomal aberrations, and hence, E7-mediated STING inhibition may be a second mechanism that promotes aneuploidy tolerance [90,91]. As a consequence of the mitotic mutator activities of the high-risk HPV E6 and E7 proteins, HPV-positive cervical cancer cells show evidence for numerical as well as structural chromosome aberrations [18].

HPV-infected epithelial cells also acquire and accumulate point mutations. An analysis of the predominating mutational pattern in HPV-positive tumors revealed a preponderance of C-to-T and C-to-G mutations at TpCpN trinucleotides, and thus, they likely represent a consequence of cytidine deamination by apolipoprotein-B mRNA editing enzyme, catalytic polypeptide like 3 (APOBEC3) [92,93]. There are seven APOBEC3 family members, all encoded in a cluster on chromosome 22. The expression pattern of the different isoforms is cell type specific, but expression increases in response to interferon [94]. Many pathogens, thus, induce expression and activity of APOBEC3 family members. APOBEC3 family members restrict invading pathogens by generating point mutations in their genomes. Deamination of cytidine residues can give rise to C-to-T and C-to-G mutations within the genome of the invading pathogen but also within the genome of the host cell. Epithelial cells express APOBEC3 family members and HPV E6 and E7 have each been reported to trigger increased expression of specific APOBEC3 family members [95]. Curiously, and in contrast to many other viruses, high-risk HPVs do not appear to have developed specific strategies to effectively counter APOBEC3 activity [96]. While HPV genomes are relatively devoid of APOBEC3 sites [97], a large percentage of the HPV16 variants that have been detected in a large-scale sequencing study of patient samples follow the consensus of APOBEC3 restriction [98]. Hence, APOBEC3 restriction may provide a mechanism for generating viral diversity that may allow for rapid adaptation to the specific genetic make-up of a human host. Host genome mutations generated by pathogen-induced APOBEC3 activity can have a dramatic effect on the physiology of the host epithelial cell and may contribute to oncogenic transformation. Consistent with this model, some of the phosphatidylinositol-4,5-bisphosphate 3-kinase catalytic subunit alpha (PIK3CA) mutants that have been detected in HPV-associated cancers may have been caused by APOBEC3 [99].

CONCLUDING REMARKS

The reprogramming of host cells by high-risk HPVs that leads to carcinogenesis is often portrayed as a one-sided act of subverting cellular tumor suppressor pathways. A more recent view is that the viral invasion of the cell triggers a concerted set of cellular defense mechanisms and many of the "oncogenic" activities that sets high-risk HPVs apart from other nononcogenic HPVs, such as RB1, TP53, and PTPN14 degradation, are driven by the need to neutralize these cellular defense responses [99]. All papillomaviruses need to reprogram terminally differentiated postmitotic epithelial cells to support viral progeny synthesis. The high-risk HPVs, however, have also evolved to successfully establish and maintain persistent infections in the proliferative compartment of squamous epithelia, a tissue that undergoes constant self-renewal and, since it functions as the main barrier to the environment, it is subject to intense surveillance by innate and adaptive immune systems and other antimicrobial defense mechanisms. It is tempting to speculate that the unique, transforming activities of this class of HPVs represents a consequence of this lifestyle.

REFERENCES

[1] Ptashne M. A genetic switch: gene control and phage lambda. Palo Alto, CA: Blackwell Press; 1986. 128 p.

[2] Koyuncu OO, MacGibeny MA, Enquist LW. Latent versus productive infection: the alpha herpesvirus switch. Future Virol 2018;13(6):431–43.

[3] Schiffman M, Castle PE, Jeronimo J, Rodriguez AC, Wacholder S. Human papillomavirus and cervical cancer. Lancet 2007;370(9590):890–907.

[4] Moody CA, Laimins LA. Human papillomavirus oncoproteins: pathways to transformation. Nat Rev Cancer 2010;10(8):550–60.

[5] Westrich JA, Warren CJ, Pyeon D. Evasion of host immune defenses by human papillomavirus. Virus Res 2017;231:21–33.

[6] Chan YK, Gack MU. Viral evasion of intracellular DNA and RNA sensing. Nat Rev Microbiol 2016;14 (6):360–73.

[7] Odendall C, Kagan JC. Activation and pathogenic manipulation of the sensors of the innate immune system. Microbes Infect 2017;19(4-5):229–37.

[8] Smola S, Trimble C, Stern PL. Human papillomavirus-driven immune deviation: challenge and novel opportunity for immunotherapy. Ther Adv Vaccines 2017;5(3):69–82.

[9] Stanley MA, Sterling JC. Host responses to infection with human papillomavirus. Curr Probl Dermatol 2014;45:58–74.

[10] Bashaw AA, Leggatt GR, Chandra J, Tuong ZK, Frazer IH. Modulation of antigen presenting cell functions during chronic HPV infection. Papillomavirus Res 2017;4:58–65.

[11] DiGiuseppe S, Bienkowska-Haba M, Guion LG, Sapp M. Cruising the cellular highways: How human papillomavirus travels from the surface to the nucleus. Virus Res 2017;231:1–9.

[12] Halec G, Scott ME, Farhat S, Darragh TM, Moscicki AB. Toll-like receptors: Important immune checkpoints in the regression of cervical intra-epithelial neoplasia 2. Int J Cancer 2018;143(11):2884–91.

[13] Scott ME, Ma Y, Farhat S, Moscicki AB. Expression of nucleic acid-sensing Toll-like receptors predicts HPV16 clearance associated with an E6-directed cell-mediated response. Int J Cancer 2015;136 (10):2402–8.

[14] Oliveira LB, Haga IR, Villa LL. Human papillomavirus (HPV) 16 E6 oncoprotein targets the Toll-like receptor pathway. J Gen Virol 2018;.

[15] Morale MG, da Silva Abjaude W, Silva AM, Villa LL, Boccardo E. HPV-transformed cells exhibit altered HMGB1-TLR4/MyD88-SARM1 signaling axis. Sci Rep 2018;8(1):3476.

[16] Lau L, Gray EE, Brunette RL, Stetson DB. DNA tumor virus oncogenes antagonize the cGAS-STING DNA-sensing pathway. Science 2015;350(6260):568–71.

[17] McBride AA. Playing with fire: consequences of human papillomavirus DNA replication adjacent to genetically unstable regions of host chromatin. Curr Opin Virol 2017;26:63–8.

[18] Duensing S, Munger K. The human papillomavirus type 16 E6 and E7 oncoproteins independently induce numerical and structural chromosome instability. Cancer Res 2002;62(23):7075–82.

[19] Moody CA, Laimins LA. Human papillomaviruses activate the ATM DNA damage pathway for viral genome amplification upon differentiation. PLoS Pathog 2009;5(10):e1000605.

[20] Mehta K, Laimins L. Human papillomaviruses preferentially recruit DNA repair factors to viral genomes for rapid repair and amplification. MBio 2018;9(1).

[21] Gluck S, Guey B, Gulen MF, Wolter K, Kang TW, Schmacke NA, et al. Innate immune sensing of cytosolic chromatin fragments through cGAS promotes senescence. Nat Cell Biol 2017;19(9):1061–70.

[22] Dou Z, Ghosh K, Vizioli MG, Zhu J, Sen P, Wangensteen KJ, et al. Cytoplasmic chromatin triggers inflammation in senescence and cancer. Nature 2017;550(7676):402–6.

[23] McLaughlin-Drubin ME, Park D, Munger K. Tumor suppressor p16INK4A is necessary for survival of cervical carcinoma cell lines. Proc Natl Acad Sci U S A 2013;110(40):16175–80.
[24] Gack MU, Kirchhofer A, Shin YC, Inn KS, Liang C, Cui S, et al. Roles of RIG-I N-terminal tandem CARD and splice variant in TRIM25-mediated antiviral signal transduction. Proc Natl Acad Sci U S A 2008;105(43):16743–8.
[25] Gack MU, Shin YC, Joo CH, Urano T, Liang C, Sun L, et al. TRIM25 RING-finger E3 ubiquitin ligase is essential for RIG-I-mediated antiviral activity. Nature 2007;446(7138):916–20.
[26] Pauli EK, Chan YK, Davis ME, Gableske S, Wang MK, Feister KF, et al. The ubiquitin-specific protease USP15 promotes RIG-I-mediated antiviral signaling by deubiquitylating TRIM25. Sci Signal 2014;7(307):ra3.
[27] Vos RM, Altreuter J, White EA, Howley PM. The ubiquitin-specific peptidase USP15 regulates human papillomavirus type 16 E6 protein stability. J Virol 2009;83(17):8885–92.
[28] Chiang C, Pauli EK, Biryukov J, Feister KF, Meng M, White EA, et al. The human papillomavirus E6 oncoprotein targets USP15 and TRIM25 to suppress RIG-I-mediated innate immune signaling. J Virol 2018;92(6).
[29] Chiang JJ, Sparrer KMJ, van Gent M, Lassig C, Huang T, Osterrieder N, et al. Viral unmasking of cellular 5S rRNA pseudogene transcripts induces RIG-I-mediated immunity. Nat Immunol 2018;19(1):53–62.
[30] Park JS, Kim EJ, Kwon HJ, Hwang ES, Namkoong SE, Um SJ. Inactivation of interferon regulatory factor-1 tumor suppressor protein by HPV E7 oncoprotein. Implication for the E7-mediated immune evasion mechanism in cervical carcinogenesis. J Biol Chem 2000;275(10):6764–9.
[31] Ronco LV, Karpova AY, Vidal M, Howley PM. Human papillomavirus 16 E6 oncoprotein binds to interferon regulatory factor-3 and inhibits its transcriptional activity. Genes Dev 1998;12(13):2061–72.
[32] Li S, Labrecque S, Gauzzi MC, Cuddihy AR, Wong AH, Pellegrini S, et al. The human papilloma virus (HPV)-18 E6 oncoprotein physically associates with Tyk2 and impairs Jak-STAT activation by interferon-alpha. Oncogene 1999;18(42):5727–37.
[33] Barnard P, McMillan NA. The human papillomavirus E7 oncoprotein abrogates signaling mediated by interferon-alpha. Virology 1999;259(2):305–13.
[34] Perea S, Lopezocejo O, Vongabain A, Arana M. Human papillomavirus type-16 (HPV-16) major transforming proteins functionally interact with interferon signaling mechanisms. Int J Oncol 1997;11(1):169–73.
[35] Doorbar J, Quint W, Banks L, Bravo IG, Stoler M, Broker TR, et al. The biology and life-cycle of human papillomaviruses. Vaccine 2012;30(Suppl. 5):F55–70.
[36] McBride AA. Mechanisms and strategies of papillomavirus replication. Biol Chem 2017;398(8):919–27.
[37] You J, Croyle JL, Nishimura A, Ozato K, Howley PM. Interaction of the bovine papillomavirus E2 protein with Brd4 tethers the viral DNA to host mitotic chromosomes. Cell 2004;117(3):349–60.
[38] McBride AA, McPhillips MG, Oliveira JG. Brd4: tethering, segregation and beyond. Trends Microbiol 2004;12(12):527–9.
[39] Ganti K, Broniarczyk J, Manoubi W, Massimi P, Mittal S, Pim D, et al. The human papillomavirus E6 PDZ binding motif: from life cycle to malignancy. Viruses 2015;7(7):3530–51.
[40] Javier RT, Rice AP. Emerging theme: cellular PDZ proteins as common targets of pathogenic viruses. J Virol 2011;85(22):11544–56.
[41] Marsh EK, Delury CP, Davies NJ, Weston CJ, Miah MAL, Banks L, et al. Mitotic control of human papillomavirus genome-containing cells is regulated by the function of the PDZ-binding motif of the E6 oncoprotein. Oncotarget 2017;8(12):19491–506.
[42] Lechler T, Fuchs E. Asymmetric cell divisions promote stratification and differentiation of mammalian skin. Nature 2005;437(7056):275–80.

[43] Thatte J, Massimi P, Thomas M, Boon SS, Banks L. The human papillomavirus E6 PDZ binding motif links DNA damage response signaling to E6 inhibition of p53 transcriptional activity. J Virol 2018;92 (16).
[44] Boxman IL, Berkhout RJ, Mulder LH, Wolkers MC, Bouwes Bavinck JN, Vermeer BJ, et al. Detection of human papillomavirus DNA in plucked hairs from renal transplant recipients and healthy volunteers. J Invest Dermatol 1997;108(5):712–15.
[45] Herfs M, Yamamoto Y, Laury A, Wang X, Nucci MR, McLaughlin-Drubin ME, et al. A discrete population of squamocolumnar junction cells implicated in the pathogenesis of cervical cancer. Proc Natl Acad Sci U S A 2012;109(26):10516–21.
[46] White EA, Munger K, Howley PM. High-risk human papillomavirus E7 proteins target PTPN14 for degradation. MBio 2016;7(5).
[47] Szalmas A, Tomaic V, Basukala O, Massimi P, Mittal S, Konya J, et al. The PTPN14 tumor suppressor is a degradation target of human papillomavirus E7. J Virol 2017;91(7).
[48] Hatterschide J, Bohidar AE, Grace M, Nulton TJ, Kim HW, Windle B, et al. PTPN14 degradation by high-risk human papillomavirus E7 limits keratinocyte differentiation and contributes to HPV-mediated oncogenesis. Proc Natl Acad Sci U S A 2019;116(14):7033–42.
[49] Strati K. Changing stem cell dynamics during papillomavirus infection: potential roles for cellular plasticity in the viral lifecycle and disease. Viruses 2017;9(8).
[50] Brehm A, Ohbo K, Zwerschke W, Botquin V, Jansen-Durr P, Scholer HR. Synergism with germ line transcription factor Oct-4: viral oncoproteins share the ability to mimic a stem cell-specific activity. Mol Cell Biol 1999;19(4):2635–43.
[51] Organista-Nava J, Gomez-Gomez Y, Ocadiz-Delgado R, Garcia-Villa E, Bonilla-Delgado J, Lagunas-Martinez A, et al. The HPV16 E7 oncoprotein increases the expression of Oct3/4 and stemness-related genes and augments cell self-renewal. Virology 2016;499:230–42.
[52] Gunasekharan VK, Li Y, Andrade J, Laimins LA. Post-transcriptional regulation of KLF4 by high-risk human papillomaviruses is necessary for the differentiation-dependent viral life cycle. PLoS Pathog 2016;12(7):e1005747.
[53] McLaughlin-Drubin ME, Crum CP, Munger K. Human papillomavirus E7 oncoprotein induces KDM6A and KDM6B histone demethylase expression and causes epigenetic reprogramming. Proc Natl Acad Sci U S A 2011;108(5):2130–5.
[54] Tyagi A, Vishnoi K, Mahata S, Verma G, Srivastava Y, Masaldan S, et al. Cervical cancer stem cells selectively overexpress HPV oncoprotein E6 that controls stemness and self-renewal through upregulation of HES1. Clin Cancer Res 2016;22(16):4170–84.
[55] Klingelhutz AJ, Foster SA, McDougall JK. Telomerase activation by the E6 gene product of human papillomavirus type 16. Nature 1996;380(6569):79–82.
[56] Kiyono T, Foster SA, Koop JI, McDougall JK, Galloway DA, Klingelhutz AJ. Both Rb/p16INK4a inactivation and telomerase activity are required to immortalize human epithelial cells. Nature 1998;396 (6706):84–8.
[57] Maurelli R, Zambruno G, Guerra L, Abbruzzese C, Dimri G, Gellini M, et al. Inactivation of p16INK4a (inhibitor of cyclin-dependent kinase 4A) immortalizes primary human keratinocytes by maintaining cells in the stem cell compartment. FASEB J 2006;20(9):1516–18.
[58] Mesri EA, Feitelson MA, Munger K. Human viral oncogenesis: a cancer hallmarks analysis. Cell Host Microbe 2014;15(3):266–82.
[59] Munger K, Jones DL. Human papillomavirus carcinogenesis: an identity crisis in the retinoblastoma tumor suppressor pathway. J Virol 2015;89(9):4708–11.
[60] Aravinthan A. Cellular senescence: a Hitchhiker's guide. Hum Cell 2015;28(2):51–64.
[61] Serrano M, Lin AW, McCurrach ME, Beach D, Lowe SW. Oncogenic ras provokes premature cell senescence associated with accumulation of p53 and p16INK4a. Cell 1997;88(5):593–602.

[62] Bennecke M, Kriegl L, Bajbouj M, Retzlaff K, Robine S, Jung A, et al. Ink4a/Arf and oncogene-induced senescence prevent tumor progression during alternative colorectal tumorigenesis. Cancer Cell 2010;18(2):135–46.
[63] Damsky WE, Bosenberg M. Melanocytic nevi and melanoma: unraveling a complex relationship. Oncogene 2017;36(42):5771–92.
[64] Agger K, Cloos PA, Rudkjaer L, Williams K, Andersen G, Christensen J, et al. The H3K27me3 demethylase JMJD3 contributes to the activation of the INK4A-ARF locus in response to oncogene- and stress-induced senescence. Genes Dev 2009;23(10):1171–6.
[65] Barradas M, Anderton E, Acosta JC, Li S, Banito A, Rodriguez-Niedenfuhr M, et al. Histone demethylase JMJD3 contributes to epigenetic control of INK4a/ARF by oncogenic RAS. Genes Dev 2009;23 (10):1177–82.
[66] Futreal PA, Barrett JC. Failure of senescent cells to phosphorylate the RB protein. Oncogene 1991;6 (7):1109–13.
[67] Dyson NJ. RB1: a prototype tumor suppressor and an enigma. Genes Dev 2016;30(13):1492–502.
[68] Boyer SN, Wazer DE, Band V. E7 protein of human papilloma virus-16 induces degradation of retinoblastoma protein through the ubiquitin-proteasome pathway. Cancer Res 1996;56(20):4620–4.
[69] Huh K, Zhou X, Hayakawa H, Cho JY, Libermann TA, Jin J, et al. Human papillomavirus type 16 E7 oncoprotein associates with the cullin 2 ubiquitin ligase complex, which contributes to degradation of the retinoblastoma tumor suppressor. J Virol 2007;81(18):9737–47.
[70] Jones DL, Munger K. Analysis of the p53-mediated G1 growth arrest pathway in cells expressing the human papillomavirus type 16 E7 oncoprotein. J Virol 1997;71(4):2905–12.
[71] Gonzalez SL, Stremlau M, He X, Basile JR, Munger K. Degradation of the retinoblastoma tumor suppressor by the human papillomavirus type 16 E7 oncoprotein is important for functional inactivation and is separable from proteasomal degradation of E7. J Virol 2001;75(16):7583–91.
[72] Keating JT, Cviko A, Riethdorf S, Riethdorf L, Quade BJ, Sun D, et al. Ki-67, cyclin E, and p16INK4 are complimentary surrogate biomarkers for human papilloma virus-related cervical neoplasia. Am J Surg Pathol 2001;25(7):884–91.
[73] von Knebel Doeberitz M, Reuschenbach M, Schmidt D, Bergeron C. Biomarkers for cervical cancer screening: the role of p16(INK4a) to highlight transforming HPV infections. Expert Rev Proteomics 2012;9(2):149–63.
[74] Wentzensen N, Fetterman B, Castle PE, Schiffman M, Wood SN, Stiemerling E, et al. p16/Ki-67 dual stain cytology for detection of cervical precancer in HPV-positive women. J Natl Cancer Inst 2015;107 (12):djv257.
[75] Arean-Cuns C, Mercado-Gutierrez M, Paniello-Alastruey I, Mallor-Gimenez F, Cordoba-Iturriagagoitia A, Lozano-Escario M, et al. Dual staining for p16/Ki67 is a more specific test than cytology for triage of HPV-positive women. Virchows Arch 2018;473(5):599–606.
[76] Ikenberg H, Bergeron C, Schmidt D, Griesser H, Alameda F, Angeloni C, et al. Screening for cervical cancer precursors with p16/Ki-67 dual-stained cytology: results of the PALMS study. J Natl Cancer Inst 2013;105(20):1550–7.
[77] Moody CA. Impact of replication stress in human papillomavirus pathogenesis. J Virol 2019;93(2).
[78] Jones DL, Thompson DA, Munger K. Destabilization of the RB tumor suppressor protein and stabilization of p53 contribute to HPV type 16 E7-induced apoptosis. Virology 1997;239(1):97–107.
[79] Werness BA, Levine AJ, Howley PM. Association of human papillomavirus types 16 and 18 E6 proteins with p53. Science 1990;248(4951):76–9.
[80] Huibregtse JM, Scheffner M, Howley PM. Cloning and expression of the cDNA for E6-AP, a protein that mediates the interaction of the human papillomavirus E6 oncoprotein with p53. Mol Cell Biol 1993;13(2):775–84.

[81] Scheffner M, Huibregtse JM, Vierstra RD, Howley PM. The HPV-16 E6 and E6-AP complex functions as a ubiquitin-protein ligase in the ubiquitination of p53. Cell 1993;75(3):495–505.
[82] Scheffner M, Munger K, Byrne JC, Howley PM. The state of the p53 and retinoblastoma genes in human cervical carcinoma cell lines. Proc Natl Acad Sci U S A 1991;88(13):5523–7.
[83] Cancer Genome Atlas N. Comprehensive genomic characterization of head and neck squamous cell carcinomas. Nature 2015;517(7536):576–82.
[84] Baldwin A, Li W, Grace M, Pearlberg J, Harlow E, Munger K, et al. Kinase requirements in human cells: II. Genetic interaction screens identify kinase requirements following HPV16 E7 expression in cancer cells. Proc Natl Acad Sci U S A 2008;105(43):16478–83.
[85] Baldwin A, Grueneberg DA, Hellner K, Sawyer J, Grace M, Li W, et al. Kinase requirements in human cells: V. Synthetic lethal interactions between p53 and the protein kinases SGK2 and PAK3. Proc Natl Acad Sci U S A 2010;107(28):12463–8.
[86] Soto DR, Barton C, Munger K, McLaughlin-Drubin ME. KDM6A addiction of cervical carcinoma cell lines is triggered by E7 and mediated by p21CIP1 suppression of replication stress. PLoS Pathog 2017;13(10):e1006661.
[87] Kiran S, Dar A, Singh SK, Lee KY, Dutta A. The deubiquitinase USP46 is essential for proliferation and tumor growth of HPV-transformed cancers. Mol Cell 2018;72(5):823–35.
[88] McClintock B. The stability of broken ends of chromosomes in *Zea mays*. Genetics 1941;26(2):234–82.
[89] Duensing S, Lee LY, Duensing A, Basile J, Piboonniyom S, Gonzalez S, et al. The human papillomavirus type 16 E6 and E7 oncoproteins cooperate to induce mitotic defects and genomic instability by uncoupling centrosome duplication from the cell division cycle. Proc Natl Acad Sci U S A 2000;97(18):10002–7.
[90] Bakhoum SF, Ngo B, Laughney AM, Cavallo JA, Murphy CJ, Ly P, et al. Chromosomal instability drives metastasis through a cytosolic DNA response. Nature 2018;553(7689):467–72.
[91] Ranoa DRE, Widau RC, Mallon S, Parekh AD, Nicolae CM, Huang X, et al. STING promotes homeostasis via regulation of cell proliferation and chromosomal stability. Cancer Res 2019;79(7):1465–79.
[92] Roberts SA, Lawrence MS, Klimczak LJ, Grimm SA, Fargo D, Stojanov P, et al. An APOBEC cytidine deaminase mutagenesis pattern is widespread in human cancers. Nat Genet 2013;45(9):970–6.
[93] Cancer Genome Atlas Research Network, Albert Einstein College of Medicine, Analytical Biological Services, Barretos Cancer Hospital, Baylor College of Medicine, Beckman Research Institute of City of Hope, et al. Integrated genomic and molecular characterization of cervical cancer. Nature 2017;543(7645):378–84.
[94] Swanton C, McGranahan N, Starrett GJ, Harris RS. APOBEC enzymes: mutagenic fuel for cancer evolution and heterogeneity. Cancer Discov 2015;5(7):704–12.
[95] Warren CJ, Westrich JA, Doorslaer KV, Pyeon D. Roles of APOBEC3A and APOBEC3B in human papillomavirus infection and disease progression. Viruses 2017;9(8).
[96] Wallace NA, Munger K. The curious case of APOBEC3 activation by cancer-associated human papillomaviruses. PLoS Pathog 2018;14(1):e1006717.
[97] Warren CJ, Van Doorslaer K, Pandey A, Espinosa JM, Pyeon D. Role of the host restriction factor APOBEC3 on papillomavirus evolution. Virus Evol 2015;1(1).
[98] Mirabello L, Yeager M, Yu K, Clifford GM, Xiao Y, Zhu B, et al. HPV16 E7 genetic conservation is critical to carcinogenesis. Cell 2017;170(6):1164–1174.e6.
[99] Henderson S, Chakravarthy A, Su X, Boshoff C, Fenton TR. APOBEC-mediated cytosine deamination links PIK3CA helical domain mutations to human papillomavirus-driven tumor development. Cell Rep 2014;7(6):1833–41.

CHAPTER

5

BIOLOGY OF THE HUMAN PAPILLOMAVIRUS LIFE CYCLE: THE BASIS FOR UNDERSTANDING THE PATHOLOGY OF PRECANCER AND CANCER

John Doorbar[1], David Jenkins[2,3], Mark H. Stoler[4] and Christine Bergeron[5]

[1]Department of Pathology, University of Cambridge, Cambridge, United Kingdom [2]Emeritus Professor of Pathology, University of Nottingham, United Kingdom [3]Consultant in Pathology to DDL Diagnostic laboratories, Rijswijk, The Netherlands [4]Professor (Emeritus) of Pathology and Clinical Gynecology, Associate Director of Surgical and Cytopathology, Department of Pathology, University of Virginia Health Health System, Charlottesville, VA, United States [5]Laboratoire Cerba, Parc d'activité "les Béthunes" 95310 Saint Ouen L'aumone, Paris, France

Papillomaviruses have been coevolving with their hosts for hundreds of millions of years, and during this time have become epithelial specialists. They complete their life cycle in the surface epithelial layers of the body, with different papillomavirus types evolving different strategies for transmission and persistence [1]. This evolutionary path has led to over 200 known human papillomavirus types (PaVE: Papillomavirus Episteme, http://pave.niaid.nih.gov/), which cause a wide range of epithelial lesions, ranging from the benign skin warts that typically afflict children, to a number of important human cancers, of which cervical cancer is numerically the most significant (Table 5.1). The vast majority of human papillomaviruses cause only in-apparent infections in normal immunocompetent individuals; however, a situation that reflects the long coevolutionary history of these viruses with humans. Coevolution and host-adaptation generally act to moderate virus pathogenicity. Thus many human papillomavirus (HPV) types from the *Beta genus* can be detected in skin swabs and plucked hairs by PCR, but cause no obvious problems for the infected individual [3]. About half of all known human papillomavirus types fall into the Beta and *Gamma genera* (see Fig. 5.1), with new members being added regularly. Although these viruses can become problematic in immunosuppressed individuals, and in individuals suffering from inherited susceptibilities, for the most part, they go unnoticed in the general population. The *Alpha papillomaviruses* have by contrast been the subject of extensive analysis, both at the clinical level and in terms of their molecular biology. This intensive research effort stems from the fact that the Alpha HPV genus contains the *high-risk HPV types* (hr HPV types) that are associated with human cancers [4], as well as the sexually transmitted *low-risk types* (lr HPV types [5]) that cause genital and respiratory papillomatosis (Fig. 5.1). Even here however our knowledge of virus biology is incomplete, and although we have a reasonable general understanding of how viral gene expression underlies

Human Papillomavirus. DOI: https://doi.org/10.1016/B978-0-12-814457-2.00005-2

Table 5.1 Cancers Caused by Human papillomaviruses [2].

Cancer Site	New Cases	Attributable to HPV	Attrib. Fraction	Men	Women
Cervix uteri	528 000	528 000	100%	–	528 000
Anus	40 000	35 000	88%	17 000	18 000
Vagina/vulva	49 000	20 000	41%	–	20 000
Penis	26 000	13 000	51%	13 000	–
Oropharynx	96 000	29 000	31%	24 000	6 000
Oral cavity/larynx	358 000	9000	2.4%	7 000	2 000
Total	1097 000	634 000	58%	61 000	574 000

the development of neoplasia and cancer (see Section *Nonproductive Infection Leads to the Development of Neoplasia* below), the specific susceptibility of the cervix to high-risk HPV-driven cancers, and the events that lead to immune-mediated disease clearance are not yet fully established. In this chapter, we will discuss our current knowledge of HPV infection and neoplastic progression, and how our understanding of viral pathogenesis underlies the management of cervical disease.

PRODUCTIVE INFECTION FACILITATES THE ASSEMBLY AND RELEASE OF VIRUS PARTICLES

GENOME MAINTENANCE AND THE REPLICATION OF HPV GENOMES IN THE EPITHELIAL BASAL LAYER

The productive life cycle of all papillomaviruses requires the infection of cells in the lowest layers of the epithelium. Once infection has occurred, the viral genome is maintained as a low copy number nuclear episome, persisting in the basal layer for months or even years. The skin serves as a physical barrier to infection, and it is generally thought that some degree of epithelial injury is necessary in order for the virus to gain access to the epithelial basal cell. A number of HPV types, however, appear to take advantage of hair follicles (*Beta* HPV types) and/or eccrine glands (*Mu* HPV types) as ''natural entry routes,'' with Beta HPV types completing their life cycle in the vicinity of the infected hair follicle. For the high risk Alpha HPV types, the transitional epithelium of the cervical transformation zone, is similarly an epithelial site where infection is thought to be facilitated (see section XX later). While lesion-morphology, and the overall course of disease is determined by the infecting HPV type and the epithelial site of infection (see Fig. 5.2), disease persistence depends on the establishment of a reservoir of infection in a long-lived epithelial cell. Within this model, there is however controversy as to whether papillomavirus persistence requires infection of a specialized type of basal cell such as an *epithelial stem cell*, or whether viral gene expression acts to modify any infected basal cell to extend its life span and to impart stem cell-like characteristics. It is likely that both situations can occur, but for the *Alpha* HPV types, current thinking suggests that a carefully regulated pattern of viral gene expression can subtly modulate the pathways that normally regulate basal cell epithelial homeostasis in order to give any infected basal

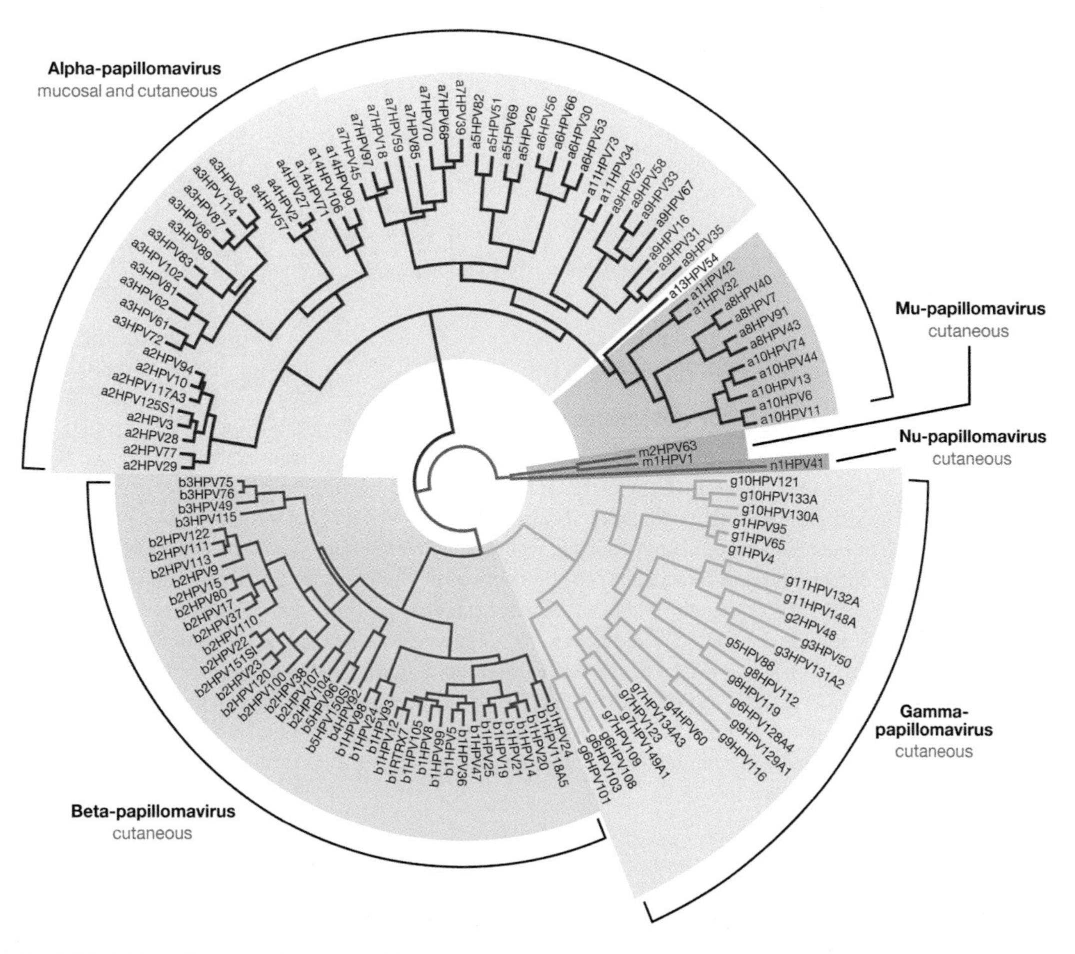

FIGURE 5.1 Diversity among human papillomaviruses.

Human papillomaviruses can be divided into five genera on the basis of their evolutionary history. Most of the Beta and Gamma HPV types (*green and blue*) cause inapparent infection of cutaneous epithelium. The Alpha genus contains the high risk HPV types (*pink*), as well as low risk HPV types, including the sexually transmitted types that cause genital warts (contained within the *orange shaded area*). Many Alpha HPV types, particularly those shaded in gray are associated with common cutaneous infections such as common warts. Within the high risk HPV group, those HPV types that are highlighted in a red font have been confirmed as human carcinogens (see Table 5.2).

cell an advantage over their uninfected neighbors. For the high-risk HPV types, this advantage appears to involve both an increase in cell proliferation, and a delay in basal cell commitment to differentiation, with the latter perhaps being sufficient to ensure the persistence of low-risk HPV types in infected basal cells following infection. To achieve this requires only a limited pattern of viral gene expression in the infected basal layer, with the minimal requirement including the E2

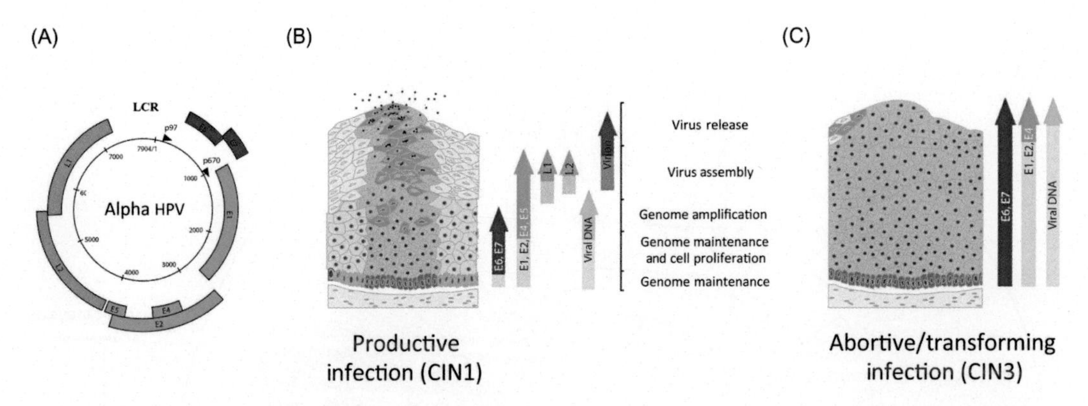

FIGURE 5.2 HPV genome organization and patterns of viral gene expression.

(A) The HPV genome is approximately 8000 base pairs in size, and typically contains eight or more open reading frames (E1,2,4,5,6,7 and L1, L2 shown). Gene expression is regulated by sequences in the LCR (*long control region*), and in the example shown (HPV16) by promoters P97 and P670. HPV types differ in their promoter positions, and also in their patterns of mRNA splicing, which along with differences on protein function, contributes to their different disease-associations. (B) Productive HPV infection is associated with a carefully regulated pattern of viral gene expression, which allows viral genome maintenance in the epithelial basal layer, viral genome amplification in the mid epithelial layers, and virus assembly and packaging toward the epithelial surface. (C) High grade neoplasia is driven by the deregulated expression of the high risk HPV E6 and E7 genes throughout the epithelial thickness. The deregulated expression of these proteins facilitates the accumulation of secondary genetic errors in the host cell genome, and also suppresses the ability of the hosts immune system to resolve infection. HPV genome integration may occur during cancer progression, facilitating the persistent expression of E6 and E7, and the loss of E2.

protein, to facilitate genome partitioning and the synchronous replication of viral and host cell genomes during cell division, and possibly also E6, to subtly retard the migration of infected basal cells into the upper epithelial layers. In fact the infected basal cell reservoir that is present during a productive infection shares similarities with the ''latently'' infected basal cells that persist following the immune-mediated lesion regression, where viral genomes also persist with only low levels of viral gene expression. As a general rule, an ordered productive infection such as occurs in benign papillomas, is characterized by low-level viral genome maintenance in the *undifferentiated cells* of the epithelial basal layer. The productive stage of the papillomavirus life cycle requires a differentiating multilayered squamous epithelium, which occurs only after the infected basal cells have divided, and infected daughter cells have entered the suprabasal differentiating cell layers.

PRODUCTIVE INFECTION AND VIRUS SYNTHESIS OCCURS IN THE DIFFERENTIATED LAYERS OF THE EPITHELIUM

For low risk *Alpha* HPV types, migration of the infected basal cell into the suprabasal layers is associated with cell cycle exit. Although suprabasal cell cycle exit is also apparent in some low-grade lesions caused by high risk HPV types, such infections are often characterized by the retention of cycling cells into the suprabasal layers, and in HSIL, into the mid and upper epithelial layers. At a pathology level, this continuous stimulation of cell cycle activity from the basal layer upwards is manifest as neoplasia, with the extent of cell cycle stimulation determining neoplastic grade. Interestingly, during the normal productive virus life cycle, the stimulation of *cell cycle re-entry* in the mid epithelial layers is necessary for viral genome amplification in preparation for packing into virus particles, and is one of the major functions of the HPV E6 and E7 proteins in the life cycle of all papillomaviruses. So how do the HPV E6 and E7 proteins normally achieve these roles? To a large extent, the cellular proteins that the virus targets for genome amplification are the same as those targeted during HPV-driven neoplasia, and include members of the retinoblastoma protein family and p53. The E6/E7 proteins of both the high and low risk papillomaviruses associate with these cellular proteins, but do it with different efficiency, and indeed, it is the activation of the common E7 target, p130 Rb, that plays a major role in driving suprabasal cell cycle entry for most HPV types. In addition to the extensively studied E6 and E7 proteins, the HPV E5 protein also contributes to successful genome amplification by constitutively activating EGFR in the absence of growth factors, in order to drive the virus infected cell to become replication competent. In this modified cellular environment in the mid epithelial layers, the viral replication proteins E1 and E2 are also expressed-binding specifically to the viral replication origin and recruiting cellular proteins necessary for viral replication and for transcriptional regulation of late events in the virus life cycle. Where it has been analyzed, papillomavirus copy number has been shown to rise from 50 copies/cell in the basal layer, to around 10,000 copies/cell in the upper layers [6]. This papillomavirus-driven replication program is accompanied by the expression of a number of other viral gene products that are involved in virus assembly and release. The first of these is E4, which is expressed at extraordinarily high levels at the time that viral genome amplification is initiated, and which eventually plays a role in virus release and transmission. The E4 protein contributes along with E5 to virus replication success, but probably plays a more important role in ensuring virus survival following release from the infected squames that are shed from the surface of infected epithelium. The L2 minor capsid protein is expressed after E4 as genome amplification is being completed in the mid epithelial layers of infected skin, where it begins the process of genome packaging. L1 expression follows, with viral genomes eventually packaged into infectious nonenveloped icosahedral virions in preparation for virus release as the surface squames are lost from the epithelial surface. Human papillomavirus particles are simple structures, with each consisting of 360 L1 proteins ordered into the 72 capsomeres that make up the virus coat. The double stranded viral genomes are packaged within these nonenveloped particles along with cellular histones. Papillomaviruses are resistant to dessication, and can persist in the environment on fomites for long periods of time, a situation that together with their easy spread by skin-skin contact explains their widespread prevalence, and their association with a diverse range of epithelial lesions in the general population.

NONPRODUCTIVE INFECTION LEADS TO THE DEVELOPMENT OF NEOPLASIA

The highly ordered patterns of viral gene expression that underlie successful productive infection are distinct from those that give rise to neoplasia. Neoplasia is a consequence of elevated viral gene expression in the epithelial basal layer, which for the high risk HPV types is facilitated at particular epithelial sites within the body. In terms of the number of individuals affected, the most important of these is the cervix, and in particular, the junctional region where the stratified epithelium of the *ectocervix* abuts the columnar epithelium of the *endocervix*. This region is referred to as the cervical transformation zone, and is a site where the normal process of metaplasia can lead to the formation of a multilayered squamous epithelium, where previously there was columnar epithelial tissue. This process can occur throughout a woman's life in response to cervical irritation, but occurs particularly at the cervical transformation zone during puberty, when normal physiological changes in the body result in endocervical epithelium becoming exposed to the acid environment of the vagina. Because the high risk HPV types are sexually transmitted, they are able to gain access to this metaplasic epithelium of the cervical transformation zone, as well as the ectocervix and the columnar epithelium of the endocervix. The basic principle underlying the vulnerability of the cervix to neoplasia and cancer is that viral gene expression is regulated differently according to epithelial site, and at the transformation zone in particular, viral gene expression becomes deregulated. The precise molecular mechanisms that underlie this process have not yet been examined, which is largely because of the lack of model systems that can be used to understand the complexity of the epithelial site. From the analysis of cervical tissue following HPV infection, however, we can reach some broad conclusions as to how this may work. In fact the concept is quite simple. At the ectocervix, the stratified epithelium supports primarily productive HPV infection, and has a cancer risk similar to that of other differentiated epithelial sites such as the vagina and vulva. At the endocervix, which consists of a columnar epithelium, there is effectively no ability of the virus to support productive infection, and instead, infection is can be manifested as high grade neoplasia especially adenocarcinoma in situ although other patterns may be seen.' At the transformation zone, a range of nonpermissive or semipermissive HPV infections occur, and these are manifest as a complex range of disease states, which for diagnostic purposes are classified as either a HSIL or an LSIL compartment. In reality, however, it appears that there is a range of deregulated HPV infection states at this site, which are characterized by changes in E6 and E7 levels and in the disruption of the virus productive cycle. It is these patterns of viral gene expression that first manifest as cervical neoplasia, and underlie the development of cancer at this site.

THE ROLE OF HPV GENE PRODUCTS IN THE PROGRESSION FROM NEOPLASIA TO CANCER

The high-risk HPV types that cause the majority of cervical cancers are contained within just one of the three broad evolutionary branches that make up the Alphapapillomavirus genus. The two other evolutionary branches contain low risk HPV types such as HPV11, which are often associated with genital infections, and a number of mainly cutaneous low risk types such as HPV2, which cause common warts and other benign lesions. The precise determinants of epithelial tropism, even within this relatively well-studied group, remain to be elucidated. It is thought, however, that the epithelial site-specific restrictions of each HPV type are not mediated at the level of virus entry and infection, but by the regulation of viral gene expression and protein function once infection has occurred. At present, we do not adequately understand how the regulation of normal epithelial homeostasis differs according to epithelial site, let alone the cell-type specific regulation of papillomavirus gene expression. Such an understanding is however central to our model of HPV-associated neoplasia. There are however a number of important biological differences among the different HPV genera that may account for their disease associations. Beta HPV types for instance restrict basal cell ''commitment to differentiation'' as a result of an association between their E6 protein and MAML [7,8], whereas the Alpha types use their E6 protein to target p53. In both the cases, the Notch pathway is disrupted, but the molecular mechanisms are different and the cell types that these viruses reside in are also different. Similarly, the Beta HPV types lack an E5 gene, which in the Alpha types contributes to effective immune evasion by suppressing HLA presentation of viral peptides to the adaptive immune system [9]. Indeed, the various immune evasion capabilities of the high risk HPV types contributes to their ability to persist as active infections in immunocompetent individuals.

To focus in further, we can also look at the high and low risk Alpha papillomaviruses, which are evolutionarily more closely related. Between these two groups there are a number of important differences that reflect the high risk HPV-associated cancer risk, as well as many similarities, which reflect a common HPV requirement to undergo productive infection in a differentiating epithelum. High risk HPV types are thought to have arisen following the acquisition of new characteristics as they became adapted to a new epithelial niche at least 300 million years ago [10]. Unlike the low risk Alpha types, they regulate their E6 and E7 genes from a single promoter, with the relative levels of E6/E7 transcripts being controlled by mRNA splicing. In general, the high risk E6 proteins more potently disrupt p53 function following infection, through both ubiquitin-mediated degradation as well as through the modulation of p53 transcription. In addition, the high risk E6 proteins possess a C-terminal motif that allows their association with a diverse family of PDZ domain proteins that are involved in maintaining cell polarity in the basal layer. Interestingly, rudimentary versions of this motif are found in a few low risk Alpha HPV types, but in the high risk viruses, a clear association with the cellular Dlg protein that is required for the normal assembly of cell junctions has been observed, and for the more problematic high risk types such as HPV16 and 18, the number of validated PDZ targets is considerable [11]. The advantage that these motifs and the extended association of high risk E7 proteins with members of the pRb family offer the high risk viruses in terms of evolutionary fitness is not yet clear, although it is apparent that these high risk HPV functions are very important in driving neoplasia and cancer when their expression becomes deregulated. Interestingly, the high risk E7 protein has other contributions to neoplasia, by disrupting chromosome segregation, and cellular gene expression through the induction of DNA methyltransferase activity [12]. Both E6 and E7 can elevate levels of Apobec 3B (A3B) in the cell, which leads to the accumulation of A3B-induced point mutations during cancer progression while the high risk E6 protein is also able to extend cellular life span by stimulating telomerase activity [4]. In general, it is the corruption of virus functions normally required for productive infection that leads to neoplasia and cancer. The E6/A3B interaction for instance is suspected to play a role in generating virus diversity during productive infection [13,14], but contributes to mutation of the host cell genome when E6 expression is deregulated. Similarly, the E6-mediated modulation of p53 function, which appears important in regulating epithelial homeostasis and maintaining the reservoir of papillomavirus infection in the epithelial basal layer, can also disrupt p53-mediated cell cycle arrest in response to DNA damage.

In addition to their effects on cell proliferation and genome integrity, the viral E6 and E7 proteins also inhibit detection of the E6/E7-expressing cell by the immune system. This involves the

E7-mediated modulation of the type 1 interferon response through the inhibition of STAT and IRF 1 (interferon response factor 1), a transcription factor involved in the induction of interferon stimulated genes [15], with E6 modulating the activity of Tyk2, another component in the interferon signaling pathway. Like E5, E7 also inhibits the MHC presentation of viral peptides, in this case by repressing LMP2 and TAP1 activity, with several viral genes acting to suppress the innate immune response, including the RIG-1-like (RLR) and STING/cGAS pathways that can detect cytoplasmic dsDNA, as well as endosomal toll-like receptor (TLR) pathways [16]. Importantly, it is the very same genes that drive cell proliferation, and which favor the accumulation of genetic errors in the host cell, that also contribute to immune evasion and persistence. It is clear from this that the deregulated expression of E6 and E7 can pose a major risk for the host, with HPV16 and a small number of related HPV types causing the majority of cervical cancers (Table 5.2).

UNDERSTANDING THE CERVIX AND OTHER VULNERABLE EPITHELIAL SITE IN THE DEVELOPMENT OF CANCER

To fully understand HPV-associated cancer, we also need to examine how the different epithelial sites where these viruses cause neoplasia are maintained, as it is at these sites that viral gene expression becomes deregulated. The most important in terms of the number of cancers is the cervix, although infection at other genital sites, and also the anus and the oropharynx, contribute to the overall HPV cancer burden. It is estimated that human papillomaviruses are responsible for just over 6% of all human cancers, with these viruses accounting for over of a third of all cancers caused by infectious agents. Although high risk HPVs can also cause infections at other epithelial sites, such as the eye ([18,19] and our unpublished data), and nail bed [20], the main sites of infection fit with their primary route of transmission being via sexual contact. Even so, the penis, vagina, and vulva are associated with a much lower incidence of cancer than the cervix, despite the fact that high risk HPV DNA is readily detectable at these sites. Although this lower cancer incidence may in part be linked to a reduced opportunity for HPV to infect the stratified cornified epithelium of the penis and the vagina, it is clear that infection does occur at both sites, with acetowhite positive areas indicating the presence of low-grade penile lesions similar to those seen at the cervix. It is generally thought that these are areas of productive HPV infection, and that productive HPV infection occurs also at the vulva and vagina. The presence of HPV particles in semen from HPV-infected men [21], and the association of HPV virions with sperm cells [22] may in fact be part of the HPV transmission strategy across these sites. The unique vulnerability of the cervix, and also the anus, is thought to stem from the presence of a transitional epithelium at these sites, which facilitates virus access to the mitotically active cells that maintain normal epithelial homeostasis at the cervix and the cervical transformation zone [23]. The transformation zone and the endocervix are in fact thought to have a fundamentally distinct pattern of cellular regulation from the ectocervix, with the process of cervical metaplasia being regulated by a special type of epithelial stem cell that can differentiate to form either the columnar cells of the endocervix, or the stratified cells of the transformation zone once metaplasia has been completed. For many years, the cervical reserve cells, which are characterized by their prominent pattern of keratin 17 expression, and by their location immediately beneath the columnar epithelial cells of the endocervix, have

Table 5.2 Association of Alpha HPV Types With Cervical Cancer [17].

Genus + Species	Type Species	Invasive Cervical Cancer	IARC Category	Squamous Cell Carcinoma	Adeno Carcinoma	Tropism
Alpha 1	HPV32		3			mucosal
	HPV42					
Alpha 2	HPV3					cutaneous
	HPV10					
	HPV28		3			
	HPV29		3			
	HPV77		3			
	HPV94					
	HPV117					
	HPV125					
Alpha 3	HPV61	0.01	3			mucosal
	HPV62		3			
	HPV72		3			
	HPV81		3	0.4		
	HPV83		3	0.4		
	HPV84		3			
	HPV86		3			
	HPV87		3			
	HPV89		3			
	HPV102					
	HPV114					
Alpha 4	HPV2		3			cutaneous
	HPV27		3			
	HPV57		3			
Alpha 5	HPV26	0.37	2B	0.22		mucosal
	HPV51	1.25	1	0.75	0.54	
	HPV69	0.08	2B			
	HPV82	0.07	2B	0.26		
Alpha 6	HPV30	0.37	2B			
	HPV53	0.26	2B	0.04		
	HPV56	0.84	1	1.09		
	HPV66	0.08	2B	0.19		
Alpha 7	HPV18	10.28	1	11.27	37.3	
	HPV39	1.67	1	0.82	0.54	
	HPV45	5.68	1	5.21	5.95	
	HPV59	1.08	1	1.05	2.16	
	HPV68	1.04	2A	0.37		
	HPV70	0.11	2B			
	HPV85		2B			
	HPV97					
Alpha 8	HPV7		3			cutaneous (mucosal)
	HPV40		3			
	HPV43					
	HPV91	0.01	3			
Alpha 9	HPV16	61.35	1	54.38	41.62	mucosal
	HPV31	3.65	1	3.82	1.08	
	HPV33	3.83	1	2.06	0.54	
	HPV35	1.94	1	1.27	1.08	
	HPV52	2.71	1	2.25		
	HPV58	2.22	1	1.72	0.54	
	HPV67	0.31	2B			
Alpha 10	HPV6	0.11	3	0.07		mucosal
	HPV11	0.02	3	0.07		
	HPV13		3			
	HPV44	0.01	3			
	HPV74	0.01	3			
Alpha 11	HPV34	0.07				mucosal
	HPV73	0.52	2B	0.49		
Alpha 12						
Alpha 13	HPV54					mucosal
Alpha 14	HPV71					
	HPV90		3			
	HPV106		3			

been considered as the ''HPV target cell'' at this site [24–27]. Although the case for the involvement of these cells in the development of cervical neoplasia is highly plausible, the regulation of high risk HPV gene expression following reserve cell infection has not yet been addressed, and there has been little substantial work to understand how reserve cell metaplasia is regulated. More recently, an alternative target cell for HPV at the cervical transformation zone, a cuboidal cell type located more precisely at the squamocolumnar junction has also been suggested to play a role in HPV-associated disease [28,29]. Although these cells stain positively for keratin 7, this marker is present in many simple epithelia cells at the cervical transformation zone, endocervix and endometrium, which has complicated the analysis of these cells. It remains possible that reserve cells originate from another cell type such as cuboidal or even the endocervical columnar cells during the metaplastic process [30]. Much of our knowledge of transitional epithelial zones has come from the analysis of the gastro-esophageal junction, which is also prone to preneoplastic metaplasia and malignancy, and which is similarly regulated by a sub columnar ''reserve cell-like'' progenitor with dual differentiation capability depending on local pH [31]. In general, such transitional epithelial sites are vulnerable to HPV-associated neoplasia and cancer, as these are sites where infection of specific types of epithelial cell or epithelial stem cells can occur. At the cervix, we have some understanding of the types of cell that may be involved, but at other sites such as the anus, our understanding is much poorer. The incidence of anal cancer in MSM, however, suggests a similar pattern of infection and disease progression, and that the incidence of HPV-associated cancer at other transitional sites will be linked to the chance of the virus gaining access to that site. Indeed, this may explain the suggested association of high risk HPV with a small number esophageal cancers and with their precursors [32,33]. In this regard, the cervix is particularly problematic in that a site of productive infection is located adjacent to a more vulnerable epithelial site where deregulated nonpermissive HPV infection is favored. We might expect therefore that the presence of a productive ectocervical CIN1 will facilitate infection of the adjacent endocervical epithelium, and indeed, it is not unusual to find both low-grade and high-grade disease in close proximity. It is likely that this also occurs at the anal transformation zone and also at the oropharynx, which can also be considered as a nonpermissive epithelial site [34,35]. Interestingly, HPV-associated squamous cell carcinomas of the oropharynx typically target the tonsillar crypts, with the stratified epithelium at the surface of the tonsil not usually involved. The crypts are normally comprised of a reticulated or ''sponge-like'' epithelium composed of both poorly differentiated or undifferentiated epithelial cells and an abundance of infiltrating lymphocytes, which allow this organ to act as a first defense against invading pathogens [36]. For a high-risk HPV types, however, this loose epithelial organization facilitates entry, and it seems likely that the poor differentiation contributes to the failure of high risk HPV types to complete their productive life cycle at this site. Furthermore, the discontinuous basal layer of the tonsillar crypt may well facilitate migration of HPV infected cells to local lymph nodes, driven by deregulated expression of the HPV E6 and E7 genes and a limited number of host cell mutations, a situation that adequately explains the absence of obvious primary disease before the detection of lymph node metastasis. Understanding the precise molecular pathways that regulate high risk HPV gene expression at this site will require further study, but it is already clear that HPV-associated oropharyngeal cancers are associated with a lower burden of host mutations, and are associated with better prognosis than those that lack HPV involvement [4]. Curiously, only a restricted number of HPV types are commonly associated with oropharyngeal cancers, with most studies finding an involvement of HPV16 in 90% or more of cases [37,38].

Although we might assume that all oropharyngeal infections are acquired directly through sexual contact, the incidence of HPV detection in oral mouthwashes in the general population is around 5% [39,40], which may suggest a productive reservoir of infection other than at the oropharynx. This would be similar to the cervix, where different epithelial sites favor either permissive/productive infection, or nonpermissive abortive infection and neoplasia.

UNDERSTANDING HPV BIOLOGY IN RELATION TO THE PATHOLOGICAL CLASSIFICATION OF HPV-ASSOCIATED CANCERS AND PRECANCER, AND THE SELECTION OF TREATMENT REGIMES

The major decisions that follow cytology-based cervical screening, the triage of HPV-positive women, and the treatment of HPV-related precancers and cancers at the cervix or other anogenital sites are reliant on microscopic, morphological interpretation, and classification of disease pathology using mainly standard tinctorial stains. For the last 30 years or so this has been based on a somewhat simplistic model of progression, from productive infection in a differentiated squamous epithelium, to a transformed undifferentiated lesion, from which invasion can arise. This approach is largely based on the biology discussed in this chapter and enshrined in the Bethesda system and its successor LAST. Although the development of cervical cancer described in Fig. 5.3 suggests multiple disease origins, the use of such diagnostic approaches have been very valuable, as detailed in the next chapter (Chapter 6: The Pathology of Cervical Precancer and Cancer and its Importance in Clinical Practice).

Making the connection between the biology of HPV and clinical decisions by simple microscopic pathology does have limitations however. One is that morphologic pathology is subjective and there is variation in interpretation of such grading between pathologists, and even following repeat examination by the same pathologist. This is inevitable as many different subjectively determined microscopic features, such as nuclear size, nuclear shape, the presence of normal and abnormal mitoses, and the position of these features in the epithelium, are involved in the diagnostic assessment, with these features being interpreted individually, and put together in different ways by different pathologists.

The simple, subjective division into high-grade and low-grade lesions does not reflect the complex differences between HPV types and their interaction with different epithelial cells [41]. There are many lesions included in the high-grade category that will resolve spontaneously but are treated on the precautionary principle, with inevitably, some negative effects of treatment. This particularly affects lesions that were previously called CIN2 as an intermediate category, particularly when seen in young women with recent HPV infection, as discussed in Chapter 9, The Natual History of Human Papillomavirus Infection in Relation to Cervical Cancer. We expect that the more detailed understanding of disease biology delineated here, to improve the classification of lesions and decisions about treatment?

One possibility is to achieve a more standardization approach to disease classification. Several studies have shown that the grading of immunohistochemical biomarkers such as p16, Ki67, MCM, and E4 is generally more reproducible than the use of conventional pathology [42,43]. Combinations of these markers, or their use alongside standard pathology analysis, may provide a way of standardizing reporting, allowing better comparison of results of clinical studies. Such

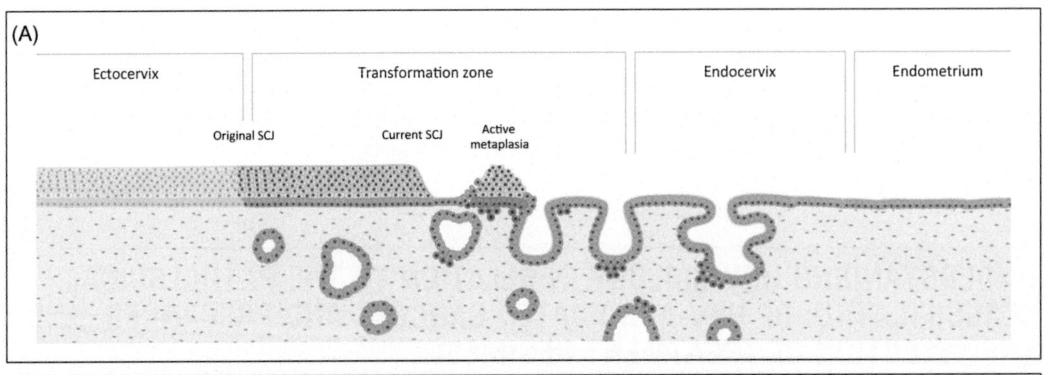

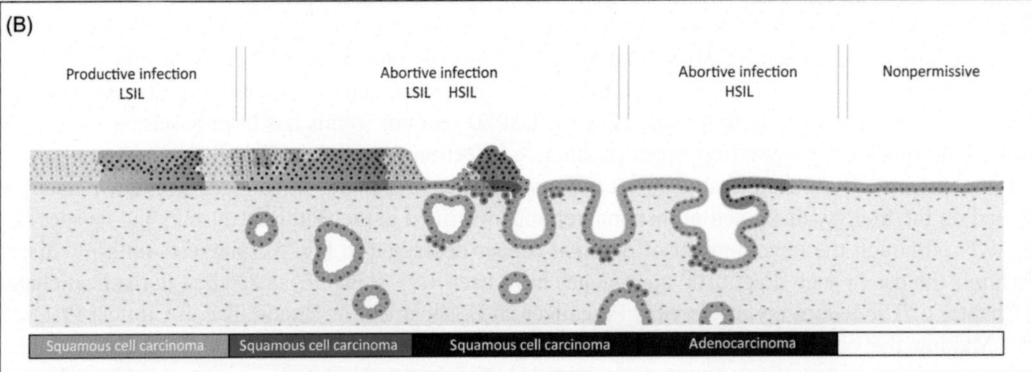

FIGURE 5.3 Diversity of HPV infection sites at the cervix, and the consequence of infection.

(A) The cervix is comprised of the conventional stratified epithelium of the ectocervix, the transformation zone, and the columnar epithelium of the endocervix, which lies adjacent to the endometrium. The reserve cells, which are shown in turquoise, lie under the columnar epithelium of the transformation zone, and play a role in normal metaplasia, a process that leads to the formation of a new stratified epithelium when required. This is shown between the original and current squamocolumnar junction (SCJ). (B) The consequence of HPV infection differs depending on the site of infection. Current thinking suggests that the ectocervix is a site where productive HPV infection and LSIL is supported, and that the other sites are associated with different level of deregulated HPV gene expression. The typical molecular phenotypes observed at these sites can be revealed using biomarkers E4 (*green*), p16 (*brown*), and MCM/Ki67 (*red*). The higher grade phenotypes observed in the transformation zone and the endocervix are potential precursors of cervical cancer (squamous cell carcinoma and adenocarcinoma), and develop more often at these sites when compared to the ectocervix.

combinations may also help to identify higher grade (CIN3) lesions within the category of high-grade squamous intraepithelial lesions, with the option of avoiding immediate treatment of the CIN2 lesions. This topic is discussed in detail in the next chapter and in chapter 17, Infection to Cancer—Finding Useful Biomarkers for Predicting Risk of Progression to Cancer.

An important issue when looking at biopsies is that in an individual patient the lesions of cervical HPV infection may be very complex and certainly not homogeneous [44]. This partly follows the three-dimensional structure of the cervix, and the likely differential susceptibility to transforming infection of cells of the endocervix, reserve cells, immature metaplasia, mature metaplastic squamous epithelium, and native ectocervix arranged in a centripetal fashion around the cervical os. Most high-grade squamous intraepithelial lesions develop in the inner part of the cervical transformation zone. Many cervical lesions consist of multiple lesions of different grades, either associated with a single HPV type or with multiple cervical HPV infections in up to 15 − 50% of precancers.

Examination of the cervix using sensitive techniques to localize the site of HPV infection to small areas of epithelium at the microscopic level such as laser capture microdissection (LCM-PCR) shows that only one HPV genotype is retrieved from one each specific lesionsal area [45]. A high-grade lesion may contain two or more HPV types in different areas, and a high-grade lesions and an adjacent low-grade lesion may show different HPV types. Studies with p16 and E4 immunohistochemistry show that CIN2 lesions (and AIN2 lesions) are biologically highly variable as to whether they show evidence of HPV productive infection. CIN3 lesions may show various degrees of superficial differentiation, with those that lack signs of obvious differentiation or evidence of HPV productive infection or life cycle completion, are more likely to show higher levels of nucleic acid methylation, which can result in the inactivation of somatic tumor suppressor genes that are associated with invasive cancer or persistent CIN3 [42]. Those CIN3 lesions that give rise to invasion are also usually large undifferentiated CIN3 lesions (carcinoma-in situ) of long duration.

A further complicating feature is that the high-grade undifferentiated lesions (CIN3) arising in areas of reserve cell hyperplasia and immature metaplasia may be quite thin lesions (Thin CIN) surrounded by columnar epithelium, which are difficult to distinguish from immature metaplasia that shows some atypia without the use of biomarkers such as p16 to indicate transforming activity of hrHPV, and cytokeratins (CK7, CK17) marking the presence of immature metaplasia [46]. Studies by LCM-PCR show that this is a particular site for HPV18 infection and transformation, and also for HPV 16, but only infrequently other hrHPV types. The finding of HPV 18 in these areas may explain the relative infrequency of CIN2/3 associated with HPV18 compared to its important role in cervical cancer, and also the late median age of finding HPV18/45 of the alpha-7 clade in high-grade precancer compared to the early median onset of invasive cervical cancer associated with these HPV types [41].

Molecular mapping of the HPV distribution in the cervix in relation to lesion grade has shown that most (90%) CIN2/3 is due to a single hrHPV type, even if multiple HPV types were present on the cervical smear, and if HPV 16 was present in the smear the probability of the high-grade lesion being due to HPV16 was 96% [44,47]. As well as being in cervical lesions, HPV DNA can be detected in normal cervical epithelium.

The important message is that precancerous and productive cervical lesions can be very complex, and routine classification by histology only provides a rough description of the lesion. Other sites affected by HPV, such as the anus and penis show similar complexity. The outcome of infection with a hrHPV is dependent partly on the HPV type and partly on the development of transforming lesions from specific cell populations at particular micro-anatomic sites. Studies of the genetic conservation of the E7 gene of HPV16 have shown that this is also important in carcinogenesis, as is integration into the human genome at certain hotspots, and the generation of somatic genomic alterations during the development of cervical cancer [48,49].

It is also important to consider that outside the cervix, in the vulva, penis and scrotum are HPV-related and non-HPV pathways that are not only distinct in behavior and relation to HPV but show very different microscopic composition. These are discussed in Chapters 12, 13 and 14 of Section 2. The glandular endocervix also presents specific issues in relation to HPV carcinogenesis from that of squamous carcinoma. The vast majority of squamous cervical carcinoma is HPV-positive, mostly hrHPV (95%–99%) with a small fraction associated with lrHPV of different types. Up to 30% of cases managed as cervical adenocarcinoma are HPV-negative when studied in large epidemiological surveys globally and in regional studies in Europe and China [50]. The reasons for this are multiple. One important problem is misdiagnosis of endometrial adenocarcinoma, especially when biomarkers are not used in diagnosis, or when there is no hysterectomy specimen for the pathologist to examine in detail. There is a small group of true endocervical adenocarcinomas, particularly common in Japan but also elsewhere in Eastern Asian populations, although rare in the USA and Western Europe. These are morphologically variable and classified as gastric-type cervical adenocarcinoma, including the pattern described as minimal deviation adenocarcinoma. These are often inherited and are due to a germline mutation in the STK11 gene [51]. In addition there are adenocarcinomas at the cervix that resemble those that occur throughout the female genital tract, which are classsified as serous and clear cell adenocarcinomas, and are not HPV-related. There is also a small group of adenocarcinomas indistinguishable morphologically from cervical adenocarcinoma in which HPV DNA cannot be found even on sensitive testing. These are mainly late stage tumors, sometimes associated with HPV elsewhere in the cervix, and show somatic mutations in major tumor associated genes. Whether these are true HPV-negative cancers or represent a late stage in tumorigenesis in which HPV has become lost is not established [52].

Cervical adenocarcinoma is thus different from squamous carcinoma. Its management requires understanding of possible alternative diagnoses and full assessment of HPV status.

CONCLUSION

Full understanding of the pathology of cervical and other HPV-related neoplasia requires awareness of the biology of HPV types in the cells of different special types that are infected.

ACKNOWLEDGEMENT

This work was supported in part, through MRC Program Grant MC_PC_13050 awarded to JD.

REFERENCES

[1] Doorbar J, Egawa N, Griffin H, Kranjec C, Murakami I. Human papillomavirus molecular biology and disease association. Rev Med Virol 2015;25(Suppl 1):2–23.

[2] Bosch FX, Broker TR, Forman D, et al. Comprehensive control of human papillomavirus infections and related diseases. Vaccine 2013;31(Suppl 8):I1–31.

[3] Quint KD, Genders RE, de Koning MN, et al. Human Beta-papillomavirus infection and keratinocyte carcinomas. J Pathol 2015;235:342–54.

[4] Schiffman M, Doorbar J, Wentzensen N, et al. Carcinogenic human papillomavirus infection. Nat Rev Dis Prim 2016;2:16086.
[5] Egawa N, Doorbar J. The low-risk papillomaviruses. Virus Res 2017;231:119–27.
[6] Maglennon GA, McIntosh P, Doorbar J. Persistence of viral DNA in the epithelial basal layer suggests a model for papillomavirus latency following immune regression. Virology 2011;414:153–63.
[7] Tan MJ, White EA, Sowa ME, et al. Cutaneous beta-human papillomavirus E6 proteins bind Mastermind-like coactivators and repress Notch signaling. PNAS 2012;.
[8] Brimer N, Lyons C, Wallberg AE, Vande Pol SB. Cutaneous papillomavirus E6 oncoproteins associate with MAML1 to repress transactivation and NOTCH signaling. Oncogene 2012;31:4639–46.
[9] Ashrafi GH, Haghshenas MR, Marchetti B, O'Brien PM, Campo MS. E5 protein of human papillomavirus type 16 selectively downregulates surface HLA class I. J Int du Cancer 2005;113:276–83.
[10] Bravo IG, de Sanjose S, Gottschling M. The clinical importance of understanding the evolution of papillomaviruses. Trends Microbiol 2010;18:432–8.
[11] Pim D, Bergant M, Boon SS, et al. Human papillomaviruses and the specificity of PDZ domain targeting. FEBS J 2012;279:3530–7.
[12] Durzynska J, Lesniewicz K, Poreba E. Human papillomaviruses in epigenetic regulations. Mutat Res Rev Mutat Res 2017;772:36–50.
[13] Hirose Y, Onuki M, Tenjimbayashi Y, et al. Within-host variations of human papillomavirus reveal APOBEC signature mutagenesis in the viral genome. J Virol 2018;92.
[14] Dube Mandishora RS, Gjotterud KS, Lagstrom S, et al. Intra-host sequence variability in human papillomavirus. Papillomavirus Res 2018;5:180–91.
[15] Kanodia S, Fahey LM, Kast WM. Mechanisms used by human papillomaviruses to escape the host immune response. Curr Cancer Drug Targets 2007;7:79–89.
[16] Hasan UA, Bates E, Takeshita F, et al. TLR9 expression and function is abolished by the cervical cancer-associated human papillomavirus type 16. J Immunol 2007;178:3186–97.
[17] Doorbar J, Quint W, Banks L, et al. The biology and life-cycle of human papillomaviruses. Vaccine 2012;30(Suppl 5):F55–70.
[18] Moyer AB, Roberts J, Olsen RJ, Chevez-Barrios P. Human papillomavirus-driven squamous lesions: high-risk genotype found in conjunctival papillomas, dysplasia, and carcinoma. Am J Dermatopathol 2018;40:486–90.
[19] Afrogheh AH, Jakobiec FA, Hammon R, et al. Evaluation for high-risk HPV in squamous cell carcinomas and precursor lesions arising in the conjunctiva and lacrimal sac. Am J Surg Pathol 2016;40:519–28.
[20] Grundmeier N HH, Weissbrich B, Lang SC, Bröcker EB, Kerstan A. High-risk human papillomavirus infection in Bowen's disease of the nail unit: report of three cases and review of the literature. Dermatology 2011;223(4):293–300. Available from: https://doi.org/10.1159/000335371 Epub 2012 Jan 24.
[21] Luttmer R DM, Snijders PJF, Jordanova ES, King AJ, Pronk DTM, Foresta C, et al. Meijer CJLM. Presence of human papillomavirus in semen of healthy men is firmly associated with HPV infections of the penile epithelium. Fertil Steril 2015;104(4):838–44.
[22] Perez-Andino J, Buck CB, Ribbeck K. Adsorption of human papillomavirus 16 to live human sperm. PLoS One 2009;4(6):e5847. Available from: https://doi.org/10.1371/journal.pone.0005847.
[23] Doorbar J, Griffin H. Refining our understanding of cervical neoplasia and its cellular origins. Papillomavirus Res 2019;7:176–9.
[24] Martens JE, Arends J, Van der Linden PJ, De Boer BA, Helmerhorst TJ. Cytokeratin 17 and p63 are markers of the HPV target cell, the cervical stem cell. Anticancer Res 2004;24:771–5.
[25] Martens JE, Smedts FM, Ploeger D, et al. Distribution pattern and marker profile show two subpopulations of reserve cells in the endocervical canal. Int J Gynecol Pathol 2009;28:381–8.

[26] Regauer S, Reich O. CK17 and p16 expression patterns distinguish (atypical) immature squamous metaplasia from high-grade cervical intraepithelial neoplasia (CIN III). Histopathology 2007;50:629–35.
[27] Reich O, Pickel H, Regauer S. Why do human papillomavirus infections induce sharply demarcated lesions of the cervix? J Low Genit Tract Dis 2008;12:8–10.
[28] Herfs M, Yamamoto Y, Laury A, et al. A discrete population of squamocolumnar junction cells implicated in the pathogenesis of cervical cancer. PNAS 2012;109:10516–21.
[29] Herfs M, Vargas SO, Yamamoto Y, et al. A novel blueprint for "top down" differentiation defines the cervical squamocolumnar junction during development, reproductive life and neoplasia. J Pathol 2012;.
[30] Herfs M, Soong TR, Delvenne P, Crum CP. Deciphering the multifactorial susceptibility of mucosal junction cells to HPV infection and related carcinogenesis. Viruses 2017;9.
[31] Jiang M, Li H, Zhang Y, et al. Transitional basal cells at the squamous-columnar junction generate Barrett's oesophagus. Nature 2017;.
[32] Rajendra S, Sharma P. Barrett esophagus and intramucosal esophageal adenocarcinoma. Hematol/Oncol Clin N Am 2017;31:409–26.
[33] Rajendra S, Xuan W, Merrett N, et al. Survival rates for patients with barrett high-grade dysplasia and esophageal adenocarcinoma with or without human papillomavirus infection. JAMA Network Open 2018;1:e181054.
[34] Egawa N, Egawa K, Griffin H, Doorbar J. Human papillomaviruses; epithelial tropisms, and the development of neoplasia. Viruses 2015;7:3863–90.
[35] Westra WH. The morphologic profile of HPV-related head and neck squamous carcinoma: implications for diagnosis, prognosis, and clinical management. Head Neck Pathol 2012;6(Suppl 1):S48–54.
[36] Perry ME. The specialised structure of crypt epithelium in the human palatine tonsil and its functional significance. J Anat 1994;185(Pt 1):111–27.
[37] Schache AG, Powell NG, Cuschieri KS, et al. HPV-related oropharynx cancer in the United Kingdom: an evolution in the understanding of disease etiology. Cancer Res 2016;76:6598–606.
[38] Liu SZ, Zandberg DP, Schumaker LM, Papadimitriou JC, Cullen KJ. Correlation of p16 expression and HPV type with survival in oropharyngeal squamous cell cancer. Oral Oncol 2015;51(862-9).
[39] Gillison ML, Broutian T, Pickard RK, et al. Prevalence of oral HPV infection in the United States, 2009-2010. J Am Med Assoc 2012;307:693–703.
[40] Pickard RK, Xiao W, Broutian TR, He X, Gillison ML. The prevalence and incidence of oral human papillomavirus infection among young men and women, aged 18-30 years. Sex Transm Dis 2012;39:559–66.
[41] Tjalma WA, Fiander A, Reich O, et al. Differences in human papillomavirus type distribution in high-grade cervical intraepithelial neoplasia and invasive cervical cancer in Europe. Int J Cancer 2013;132:854–67.
[42] Zummeren MV, Kremer WW, Leeman A, et al. HPV E4 expression and DNA hypermethylation of CADM1, MAL, and miR124-2 genes in cervical cancer and precursor lesions. Mod Pathol 2018;31:1842–50.
[43] van Zummeren M, Leeman A, Kremer WW, et al. Three-tiered score for Ki-67 andp16(ink4a) improves accuracy and reproducibility of grading CIN lesions. J Clin Pathol 2018;71:981–8.
[44] van der Marel J, Berkhof J, Ordi J, et al. Attributing oncogenic human papillomavirus genotypes to high-grade cervical neoplasia: which type causes the lesion? Am J Surg Pathol 2015;39:496–504.
[45] Quint W, Jenkins D, Molijn A, et al. One virus, one lesion-individual components of CIN lesions contain a specific HPV type. J Pathol 2012;227:62–71.
[46] van der Marel J, van Baars R, Alonso I, et al. Oncogenic human papillomavirus-infected immature metaplastic cells and cervical neoplasia. Am J Surg Pathol 2014;38:470–9.
[47] van der Marel J, van Baars R, Quint WG, et al. The impact of human papillomavirus genotype on colposcopic appearance: a cross-sectional analysis. BJOG 2014;121:1117–26.

[48] Mirabello L, Yeager M, Yu K, et al. HPV16 E7 genetic conservation is critical to carcinogenesis. Cell 2017;170:1164–74e6.
[49] Hu Z, Zhu D, Wang W, et al. Genome-wide profiling of HPV integration in cervical cancer identifies clustered genomic hot spots and a potential microhomology-mediated integration mechanism. Nat Genet 2015;47:158–63.
[50] Molijn A, Jenkins D, Chen W, et al. The complex relationship between human papillomavirus and cervical adenocarcinoma. Int J Cancer 2016;138:409–16.
[51] Pirog EC, Park KJ, Kiyokawa T, et al. Gastric-type adenocarcinoma of the cervix: tumor with wide range of histologic appearances. Adv Anat Pathol 2019;26:1–12.
[52] Ojesina AI, Lichtenstein L, Freeman SS, et al. Landscape of genomic alterations in cervical carcinomas. Nature 2014;506:371–5.

CHAPTER

THE PATHOLOGY OF CERVICAL PRECANCER AND CANCER AND ITS IMPORTANCE IN CLINICAL PRACTICE

6

Mark H. Stoler[1], David Jenkins[2,3] and Christine Bergeron[4]

[1]Professor (Emeritus) of Pathology and Clinical Gynecology, Associate Director of Surgical and Cytopathology, Department of Pathology, University of Virginia Health Health System, Charlottesville, VA, United States [2]Emeritus Professor of Pathology, University of Nottingham, United Kingdom [3]Consultant in Pathology to DDL Diagnostic laboratories, Rijswijk, The Netherlands [4]Department of Pathology, Laboratoire Cerba, Cergy Pontoise, France

INTRODUCTION: FRAMING THE PROBLEM

In screened populations, the diagnosis of cervical squamous intraepithelial lesions [aka cervical intraepithelial neoplasias (CINs), cervical dysplasias] impacts a significant proportion of the population. Screening programs to detect and eradicate the subset of these lesions that are precancerous are a success story in preventive medicine. Recent decades have seen a rapid expansion in our understanding of cervical carcinogenesis (Chapter 2: Linking Human Papillomavirus to Human Cancer and Understanding Its Carcinogenic Mechanisms, Chapter 4: Finding How Human Papillomavirus Alter the Biochemistry and Identity of Infected Epithelial Cells, and Chapter 5: Biology of the Human Papillomavirus Life Cycle: The Basis for Understanding the Pathology of Precancer and Cancer) leading to the development of molecular screening strategies (Chapter 16: Primary Screening by Human Papillomavirus Testing: Development, Implementation, and Perspectives), new diagnostic biomarkers (Chapter 17: Infection to Cancer—Finding Useful Biomarkers for Predicting Risk of Progression to Cancer) and remarkably effective vaccines for the prevention of cancer (Chapter 19: Clinical Trials of Human Papillomaviruses Vaccines and Chapter 20: The Key Role of Mathematical Modeling and Health Economics in the Public Health Transitions in Cervical Cancer Prevention). Yet, pathologists, clinicians, and patients are still confronted with imperfection and imprecision in our systems. Morphologic pathology is somewhat subjective with an inherent degree of interpretive variability and error. Part of what follows reflects an attempt to use our scientific knowledge to improve diagnostic reproducibility and objectivity while improving patient care.

TERMINOLOGY: HISTORICAL PERSPECTIVE

Uterine cervical squamous intraepithelial lesions form the model for intraepithelial neoplasia throughout the body. Here "lesion" will be first used generically to describe an abnormality

Human Papillomavirus. DOI: https://doi.org/10.1016/B978-0-12-814457-2.00006-4

potentially related to cancer, that is, a precancer. But as will be seen, the term becomes one of formal usage in the current classification of cervical neoplasia.

The cervix is accessible and treatable, unlike many deep-seated cancers, facilitating serial observation and hence natural history. Thus, cervical neoplasia was initially recognized in the early 1900s, when histologic descriptors such as "surface carcinoma," "intraepithelial carcinoma," and "carcinoma in situ (CIS)" were applied to histologic lesions that bore the cytologic features of malignancy, but that had yet to invade through their basement membrane, the hallmark of the transition to epithelial cancer. At that time, these diagnoses corresponded with a fundamental clinical decision: a diagnosis of CIS meant hysterectomy, while the uterus and cervix were retained in cases that failed to meet CIS criteria. The mid-20th century and the process of cytology–histology correlation led to the description of lesions with less marked severity than CIS, ultimately leading to the application of the term "dysplasia" by Reagan and colleagues, who went on to usher in gradations of mild, moderate, and severe dysplasia. Although the diagnostic interface between CIS and severe dysplasia remained somewhat murky, the former continued to provoke complete hysterectomy while the latter was more often conservatively treated with just conization.

The next significant development in the characterization of squamous intraepithelial lesions came with Koss and Durfee's description of koilocytes in 1956. They derived this term from the Greek word for cave due to the "cave-like" vacuoles that surrounded the enlarged nuclei of these cells, and noted the morphologic homology between these cells and the mild dysplasia depicted in Reagan's system. It took another two decades for this morphology to be ascribed to human papillomavirus (HPV) infection by Miesels and Fortin, with confirmatory electron microscopy studies showing intranuclear virions in these cells shortly thereafter.

The concept that cervical dysplasias were precursors for invasive carcinoma was strengthened by the work of Ralph Richart, one of the first pathologist-colposcopists, who in 1969 suggested that the continuum of mild, moderate, and severe dysplasias all imparted some risk of progression to carcinoma. Recognition of this shared risk led nomenclature to change to the cervical intraepithelial neoplasia (CIN) system, with mild dysplasia equating with CIN1, moderate dysplasia CIN2, and severe dysplasia and CIS being collapsed into CIN3. This conceptual shift was accompanied by a shift in treatment practices, with local resection supplanting hysterectomy for all grades of CIN, including those formerly termed CIS. The end of the 20th century saw the advent of molecular biology and an explosion in our understanding of cervical carcinogenesis and its relationship to human papillomavirus. Notable breakthroughs included the first demonstration of high-risk HPV DNA in cervical cancer cell lines by Boshart et al. while working in the laboratory of Harald zur Hauzen, who was awarded the Nobel Prize in 2008 for the relationship of HPV to cervical carcinogenesis. Subsequently, Crum and colleagues were the first to identify HPV 16 in CIN.

As molecular evidence mounted and was carefully correlated with epidemiologic studies, it became clear that CIN1 (e.g., mild dysplasia, usually with koilocytes) represented the histologic correlate for productive HPV infection, while CIN2 (at least for some) but definitely CIN3/CIS were identified as a morphologic indication of HPV oncogene-induced cell transformation. The "at least" refers to the interpretive variability issue to be discussed later in this chapter. Natural history studies demonstrated CIN1 lesions regress in the majority of instances (like a wart) whereas CIN2/CIN3/CIS lesions showed much higher rates of persistence and progression. This understanding led to the return of a binary risk-based managerial approach to cervical pathology: CIN1 lesions were considered low-grade squamous intraepithelial lesions (LSIL) and managed with observation, whereas CIN2/CIN3/CIS lesions were lumped together as high-grade squamous intraepithelial lesions (HSIL) and warranted resection.

This two-tiered risk schema informed the Bethesda Classification System for Cervical Cytology, first introduced in 1988 and refined three times, most recently in 2014. In 2012, the Lower Anogenital Squamous Terminology (LAST) project further advocated for the use of LSIL/HSIL terminology not only in the uterine cervix, but also elsewhere in the male and female genital tracts, as did the 4th edition of the World Health Organization's text on gynecologic neoplasia. Thus today, we have a unified, biologically based terminology for both cytology and histology that extends to the whole spectrum of cervical neoplasia and helps to guide management.

DIAGNOSTIC PROCESS AND SPECIMEN TYPES

Cervical squamous neoplasia is typically first encountered in a screening cytology sample (Pap smear). A diagnosis of squamous intraepithelial lesion by cytology prompts a range of interventions depending upon the degree of dysplasia and the patient's age, HPV status, and gravidity; the most up-do-date management guidelines in the United States are available through the American Society for Colposcopy and Cervical Pathology (ASCCP). Yet despite all our progress, what can be done to patients is fairly limited. Follow-up actions include watchful waiting and repeat screening, triage to colposcopy and cervical biopsy, with the latter sometimes prompting a larger excision [either loop electrosurgical excision (LEEP) or cone biopsy] or, occasionally, hysterectomy all with the aim of eradicating precancer before the development of invasive cancer. Much of the discussion in this book and in the greater scientific literature centers on how to balance the benefit of screening and precancer detection versus the harms of being identified as being at risk for cancer and the harms of treatment.

CYTOLOGIC SPECIMENS ("PAP SMEARS")

The term "Pap smear" is used colloquially to describe all manner of cervical cytology samplings. A more apt generic term might be just "cervical cytology" or perhaps even better the "Pap test," as this term allows for a variety of collection methods and still retains the homage to Dr. George Papanicolaou, the anatomist who first reported malignant cells in cell smears on slides from gynecologic samples. Technically, a "Pap smear" refers to a conventional smear collected directly from the cervix, typically using a specialized spatula alone or in tandem with an endocervical sampling device (historically a cotton-tipped applicator but today more often a patented "broom" or "brush"). The collected material is then transferred directly a glass slide and smeared out. Conventional smears typically contain between 50,000 and 300,000 cells with a minimum of 8000–12,000 well-visualized cells required for specimen adequacy. Although in optimal conditions sensitivity for CIN can be robust, it remains imperfect and DNA analysis has shown that the cells represented on a direct smear constitute a small minority (as little as 5%) of the total cells removed from the patient, with practitioner skill/technique, anatomy, and technical preparation considerably influencing specimen sensitivity. Furthermore, interpretation can be hindered significantly by excessive smear thickness, obscuring blood, and variable fixation.

Given the propensity for such variables to confound conventional smear interpretation, many cytopathologists prefer more standardized preparation methods, so-called LBC samples for liquid based cytology, where the cervical sample is suspended in a preservative fluid. When compared to conventional smears, liquid-based specimens have equivalent or better sampling sensitivity, more uniform fixation and staining, and decreased obscuring background. Furthermore the LBC approach allows additional sampling and testing, for example, HPV testing, obviating the need for the collection of multiple samples and thereby minimizing sampling variability. The most commonly employed LBC techniques are ThinPrep (Hologic) and SurePath (BD).

ThinPrep samples are collected using either a broom-type device or a plastic spatula in combination with an endocervical brush. The sampling apparatus is then swished and stored in a vial containing methanol-based preservative solution which lyses red blood cells. Collection vials are loaded into a patented instrument which disperses cells and collects them on a 20 mm polycarbonate filter, which is then transferred onto a glass slide. Generally, only a small proportion of the specimen is used to create a ThinPrep slide, therefore residual material remains available for molecular diagnostic testing, additional slides preparation, and/or cell block preparation. Several studies have shown improved detection of dysplasia in ThinPrep slides when compared to conventional smears.

In contrast to ThinPrep, SurePath enlists an ethanol-based preservative and the collection device is snipped off and included in the vial. The specimen is vortexed and syringed through a small opening to disaggregate large cell clumps. Next, it is centrifuged through a density gradient which eliminates red blood cell and some white blood cells. The centrifuged pellet is then resuspended and centrifuged again. Finally, the centrifuge tube is transferred to a staining instrument which samples the pellet and settles the cells onto a cationic polyelectrolyte-coated slide. Although some studies have shown improved detection of LSIL by SurePath as compared to conventional smears, a significant difference in the detection of HSIL has not been demonstrated.

Further to this point, during the preparation of the *European Guidelines for Quality Assurance in Cervical Cancer Screening*, a meta-analysis on test characteristics of LBC and conventional cytology (CP) was prepared. Results pooled from studies with concomitant testing showed nearly equal detection rates of HSIL and positive predictive value for CIN2 + in CP and LBC. However, in two-cohort studies, detection rates of HSIL were substantially and statistically significantly higher in LBC. The sensitivity and specificity of LBC at ASC-US + and LSIL + for CIN2 + pooled from the most stringent studies was never significantly different from CP. In two-cohort studies, less unsatisfactory smears were found in ThinPrep and SurePath smears, respectively. Overall, interpretation of LBC required 30% less time to interpret than CP. It was concluded that no evidence is available to claim higher accuracy of LBC to predict histologically confirmed CIN2 +, but recognized that LBC improves the quality and speed of interpretation, and offers the important possibility of additional molecular testing from a single sample thereby simplifying clinical sampling and paving the way for modern screening where both molecular and morphologic techniques can be applied to the same sample to minimize sampling variability.

CERVICAL BIOPSY

Surgical biopsies are typically small, unoriented portions of tissue. While most often directed at colposcopically visible lesions, they may also be either random or systematic quadrant samplings.

The sensitivity of sampling may be significantly influenced by operator skill and patient anatomy, and the absence of a lesion on biopsy does little to assuage concern prompted by positive cytology. Indeed, while colposcopic biopsy has long been considered the diagnostic "gold standard" by many clinicians, it may miss roughly 25% to up to 50% of cytologically detected intraepithelial lesions. This makes sense, as surgical biopsies collect only a small focus of the epithelium whereas cytologic samples include cellular representations of much broader areas. This issue of sampling discrepancy is also of significance in discussions regarding the utility (and limitations) of prognostic markers in cervical specimens, discussed later in this chapter.

CONE BIOPSY

These excisional specimens encompass the transformation zone and serve the dual purposes of diagnosis and therapy. They are performed using a scalpel (hence the term "cold-knife cone"). Specimen size varies considerably based on the cervical anatomy, surgical technique, and patient age and interest in preserving fertility. Generally, a cold-knife cone removes more tissue and is often used to get better assurance of complete excision.

Loop Electrosurgical Excision Procedure (LEEP), LEEP has largely supplanted traditional conization as the excisional method of choice for cervical dysplasia. It is similar to a traditional or "cold-knife" cone, but instead of an unheated scalpel it relies on a hot wire-shaped loop or similar device carrying an electrical current to slice and simultaneously cauterize the cervical tissue. The advantage of conservation of cervical substance while still providing definitive pathology and the ease of excision with less bleeding all in an office setting has made LEEP the current treatment standard of care.

HYSTERECTOMY

Hysterectomy typically involves the en bloc resection of the uterus and cervix although the specimen type varies somewhat based on the underlying pathology. Radical hysterectomy with pelvic lymph node resection is the intervention of choice for the treatment of invasive cervical cancer in patients who are surgical candidates (typically those with disease stage ≤ 1B, or tumors <4 cm in size). Patients with higher stage disease are often subjected to neoadjuvant chemotherapy and/or radiation. Squamous intraepithelial lesions may be encountered overlying and/or adjacent to areas of invasion in patients with squamous cell carcinoma. Dysplasia can also be seen without associated invasive malignancy as the decision to perform hysterectomy may be spurred by recalcitrant dysplasia, particularly when childbearing is not an issue. Finally, squamous dysplasia may be an incidental finding in hysterectomies performed for benign lesions such as leiomyomas or noncervical neoplastic reasons.

GROSS MORPHOLOGY—COLPOSCOPIC FEATURES

With the exception of exophytic lesions such as condyloma accuminatum, cervical squamous intraepithelial lesions are not typically visible to the naked eye. They may be clinically identified using colposcopic magnification and the sequential application of normal saline, 3%–5% dilute glacial acetic acid, and in some cases Lugol's iodine. Lesions may be demarcated from the background normal cervix based on the color tone/intensity of acetowhite areas, the margins and surface contour of acetowhite areas, and vascular features and color changes following acetic acid and/or iodine application.

The initial application of normal saline aids in the identification of vasculature and demarcates the transition zone borders. Subsequent application of acetic acid highlights intraepithelial lesions: LSIL are typically thin, smooth acetowhite lesions with irregular but well-demarcated margins, whereas HSIL are more often thick and dense with an irregular surface and raised or rolled margins and abnormal vascular patterns.

In the final (but for some optional) step, iodine solution is taken up by glycogen-rich normal squamous and mature metaplastic cells but not by columnar cells or intraepithelial lesions, further delineating CIN from the background cervix. Directed biopsies can then be collected using information gleaned from this colposcopic exam. That said, subjective colposcopic impressions are imperfect correlates for microscopic findings. Hence, discordance between the colposcopically anticipated grade and the ultimate diagnosis is not surprising.

MORPHOLOGY (HISTOLOGY AND CYTOLOGY)

Under the WHO classification, cervical dysplasia manifests microscopically either as LSIL (CIN1) or HSIL (CIN2/3). These two basic morphologic patterns reflect the status of HPV infection in the involved tissues. Although some assume that CIN1, CIN2, and CIN3 represent a continuous spectrum, linear progression through each level of dysplasia does not necessarily occur: varying degrees of dysplasia may occur in tandem with progression of LSIL to HSIL within the same biopsy, or they can occur independently as not all cases of LSIL are obligate precursors to HSIL and a "direct-to-HSIL" pathway is thought to exist. Furthermore, the issues of interpretive variability and sampling issues confound the simplistic but attractive idea of linear progression

NORMAL SQUAMOUS EPITHELIUM

To understand the morphology of cervical neoplasia, it is important to appreciate the maturation pattern of normal squamous epithelium. Squamous epithelium lines the ectocervix with a transition to glandular epithelium in the endocervix. In normal squamous epithelia, cell division is restricted to basal and parabasal cells. These proliferative basal layers are responsible for populating the full thickness of the epithelium, with maturation and functional refinement (but not cell division) occurring as the cells become more superficial. Maturation manifests in a decrease in nuclear size and an increase in cytoplasmic volume, often with acquisition of cytoplasmic glycogen. The maturing cytoplasm stains differentially on Pap specimens with intermediate cells derived from the middle of the epithelium bearing pale blue cytoplasm and the most superficial and mature cells bearing abundant "orangophilic" cytoplasm. On H&E-stained slides, maturation manifests in decreasing nuclear size with concomitant increase in eosinophilic, keratin-enriched cytoplasm. Perinuclear clearing is often present due to the presence of cytoplasmic glycogen, and should not be mistaken for koilocytic halos of a productive HPV infection.

The degree of maturation is influenced by the amount of estrogen available either endogenously or exogenously. Varying degrees of immaturity or atrophy may be seen approaching and after menopause, during pregnancy and lactation, and in the setting of hormonal contraceptives.

Negative for Intraepithelial Lesion or Malignancy

In cytology a normal smear is categorized as NILM. Table I summarizes the Third Edition of the Bethesda System. This category is used for all cervical samples that fail to even suggest the possibility of cervical neoplasia. Many unrelated infectious agents or reactive patterns are categorized under NILM (Tables 6.1 and 6.2).

Table 6.1 Modified/Simplified Bethesda System for Reporting Cervical Cytology, 3rd Ed.

- Specimen type
- Specimen adequacy
- Interpretation/result
  - Negative for intraepithelial lesion or malignancy
    - Nonneoplastic cellular variations
    - Organisms
    - Other
  - Epithelial cell abnormalities
    - Squamous cell
      - ASC-US
      - ASC-H (cannot exclude HSIL)
      - LSIL (encompassing HPV/mild dysplasia/CIN1)
      - HSIL (encompassing moderate and severe dysplasia, CIN2, CIN3/CIS
      - Squamous cell carcinoma
    - Glandular cell
      - Atypical
        - Endocervical NOS
        - Endometrial NOS
        - Glandular cell NOS
      - Atypical
        - Endocervical favor neoplastic
        - Endometrial favor neoplastic
      - Endocervical adenocarcinoma In Situ
      - Adenocarcinoma
    - Other malignant neoplasms
      - Neuroendocrine carcinoma
      - Other

Table 6.2 Modified/Simplified WHO 2014 Classification of Cervical Neoplasms.

Epithelial tumors
Squamous cell tumors and precursors
Squamous intraepithelial lesions
Low-grade squamous intraepithelial lesion (CIN1)
High-grade squamous intraepithelial lesion (CIN2)
High-grade squamous intraepithelial lesion (CIN3)
Squamous cell carcinoma, NOS
Keratinizing
Nonkeratinizing
Other squamous variants
Benign squamous lesions
Condyloma acuminatum (LSIL)
Other
Glandular tumors and precursors
Adenocarcinoma in situ
Adenocarcinoma
Endocervical adenocarcinoma, usual type
Mucinous carcinoma
Variants
Endometrioid adenocarcinoma
Clear cell carcinoma
Serous carcinoma
Other rare variants
Benign glandular tumors
Other epithelial tumors
Adenosquamous carcinoma
Undifferentiated carcinoma
Other
Neuroendocrine tumors
Low-grade neuroendocrine tumors
High-Grade neuroendocrine carcinomas
Small cell neuroendocrine carcinoma
Large cell neuroendocrine carcinoma

Cervical Cytology's Special Case of Equivocal Morphology

Atypical Squamous Cells

Following numerous discussions, based on the practicality of maintaining or not an invalid category, it has been decided to keep this category, which is associated with approximately 10–20% of high grade squamous intraepithelial lesions (HSIL)/CIN2 + from biopsies. The general term for this category is no longer "atypical cells of undetermined significance" (ASCUS). It is replaced by the term "atypical squamous cells." The more specific term "atypical squamous cells of undetermined significance" (ASC-US) is used for abnormalities suggesting a low-grade intraepithelial squamous lesion, which has not been confirmed, or for nonspecific atypical cells. Not more than 3%–5% of the smears should have this designation. Atypical findings of undetermined significance associated with inflammation come out of this group and should now be included among normal smears. The term "atypical squamous cells cannot exclude high grade squamous intraepithelial squamous lesion" (ASC-H) is proposed for those unconfirmed although suspected of high-grade intraepithelial squamous lesions (HSIL). This term should apply to 5%–10% of atypical squamous cell alterations (i.e., nor more than 0.3%–0.5% of all smears) and are associated with a HSIL confirmed through biopsy taken by colposcopy in about 50% of cases.

Epithelial Glandular Cells Abnormalities

Sampling of the Endocervical Epithelium

Cytological sampling can be performed both for conventional smears using the cytobrush and for liquid based cytology using the Cervexbrush and its variants. Improvement of cytology brushes allows better sampling of the endocervical epithelium. Endocervical cells sampling can be difficult because adenocarcinoma in situ may have a multifocal development and endocervical glands can be partially or totally involved. If abnormal endocervical cells do not reach the superficial part of the glandular structures, they may not shed on the smears. This explains why sometimes adenocarcinoma in situ is diagnosed late on the cone biopsy.

LOW-GRADE SQUAMOUS INTRAEPITHELIAL LESIONS/CIN1

In LSIL, HPV infects all levels of the epithelium; however, productive viral gene expression is restricted to those cells that have begun to mature. In the suprabasal zone, viral gene expression is restricted exclusively to the early viral genes. Further up in the epithelium all viral genes are induced and ultimately viral DNA is synthesized, leading to the production and assembly of virions in the most superficial cell layers. LSIL includes both flat lesions and exophytic lesions conventionally referred to as condylomata. Some pathologists have attempted to divide LSIL into lesions showing koilocytosis without atypia, condylomata, and CIN1, but these distinctions do not appear biologically relevant or clinically reproducible and are not recommended by LAST criteria. LSIL can be due to either high-risk or low-risk HPV subtypes, with high-risk HPV accounting for the majority (80%–85%) of cases at the cervix.

Low-Grade Squamous Intraepithelial Lesions: Histology

LSIL is morphologically characterized by koilocytic atypia. Koilocytes contain enlarged and irregular nuclei, sometimes with binucleation, surrounded by a region of cytoplasmic clearing. The area of cleared cytoplasm shows an irregular edge (a "ribbon-like" or "calligraphy pen" border due to its vacillating thickness). The nuclear chromatin is coarse and irregular with either no nucleoli or only tiny, indistinct nucleoli.

Histologic samples of LSIL may show a proliferation of basal-like cells with occasional mitoses; however; this proliferation is by definition limited to the lower-third of the epithelium. Although some descriptions of LSIL imply that the cytologic atypia ends in the lower third, it is critical to note that full-thickness atypia is indeed present in LSIL: enlarged, hyperchromatic, and irregular nuclei percolate throughout the upper portions of the epithelium. Indeed, that is why cytologic specimens, which typically sample only the superficial layers of epithelium, remain such an effective predictor of LSIL histology. However, in contrast to HSIL, the surface atypia of LSIL is koiolocytic or maturing in nature, with retention of abundant cytoplasm and a relatively low/normal N:C ratio.

In straightforward cases, LSIL is readily and reproducibly classified, but many cases present diagnostic difficulties. In particular, the CIN1/CIN2 interface can be challenging.

Cytologically, LSIL corresponds to the HPV productive infection and is characterized by the presence of koilocytosis. LSIL (2%–4% of the samples) regress spontaneously in at least half of cases, especially in young patients. It is critical to note that most true LSIL cytology samples are due to high-risk HPV types (80%–85%) the remainder are due to low risk types.

HIGH-GRADE INTRAEPITHELIAL SQUAMOUS LESION: CIN2

HSIL/CIN2 Histology

On histologic sections, the extension of atypical basaloid high N:C ratio cells and mitotic activity into the middle of the epithelium raises concern for moderate dysplasia, or CIN2, which represents the lower end of the HSIL spectrum and an area of significant diagnostic difficulty. This difficulty is highlighted by the fact that CIN2 has the lowest interobserver reproducibility of all cervical diagnoses. A common explanation for the over-diagnosis of CIN2 is the tendency of some observers to over interpret high-riding nuclear atypia without attention to the associated cytoplasm. It is therefore worth reemphasizing: full-thickness nuclear atypia is allowable and expected for CIN I. It is the proliferation of high N:C ratio, basal-like cells, often with associated mitotic figures, into the middle or upper third of the epithelium that warrants consideration for a HSIL diagnosis. The more extensive the expansion of the cellular component the more certain the diagnosis of HSIL. Yet, assessing the level of atypia and proliferative activity can also be complicated by tangential sectioning and epithelial sloughing; therefore, a confident diagnosis of CIN2 requires clear visualization of the epithelium from base to surface. If orientation is suboptimal but CIN2 remains on the differential, level sections should be considered. If ambiguity remains, a diagnosis of "dysplasia" can be rendered and accompanied by a note explaining the issue. However, such equivocal, ungraded reads can generate understandable frustration for our clinical colleagues and patients and should therefore be used very sparingly (Figs. 6.1–6.9).

Given the poor interobserver reproducibility for CIN2, the frustration generated by equivocal reads and the importance of managing HSIL, LAST recommends the use of ancillary studies (specifically, p16 immunohistochemistry) for all cases of suspected CIN2. Recent work suggests that enlisting p16 downgrades roughly 1/3 of CIN2 lesions, preventing unnecessary follow-up in a significant subset of patients. Patterns of p16 interpretation are further discussed in the "Biomarkers" section of this chapter. It is also notable that LAST emphasizes that it is the distinction between LSIL versus HSIL that guides management. Separating CIN2 from CIN3, even with biomarkers, is only relevant in ASCCP guidelines for the management of women under age 25 with small lesions, but in many European countries screening does not begin until this age. Hence in many laboratories CIN2 and CIN3 are combined ad HSIL (CIN2-3 or CIN2/3) for diagnostic purposes.

High-Grade Intraepithelial Squamous Lesion: CIN3

CIN3 represents the highest grade of dysplasia and denotes near or complete repopulation of the epithelium by a clonal expansion of proliferative, high-N:C ratio basaloid cells. As previously

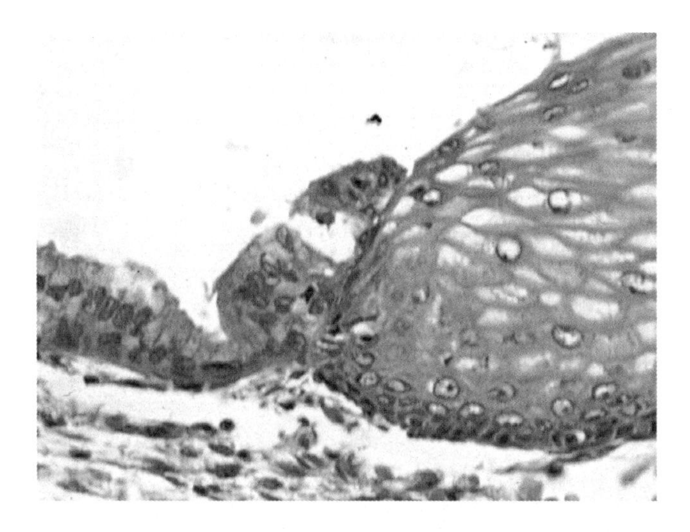

FIGURE 6.1

The cervical transition zone, demonstrating a convergence of normal squamous and glandular mucosa.

mentioned, historic attempts to segregate CIN3 from CIS proved poorly reproducible and biologically inconsequential; therefore, the latter term has been discarded and these cases collapsed into CIN3. Often the constituent cells of CIN3 bear nuclei that are smaller and less strikingly pleomorphic than the cells observed in LSIL, and are instead marked by their monotony a manifestation of their clonality.

HSIL/CIN3: Histology

On histologic sections, this monotony translates into an epithelium that essentially shows no distinction between the superficial and base layers: in the most classic cases one could imagine flipping the image over entirely, so that the superficial cells lined the base and vice versa, without changing the appearance. Mitotic figures are commonly present and extend into the upper third of the epithelium. This impressive appearance leads to relatively high interobserver reproducibility in the diagnosis of CIN3 when compared to CIN2, although benign mimics such as squamous metaplasia and atrophy can lead to diagnostic difficulty. It is critical to note that while there is remarkable monotony within a CIN3 lesion that lesion should differ significantly from the background squamous epithelium. The presence of a morphologically distinct clonal population is a hallmark of high-grade dysplasia. If the entire epithelium is lined by a relatively uniform basaloid, high N:C ratio cells, the differential of atrophy should be considered.

Given the above, cytologic HSIL represents transforming HPV infection and is characterized by the presence of abnormal basal-like cells. In some labs HSIL is still split into moderate versus severe dysplasia based on the degree of maturation, N:C ratio, etc. of the cells in question. HSIL (0.5%–1% of smears) are associated with persistent viral infection and viral persistence is highly predictive of a high-grade lesion being detectable on biopsy. It is critical to note that an HSIL cytology is not negated by a negative biopsy. The variability in colposcopic accuracy makes any

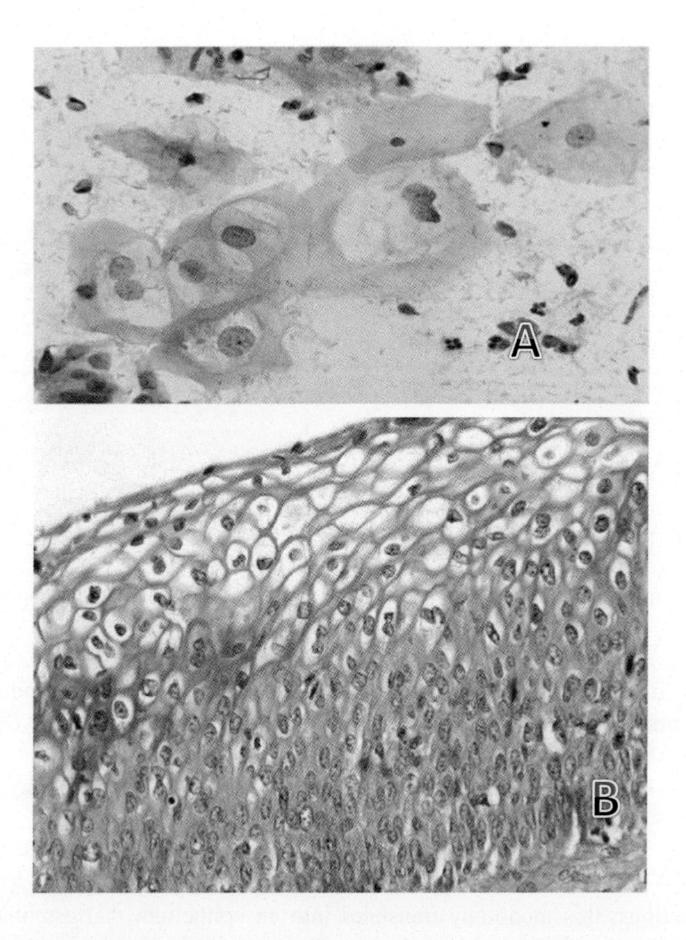

FIGURE 6.2

LSIL cytology A/Histology B. Pap Cytology of Koilocytes in background of normal cells (A). Koilocytic atypia extends into the superficial epithelium, and occasional binucleate cells are seen. Nuclear enlargement is marked, often exceeding the nuclear size of basal cells; however, cytoplasm remains abundant (B).

patient with HSIL cytology worthy of continued surveillance or even treatment given its high positive predictive value

HIGH-GRADE INTRAEPITHELIAL SQUAMOUS LESION VARIANTS

Thin High-Grade Intraepithelial Squamous Lesion

Thin HSIL is an immature squamous epithelial lesion typically measuring fewer than 10 cells in thickness. These lesions may be difficult to differentiate from atrophy or repair, and their diagnosis can be aided by the use of ancillary biomarkers such as p16.

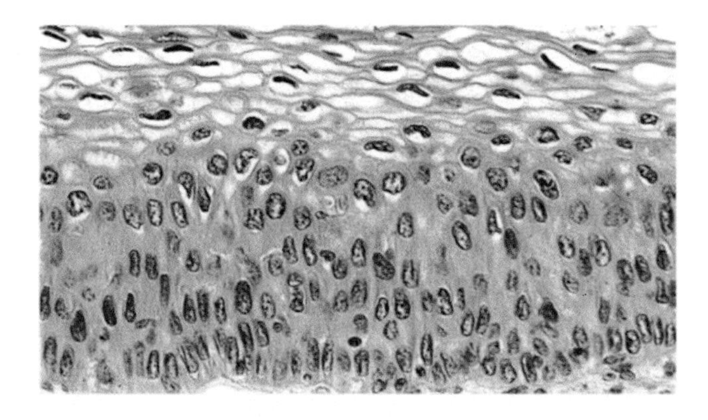

FIGURE 6.3

HSIL (CIN2) histology. This case shows basaloid cells and mitotic activity percolating through the middle of the epithelium; however, the surface retains some degree of maturation with some preservation of the nuclear: cytoplasmic ratio.

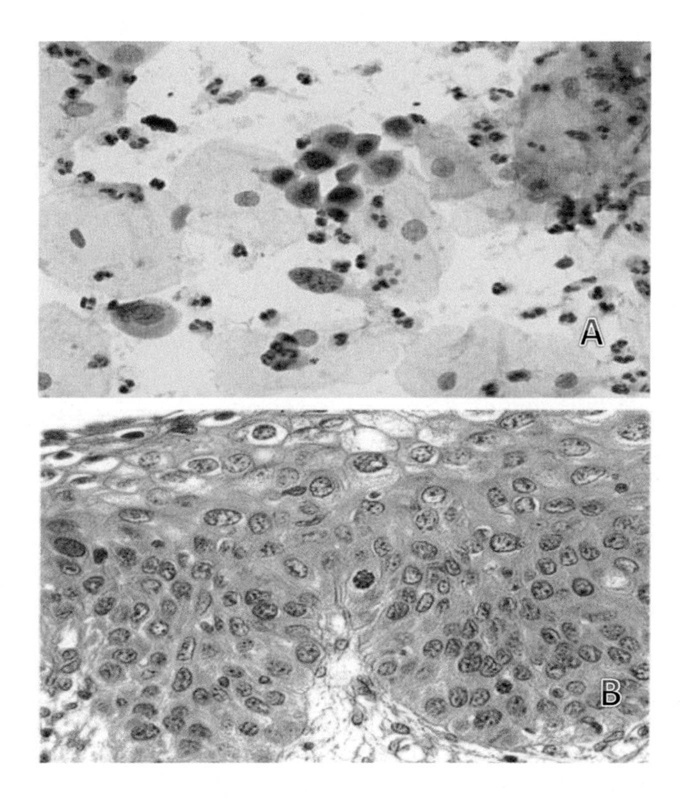

FIGURE 6.4

HSIL (CIN3) cytology A/Histology B. Pap cytology of HSIL (A). The epithelium is completely repopulated by basaloid cells with nuclear enlargement, hyperchromasia, and irregular nuclear outlines. Cytoplasm is scant and uniformly distributed across the thickness of the epithelium. Mitotic figures are readily identifiable and extend to the surface (B).

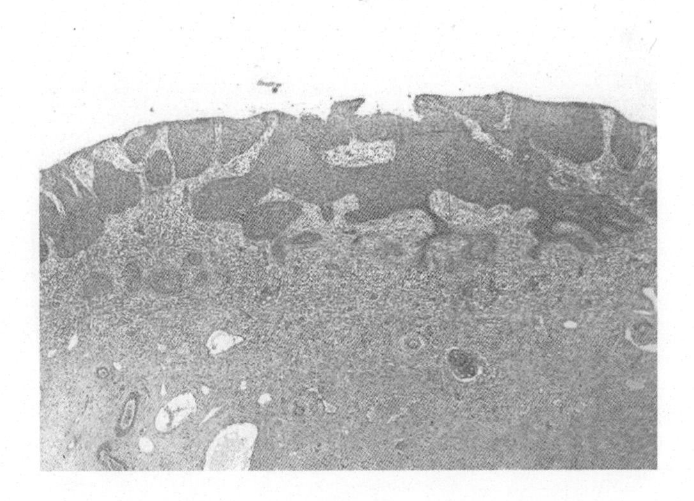

FIGURE 6.5

HSIL with invasive squamous cell carcinoma. This specimen showed extensive HSIL However, several foci of jagged, irregular squamous nests were seen underlying the in situ lesion demonstrating invasion.

Keratinizing High-Grade Intraepithelial Squamous Lesion

Keratinizing HSIL has an atypical surface layer bearing dyskeratotic and pleomorphic nuclei. These lesions are often encountered in the ectocervix and are reminiscent of the dysplasia seen in high-risk HPV-related cancers outside the cervix, such as the vulva and penis. Care must be taken not to underestimate these lesions as LSIL based on the presence of moderate to abundant keratin-rich cytoplasm. Keratinizing HSIL can also be found in lesions which are clinically condylomas, and in these instances the HSIL dictates the prognosis.

Papillary High-Grade Intraepithelial Squamous Lesion

Papillary HSIL (e.g., "noninvasive papillary squamo-transitional carcinoma" or "papillary squamous carcinoma in situ") is reminiscent of urothelial neoplasia. It is a diagnosis contingent on complete excision with exclusion of stromal invasion.

Microscopic Mimics

There are several important mimics to consider in the assessment of cervical dysplasia, including reactive atypia, basal cell hyperplasia, squamous metaplasia, and atrophy. These benign lesions are very much a "part of the problem" of distinguishing precancer that needs treatment from a benign mimic. The judicious use of biomarkers helps to dichotomously solve this common clinical problem.

SQUAMOUS CELL CARCINOMA

Approximately, 85% of cervical cancers are squamous cell carcinomas. The remainder are mostly adenocarcinomas and a rare but deadly variant, neuroendocrine carcinoma. In the cervix, more than 95% of all

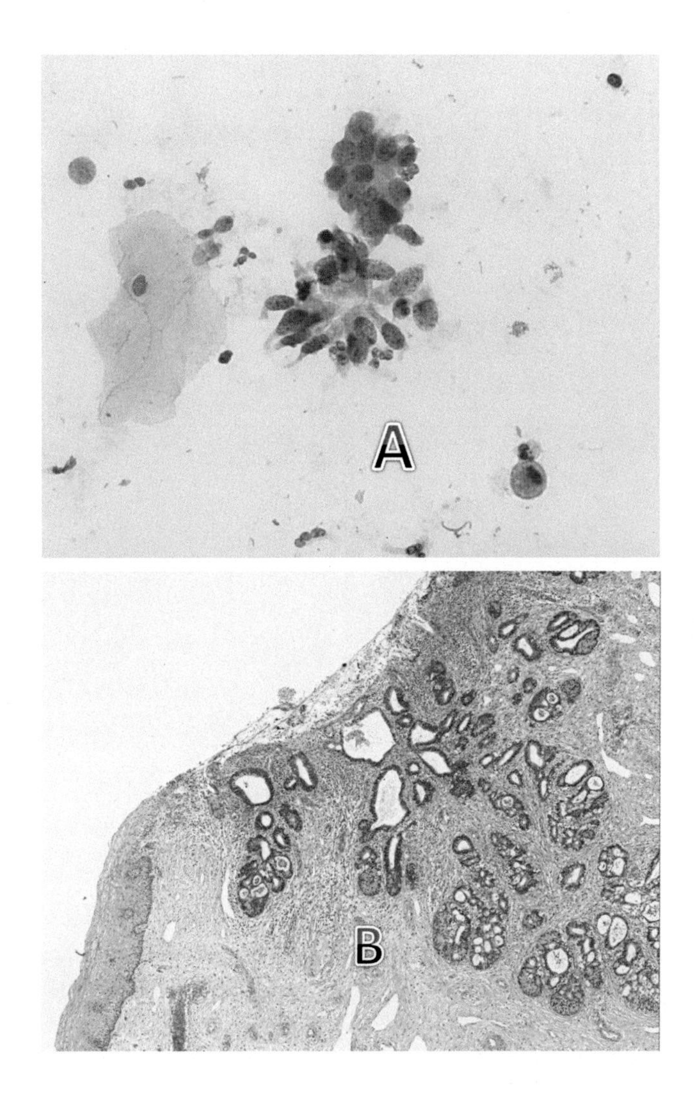

FIGURE 6.6

AIS cytology A/histology B. Pap cytology with abnormal glandular cells arranged in rosettes (A). Leep conisation showing an AIS localized around the squamocolummar junction (B).

carcinomas regardless of cell type are caused by HPV. Squamous cell carcinoma warrants brief discussion here for direct comparison with SIL. On histologic sections the differentiation of cervical squamous dysplasia from invasive squamous cell carcinoma is usually straightforward but can be quite difficult in cases with only superficial invasion. Tangential sectioning of a tongue of dysplastic epithelium must be excluded in cases with suspected invasion which show only rare infiltrative-appearing squamous nests lying close to the surface. The presence of a stromal desmoplastic or inflammatory response provides helpful support for a diagnosis of invasion when present. Occasional cases with dysplasia with extensive

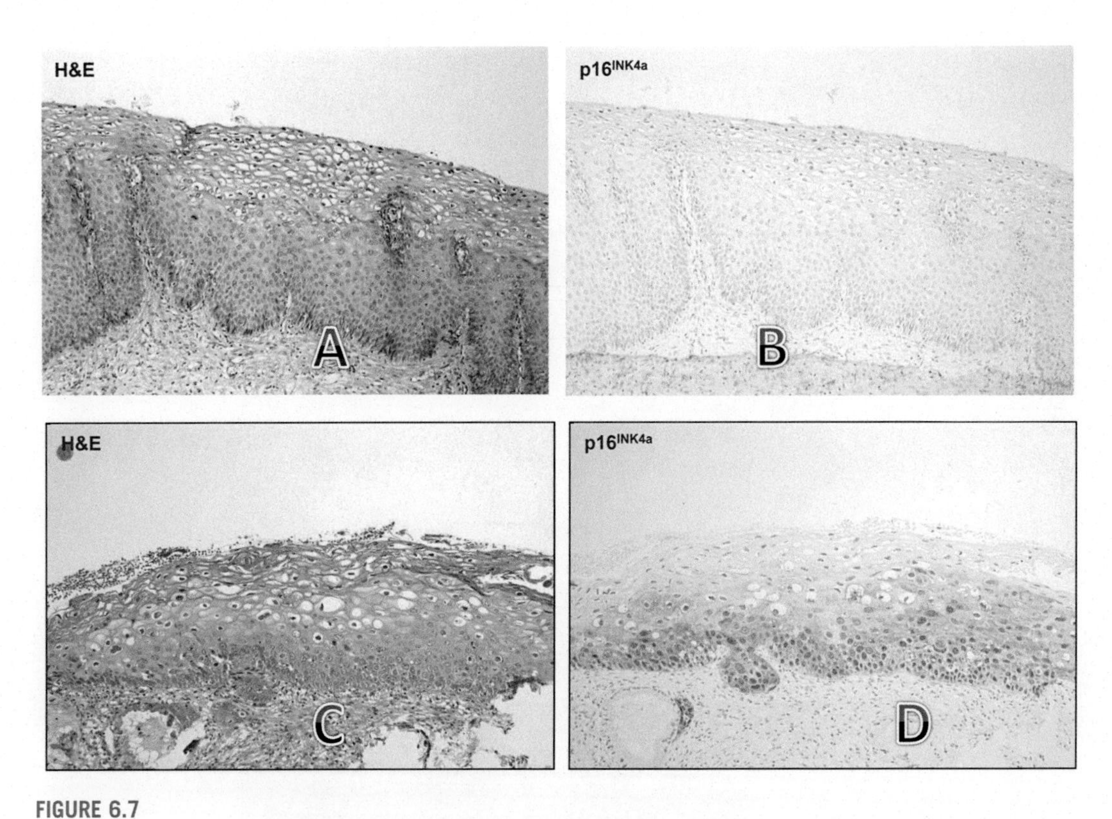

FIGURE 6.7

p16 in LSIL. LSIL (A and C) shows variable staining with p16, with some cases showing completely negative (B) while some are positive usually in the lower third (D).

endocervical gland extension may also present difficulty. The presence of squamous nests with jagged, irregular borders raises considerable concern for invasion, as do nests located outside of the normal pattern of endocervical glands. Paradoxical squamous maturation with accumulation of eosinophilic cytoplasm and keratin pearl formation at the epithelial stromal should prompt a careful search for invasion.

GLANDULAR NEOPLASIA

The focus of the HPV story centers on squamous neoplasia. However as noted above roughly 15% of cervical cancers are adenocarcinomas and hence, a few comments are warranted. Cytology has shown little success in preventing cervical adenocarcinoma despite the fact that there is a histologic precursor called adenocarcinoma in situ. This lesion is detectable cytologically but is confounded by sampling issues and interpretive mimics that perhaps explain the failure of cytology to impact the epidemiology for this form of cancer. The good news is HPV testing because of its superior sensitivity may change this and even more importantly HPV vaccination may equally eradicate

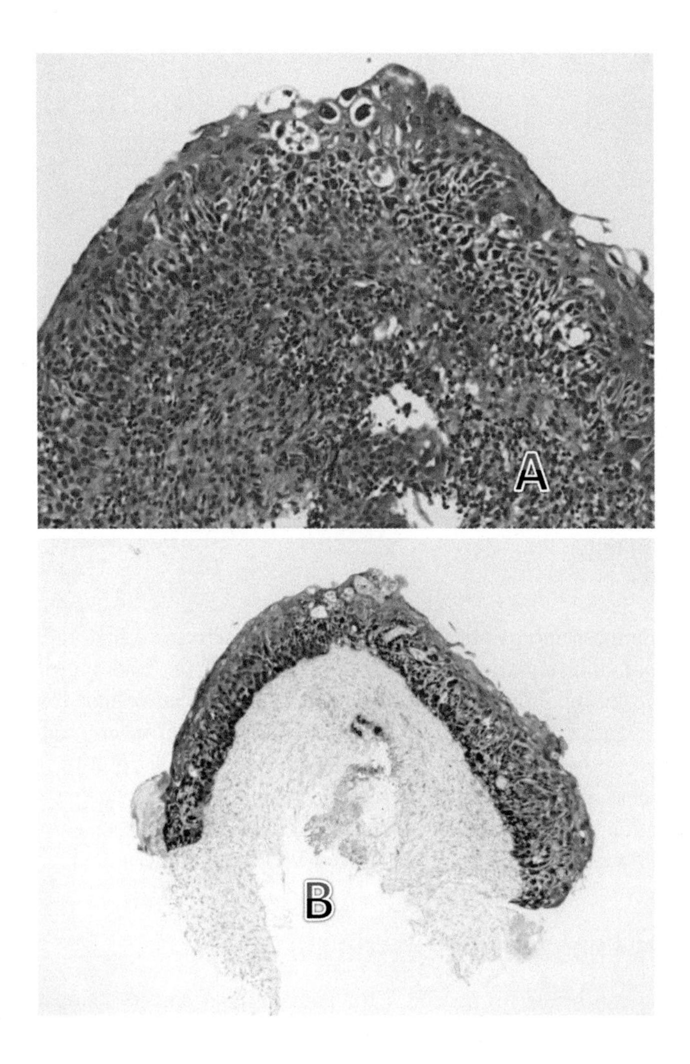

FIGURE 6.8

p16 in HSIL versus reactive/metaplastic. p16 immunostaining is also of value for confirming HSIL in cases with a differential of reactive, metaplastic changes. Although very mitotically active, this HSIL case has a somewhat metaplastic appearance; furthermore, the background is inflammatory and nucleoli are focally prominent, invoking the possibility of a reactive process (A). However, the strong diffuse nuclear p16 positivity extending throughout the epithelium permits unequivocal assignment as HSIL (B).

these lesions since even more than squamous cancer the spectrum of HPV types in AIS and adenocarcinoma is narrower. HPV 18 is relatively more common in adenocarcinomas compared to HPV16. Moreover, of the nonsquamous cancers of the cervix, more than 85%–90% of them are also HPV associated and therefore, both HPV screening and HPV vaccination can have major impact in reducing the incidence of these tumors as well.

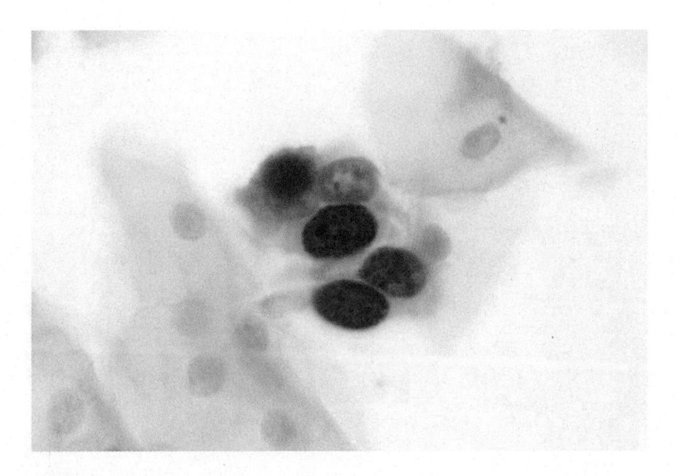

FIGURE 6.9

Dual stain cytology of an HSIL.

Histologically, adenocarcinomas of the cervix have a spectrum of histologic patterns correlating with the cell's type of differentiation. Most are endocervical, but endometrioid, papillary, serous, clear cell, and gastric morphologies are all seen. HPV can be pathogenic for almost all types and those that are HPV negative are most often lower uterine segment tumors rather than true cervical cancers. The rare exceptions like gastric type adenocarcinoma are virtually never HPV-associated. HPV associated cancers almost never have p53 mutations. The HPV negative tumors often do, a useful direct application of molecular biology to diagnostics and furthermore, gastric-type adenocarcinoma is frequently associated with a germline STK11 mutation.

SAMPLING OF THE ENDOCERVICAL EPITHELIUM

Cytological sampling can be performed both for conventional smears using the cytobrush and for liquid based cytology using the Cervexbrush and its variants. Improvement of cytology brushes allows better sampling of the endocervical epithelium. Endocervical sampling of glandular precancer can be problematic because adenocarcinoma in situ may have a multifocal development and endocervical glands can be partially or totally involved. If abnormal endocervical cells do not reach the superficial part of the glandular structures they may not available for sampling.

ATYPICAL GLANDULAR CELLS

The term "atypical endocervical, endometrial or glandular cells" replaces the generic term "atypical glandular cells of undetermined significance" (AGUS) in the current Bethesda System. The nature of glandular, endocervical, or endometrial cells should be noted if possible by the pathologist; this will permit a more appropriate diagnostic approach, cervical biopsy, and examination of the endocervix to identify endocervical lesion, or endometrial biopsy to identify an endometrial lesion.

Glandular cells abnormalities are rare and represent less than 0.1–0.5% of all smears and less than 5% of abnormal smears

ENDOCERVICAL ADENOCARCINOMA IN SITU

''Adenocarcinoma in situ" is the category, which corresponds to specific morphological abnormalities that correlate with histologic AIS, the precancer of adenocarcinoma. This entity is characterized by specific morphological features, such as the radial arrangement of nuclei in the periphery, like "at the end of the feathers of a bird's wing" *(feathering of cells)*, images of nuclei palisading or rosettes without tumoral diathesis. Making this diagnosis allows for a more aggressive diagnostic approach such as a conization rather than just a biopsy.

ADENOCARCINOMA

On cytology, invasive adenocarcinoma is characterized by the presence of atypical glandular cells showing nuclear changes comprising enlargement, hyperchromasia, coarse granulation of chromatin and large, sometimes irregular multiple nucleoli. The abnormal cells often form spherical or oval clusters and the smear background often shows blood, necrosis, and cell debris. The TBS ask if possible, to specify if the adenocarcinoma is endocervical, endometrial, or extrauterine.

USING OUR MOLECULAR UNDERSTANDING OF DISEASE PROGRESSION TO IMPROVE DIAGNOSIS: DETECTION OF HPV DNA AND BIOMARKERS OF INFECTION

BIOMARKERS

The subject of biomarkers is discussed in detail in Chapter 17: Infection to Cancer—Finding Useful Biomarkers for Predicting Risk of Progression to Cancer). In this chapter the use of immunohistochemical markers in biopsy pathology and cytology is discussed.

p16 immunohistochemistry is the most widely enlisted biomarker in the uterine cervix and in the HPV-related neoplasia in general. p16 protein is thought to accumulate in the setting of transcriptionally active hrHPV infection as a response to unchecked proliferation due to the viral E7 oncoprotein's interference with the tumor suppressor protein retinoblastoma (RB). As such, p16 is not a globally specific marker of HPV infection; it can be overexpressed in a variety of non-HPV related malignancies such as myometrial leiomyosarcoma. However, it has strong specificity for the presence of a high-risk HPV-driven lesion in the epithelia of the lower anogenital tract and the oropharynx.

p16's utility in these locales is chiefly in the diagnosis of high-grade dysplasia. It is most often performed on tissue sections but has occasional (and potentially growing) utility in cytology preparations. Although some early studies suggested that p16 may be able to differentiate between LSIL and its benign benign/reactive mimics that has not borne out in subsequent works which showed unreliable p16 marking in histologically unequivocal CIN1 and frequent patchy, blush-like staining in morphologically benign epithelia. Such focal, predominantly cytoplasmic staining should be considered negative. In contrast, the presence of confluent groups (e.g., >5–6 cell in a row or so-called block positivity) of cervical epithelial cells with nuclear and/or nuclear and cytoplasmic p16 positivity that in histologic section also extends upward at least 1/3rd of the epithelial thickness can greatly aid in confirming dysplasia in cases of possible HSIL with a differential of atrophy or metaplasia. Biologically, high grade lesions, for example, true precancers are virtually always p16 positive. Ergo, The LAST recommendations therefore advocate for p16 application in all cases of suspected CIN2 as a way of minimizing CIN2 cases as well as cases with a differential diagnosis of CIN3 versus benign (atrophy, squamous metaplasia, etc.).

Although the diagnostic value of p16 immunohistochemistry in the uterine cervix is well-established in these scenarios (e.g., diagnosis of CIN2 and CIN3 vs mimics), p16 falters when it comes to prognostication. LSIL lesions are ~40%–50% p16 positive. Thus, LAST recommends using p16 only in cases where the pathologist is on the fence between calling a lesion CIN2 versus CIN1. Recent work—including a recent study of LSIL/CIN1 expert-adjudicated cases—have not supported the prognostic value of p16 in CIN1, and at this point p16 is not considered a reliable prognostic marker in LSIL cases.

Ki67

The proliferative marker Ki67 has long been enlisted as an ancillary test in the diagnosis of cervical neoplasia. It has some value in limited diagnostic samples, where its negative predictive value can prevent unnecessary colposcopy and may be useful in conjunction with p16 for morphologically equivocal cases. It has no clear utility; however, in the triage of morphologic CIN1 cases. It occasionally can be used as a "tie breaker" if the p16 is negative but the morphology in the pathologist's opinion has to be HSIL.

Cytokeratin 7 and Other Squamocolumnar Junctional Markers

Cytokeratin 7 (CK7) has appeared more recently as a putative prognostic marker in tissue sections of CIN; its role in cytologic specimens has not been investigated. CK7 is thought to mark a biologically distinct squamocolumnar junctional (SCJ) cell population which represents the putative origin of HSIL. Herf, Crum, and colleagues have studied this cellular subset and regard CK7 as marking the subset more likely to progress to HSIL.

In Situ Hybridization (ISH)

HPV DNA In Situ Hybridization

HPV DNA ISH has historically shown robust specificity but less-than-optimal sensitivity for the detection of high-risk HPV, with poor detection in cases with low viral copy numbers. It also performs inferiorly to p16 in differentiating HSIL versus benign mimics. Nevertheless, HPV DNA ISH had value, particularly in the minimization of false positive CIN1 results.

HPV RNA ISH can identify transcriptionally-active HPV in the context of neoplastic morphology. The last few years have seen the emergence of HPV RNA ISH assays that can be performed on formalin-fixed, paraffin-embedded tissue such assays and multiple investigations of their sensitivity and specificity has shown great promise. In addition, some studies of HPV E6/E7 RNA ISH patterns may help stratify CIN. While ISH assays are attractive especially because of their specificity, there will always be a tension between more complex ISH and the ease of p16 IHC that can be viewed as an assay capturing the activity of any high-risk HPV E7 expression.

BIOMARKERS ON CYTOLOGY SAMPLES

TRIAGE OF HPV-POSITIVE WOMEN

Randomized controlled trials of HPV testing have repeatedly shown more sensitive detection of HSIL compared to cytology and is more effective in preventing invasive cervical cancer. However, directly referring to colposcopy all HPV-positive women results in a marked increase in the number of colposcopies needed to detect a precancerous lesion. Therefore, methods are needed for selecting, among HPV-positive women, those who have very low probability of carrying a colposcopy-detectable precancerous lesion and therefore not needing immediate colposcopy. Supplements to European guidelines have recommended the HPV primary screening alone with triaging positives with cytology and referring immediately to colposcopy HPV-positive women with abnormal cytology (ASC-US or more severe) and retesting HPV-positive cytological negative women after 1 year. Women positive at that time for either test are referred to colposcopy. Cytology informed of HPV positivity is more sensitive than blind cytology. Screening programs with informed cytology triage are expected to perform better than predicted by trials and could possibly allow longer intervals before retesting HPV-positive women with normal cytology. Alternative triage strategies like combining genotyping (16/18 only vs extended genotyping) with cytology, p16/ki67 dual stain IHC, or methylation analyses are all under active evaluation for optimization of the balance between immediate referral versus deferred assessment of HPV positive women.

P16/KI-67 DUAL STAIN

When compared to molecular HPV tests, immunochemical detection of p16-stained cells demonstrates a significantly improved specificity with remarkably good relative sensitivity. However, p16 single-staining immunocytochemistry protocols required the morphologic interpretation of immunoreactive cells to distinguish between $p16^{INK4a}$-positive cells showing intraepithelial lesions and those cervical cells or contaminating endometrial cells that normally overexpress p16. Using a combination of antibodies to detect p16 and the cell cycle progression marker Ki67 identifies, in the context of cervical neoplasia, HPV-transformed cervical cells. The clinical performance of this approach has been evaluated in the triage of ASC-US and LSIL cytology results and more recently in HPV-primary screening. In all settings, dual-stain demonstrated superior sensitivity and specificity for HSIL detection. The low risk of cervical precancer in p16/Ki-67 negative women permits safe extension of follow-up intervals for at least 3 years.

CLINICAL COURSE AND MANAGEMENT OF CERVICAL NEOPLASIA

Screening and triage practices are discussed in other chapters. The most up-to-date US guidelines for the management of cervical squamous lesion can be obtained from the ASCCP or IARC.

However, some nuances of these practices warrant discussion here within the context of the behavior of squamous intraepithelial lesions as well as the diagnostic and prognostic uncertainty that sometimes attends their diagnosis.

LOW-GRADE SQUAMOUS INTRAEPITHELIAL LESIONS BEHAVIOR/MANAGEMENT

Current clinical management of LSIL is driven largely by the small percentage (~10%) of cases that "progress" to HSIL. The vast majority of LSIL cases will resolve without excisional intervention within approximately 12 months although it is notable that teasing out true regression from removal due to "therapeutic biopsy" can be difficult. However, with the exception of very young women it is not acceptable to leave LSIL patients unmonitored for long durations because of the subset who will go on to develop HSIL lesions. It is unclear whether these represent true progression versus initially unsampled lesions, and it bears reinforcing that an LSIL surgical biopsy result should never simply negate a HSIL or ASC-H cytologic result.

HIGH-GRADE INTRAEPITHELIAL SQUAMOUS LESION BEHAVIOR/MANAGEMENT

The management of HSIL is guided by its relatively high propensity to eventually progress to invasive cancer. While HSIL more often occurs in older patients when compared to LSIL, this does not necessarily mean that HSIL invariably takes a long time to develop. HSIL has been documented within 1–2 years of initial HPV infection among adolescents, suggesting that it can manifest relatively rapidly. HSIL is estimated to progress to invasive carcinoma at a rate of 0.5%–1.0% per year, with invasive cancers diagnosed up to two decades later—and on average, one decade later—than in situ lesions. Although a large proportion of untreated HSIL are likely to progress to invasive cancer, HSIL nevertheless represents a nonobligate precursor to invasive cancer, with reported regression rates ranging from 30% to 50%. However, these numbers may be considerably inflated by the potential for therapeutic biopsy in a large proportion of these presumably "regressed" cases, as well as contamination by CIN1 cases misclassified as CIN2.

Another somewhat controversial issue in the management of HSIL is whether or not CIN2 and CIN3 should ever be treated differently. Although our understanding of their biologic underpinnings and behavior supports the collapse of CIN2 and CIN3 into a single managerial category (HSIL), differentiation between moderate and severe dysplasia sometimes becomes important in women with HSIL who are willing to accept some risk of progression in order to maintain fertility. There is evidence that LEEP increases rates of miscarriage and preterm labor (although these risks are lower than was previously thought), therefore one could argue for less aggressive excision in women who wish to maximize childbearing potential. In such cases, less aggressive interventions may be allowable for a diagnosis of CIN2. However, such differential management for CIN2 and CIN3 must be performed with awareness of the considerable interpretative variability that attends the classification of these lesions. The potential for using a marker of productive infection to identify early from late, more complete transforming infection is discussed with other potential biomarkers of the extent of transformation in Chapter 17: Infection to Cancer—Finding Useful Biomarkers for Predicting Risk of Progression to Cancer.

FURTHER READING

Reagan JW, Hamonic MJ. The cellular pathology in carcinoma in situ: a cytohistopathological correlation. Cancer 1956;9(2):385–402.

Koss LG, Durfee GR. Unusual patterns of squamous epithelium of the uterine cervix: cytologic and pathologic study of koilocytotic atypia. Ann N Y Acad Sci 1956;63(6):1245–61.

Meisels A, Fortin R, Roy M. Condylomatous lesions of the cervix. II. cytologic, colposcopic and histopathologic study. Acta Cytol 1977;21(3):379–90.

Hills E, Laverty CR. Electron microscopic detection of papilloma virus particles in selected koilocytotic cells in a routine cervical smear. Acta Cytol 1979;23(1):53–6.

Richart RM. A theory of cervical carcinogenesis. Obstet Gynecol Surv 1969;24(7 Pt 2):874–9.

Boshart M, Gissmann L, Ikenberg H, Kleinheinz A, Scheurlen W, zur Hausen H. A new type of papillomavirus DNA, its presence in genital cancer biopsies and in cell lines derived from cervical cancer. EMBO J 1984;3(5):1151–7.

Crum CP, Ikenberg H, Richart RM, Gissman L. Human papillomavirus type 16 and early cervical neoplasia. N Engl J Med 1984;310(14):880–3. Available from: https://doi.org/10.1056/NEJM198404053101403.

Doorbar J, Quint W, Banks L, et al. The biology and life-cycle of human papillomaviruses. Vaccine 2012;30(Suppl 5):F55–70. Available from: https://doi.org/10.1016/j.vaccine.2012.06.083.

Darragh TM, Colgan TJ, Thomas Cox J, et al. The Lower Anogenital Squamous Terminology standardization project for HPV-associated lesions: background and consensus recommendations from the College of American Pathologists and the American Society for Colposcopy and Cervical Pathology. Int J Gynecol Pathol 2013;32(1):76–115. Available from: https://doi.org/10.1097/PGP.0b013e31826916c7.

Nayar R, Wilbur D. The Bethesda system for reporting cervical cytology. Definitions, criteria, and explanatory notes. 3rd ed Springer; 2015.

Stoler M, Bergeron C, Colgan TJ, Ferenczy A, Herrington CS, Loening T, et al. Squamous cell tumors and precursors Chapter 7 In: Kurman RJ, Carcangiu ML, Herrington CS, Young CS, editors. World Health Organization classification of tumours (WHO) Classification of tumours of female reproductive organs. Published by International Agency for Research on Cancer (IARC); 2014. p. 1–12.

Saslow D, Solomon D, Lawson HW, et al. American Cancer Society, American Society for Colposcopy and Cervical Pathology, and AmericanSociety for Clinical Pathology screening guidelines for the prevention and early detection of cervical cancer. Am J Clin Pathol. 2012;137(4):516–42. Available from: https://doi.org/10.1309/AJCPTGD94EVRSJCG.

Arbyn M, Bergeron C, Klinkhamer P, Martin-Hirsch P, Siebers AG, Bulten J. Liquid compared with conventional cervical cytology: a systematic review and meta-analysis. Obstet Gynecol 2008;111(1):167–77. Available from: https://doi.org/10.1097/01.AOG.0000296488.85807.b3.

Scottish Cervical Screening Programme: Steering group report on the feasibility of introducing liquid-based cytology, January 2002. http://www.show.scot.nhs.uk

National Institute for Clinical Excellence (NICE). Guidance on the use of liquid-based cytology for cervical screening, October 2003. http://www.nice.org.uk

Stoler MH, Schiffman M. Atypical Squamous Cells of Undetermined Significance-Low-grade Squamous Intraepithelial Lesion Triage Study (ALTS) Group. Interobserver reproducibility of cervical cytologic and histologic interpretations: realistic estimates from the ASCUS-LSIL triage study. JAMA 2001;285 (11):1500–5. Available from: https://doi.org/jto10000 [pii].

Stoler MH, Vichnin MD, Ferenczy A, et al. The accuracy of colposcopic biopsy: analyses from the placebo arm of the Gardasil clinical trials. Int J Cancer 2011;128:1354–62.

Guido RS, Jeronimo J, Schiffman M, Solomon D, ALTS Group. The distribution of neoplasia arising on the cervix: results from the ALTS trial. Am J Obstet Gynecol 2005;193(4):1331–7. Available from: https://doi.org/S0002-9378(05)00651-4.

Galgano MT, Castle PE, Atkins KA, Brix WK, Nassau SR, Stoler MH. Using biomarkers as objective standards in the diagnosis of cervical biopsies. Am J Surg Pathol 2010;34(8):1077–87. Available from: https://doi.org/10.1097/PAS.0b013e3181e8b2c4.

Bergeron C, Ronco G, Reuschenbach M, Arbyn M, Stoler M, Von Knebel Doeberitz M. The clinical impact of using p16 immunochemistry in cervical histopathology and cytology: an update of recent developments, Minireview. IJC 2014; Online 17 April.

Mills AM, Paquette C, Castle PE, Stoler MH. Risk stratification by p16 immunostaining of CIN1 biopsies: a retrospective study of patients from the quadrivalent HPV vaccine trials. Am J Surg Pathol 2015;39(5):611–17. Available from: https://doi.org/10.1097/PAS.0000000000000374.

Kisser A, Zechmeister-Koss I. A systematic review of p16/ki-67 immuno-testing for triage of low grade cervical cytology. BJOG 2015;122(1):64–70. Available from: https://doi.org/10.1111/1471-0528.13076.

Herfs M, Parra-Herran C, Howitt BE, et al. Cervical squamocolumnar junction-specific markers define distinct, clinically relevant subsets of low-grade squamous intraepithelial lesions. Am J Surg Pathol 2013;37 (9):1311–18. Available from: https://doi.org/10.1097/PAS.0b013e3182989ee2.

Paquette C, Mills AM, Stoler MH. Predictive value of cytokeratin 7 immunohistochemistry in cervical low-grade squamous intraepithelial lesion as a marker for risk of progression to a high-grade lesion. Am J Surg Pathol 2015;. Available from: https://doi.org/10.1097/PAS.0000000000000548.

Zhang W, Kapadia M, Sugarman M, et al. Adjunctive HPV in-situ hybridization (ISH) assay as an aid in the diagnosis of cervical intraepithelial neoplasia in cervical tissue specimens: An analytical and functional characterization. Int J Gynecol Pathol 2012;31(6):588–95. Available from: https://doi.org/10.1097/PGP.0b013e318254349a.

Evans MF, Peng Z, Clark KM, et al. HPV E6/E7 RNA in situ hybridization signal patterns as biomarkers of three-tier cervical intraepithelial neoplasia grade. PLoS One 2014;9(3):e91142. Available from: https://doi.org/10.1371/journal.pone.0091142.

Mills AM, Coppock JD, Willis BC, Stoler MH. HPV E6/E7 mRNA In Situ Hybridization in the Diagnosis of Cervical Low-grade Squamous Intraepithelial Lesions (LSIL). Am J Surg Pathol 2018;42:192–200.

Mills AM, Dirks DC, Poulter MD, Mills SE, Stoler MH. HR-HPV E6/E7 mRNA in situ hybridization: validation against PCR, DNA in situ hybridization, and p16 immunohistochemistry in 102 samples of cervical, vulvar, anal, and head and neck neoplasia. Am J Surg Pathol 2017;41:607–15.

Katki HA, Gage JC, Schiffman M, et al. Follow-up testing after colposcopy: five-year risk of CIN 2+ after a colposcopic diagnosis of CIN 1 or less. J Low Genit Tract Dis 2013;17(5 Suppl 1):S69–77. Available from: https://doi.org/10.1097/LGT.0b013e31828543b1.

Cox JT, Schiffman M, Solomon D. ASCUS-LSIL Triage Study (ALTS) Group. Prospective follow-up suggests similar risk of subsequent cervical intraepithelial neoplasia grade 2 or 3 among women with cervical intraepithelial neoplasia grade 1 or negative colposcopy and directed biopsy. Am J Obstet Gynecol 2003;188(6):1406–12. Available from: https://doi.org/S0002937803004174.

Bansal N, Wright JD, Cohen CJ, Herzog TJ. Natural history of established low grade cervical intraepithelial (CIN 1) lesions. Anticancer Res 2008;28(3B):1763–6.

McCredie MR, Sharples KJ, Paul C, et al. Natural history of cervical neoplasia and risk of invasive cancer in women with cervical intraepithelial neoplasia 3: a retrospective cohort study. Lancet Oncol 2008;9(5):425–34. Available from: https://doi.org/10.1016/S1470-2045(08)70103-7.

International Agency for Research on Cancer World Health Organization. Cervix cancer screening. IARC Handbooks of cancer prevention, vol. 10. Lyon: IARC; 2005.

European guidelines for quality assurance in cervical cancer screening. In: Arbyn M, Antilla M, Jordan J, Ronco G, Schenck U, Segnan N, Wiener HG, Herbert A, Daniel J, von Karsa L, editors. IARC Lyon; 2008.

American Society for Colposcopy and Cervical Pathology, JLGTD 2012, Supplement 16, pp. 205–42.

Arbyn M, Ronco G, Anttila A, Meijer CJ, Poljak M, Ogilvie G, et al. Evidence regarding human papillomavirus testing in secondary prevention of cervical cancer. Vaccine 2012;30(Suppl 5):F88–99.

Ronco G, Dillner J, Elfstrom KM, Tunesi S, Snijders PJ, Arbyn M, et al. Efficacy of HPV-based screening for prevention of invasive cervical cancer: follow-up of four European randomised controlled trials. Lancet 2014;383:524–32.

Ronco G, Arbyn M, Meijer CJLM, Snijders PJF, Cuzick J. S1 Screening for cervical cancer with primary testing for human papillomavirus. In Supplements to EU guidelines for quality assurance in cervical cancer screening. In: Anttila A, Arbyn M, De Vuyst H, Dillner J, Dillner L, Franceschi S, Patnick J, Ronco G, Segnan N, Suonio E, Tornberg S, von Karsa L, editors; 2015.

Clarke MA, Cheung LC, Castle PE, Schiffman M, Tokugawa D, Poitras N, et al. Five-year risk of cervical precancer following p16/Ki-67 dual-stain triage of HPV-positive women. JAMA Oncol 2018;. Available from: https://doi.org/10.1001/jamaoncol.2018.42 Published online October 11.

Bergeron C, Giorgi-Rossi P, Cas F, Schiboni ML, Ghiringhello B, Dalla Palma P, et al. Informed cytology for triaging HPV-positive women: substudy nested in the NTCC randomized controlled trial. J Natl Cancer Inst 2015;1–5.

CHAPTER

7 DEVELOPING AND STANDARDIZING HUMAN PAPILLOMAVIRUS TESTS

Attila Lorincz[1], Cosette Marie Wheeler[2], Kate Cuschieri[3], Daan Geraets[4], Chris J.L.M. Meijer[5] and Wim Quint[4]

[1]Wolfson Institute of Preventive Medicine, Queen Mary University of London, London, United Kingdom [2]University of New Mexico Comprehensive Cancer Center, Albuquerque, NM, United States [3]Scottish HPV Reference Laboratory, Royal Infirmary of Edinburgh, United Kingdom [4]DDL Diagnostic Laboratory, Rijswijk, The Netherlands [5]Amsterdam UMC, Vrije Universiteit Amsterdam, Pathology, Cancer Center Amsterdam, Amsterdam, The Netherlands

HISTORICAL DEVELOPMENT

DISCOVERY OF HUMAN PAPILLOMAVIRUS

Methods to detect human papillomavirus (HPV) evolved greatly during the last 70 years. In the 1940s, there was no knowledge of different HPV genomes and understanding of the viruses was rudimentary. Extracts of warts filtered through pores too small for bacteria to pass through were known to produce new warts on the scarified skin of test animals and crystalline arrays of HPV in epithelial extracts were first visualized by electron microscopy in 1949 [1].

HPV were also detectable by immunohistochemical (IHC) peroxidase staining with viral capsid antibodies in tissue sections. The IHC method was highly regarded at the time as a promising diagnostic test. It was seen as complementary to the morphological visualization of the koilocyte, which was thought to be the pathognomonic feature of papillomavirus infection [2], although we now know that other infections may cause koilocytosis. In the end neither microscopy nor immunostaining for HPV proteins was proven to be sufficiently sensitive and specific for clinical use and these approaches were unable to distinguish different HPV types that would subsequently be identified. It was nucleic acid technology that revolutionized HPV basic and applied research and produced today's essential tools for cancer prevention.

In the mid-1970s, Dr. Gerard Orth and colleagues isolated an HPV genome from plantar warts and roughly characterized it by low-resolution restriction enzyme mapping [3]. This led to the recognition of HPV genomes with different restriction maps and the realization that there may be a vast array of different HPV genotypes.

A pivotal nucleic acid-based technology, low stringency Southern blot hybridization (Fig. 7.1) used prominently in the 1980s revealed the true diversity of HPV genomes in human specimens. Most studies were done in women harboring different kinds of lesions including warty condylomas, flat lesions with different histopathological appearances, cervical and other anogenital preinvasive

Human Papillomavirus. DOI: https://doi.org/10.1016/B978-0-12-814457-2.00007-6

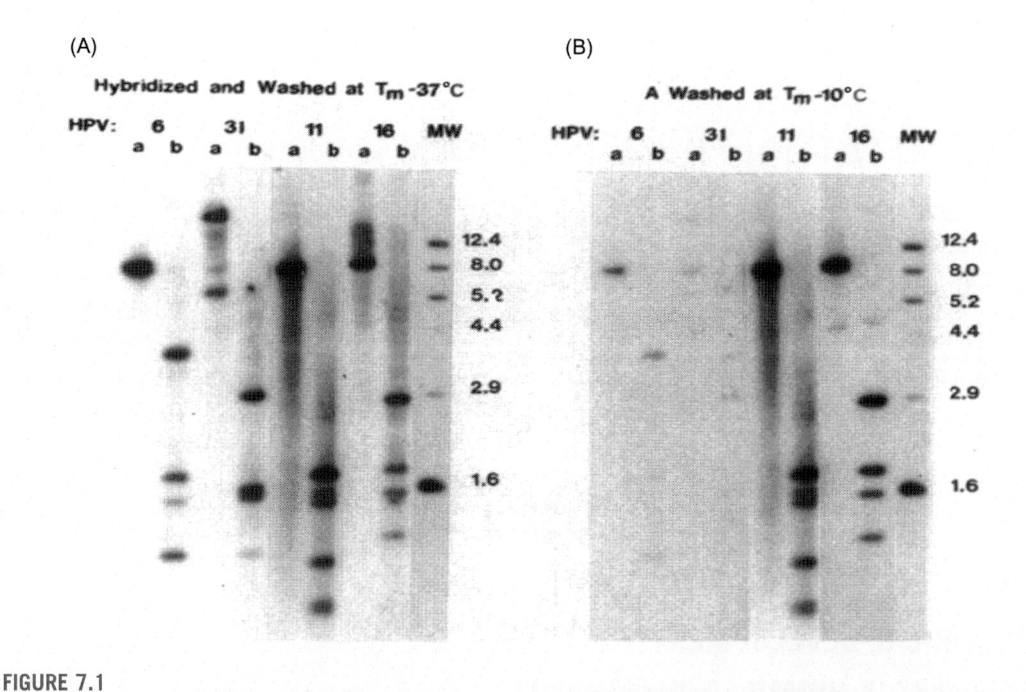

FIGURE 7.1

Photo of a low stringency (T_m: −37°C) and a high stringency (T_m: −10°C) Southern blot autoradiograph showing the restriction enzyme fragments of different HPV types and the molecular weight (MW) ladder.

and cancerous lesions and lesions from individuals with chronic genetic diseases, such as epidermodysplasia verruciformis. Southern blotting and, to a lesser extent, dot blot testing demonstrated that a significant and rich diversity of HPV DNAs were present across these various lesion types. Numerous papillomavirus genotypes were cloned from various tissues in order to codify and interrogate the varied primary HPV genome sequences [4].

CLONING OF HUMAN PAPILLOMAVIRUS

HPV genomes continue to be numbered according to chronology of discovery, starting with HPV1. The recognition of a new HPV type initially required that the cloned DNAs: (1) represent an entire individual genome; (2) be characterized by restriction enzyme mapping; (3) show <90% homology to any other previously characterized type(s); and (4) be submitted to a reference center for verification and issue of a new HPV number. The initial reference center was based with the Orth HPV team at Institute Pasteur in Paris France. An HPV reference center was then established in Heidelberg in 1985. The Heidelberg center served the papillomavirus community under the leadership of Dr. Ethel-Michele deVilliers until 2012. Currently the international HPV reference center is based at the Karolinska Institute in Uppsala, Sweden (https://ki.se/en/labmed/international-hpv-reference-center). In the earlier days of cloning and interrogation, HPV type 6, discovered by Dr. Lutz Gissmann of the Harald zur Hausen team, was of particular interest given its dominance in anogenital warts and respiratory papillomas [5]. Dr. Matthias Durst and the zur Hausen team then made a truly fundamental discovery in 1983 cloning and characterizing the first of two key HPV genomes from cervical carcinoma samples obtained in Germany and Brazil. They found that a majority of cervical cancer specimens had detectable DNA of HPV16 and to a lesser extent HPV18 [6,7].

For a time in the mid-to-late 1980s, there was a major emphasis on only HPV16 and HPV18, with some researchers believing that these two HPVs were responsible for all cervical cancers and that more sensitive methods would reveal that all cervical cancers were associated with these genotypes. However, low stringency Southern blotting of restriction enzyme digests using extracted DNA from a large series of geographically diverse cervical specimens demonstrated the existence of many other oncogenic HPV types beyond HPV16 and HPV18. These "novel" HPV genomes were observed in cancers, precancers and condylomas, as well as in morphologically normal anogenital tissues [8].

Soon after HPV16 and HPV18 were cloned, various researchers saw an opportunity to create diagnostic tests using DNA methodology. The new tests were imagined as supplementary or possibly replacements to cytological screening, which was in a sense a "secondary" test of HPV infection based on morphological interpretation of virus-driven cytopathology. Development of HPV diagnostic tests required the cloning and characterization of the multitude of HPV genomes. Thus there arose a fast-paced but amicable race between different research teams to identify and clone all remaining important anogenital HPV types. These endeavors led to the establishment of a panel of 13 HPVs representing the types most relevant to the risk of developing cervical cancer, which to this day remains the core set of high-risk HPVs (hrHPVs) that must be detected by all HPV DNA tests relevant for cervical screening and disease management [9,10] (Table 7.1). In summary, the comprehensive cloning of HPV types associated with cervical and other anogenital cancers was achieved predominantly by research teams in Germany (zur Hausen et al.), France (Orth et al.) and the United States (Lorincz et al.). Notable is the fact that although a complete set of hrHPV DNA genomes were available and fully characterized at the close of the 1980s, it took another 10 years for the broader clinical community to show interest in HPV DNA diagnostics and a full 20 years before a globally competitive commercial marketplace evolved.

FIRST WAVE OF HUMAN PAPILLOMAVIRUS DNA TESTS

Much of the early epidemiological research on HPV infection and disease-association was done with high stringency DNA hybridization methods, which allowed a more accurate determination of HPV types in different lesions. Hybridization studies preceded polymerase chain reaction (PCR) assays by almost 10 years [6–9]. In high stringency hybridization, each HPV type must be present via detection with a cocktail of probes or through individual probe–target detection. The method required purified DNA, which was then fragmented into characteristic patterns by various restriction enzymes, often *BamH1* and *Pst1*. The restricted DNA fragments were then separated according to their size by gel electrophoresis and transferred (blotted) onto nitrocellulose or nylon filter membranes for hybridization to radiolabeled probes. The advantage of Southern blotting was that the characteristic DNA banding patterns as well as the hybridization signals, combined to give very accurate information on HPV types, even in samples with multiple infections. However, while technically robust, the process was both labor intensive and low throughput (Fig. 7.1).

In dot blot methods, the pure or crude liberated DNA was blotted directly onto the filters. These methods were less demanding technically but did not give banding information and thus were more

Table 7.1 Brief Overview of a Selection of HPV Tests, Summarizing Their Technical Characteristics (HPVs Targeted, Concurrent Genotyping Capability, Technology, Target, and Internal Control), and Main Indicated Use

HPV Test *Registration*	HPVs Detected[a]	Concurrent Genotyping	Technology	Target	Internal Control: Endogenous/ Exogenous	Main Indicated Use
Hybrid Capture 2 *FDA*	13 hrHPVs	None	Signal amplification by hybrid capture	DNA	No/No	Screening triage after cytology test-of-cure
GP5 + /6 + -EIA *CE-IVD*	14 hrHPVs	None	Broad-spectrum PCR and probe hybridization	DNA, L1, ~150 bp	Yes/No	Screening triage after cytology test-of-cure
GP5 + /6 + LMNX (LQ Test)[b] *CE-IVD*	14 hrHPVs and 4 other HPVs	Full				
Abbott RealTime HR HPV *CE-IVD*	14 hrHPVs	HPV16 and 18	Broad−spectrum qPCR	DNA, L1, ~150 bp	Yes/No	Screening
APTIMA *FDA*	14 hrHPVs	None	Transcription-mediated amplification (TMA) and probe hybridization	mRNA, E6/ E7[c]	No/Yes	Screening (with cytology, Among >30 years) triage after cytology (among >21 years)
Onclarity *FDA*	14 hrHPVs	HPV16, 18, 31, 45, 51, and 52[d]	Three separate broad−spectrum qPCR	DNA, E6/ E7[d]	Yes/No	Screening triage after cytology
Cobas *FDA*	14 hrHPVs	HPV16 and 18	Broad−spectrum qPCR	DNA, L1, ~200 bp	Yes/No	Screening (with or without cytology) Triage after cytology
Xpert CE-IVD	14 hrHPVs	HPV16, 18, 18/45	Broad−spectrum qPCR	E6/E7 DNA	Yes/No	Screening (with cytology triage) cytology with reflex HPV (ASCUS Triage) cotesting
PGMY Linear Array *CE*	14 hrHPVs and 23 other HPVs	Full	Broad−spectrum PCR and probe hybridization	DNA, L1, ~450 bp	Yes/No	Not specified. Used in epidemiological and surveillance studies. Not clinically validated
SPF_{10} DEIA (version 1) *CE*	14 hrHPVs and at least 55 other HPVs	None	Broad−spectrum PCR and probe hybridization	DNA, L1, ~65 bp	No/No	Vaccine trials, surveillance and epidemiology
SPF10-DEIA-LiPA25 system (version1) *CE*	14 hrHPVs and 11 other HPVs	Full				

[a]*Unless otherwise specified, 13 hrHPVs refers to types 16, 18, 31, 33, 35, 39, 45, 51, 52, 56, 58, 59, and 68; 14 hrHPVs refers to the aforementioned 13 hrHPV types plus HPV66, which is now regarded as a probable low risk HPV type.*
[b]*The GP5 + /6 + EIA and LMNX Genotyping kit HPV GP (GP5 + /6 + LMNX; LBP) were previously marketed as the Digene Inc. HPV Genotyping LQ Test (LQ Test, Qiagen).*
[c]*Size of region(s) targeted by primers not specified in kit manual or literature.*
[d]*Onclarity detects additional hrHPVs as three groups, that is, 33/58, 56/59/66, and 35/39/68.*
CE, Conformité Européene/European Conformity*; DEIA,* DNA Enzyme ImmunoAssay*; IVD,* In vitro diagnostic*; LMNX,* Luminex*; mRNA,* messenger RNA*; PCR,* quantitative PCR.

(A)

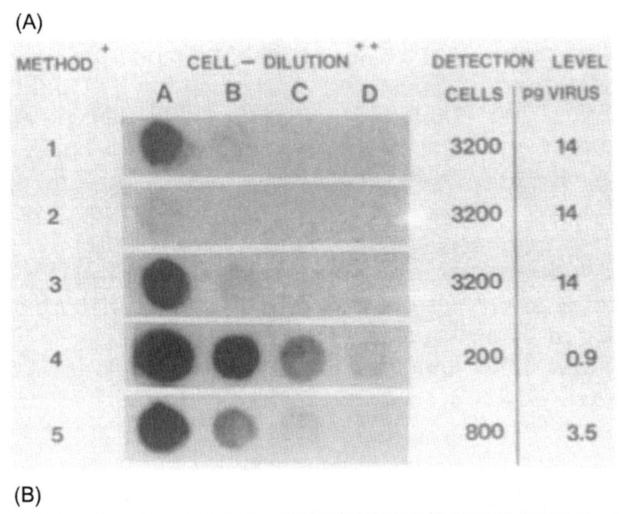

(B)

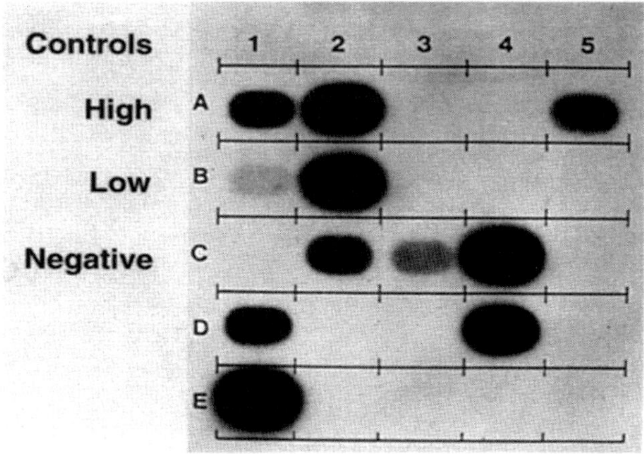

FIGURE 7.2

(A) Example of a research dot spot showing HPV16 detection in CaSki cells with different DNA denaturation methods. The numbers 1-5 correspond to the different denaturation methods. The CaSki cells were processed on 78-mm^2 spots. Numbers of CaSki cells spotted: A, 12,800; B, 3200; C, 800; D, 200 [11]. (B) Example of the ViraPap dot blot.

prone to false positives and low analytical specificity for individual HPV types, especially in samples associated with multiple infection (Fig. 7.2) [11].

Contemporaneously with filter hybridization methods, tissue in situ hybridization (ISH) was a popular technique, especially with pathologists [12]. In DNA ISH, thin sections (typically 3–5 μm) of biopsied tissues, or cell spreads deposited onto slides, were permeabilized to allow access of the cellular DNA to labeled probes. After hybridization of the probes and washing away of unbound DNA the labeled nuclei were visualized by either radiographic methods (e.g., silver deposition), by

HPV 16 E6/E7 mRNA expression in high-grade CIN and cervical carcinoma

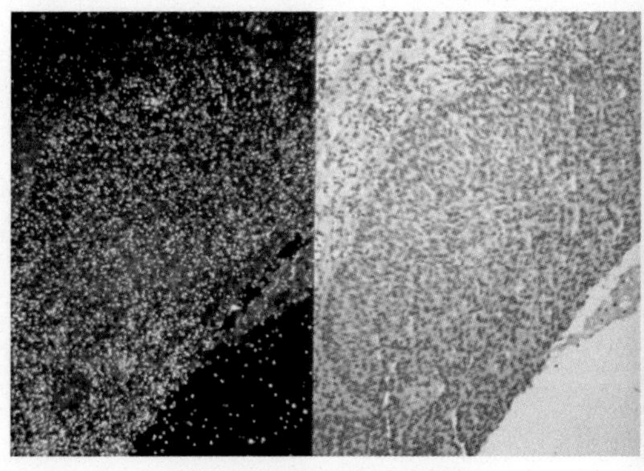

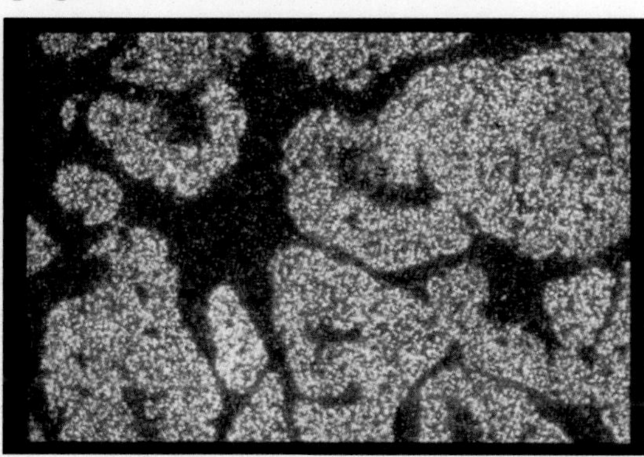

(A) HPV 16 positive CIN3 (B) HPV16 positive cervix carcinoma

FIGURE 7.3

Example of an ISH test. Nonradioactive RNA in-situ hybridization (RISH) staining (left panel) and H&E stain (right panel) using an HPV16-E7 antisense probe visualized by gold/silver enhancement and confocal laser scanning microscopy (CLSM) on CIN3 lesion (A) and cervix carcinoma (B) both HPV16 positive as determined by PCR, [13], showing that HPV16 is transcriptionally active.

enzymatic deposition of insoluble colored dye or by fluorescent microscopy, depending on the nature of the probes used (Fig. 7.3). While Southern and dot blotting are not generally used for diagnostic work, tissue ISH is still used as an adjunct to histopathology as it can reveal the cellular locations of target DNAs in relation to proteins in surrounding tissue and thus provide useful morphological information. For example, ISH demonstrated that HPV was transcriptionally active using RNA ISH (Fig. 7.3) [13].

SPECTRUM OF ANOGENITAL HUMAN PAPILLOMAVIRUS TYPES

Hybridization methods permitted the initial separation of the panoply of anogenital HPVs into three groups: (1) hrHPV types 16, 18, and 45; (2) intermediate risk HPV types 31, 33, 35, 39, 51, 52, 56, 58, 59, and 68; and (3) low-risk HPV types 6, 11, 42, 43, and 44 [9]. The initial three groups of HPV were soon collapsed into high-risk (hr) and low-risk (lr) groups: to allow for simpler algorithms. The definition of "high"- and "low"-risk HPV is helpful when considering assay design and communications to support HPV-based interventions. However, in a purist sense, these groupings are not absolute and natural history studies show that instead of simplistic HPV groups there is actually a continuum of distinct viral characteristics and risk distributions. Furthermore, the magnitude of risk of any given HPV genotype is affected by the end point used (i.e., whether CIN3 or cancer).

FIRST COMMERCIAL HUMAN PAPILLOMAVIRUS DNA TESTS

Southern blot methods were never commercialized for HPV testing. Only one dot blot hybridization method, ViraPap, was ever manufactured by Digene Inc., (now part of Qiagen GMBH, Hilden

Germany) with rather limited market success. This and a related genotyping dot blot test called ViraType were approved by the US Food and Drug Administration (FDA) for testing of female cervical clinical specimens in 1988. A limitation of ViraPap and ViraType was that they detected only five hrHPV types. While the tests were ground-breaking in their routine and standardized manufacture, their relative ease of use, and for being the first mass-marketed clinical tests for HPV DNA, they had some serious flaws. The combination of low-, intermediate-, and high-risk HPV types in the same cocktail, the use of short-lived radioactively labeled probes (P^{32}), and the incomplete set of hrHPV probes led to the discontinuation of these tests. ViraPap was soon replaced by Hybrid Capture 1 (Digene) and in 1996 by a much more sensitive microplate-based hybridization and signal amplification test, Hybrid Capture 2 (HC2; Digene). HC2, in particular had 13 unlabeled hrHPV RNA probes, without HPV6 and HPV11, and thus mitigated some failings of ViraPap.

Hybrid Capture 2 (HC2) is an exemplar of a successful signal amplification DNA test, both in terms of its market success and in the huge numbers of women screened worldwide. HC2 also lead the way for most other HPV DNA tests, for example, the probe cocktails and analytical cut-off limits of all current HPV tests with regulatory approvals are based on the HC2 endpoint design. Details of the HC2 test methodology were first published in 1996 [14]. HC2 was approved in 1999 by the US FDA for triage of cytology with the result of atypical squamous cells of unknown significance (ASC-US) and for general population screening in combination with Papanicolaou (Pap) cytology in women aged 30 years and older in 2003. In the early 1990s when HC2 was conceived and was being developed, there were many competing tests in development, including several PCR-based HPV assays.

Interestingly, HC2 was almost not advanced toward commercialization as there was a vigorous debate in the early 1990s at Digene Inc. on whether to opt for a target- or a signal-based amplification HPV test. Given the importance of the US and European markets, it was crucial to develop routinely applicable HPV test methods suitable for service labs and clearly PCR was technology that could achieve this. However, a strong headwind against PCR-based HPV assays was the fact that the royalties demanded by the owners of patents use of PCR in commercial applications were close to 20% (remarkably high for diagnostics). Thus the decision went in favor of HC2 where Digene Inc. held patents for HPV types that were a barrier to entry by other diagnostic companies. It is probable that without patents HPV DNA testing would not have evolved as it did due to the additional risks of the pioneer losing an expensive multiyear development effort to a big powerful competitor as soon as the market became interesting. The clear winner in the HPV diagnostics space over the longer term, was PCR; however, it took almost an additional two decades to commercialize and obtain US FDA approval of the first PCR-based HPV test.

Some of the critical decisions in the design of HC2 related to the 96-well microplate chemiluminescent signal format, which was quite novel at the time and that required the development of a special luminometer. Another critical decision was the HPV type composition of the 13 hrHPV RNA probe cocktail, which in the early 1990s, had not been defined by any consensus group and was based on a single large study [9]. The choice of the HC2 test cut-off was 100,000 HPV genomes per milliliter (mL) of original clinical specimen, also an important decision.

It was necessary to achieve an optimal combination of sensitivity and specificity as provided by the set of standard HC2 DNA positive controls equal to 1 pg of target per HC2 assay. HC2 was comprehensively validated in subsequent years and underpinned many large randomized controled trials (RCT) of HPV testing globally. Thousands of peer-reviewed papers employing HC2 have been published and tens of millions of women worldwide have been screened with this assay. HC2 remains in use today and indeed has been employed extensively as a gold-standard assay on which to compare the performance of new HPV tests as described in the well-established Meijer guidelines for HPV DNA test requirements for use in screening [15].

TARGET AMPLIFICATION FOR DETECTION OF HUMAN PAPILLOMAVIRUS DNA

Although PCR was invented in the mid-1980s it did not come into widespread use in HPV research until the early 1990s. The first set of broad-spectrum HPV PCR primers to achieve success in research were the molecularly degenerate MY09/MY011 primers developed by Dr. Michelle Manos and her team at Cetus, however, these proved difficult to make reproducibly [16]. Later, the availability of nearly 70 HPV genome sequences supported the PGMY09/11 primers targeting the same region as MY09/MY11 but as consensus primers, thus the manufacturing challenges were simplified [17]. At about the same time, Drs. Jan Walboomers, Chris Meijer, Adriaan van den Brule, and Peter Snijders developed the consensus GP5/GP6 primers [18] which proved to be very useful. Another set of established and useful primers are the SPF10 primers developed by Dr. Wim Quint and his team at DDL Diagnostic Laboratory [19].

The amplicon lengths that each of the above respective primer sets produce vary and this can make them more or less suited to particular applications; for example, the SPF10 PCR assay amplifies the smallest DNA target making it particularly useful for epidemiological studies where archived and diverse sample types can make amplification of fragmented DNA more challenging.

BROAD-SPECTRUM POLYMERASE CHAIN REACTION

Many different HPV primer systems were developed, with a number targeting the highly conserved L1 capsid gene region. The degree of heterogeneity of the viral genome enables two approaches for amplification of viral DNA by PCR primers, that is, broad-spectrum (degenerate or consensus) primers or type-specific (TS) primers. Broad-spectrum primers target relatively well-conserved genomic sequences (often L1), enabling the simultaneous amplification of a wide diversity of HPV types. A technical limitation of all broad-spectrum PCR-based assays is the underestimation of the prevalence of genotypes present in low concentrations within multiple infections, also known as masking. Broad-spectrum PCR primers do not necessarily have the same analytical sensitivity and specificity for each HPV genotype and amplification efficiency might differ among individual genotypes. Spiking experiments with plasmid mixtures of different HPV genotypes have shown that a competitive effect occurs in mixed infections when one genotype is present in a much lower concentration than another (PCR competition) [20].

TYPE-SPECIFIC POLYMERASE CHAIN REACTION

As opposed to broad-spectrum PCR, TS PCR primers target HPV sequences that are specific for a single genotype. These primers permit highly sensitive and specific identification of HPVs. Although TS PCRs can be designed to target any region in the HPV genome, many TS PCR assays amplify regions in the E6 or E7 open reading frame (ORF). These PCR target regions have unique coding sequences enabling greater ease of TS primer design and are not interrupted by viral integration, since over-expression of E6 and E7 is required for transforming HPV infections.

An important advantage of TS PCRs is that they are less prone to underestimate the prevalence of HPV types that have low viral loads in multiple infections. However, TS PCRs also have some limitations. Unknown variations in the primer target sequences could cause false-negative results. In addition, the requirement to deliver a separate TS PCR for each HPV potentially within a sample is highly laborious and would require a substantial quantity of sample. This issue can be mitigated by the development of multiplex type-specific (MPTS) PCRs, where multiple TS primer sets are combined in a single PCR reaction for simultaneous amplification of a defined set of HPVs.

DETECTION OF HUMAN PAPILLOMAVIRUS mRNA

The detection of hrHPV mRNA encoding viral oncoproteins E6 and E7 instead of DNA has the possibility of offering better distinction between productive (low expression of E6/E7) and transforming (high expression of E6/E7) infections. Overexpression of E6 and E7 is required for malignant transformation in HPV-related cancers. To this end, commercial assays targeting the E6 and E7 genome regions have been developed, largely based on transcription-mediated amplification (TMA) involving initial reverse-transcription of DNA [21].

POLYMERASE CHAIN REACTION READ-OUT METHODS

Amplification products can be detected by hybridization to a mixture of probes in a DNA enzyme immuno-assay (DEIA/EIA) in a microtiter well plate. In addition, reverse hybridization (RH) can be performed by separate genotype-specific probes immobilized on carriers such as nylon membrane, nitrocellulose strips, microsphere beads, or DNA chips. Read-out is typically achieved through labeled or tagged probes or targets and enzyme- or fluorochrome detection and generally performed in a real-time format as opposed to a conventional end-point PCR—particularly for high throughput tests. Read-out systems of PCR products can be designed to recognize a range of HPV types simultaneously (detection), individually (full genotyping), or as a combination of detection and genotyping (partial genotyping).

DESIGN AND PERFORMANCE ASSESSMENT OF HUMAN PAPILLOMAVIRUS TESTS

INTENDED USE

The design of an HPV DNA test should match its intended use. HPV tests designed for epidemiologic purposes, for example, disease association studies, vaccine monitoring and efficacy trials, and surveillance of HPV prevalence, require a high analytical sensitivity and specificity and the capacity to individually identify a range of hrHPVs in diverse bio-specimen types, including archival specimens. In contrast, within the clinical setting, HPV tests should detect hrHPV infections mainly associated with clinically meaningful disease, that is, cervical intraepithelial neoplasia grade 2 or worse (CIN2 + , clinical sensitivity), limiting the detection of transient HPV infections not associated with CIN2 + (clinical specificity) [15].

In addition to primary cervical cancer screening, two other main, evidence-based indications for hrHPV testing have been identified: (1) "test-of-cure" in women treated for preinvasive cervical lesions, and (2) as a triage test for women with equivocal and low-grade cytology; alternatively designated by The Bethesda System terminology as ASC-US and low-grade squamous intraepithelial lesions (LSIL), respectively. In both situations, the HPV test is used to inform decisions on referral to colposcopy or more conservative management. All these indications take advantage of the high sensitivity and negative predictive value of HPV tests

CLINICAL SPECIMEN, COLLECTION, AND PROCESSING

A range of clinical specimens have been used for HPV analysis, for example, cervical cells collected by lavage and various brushes and swabs that were stored in liquid-based medium or cervicovaginal specimens self-collected by lavage, swab, brush and urine [22]. It is important to stress that the yield and quality of the specimen for HPV testing is influenced by the methodology used for each individual aspect, including collection, storage, and processing. Thus all elements of the preanalytical and analytical process determine the outcome of the HPV test. Proper control and validation of the whole process is essential, particularly if the test is to be used for cervical screening or diagnosis to inform patient management.

TARGET SELECTION AND AMPLIFICATION

Most commercial and in-house tests that have been developed for HPV detection and genotyping are based on PCR amplification of HPV DNA. These assays target one or more regions within the HPV genome that are sufficient for accurate recognition of one or more previously designated HPV genotypes. Misrecognition or inaccurate classification, due to the presence of unclassified HPV types or genetic variants of designated HPV genotypes, is inherently possible when using any HPV DNA assay. The regions in the viral genome commonly targeted by a selection of HPV DNA PCR-based assays are shown in Fig. 7.4.

HUMAN PAPILLOMAVIRUS DNA DETECTION TESTS WITH PARTIAL AND FULL SPECTRUM GENOTYPING

Increasingly HPV tests offer concurrent identification of a limited number of individual HPVs in addition to a group or multiple groups of pooled hrHPVs (designated as partial genotyping). Partial genotyping has largely been restricted to HPV16 and HPV18, with the remaining hrHPVs detected as a single group (pool). The rationale for this approach is based on definitive evidence that HPV16 and HPV18 (particularly HPV16) together account for the largest attributable fraction of cervical cancer worldwide [9,10].

HPV tests with full genotyping capacity distinguish all individual hrHPV genotypes and some of these assays also offer detection of several lrHPV genotypes. These assays can address the epidemiological burden of HPV infections and the efficacy and real-world impact of vaccines in diverse population settings. In addition, full genotyping can be useful when assessing new sampling devices

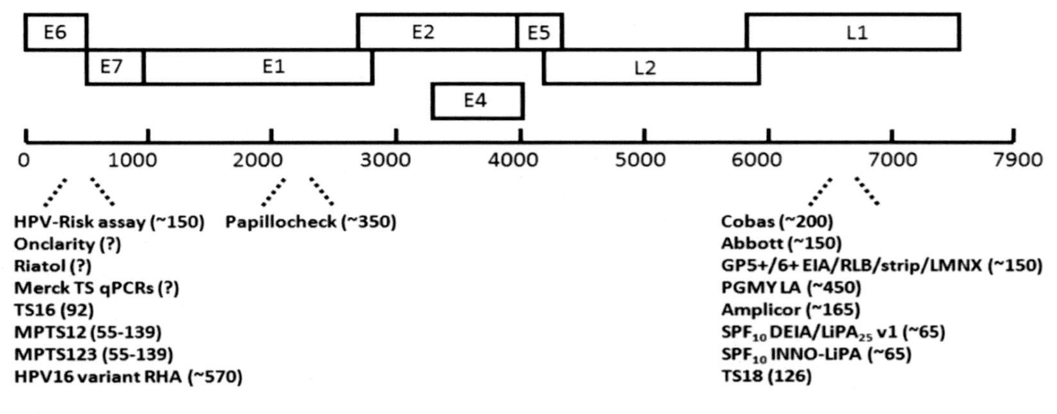

FIGURE 7.4

Schematic representation of primer target regions for selected type-specific (TS) and broad-spectrum DNA-based PCR assays. The approximate fragment lengths (in nucleotides) are shown in parenthesis. The HPV16 genome is shown as the map reference [23].

and specimen collection media; however, current evidence does not support a clinical utility for full genotyping.

PERFORMANCE ASSESSMENT OF HPV TESTS AND REQUIREMENTS FOR ONGOING QUALITY ASSURANCE

ANALYTICAL VERSUS CLINICAL PERFORMANCE

As described above, there are now a multitude of HPV assay-approaches which vary in their intrinsic chemistry and their range of HPV genotype detection. This variety of HPV tests offers more choice to suit desired applications whether that be research, epidemiology or screening and clinical services. Irrespective of the application, any HPV assay should deliver high technical and analytical performance. The nonclinical performance assessment of HPV tests includes analytical sensitivity (limit-of-detection), analytical specificity (interference), and precision and accuracy (comparison of methods). The analytical accuracy of a novel HPV test can be evaluated by comparison with an established reference or "gold standard" as the comparator test, using panels of quantitated artificial samples, for example, plasmids cloned with HPV target sequences, and/or clinical specimens for which the HPV test was designed.

However, an analytically validated HPV test should not be used in clinical practice without explicit prior clinical validation, since analytical and clinical accuracy are nonsynonymous. The clinical relevance of a novel hrHPV test for primary cervical cancer screening is best supported by data generated through longitudinal RCT. Realistically RCT are not feasible for most new commercial entrants and instead sets of representative clinical samples with relevant endpoints (most importantly CIN2 + versus normal/CIN1) from clinical studies can support robust validation as can validation-metrics where the candidate test is compared to the performance of a gold standard previously validated in the context of a RCT. International guidelines for panel composition and criteria for clinical accuracy and test reproducibility have been formulated by Dr. Chris Meijer and colleagues [15].

PROFICIENCY PANELS

Laboratories performing hrHPV testing must comply with a raft of quality assurance (QA) process measures, such as participation in accredited external quality assessment schemes, in-house validation, and verification. External accreditation for clinical molecular testing [e.g., by the College of American Pathologists (CAP) and the United Kingdom Accreditation Service (UKAS)] necessitates that a laboratory complies with several quality process measures including internal verification of individual assay results and participation in external QA (EQA) schemes. There are a number of HPV EQA schemes available as described in a recent review which vary in their scope, frequency and scoring remit (i.e., the minority score on TS performance) [24].

INTERNAL/ENDOGENOUS CONTROL

Many HPV tests have been designed with an internal control (IC) to aid laboratories in assessing specimen quality and/or monitoring of sample processing to mitigate against false-negative HPV results. A variety of commercially available tests have primers that can amplify a conserved fragment of the human genome extracted from cells present in the collected cervical specimen. An IC usually comprises a spiked exogenous nucleotide sequence or an endogenous target (e.g., the beta-globin housekeeping gene) that is simultaneously extracted and amplified with HPV DNA.

If a specimen does not allow amplification of the target HPV DNA and the IC, this could indicate inhibition of amplification or failure of DNA extraction. A positive endogenous human target indicates presence of human cells and is therefore often interpreted as an accurately collected cervical specimen. However, we should be careful in relying on this endogenous IC for the following reasons: (1) An IC for a human target can be positive due to presence of any type of human cells (e.g., inflammatory or irrelevant human skin cells), which may then provide a false reassurance that cervical epithelial cells from the squamo-columnar junction have been sampled. (2) There is little correlation between the HPV viral load and the number of cells in a given cervical swab; thus a numerical correlate of adequacy for HPV testing does not exist. (3) The internal assay control may have an amplification advantage and thus may not accurately test for inhibition of HPV amplification. Certain assays incorporate an exogenous IC template at a fixed and relatively low concentration to monitor inhibition, although this does not address the issue of acellular samples. Notably, inhibition of amplification and/or extraction failure in cervical samples has become a rare event with contemporary DNA extraction methods. Furthermore, it could be argued that while it is not unfeasible that a sample which contains no sample/cells reaches a laboratory, the frequency of this is very rare and this is emphasized by the longitudinal data on HC2 which does not incorporate an IC and which nonetheless shows high cross sectional and longitudinal sensitivity for high-grade cervical disease.

While helpful to an extent, arguably the endogenous IC for human DNA is an imperfect measure of sample/sampling adequacy. True "false" negative rate/test performance is more comprehensively derived from longitudinal assessment of assay performance relative to disease.

BRIEF SUMMARY OF SELECTED COMMERCIAL HUMAN PAPILLOMAVIRUS TESTS

There are more than 190 HPV tests that have been described. However, some HPV tests have played essential roles in the process of adoption of hrHPV testing in cervical cancer screening and in establishing the current HPV diagnostic landscape, these tests deserve special mention. A more comprehensive listing of the multitude of HPV tests can be found in a recent review [23].

THE HYBRID CAPTURE 2 HUMAN PAPILLOMAVIRUS DNA TEST

The Hybrid Capture 2 HPV DNA Test (HC2; Qiagen, Hilden, Germany) has been the most frequently used HPV test in the world. HC2 uses liquid-based chemiluminescent signal amplification of hybridized target DNA for the simultaneous detection of 13 hrHPVs [14]. The assay has no IC for a human DNA target. The RNA probes can cross-hybridize with several other lrHPVs [25]. HC2 is intended for (1) primary cervical cancer screening, (2) triage of women with equivocal cytology results, and (3) test-of-cure following treatment. HC2 has been evaluated in many longitudinal randomized, controlled studies and is therefore considered a reference test for validation of novel hrHPV assays [15].

GP5 + /6 + SYSTEM

The GP5 + /6 + -PCR-EIA was originally developed as an in-house test by the Walboomers and Meijer team and has now become commercially available (EIA kit GP HR; LBP, Rijswijk, The Netherlands). The GP5 + /6 + primers amplify a region of approximately 150 bp in the highly conserved HPV L1 ORF. Amplification of a broad-spectrum of HPV genotypes using only two primers is achieved by a relatively low annealing temperature. A cocktail of probes specific for 13 hrHPVs and HPV66 (now regarded as probably low risk) hybridizes with the GP5 + /6 + amplification products in an EIA format, providing a qualitative result [26]. Similarly to HC2, the GP5 + /6 + -PCR-EIA has been evaluated in many longitudinal, randomized controlled studies and is therefore considered a reference assay for HPV testing in cervical cancer screening [15].

The LMNX Genotyping Kit HPV GP (GP5 + /6 + LMNX, LBP); previously marketed as the Digene HPV Genotyping LQ Test, provides identification of the same 14 HPVs, but read-out is performed using a bead-based xMAP technology on a Luminex platform. This platform is more suitable for high-throughput testing, enabling the GP5 + /6 + LMNX to be used as a reflex genotyping test and also as a stand-alone test for hrHPV detection and concurrent genotyping, if desired [27].

SPF10/DEIA/LIPA$_{25}$ SYSTEM

The SPF10-PCR-DEIA (Labo Bio-medical Products, Rijswijk, The Netherlands) was developed around 18 years ago and is usually combined in a test algorithm with the LiPA25 reverse hybridization strip version 1 (Labo Bio-medical Products) [28]. This algorithm was designed to have high analytical sensitivity and specificity. The SPF10 primers are non-degenerate and amplify a 65-bp fragment in the L1 ORF of a broad-spectrum of at least 69 mucosal and cutaneous HPV types. Qualitative detection of HPV is performed in a DEIA with conserved, universal HPV probes. A control target for the housekeeping gene beta-globin can be separately amplified and detected. The short region of only 65 bp targeted by SPF10 primers is particularly sensitive for amplification in FFPE cervical biopsy specimens, in which DNA is often poorly preserved. The sensitivity of the SPF10-PCR-DEIA is ideal for epidemiological research but is too high for application in clinical settings [29]. This assay may eventually find clinical utility in vaccinated women as a test of HPV presumptive suppression or clearance.

The LiPA25 version 1 (Labo Bio-medical Products) is used for identification of 25 individual HPV genotypes by reverse hybridization to SPF10 amplimers with genotype-specific probes immobilized on a reverse hybridization strip. In a combined test algorithm, SPF10 amplimers that are positive by the DEIA (Labo Biomedical Products) can be used directly for LiPA25, eliminating the need for a separate PCR reaction [19].

THE LINEAR ARRAY HUMAN PAPILLOMAVIRUS GENOTYPING SYSTEM

The LINEAR ARRAY HPV Genotyping Test (LA; Roche Diagnostics) is a reverse hybridization strip test with immobilized probes enabling the recognition of 37 anogenital HPVs and an IC target (beta-globin). Consensus PGMY primers are used for broad-spectrum amplification of a viral DNA fragment of ~450 bp in the L1 ORF [30]. The PGMY primer system was first reported as the Roche Line Blot Assay (LBA) that detected 27 distinct HPV genotypes [31]. Later the LBA assay was extended to detect 38 individual HPV genotypes [31]. Manufacturing of the LA HPV genotyping test is planned for discontinuation in 2020.

THE COBAS 4800 HUMAN PAPILLOMAVIRUS TEST (ROCHE COBAS)

The Cobas test (Roche Diagnostics) enables the qualitative, simultaneous detection of 14 HPVs using qPCR amplification of a region in L1. This test provides individual genotyping of HPV16 and HPV18, if desired, and distinction of a pool of non-16/18 or "other" HPVs. A beta-globin IC target is included in the same PCR reaction. In recent years, its clinical value has been supported by the ATHENA trial [32].

THE APTIMA HUMAN PAPILLOMAVIRUS ASSAY

The APTIMA HPV assay (APTIMA; Hologic GenProbe, San Diego, CA, USA) is an HPV assay designed for pooled detection of E6/E7 mRNA from 14 HPVs. APTIMA is based on target capture after cell lysis, with subsequent TMA and probe hybridization for detection of E6/E7 mRNA expression of 14 HPV genotypes. APTIMA met the cross-sectional clinical and reproducibility criteria of the international guidelines for HPV test requirements for cervical screening [15,21]. An

APTIMA genotyping assay for TS detection of HPV16 and for combined detection of HPV18 and HPV45 received approval from the US FDA in 2012 for application in women with APTIMA HPV-positive test results.

THE BD ONCLARITY HUMAN PAPILLOMAVIRUS ASSAY

The BD Onclarity HPV Assay (Onclarity; Becton Dickinson, Sparks, MD, USA) is an E6/E7-based qPCR, enabling detection of 14 HPVs in three separate reactions. Concurrent individual genotyping is currently offered for six hrHPV types: 16, 18, 31, 45, 51, and 52, while the remaining HPVs are detected as three separate groups (i.e., HPV33/58, HPV56/59/66, and HPV35/39/68). The test also detects human beta-globin as an endogenous IC. The Onclarity test has been clinically validated according to international guidelines as a primary screen and as a triage of low-grade abnormalities. The test was approved for use limited to SurePath cytology media by the US FDA in 2018.

THE XPERT HUMAN PAPILLOMAVIRUS ASSAY

The Xpert HPV test is a cartridge-based assay, with the cartridge acting as a self-contained unit for extraction and amplification. It has been clinically validated for primary screening and triage of low-grade abnormalities [33]. While levels of automation do exist in this test it is different from the before-mentioned clinically validated assays given that it is less reliant on batch testing and has capabilities as a "point-of-care" test, which may be of use to address specific healthcare delivery needs including delivery of care in rural settings as well as in low- and middle-income countries, where the required infrastructure for large service laboratories is more challenging.

FUTURE PERSPECTIVES

VACCINOLOGY

Future trials of licensed and next-generation HPV vaccines for which efficacy is demonstrated via protection against high-grade lesions associated with vaccine- and nonvaccine-targeted genotypes are supported by the application of highly sensitive genotyping tests. TS PCR-based technology has been used in efficacy studies of the quadrivalent and bivalent vaccines, as a stand-alone assay or used in a combined testing algorithm with a sensitive broad-spectrum PCR. A higher analytical sensitivity was achieved by using TS PCR and broad-spectrum PCR separately under a defined algorithm, supporting the advantages of combining both approaches [20]. A direct analytical comparison of the quantitative TS PCR assays used for the quadrivalent vaccine versus the combined algorithm of TS and broad-spectrum PCR used for the bivalent vaccine has not been performed. A thorough insight into their TS analytical characteristics, particularly in multiple infections, would be informative to determine which HPV test systems should be used in future vaccine trials.

HUMAN PAPILLOMAVIRUS SURVEILLANCE

Knowledge on TS HPV prevalence pre- and post-vaccination is required for surveillance of effects on HPV genotype diversity in nations that have implemented a vaccination program. In addition to the surveillance of CIN2 + prevalence associated with a diverse spectrum of HPVs, tests with high analytical sensitivity should be used for the longitudinal monitoring of HPV TS prevalence to estimate decreases in specific HPV genotypes and any unanticipated increases, the latter potentially resulting from genotype replacement or unmasking. When vaccinated cohorts reach the age of the first screening round, the effect of vaccination on type-specific prevalence may be monitored in "real-time" within organized screening programs. This will be more straightforward to perform in countries that have or will implement primary HPV-based screening and in programs where the HPV test used for screening has genotyping capability.

HUMAN PAPILLOMAVIRUS-BASED SCREENING

HPV-based screening is more sensitive than cytology in detecting CIN2 + and a negative HPV test provides greater and longer reassurance against CIN2 + (see Chapter 17: Infection to Cancer—Finding Useful Biomarkers for Predicting Risk of Progression to Cancer). Primary hrHPV screening is at least as effective as cytology when comparing the same screening intervals. Furthermore, HPV testing is more reproducible, can be used in self-sampling strategies and can help reduce disparities by increasing screening uptake among unscreened or underscreened women. Importantly, the performance (sensitivity, specificity positive and negative predictive values) of primary HPV screening is expected to be less affected than cytology in countries where cohorts of vaccinated women are now moving into screening. Primary HPV screening has been implemented in several settings, but adoption of HPV testing as a primary screening modality has encountered some resistance. Central arguments against HPV-based screening have focused on its lower specificity and the potential harms such as increased colposcopy and treatment for nonneoplastic lesions, the lack of well-defined and evaluated strategies to manage hrHPV-positive women, and inadequate information to define appropriate screening intervals for women who are hrHPV-negative. It is likely that as more data emerges from trials and population-based real-world data where multiple rounds of screening have been done that greater support for primary HPV screening will be achieved. Another key aspect of HPV screening implementation is triage of hrHPV-positive women, which is covered briefly below.

CERVICAL SCREENING IN VACCINATED WOMEN

While the prophylactic HPV vaccines have been shown to exert massive impact on infection and disease at the population level, not all women will be comprehensively protected from cervical cancer given the first HPV vaccines do not cover all hrHPV types and many women, particularly older women and those in low- and middle-income countries will not have been vaccinated. A form of screening is therefore required to take into account increasingly mixed populations of vaccinated and unvaccinated women. As vaccinated young women reach the age of screening, the decrease in cervical lesion prevalence will adversely impact HPV test performance, particularly the predictive values. Evaluation of the impact of prophylactic HPV vaccination on cervical precursor lesions

(CIN2 +) will help inform the optimal age of screening initiation and whether screening intervals can be extended. These changes might help retain cost-effectiveness of HPV-based screening of vaccinated populations, but the optimal strategy remains to be determined. In theory, around 70% of all cervical cancer cases could be prevented by the licensed bivalent and quadrivalent vaccines and this proportion may increase to around 80% through cross-protection and 90% or more with the nonavalent vaccine.

FULL MOLECULAR-BASED SCREENING

In active hrHPV-based primary screening programs cytology and/or limited genotyping have been implemented for the triage of hrHPV-positive women. However, as opposed to cytology, objective biomarkers, including those which incorporate methylation targets have the potential to reduce unnecessary referrals through specific detection of invasive cancers and advanced transforming CIN with a greater short-term risk of progression [34]. Biomarker tests also have the advantage of being unaffected by the vagaries of subjective interpretation (which cytology relies on) and are also amenable to self-collected cervicovaginal samples. These qualities support the aspiration towards an "objective," fully molecular-based screening strategy of hrHPV testing and biomarker triage. This is a realistic trajectory for the next phase in cervical cancer screening programs.

NEXT GENERATION SEQUENCING

The evolution and application of next generation sequencing (NGS) has moved exponentially in the last 5 years. Although NGS is relatively expensive compared to standard PCR, there are many strong advantages to NGS, and NGS platforms are becoming an increasing feature of service laboratories. On a per-nucleotide basis, NGS costs less than any existing HPV DNA test; however, NGS produces a huge amount of data, whereas the region that needs to be sequenced to definitively identify and distinguish HPV genotypes by PCR is very small, therefore for routine testing/typing NGS may be overengineered.

Additionally, the massive resolving power of NGS make it reconcilable to the future of laboratory diagnostics which will demand a more holistic depiction of risk potentially using various informative aspects of the microbial and host genome. In time, the information provided by qualitative HPV testing may be insufficient for a definitive determination of patient risk. Furthermore, NGS instrumentation, consumables, and bioinformatics pipelines are becoming less expensive and more robust. We may expect that in a relatively short span of time, NGS will become competitive with existing HPV DNA tests. Moreover, NGS may be suitable for elucidating the genotype–phenotype relationships between HPV genotypes or variants and clinical outcomes, such as cervical cancer. The identification of specific nucleotides or combinations of nucleotides associated with disease could provide insight into the molecular basis of HPV-associated malignancies. Elucidating the genetic basis of pathogenesis and of virus–host interactions might help reveal the reasons for differences in carcinogenic potential between closely related HPV genotypes and intratypic variant lineages within alpha species 5, 6, 7, 9, and 11.

OVERVIEW AND SUMMARY

An ability to routinely test for HPV was one of the longer and more complex diagnostic development paths spanning 50 years. It was the work product of many different investigators and required numerous seminal discoveries, one of which deserved the Nobel Prize in Medicine. Initial studies of the virus by microscopy and immunochemistry led to better performing but more complicated DNA tests. Almost 70 different HPV genomes were cloned before we arrived at the final set of HPV probes for cervical cancer screening. DNA tests based on radioactively labeled probes led to colorimetric assays, to signal amplified tests and finally to target amplified tests. However the acceptance and translation of HPV technologies has been long and complex; we must strive to be more efficient in the future. Despite the sunk costs, energy and resources expended on existing HPV tests, they cannot preclude the rapidly approaching next paradigm shift to total molecular "risk profiling" of HPV-infected individuals. Cervical cancer is essentially a fully preventable disease and yet it has not been reduced worldwide. Proper and enthusiastic implementation as well as integration of screening and vaccination has great potential. The time has come to grasp the real solutions which lie in the political will to solve the problem and to get moving.

REFERENCES

[1] Strauss MJ, Shaw EW, et al. Crystalline virus-like particles from skin papillomas characterized by intranuclear inclusion bodies. Proc Soc Exp Biol Med. 1949;72(1):46–50.

[2] Kurman RJ, Shah KH, Lancaster WD, Jenson AB. Immunoperoxidase localization of papillomavirus antigens in cervical dysplasia and vulvar condylomas. Am J Obstet Gynecol 1981;140(8):931–5.

[3] Favre M, Orth G, Croissant O, Yaniv M. Human papillomavirus DNA: physical map. Proc Natl Acad Sci USA 1975;72(12):4810–14.

[4] Bernard HU, Burk RD, Chen Z, van Doorslaer K, zur Hausen H, de Villiers EM. Classification of papillomaviruses (PVs) based on 189 PV types and proposal of taxonomic amendments. Virology 2010;401 (1):70–9.

[5] Gissmann L, deVilliers EM, zur Hausen H. Analysis of human genital warts (condylomata acuminata) and other genital tumors for human papillomavirus type 6 DNA. Int J Cancer 1982;29(2):143–6.

[6] Durst M, Gissmann L, Ikenberg H, zur Hausen H. A papillomavirus DNA from a cervical carcinoma and its prevalence in cancer biopsy samples from different geographic regions. Proc Natl Acad Sci USA 1983;80(12):3812–15.

[7] Boshart M, Gissmann L, Ikenberg H, Kleinheinz A, Scheurlen W, zur Hausen H. A new type of papillomavirus DNA, its presence in genital cancer biopsies and in cell lines derived from cervical cancer. EMBO J 1984;3(5):1151–7.

[8] Lorincz AT, Lancaster WD, Temple GF. Cloning and characterization of the DNA of a new human papillomavirus from a woman with dysplasia of the uterine cervix. J Virol 1986;58(1):225–9.

[9] Lorincz AT, Reid R, Jenson AB, Greenberg MD, Lancaster W, Kurman RJ. Human papillomavirus infection of the cervix: relative risk associations of 15 common anogenital types. Obstet Gynecol 1992;79(3):328–37.

[10] Munoz N, Bosch FX, de Sanjose S, Herrero R, Castellsague X, Shah KV, et al. Epidemiologic classification of human papillomavirus types associated with cervical cancer. N Engl J Med 2003;348(6):518–27.

[11] Melchers WJ, Herbrink P, Walboomers JM, Meijer CJ, vd Drift H, Lindeman J, et al. Optimization of human papillomavirus genotype detection in cervical scrapes by a modified filter in situ hybridization test. J Clin Microbiol 1989;27(1):106–10.

[12] Walboomers JM, Melchers WJ, Mullink H, Meijer CJ, Struyk A, Quint WG, et al. Sensitivity of in situ detection with biotinylated probes of human papilloma virus type 16 DNA in frozen tissue sections of squamous cell carcinomas of the cervix. Am J Pathol 1988;131(3):587–94.

[13] van den Brule AJ, Cromme FV, Snijders PJ, Smit L, Oudejans CB, Baak JP, et al. Nonradioactive RNA in situ hybridization detection of human papillomavirus 16-E7 transcripts in squamous cell carcinomas of the uterine cervix using confocal laser scan microscopy. Am J Pathol 1991;139(5):1037–45.

[14] Lorincz AT. Hybrid Capture method for detection of human papillomavirus DNA in clinical specimens: a tool for clinical management of equivocal Pap smears and for population screening. J Obstet Gynaecol Res 1996;22(6):629–36.

[15] Meijer CJ, Berkhof J, Castle PE, Hesselink AT, Franco EL, Ronco G, et al. Guidelines for human papillomavirus DNA test requirements for primary cervical cancer screening in women 30 years and older. Int J Cancer 2009;124(3):516–20.

[16] Manos M, Lee K, Greer C, Waldman J, Kiviat N, Holmes K, et al. Looking for human papillomavirus type 16 by PCR. Lancet 1990;335(8691):734.

[17] Gravitt PE, Peyton CL, Alessi TQ, Wheeler CM, Coutlee F, Hildesheim A, et al. Improved amplification of genital human papillomaviruses. J Clin Microbiol 2000;38(1):357–61.

[18] van den Brule AJ, Snijders PJ, Raaphorst PM, Schrijnemakers HF, Delius H, Gissmann L, et al. General primer polymerase chain reaction in combination with sequence analysis for identification of potentially novel human papillomavirus genotypes in cervical lesions. J Clin Microbiol 1992;30(7):1716–21.

[19] Kleter B, van Doorn LJ, ter Schegget J, Schrauwen L, van Krimpen K, Burger M, et al. Novel short-fragment PCR assay for highly sensitive broad-spectrum detection of anogenital human papillomaviruses. Am J Pathol 1998;153(6):1731–9.

[20] van Doorn LJ, Molijn A, Kleter B, Quint W, Colau B. Highly effective detection of human papillomavirus 16 and 18 DNA by a testing algorithm combining broad-spectrum and type-specific PCR. J Clin Microbiol 2006;44(9):3292–8.

[21] Heideman DA, Hesselink AT, van Kemenade FJ, Iftner T, Berkhof J, Topal F, et al. The Aptima HPV assay fulfills the cross-sectional clinical and reproducibility criteria of international guidelines for human papillomavirus test requirements for cervical screening. J Clin Microbiol 2013;51(11):3653–7.

[22] Arbyn M, Verdoodt F, Snijders PJ, Verhoef VM, Suonio E, Dillner L, et al. Accuracy of human papillomavirus testing on self-collected versus clinician-collected samples: a meta-analysis. Lancet Oncol 2014;15(2):172–83.

[23] Poljak M, Kocjan BJ, Ostrbenk A, Seme K. Commercially available molecular tests for human papillomaviruses (HPV): 2015 update. J Clin Virol 2016;76(Suppl 1):S3–13.

[24] Carozzi FM, Del Mistro A, Cuschieri K, Frayle H, Sani C, Burroni E. HPV testing for primary cervical screening: laboratory issues and evolving requirements for robust quality assurance. J Clin Virol 2016;76(Suppl 1):S22–8.

[25] Poljak M, Marin IJ, Seme K, Vince A. Hybrid Capture II HPV test detects at least 15 human papillomavirus genotypes not included in its current high-risk probe cocktail. J Clin Virol 2002;25(Suppl 3): S89–97.

[26] Snijders PJ, van den Brule AJ, Jacobs MV, Pol RP, Meijer CJ. HPV DNA detection and typing in cervical scrapes. Methods Mol Med 2005;119:101–14.

[27] Geraets DT, Cuschieri K, de Koning MN, van Doorn LJ, Snijders PJ, Meijer CJ, et al. Clinical evaluation of a GP5 + /6 + -based luminex assay having full high-risk human papillomavirus genotyping capability and an internal control. J Clin Microbiol 2014;52(11):3996–4002.

[28] Kleter B, van Doorn LJ, Schrauwen L, Molijn A, Sastrowijoto S, ter Schegget J, et al. Development and clinical evaluation of a highly sensitive PCR-reverse hybridization line probe assay for detection and identification of anogenital human papillomavirus. J Clin Microbiol 1999;37(8):2508–17.
[29] Hesselink AT, van Ham MA, Heideman DA, Groothuismink ZM, Rozendaal L, Berkhof J, et al. Comparison of GP5 + /6 + -PCR and SPF10-line blot assays for detection of high-risk human papillomavirus in samples from women with normal cytology results who develop grade 3 cervical intraepithelial neoplasia. J Clin Microbiol 2008;46(10):3215–21.
[30] Coutlee F, Rouleau D, Petignat P, Ghattas G, Kornegay JR, Schlag P, et al. Enhanced detection and typing of human papillomavirus (HPV) DNA in anogenital samples with PGMY primers and the Linear array HPV genotyping test. J Clin Microbiol 2006;44(6):1998–2006.
[31] Peyton CL, Gravitt PE, Hunt WC, Hundley RS, Zhao M, Apple RJ, et al. Determinants of genital human papillomavirus detection in a US population. J Infect Dis 2001;183(11):1554–64.
[32] Wright TC, Stoler MH, Behrens CM, Sharma A, Zhang G, Wright TL. Primary cervical cancer screening with human papillomavirus: end of study results from the ATHENA study using HPV as the first-line screening test. Gynecol Oncol 2015;136(2):189–97.
[33] Castle PE, Smith KM, Davis TE, Schmeler KM, Ferris DG, Savage AH, et al. Reliability of the Xpert HPV assay to detect high-risk human papillomavirus DNA in a colposcopy referral population. Am J Clin Pathol 2015;143(1):126–33.
[34] Verhoef VM, Bosgraaf RP, van Kemenade FJ, Rozendaal L, Heideman DA, Hesselink AT, et al. Triage by methylation-marker testing versus cytology in women who test HPV-positive on self-collected cervicovaginal specimens (PROHTECT-3): a randomised controlled non-inferiority trial. Lancet Oncol 2014;15(3):315–22.

CHAPTER

8 PROVING THE CAUSAL ROLE OF HUMAN PAPILLOMAVIRUS IN CERVICAL CANCER: A TALE OF MULTIDISCIPLINARY SCIENCE*

Eduardo L. Franco[1] and F. Xavier Bosch[2]

[1]McGill University, Montreal, QC, Canada [2]Catalan institute of Oncology (ICO), Barcelona, Spain

AVOIDING A CLASH BETWEEN DIFFERENT LINES OF SCIENTIFIC EVIDENCE

The mood in the room was somber yet you could feel the excitement. The brisk late November air outside of the plenary room at the Institute of Hygiene and Epidemiology in Brussels was not in the mind of the dozens of people[1] inside because the business at hand required a lot of concentration. The International Agency for Research on Cancer (IARC), the main cancer research agency of the World Health Organization, had convened a meeting to examine the state of the science on human papillomavirus (HPV) infection and cervical cancer by summoning to Belgium a representative group of scientists from the various domains related to epidemiologic research on HPV and cervical cancer. The meeting had been convened by Nubia Muñoz, at the time the lead scientist at IARC's Unit of Field and Intervention Studies. It was the second time that the IARC was trying to build consensus on this topic.[2] In addition to FXB, the coconveners in 1991 were Keerti Shah, a professor of molecular pathology at Johns Hopkins University in Baltimore, and André Meheus, a policy expert on sexually transmitted infections based at the World Health Organization in Geneva, at the time. The Brussels workshop brought together epidemiologists, virologists, pathologists, molecular biologists, clinicians, and sexually transmitted disease experts. Throughout the three-and-a-half-day workshop, ELF and FXB sat next to each other on the inner bench in the ring-shaped room. It was the first time we met. FXB was at the IARC and ELF was at the Université du

*This chapter is dedicated to our dear colleague Nubia Muñoz, whose intellect, boundless energy, and vision have inspired legions of public health scientists since 1970.

[1]A. Schneider, A. VandenBrule, A. Singer, A. Meheus, A. Lorincz, C. Critchlow, C. Lacey, C. Meijer, D. Galloway, E. Guerrero, E. Franco, E-M de Villiers, F. Judson, H. zur Hausen, J. Walboomers, J. Daling, K. Sherman, K. Shah, L. Koutsky, L. Brinton, M. Schiffman, M. Manos, N. Aristizabal, N. Kiviat, N. Day, N. Muñoz, P. Kenemans, P. Alonso de Ruiz, P. Gravitt, P. Snijders, R. Barrasso, R. Kurman, S. de Sanjosé, S. Kjaer, T. Helmerhorst, V. Beral, X. Castellsagué, F.X. Bosch.

[2]It had been done in March 1988 in a meeting convened by Nubia Muñoz, FXB, and O.M. Jensen in Copenhagen, when the science was less clearcut; the uncertainty captured in the last sentence of a commentary by Nubia Munoz, Xavier Bosch, and John Kaldor as "...*while experimental data suggest an oncogenic potential for HPV, the epidemiological evidence implicating it as a cause of cervical neoplasia is still rather limited.*" [1]

Human Papillomavirus. DOI: https://doi.org/10.1016/B978-0-12-814457-2.00008-8

Québec then. We all had scheduled presentations of our findings. The overall goal was to see if there was consensus within the community for the notion that HPV infection caused cervical cancer. The solemnity of the room and its circular seating arrangement was perfect for encouraging scholarly debate. There was plenty of it to fill the duration of the workshop and the evenings outside of the venue.

For those of us in Brussels it felt like the culmination of a long-drawn pursuit for the elusive sexually transmitted cause of cervical cancer. It had been known since the 1950s that cervical cancer risk was consistently higher among women with multiple sexual partners [2]. Likewise, it had been known since the late 1970s that wives of men who developed penile cancer were at higher risk of developing cervical cancer than those married to men who never developed penile cancer [3]. We and others had also carried out research that showed that the incidence of cervical and penile cancers were strongly correlated across countries or regions in the same country, much like pairs of cancers that originate from the same cause, for example, tobacco-associated lung and laryngeal cancers [4–6]. With this compelling knowledge base, Richard Doll and Richard Peto had stated prophetically in 1981 that "*the present evidence strongly suggests that one of the primary causes of the disease is an agent passed between partners in intercourse, quite possibly a virus. If this is indeed so, it may eventually be possible to protect against the disease by immunization.*" [7] Undeniably, the epidemiologic evidence was overwhelming for a sexually transmitted disease model for cervical cancer.

The nonepidemiologists in the room held a generally favorable view and felt ready to declare the causal relationship with HPV infection as established. Some even felt genuinely annoyed with the delay in taking action.[3] Not quite as sparring opponents, but with a distinct sobriety in looking at the evidence, were the epidemiologists in the Brussels meeting. They were not a block but had varying degrees of skepticism. After all, they had been burned before, with the attention given to so many microorganisms as putative causal agents for cervical cancer.[4] The science did not pan out for any of them, despite the advocacy of those doing research on these agents. There was also the intellectual challenge of countering the lucid arguments of Leopold Koss, a towering cytopathology authority who urged the community to consider that "*the activation of the virus and its ability to interact with cervical epithelium is likely to be due to "patient factors" rather than the presence of the virus per se*" [9].[5]

---

[3]Harald zur Hausen spoke with confidence that the community was taking much too long to consider the voluminous evidence placing HPV as the culprit. Chris Meijer and Jan Walboomers even had a well-defined plan for dramatically changing cervical cancer screening; they believed that testing for HPV would be better in finding precancerous cervical lesions than with Papanicolaou (Pap) cytology, the 40-year-old paradigm at the time.

[4]The list was long for infectious agents: Herpes simplex virus 2, *Chlamydia trachomatis, Neisseria gonorrhoeae, Trichomonas vaginalis, Treponema pallidum*, and *Cytomegalovirus*. Even a protein in sperm cells had at one point been considered a putative carcinogen [8].

[5]Koss coined the term koilocytosis in 1956 to denote abnormal epithelial cells with a perinuclear halo in cervical precancerous lesions [10]. Later, koilocytosis became the main telltale sign that a cell was infected with HPV. Koss writings tended to acknowledge that cervical precancerous lesions and cancer were associated with HPV but generally downplayed the importance of this virus' etiologic role [9,11,12]. His defiance of the HPV hypothesis continued well into the late 1990s [13].

Despite the tension, the Brussels workshop reached more sanguine conclusions than the previous one held by the IARC in Copenhagen in 1988. The science on HPV and cervical cancer was not all coherent but at least we were beginning to understand where the facts disagreed with expectation. The molecular biologists and virologists were telling the epidemiologists that they were confident this time. A causal framework was emerging, propelled by Harald zur Hausen's and his colleagues' decade-long work [14], to the effect that human genital cancer would occur as the result of deficient intracellular control of persisting HPV genomes in infected cells. Strong evidence had just been obtained in the laboratories of Peter Howley and Lou Laimins, in the United States, and that of Lionel Crawford, in the United Kingdom, to the effect that HPV oncogenes, such as E6 and E7, had the ability to interfere with the function of the *TP53* and Retinoblastoma (*Rb*) tumor suppressor genes [15–18]. The momentum was thus clearly strong in favor of HPV as a causative agent of cervical cancer.

Before the HPV era, the most likely etiologic candidate had been the Herpes simplex virus type 2 (HSV-2). The epidemiologic evidence was more or less supportive of a role for HSV-2 as the cause of cervical cancer, as the patients affected by this disease tended to have higher titers of HSV-2 antibodies than comparable women of the same age who did not have the disease. Owing to the work of André Nahmias, Bernard Roizman, William Rawls, and Irving Kessler beginning in the late 1960s, the HSV-2 etiological hypothesis had gained traction for more than a decade [19–21]. In fact, gynecology textbooks of the 1970s and early 1980s tended to refer to cervical cancer as a disease etiologically linked to HSV-2 infection. However, the HSV-2 hypothesis was losing steam. A prospective epidemiologic study by Vladimir Vonka initiated in 1973 and published in 1984 did not support an association between antibodies to HSV-2 and subsequent occurrence of cervical cancer or its precancerous[6] lesions [22]. Even the Nahmias–Roizman team stopped upholding their long-held HSV-cervical cancer hypothesis [23]. But old ideas die hard; Irving Kessler held on to his belief "*of the role of the herpes genitalis virus as a necessary, though not sufficient, causal factor*" well into the late 1980s [24].

LEARNING FROM ERRORS

That there was no point insisting on linking HSV-2 and cervical cancer was clear to cancer epidemiologists at the time. The association was mostly a consequence of the fact that both HSV-2 infection and cervical cancer have their roots in sexual transmission, albeit cervical cancer is a much more downstream consequence of sexual behavior than HSV-2. Likewise, that the mechanistic and biological evidence supporting a role for HPV was strong was also not in question; it was in fact overwhelming. Therefore why were epidemiologists timid in fully embracing the association as a

[6]Throughout this chapter, this term is used to denote the different stages (more commonly denoted as grades) of dysplastic, i.e., malignant growth still confined within the cervical epithelium. These lesions have been designated as cervical intraepithelial neoplasia (grades 1, 2, and 3 for mild, moderate, and severe dysplasia, respectively) or more recently as squamous intraepithelial lesions (low or high grade). Once a high-grade lesion traverses the basement membrane that serves as boundary between the cervical epithelium and the adjacent connective tissue, the lesion is then considered invasive, that is, cancer.

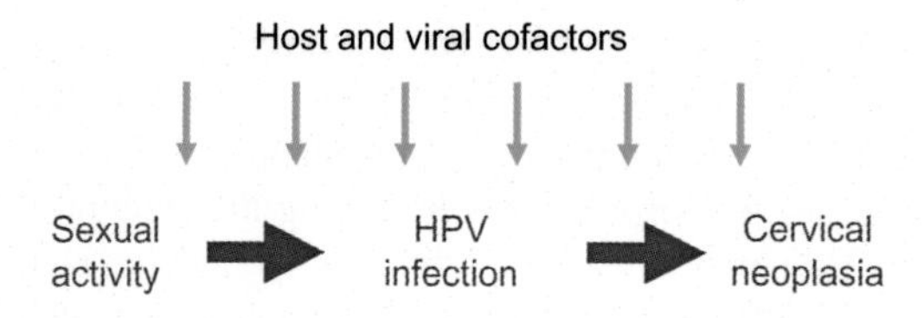

FIGURE 8.1

Schematic of the sexually transmitted disease model that describes the causal relations that lead to cervical cancer. From the onset of sexual activity leading to acquisition of HPV infection and then to development of cervical cancer two or more decades may elapse in women who develop this disease. The model is an oversimplification. It omits the precancerous steps that are amenable to be detected via screening with Papanicolaou (Pap) cytology and molecular HPV testing, and eventually treated. The arrows indicate causal relations that can be measured statistically in epidemiologic studies. The extent of progression is affected by host and viral cofactors, such as smoking, prolonged oral contraceptive use, parity, nutrition, immunosuppression, other sexually transmitted infections such as HSV-2 and *Chlamydia*, HPV type and intratypic variant. Preventive strategies, such as condom use, HPV vaccination, and screening followed by ablative or excisional treatment can arrest disease development at different steps in the model.

true etiological mechanism? To a large extent, the mindset of the community around the time of the Brussels meeting was one of puzzlement. To understand why, we must examine the epidemiologic findings that had just been published in light of the etiological model for cervical cancer shown in Fig. 8.1. It portrays cervical cancer as the logical end result of two prior steps: one remote or distal, the woman's sexual activity that eventually places her at risk of acquiring cervical HPV infection from one or more sexual partners, which is then the intermediate or proximal step, which may lead to cervical cancer after a certain period of time. No step is guaranteed to happen following a previous one. What the model's arrows denote are causal relations explained by a higher risk of the subsequent step if the condition in the prior step is acquired. The ancillary arrows stemming from host and viral cofactors indicate that the underlying risk relations in the main causal pathway can be affected by additional variables.

Molecular epidemiologic studies were launched in the mid-1980s trying to address the causal relations in the model in Fig. 8.1, but the findings were not exactly what we expected. Independent studies led by Susanne Kjaer in Denmark and Greenland, Ethel-Michele de Villiers in Germany, William Reeves and Louise Brinton in multiple Latin American countries (Mexico, Costa Rica, Panama, and Colombia), and Luisa Villa and ELF in different regions in Brazil (São Paulo and Recife), all came to surprising conclusions, albeit perfectly coherent with each other's findings [25–28]. While the Kjaer and the Villa-Franco cross-sectional studies examined the link between the first two steps of the etiological model in Fig. 8.1, the Reeves–Brinton case-control investigation looked at the link between the second and last steps. The de Villiers study also used a cross-sectional study design to examine the link between HPV and cancer. These teams all used epidemiological study designs that permitted measuring the strength and magnitude of the statistical associations that underlie the links between the steps above. These studies included thousands of women each and employed the same state-of-the-art molecular technique to detect the presence of HPV in exfoliated cervical samples. This technique, developed in Germany [29] was remarkably simple and easy to deploy in large-scale investigations. It became known as filter in situ hybridization (FISH).[7]

What were the surprising conclusions from the above studies? They largely failed to support the validity of the causal relations in Fig. 8.1, first by not finding that HPV infection risk increased appreciably with the lifetime number of sexual partners reported by the female participants, and second, by finding at most weak associations between HPV and cervical cancer. Alas, the culmination of the Reeves–Brinton investigation, published in the prestigious *New England Journal of Medicine* in 1989, indicated that the underlying relative risk estimate[8] for the association between HPVs 16 and 18[9] in combination and cervical cancer risk was a paltry 2.5.[10] In other words, assuming that the causal relation is not in dispute, HPV 16/18 infection seemed to merely increase risk of this disease by 150%. It was an anti-climactic conclusion for what was then the largest epidemiologic investigation of cervical cancer ever conducted, involving massive resources in four different countries. How was it possible that the all-too powerful carcinogenic role ascribed by molecular biologists to HPV could be translated in real terms to such a low level of risk?[11]

Also disturbingly, the remote relation between sexual activity and cervical cancer did not seem to be mediated by HPV infection, which further challenged the validity of model in Fig. 8.1. This state of affairs was further compounded by the remarkably different prevalence rates of cervical HPV infection in populations with more or less the same incidence of cervical cancer. Further aggravating the confusion was the advent of the polymerase chain reaction (PCR) era in the late 1980s. This technological breakthrough permitted target amplification of HPV sequences, thus greatly increasing our ability to detect HPV in biological samples. The first study that used PCR to detect HPV type 16 in asymptomatic women [30] had been prompted by earlier claims that the FISH technique was not sufficiently sensitive to detect HPV DNA. It found that 84% of women with normal cervical cytology were positive for HPV16,[12] which contrasted with the range of 2%–40%, up until then obtained with FISH in similar populations. Since PCR is much more sensitive

---

[7]Not to be confused with fluorescence in situ hybridization, a technique used in cytogenetics to locate specific DNA sequences in chromosomes. Before FISH, Southern blot was the gold standard to detect HPV DNA in exfoliated cells or tumor samples. However, Southern blot was laborious and not amenable to large-scale use in epidemiologic investigations; hence, the optimism about FISH among HPV epidemiologists in the mid-1980s.

[8]Epidemiologists refer generically to the "relative risk" (RR) as the numerical value that indicates the magnitude of risk associated with an exposure under investigation. It denotes a multiple of the risk in relation to the unexposed category. In a cohort study, the relative risk is estimated via the risk ratio (also conveniently abbreviated as RR). In case-control studies, such as those mentioned in this chapter, one estimates the relative risk via a different calculation called the odds ratio. Ultimately, assuming that a causal relation is real, they measure the same theoretical quantity, that is, the magnitude of risk for a particular exposure in relation to the unexposed state. The RR has an inherent knowledge translation meaning. For instance, epidemiological studies have determined that ever smoking cigarettes is associated with a 10- to 20-fold increase in risk of lung cancer relative to never smoking. This is a powerful message that is central to the mission of tobacco control.

[9]The two genotypes (i.e., strains) of HPV that had been implicated as of highest carcinogenic potential.

[10]Calculated from table 1 in the paper.

[11]A 2.5-fold or, equivalently, a 150% increase in risk may seem high to those not accustomed to seeing the strong associations that have become the mainstay of cancer prevention. Contrast this level of risk with that described above for smoking in lung cancer: 10- to 20-fold, which represents increases in risk of 900%–1900%.

[12]Tidy et al. concluded that "*Epidemiological studies on an etiological role for HPV16 in cervical neoplasia may now have to be altered to take account of the very high rate of infection in the general population.*" [30]

than FISH[13] the apparent universal ubiquity of HPV16 was cause for concern. How can a virus that is apparently present in the vast majority of women's cervices be the cause of cancer? That uncertainty played well in the hands of those who wished to look for other putative causes of cervical cancer and felt that the pursuit of the HPV hypothesis was destined to failure.

A NEW BEGINNING FOR MOLECULAR EPIDEMIOLOGY

Such was the embarrassing state of affairs for the epidemiology of cervical cancer in the late 1980s. Luckily, the community quickly regrouped and learned from its mistakes [31,32], the first of which was to have relied on the dubious validity of FISH to serve as the technological mainstay of complex molecular epidemiologic investigations in different continents that ended up costing many millions of dollars. FISH lacked sufficient sensitivity and specificity to provide reliable results in epidemiologic studies but was also prone to subjective interpretation.

The aforementioned initial adoption of the more sensitive PCR technique did not solve the problem but showed how easy it was to get blindsided by the problems with a technology that multiplies a target DNA sequence by making millions of copies of it. PCR requires separating pre and postamplification areas in the laboratory and adopting stringent safeguards to avoid contaminating clinical samples with aerosolized amplified products that remain from a previous set of assays.[14] Contamination leads to false-positive results, which adversely affect the specificity of the assay.

Fully cognizant of the importance of avoiding the pitfalls of measurement error in HPV assays the epidemiologic community turned the page on the earlier incoherent findings and thus the full extent of the causal relations in Fig. 8.1 could be validated. The 1990s began with a string of robust studies that dotted all the i's and crossed all the t's in fitting HPV squarely as the central piece in the sexually transmitted disease model for cervical cancer [34,36,37]. Nubia Muñoz, FXB, and their associates completed their case-control study of cervical cancer in Colombia and Spain in 1992 by showing for the first time that relative risks for the association between HPV and cervical cancer was in the double digits [36]. From that point on, progress came at a fast pace with no misgivings; case-control and cohort studies using properly executed laboratory techniques based on PCR demonstrated that infection by certain genotypes of HPV is one the strongest cancer risk factors ever found in cancer prevention [38].

By 2003, the IARC case-control study of HPV and cervical cancer, led by Nubia Muñoz and FXB had been completed and appeared on the pages of the same *New England Journal of Medicine* that published the results of the Reeves–Brinton study in 1989. However, this time relative risks reached the upper triple digits [39], thus making the HPV–cervical cancer association the strongest ever recorded in cancer prevention science. How it evolved in about 15 years from a weak association with RRs in low single digits to one in the high triple digits is a story that shows

[13]FISH can detect between 1000 and 10,000 HPV-infected cells, each with its own variable viral load, whereas PCR can detect as low as 10 molecules.

[14]In fact, the initial study of Tidy et al. [30] was retracted with the authors' candid admission that sample contamination by HPV16 DNA from a previous PCR assay run was what caused the near universal ubiquity of HPV16 that they reported for clinical samples [33]. Tidy et al.'s honesty was very much appreciated. Subsequent uses of PCR assays for HPV became increasingly better at avoiding misclassification of HPV status in cervical and other samples.

the self-correcting nature of science [40]. As the technological experience for detecting HPV infection improved from the late 1980s to the mid-2000s the validity and full extent of the model relations shown in Fig. 8.1 became axiomatic, undisputed scientific truths. We hasten to add that the earlier studies using FISH or early implementation of PCR were not wrong; they reflected the proper application of scientific methods that were available at the time they were conducted and represented "state-of-the-art" methodology. The knowledge translation from that remarkable journey of discovery by learning from past mistakes is worth being told to students of epidemiology and public health as a demonstration of the value of working in a multidisciplinary environment to avert or minimize the damaging effects of measurement error.

THE BIRTH OF A NEW PARADIGM IN CANCER PREVENTION

As mentioned above, the 1990s brought coherence and fast progress on HPV and cervical cancer research. The 1991 Brussels meeting served as a stepping stone for a series of well-orchestrated policy and technical meetings that greatly accelerated progress towards preventing cervical cancer via HPV-based technologies. This progress came in two fronts: as new research knowledge on HPV and in policies or high-level decisions of universal impact on cervical cancer control. These two fronts are shown in Fig. 8.2 as the annual frequency of scientific papers on the topic of HPV infection overlaid with the policy or historical milestones. The various domains of public health research and policy milestones are shown in Fig. 8.3 as a comparative timeline.

How did it all start? Two independent teams, that of Alexander Meisels and Roger Fortin in Canada (Quebec), and Esko Purola and Eeva Savia in Finland [41,42] suggested in 1976 and 1977, respectively, that viral alterations consistent with condylomatous atypia[15] were the underlying characteristic of cervical precancerous lesions. As early as in 1974–76, through the auspices of her IARC unit, Nubia Muñoz conducted surveys in Brazil, Colombia, and Uganda to collect cervical cancer specimens. At the time, the most promising microbial candidates were HSV-2, cytomegalovirus, and HPV [44]. Intrigued by observations she made jointly with Adonis de Carvalho, a pathologist in Recife, Brazil, about the common cooccurrence of giant condylomas and cervical cancer, she collected tissue samples and sent them to Gerard Orth in Paris and Harald zur Hausen in Germany to look for HPV and HSV-2. At that time Orth was the authority on HPV and zur Hausen on HSV-2 and Epstein-Barr virus. Orth analyzed Muñoz's specimens with electron microscopy, the only technique available at that time to identify HPV and found a few viral particles only in the condylomas, as expected. zur Hausen could not detect HSV-2 DNA in the cervical cancer specimens. Since the results were negative she did not publish the findings, but a summary of this exploratory work appeared in the IARC Annual Reports of that period.

The ability to have an adequate supply of biological samples from high-risk areas greatly assisted the discoveries that ensued in zur Hausen's laboratory with the collaboration of Lutz Gissmann and Mathias Dürst. Using DNA hybridization techniques, they were able to demonstrate the genomic heterogeneity of HPVs and find DNA sequences of different HPV types in cervical cancer specimens [45,46]. zur Hausen's Heidelberg lab became a hub for training scientists on the

[15]See note about koilocytosis, the term coined by Leopold Koss. These two papers, published a few months apart in Acta Cytologica, are considered by the current editor, Dr. Kari Syrjanen, to be "*the two most influential studies ever published in this journal*" [43].

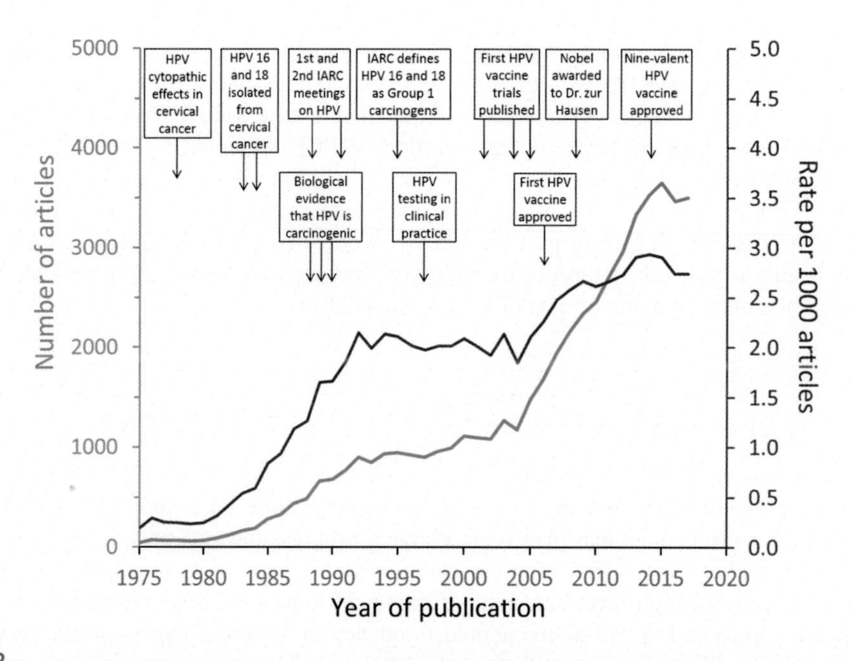

FIGURE 8.2

Double plot of the number of annual articles on HPV research curated in the National Library of Medicine's PubMed database. Blue curve and left *Y* axis: actual number of articles. Red curve and right *Y* axis: rate of HPV articles per 1000 papers. The latter value corrects for the secular growth trend in the biomedical literature. While there is a clear trend for an absolute increase in research output on HPV, this increase is real and not a consequence of the volume of biomedical research alone. The rate of HPV papers relative to the entire biomedical research output increased in real terms from about 0.25 per 1000 before 1980 to over 2 per 1000 between 1992 and 2004. Following the publication of the succession of vaccination trials the rate increased to the present level of nearly 3 per 1000 (over 3000 papers per year today from less than 70 per year in the 1970s). Historical scientific and policy milestones are shown (arrows point to the year).

hybridization techniques and as a source of the reference HPV types to be used in myriad studies that were published subsequently. By the mid-1980s it was already known that HPV types 16 and 18 were dominant findings in cervical cancer (Fig. 8.3).

From about 50 papers a year in the mid-1970s the research output on HPV increased by more than 10-fold in 1990, the crucial tipping point when the momentum clearly shifted towards the HPV hypothesis. The IARC meetings in Copenhagen (1988) and Brussels (1991) meetings, and their associated IARC publications [35,47], paved the way for the 1995 IARC HPV monograph meeting[16], which unequivocally established HPVs 16 and 18 as definite carcinogens and HPVs 31

[16]These meetings are multidisciplinary and bring experts in all technological domains of candidate physical, chemical, or biological agents (or their circumstantial exposures) to sift through the entire scientific literature on the topic at hand. At the end of 2 weeks the panel experts reach conclusions with respect to classifying an agent (or its mixtures or circumstances) as frankly carcinogenic (group 1), probably carcinogenic (group 2A), possibly carcinogenic (group 2B), not classifiable (group 3), or probably not carcinogenic (group 4).

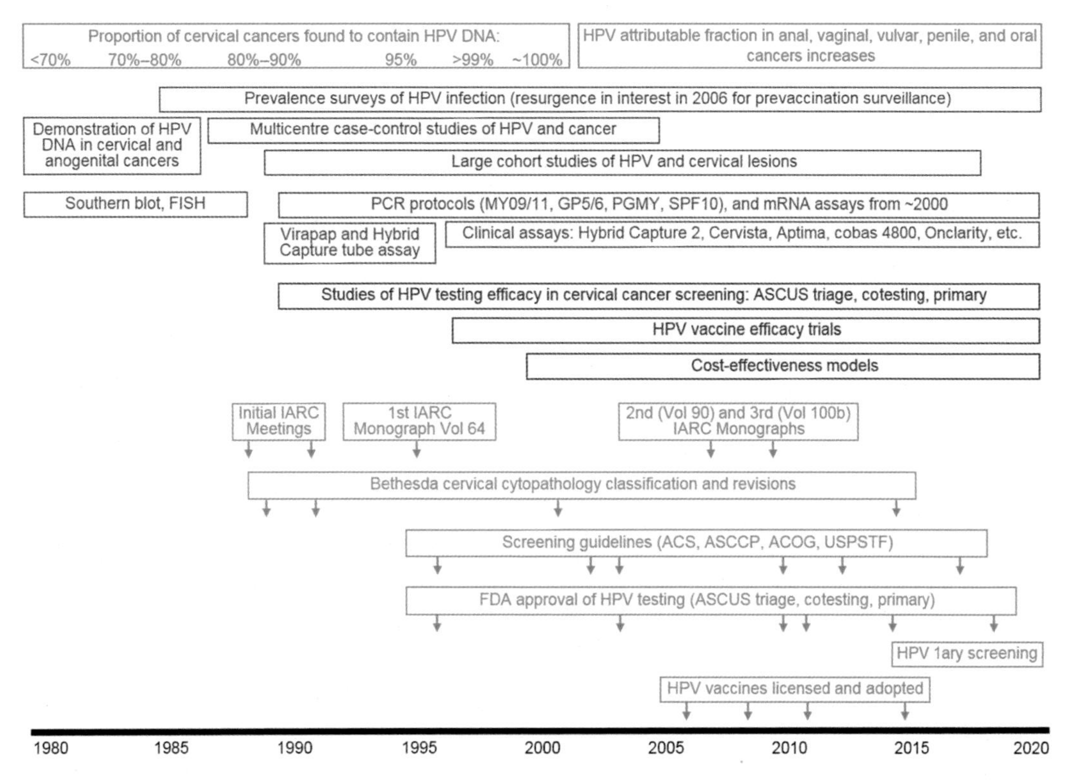

FIGURE 8.3

Forty-year timeline of public health research and policy on HPV and cervical cancer. The domains of research are shown in different colors. Prevailing diagnostic methods in different eras are shown in brown below the periods in which different types of molecular epidemiologic studies were conducted (purple). As diagnostic methods evolved and the epidemiologic knowledge base improved, the fraction of cervical cancers that were shown to contain HPV DNA (green, top) increased gradually until 1999 when HPV became widely accepted as the necessary cause of cervical cancer. The attention regarding the HPV etiologic fraction is now focused on other HPV-associated cancers. Prevention studies and related health economics research are shown in red. Important policy implementation, guidelines, or consensus milestones are shown in blue with the corresponding arrows indicating the year of adoption. *ACS*, American Cancer Society; *ASCCP*, American Society for Colposcopy and Cervical Pathology; *ACOG*, American College of Obstetrics and Gynecology; *ASCUS*, atypical squamous cells of undetermined significance; *FISH*, filter in situ hybridization; *HPV*, human papillomavirus; *IARC*, International Agency for Research on Cancer; *PCR*, polymerase chain reaction; *USPSTF*, US Preventive Services Task Force.

and 33 as probable carcinogens. It also concluded that other HPV types were possible carcinogens [48]. By 1999, the fraction of cervical cancers known to contain traces of HPV DNA had increased to just under 100% [49] (Fig. 8.3). By then there was widespread acceptance to the notion that cervical cancer was the first human cancer proven to have a necessary cause, that is, HPV infection. Proving absolutes in the biomedical sciences is a virtual impossibility. There are limitations to what can be demonstrated at the boundaries of epidemiologic scrutiny [50]. However, for all practical purposes, the proof that nearly all cervical cancers begin via a carcinogenic route triggered by HPV infection is good enough as a scientific rationale to compel immediate public health action.

MOVING FROM UNDERSTANDING ETIOLOGY TO IMPROVING CERVICAL CANCER SCREENING

While technical discussions about whether or not HPV was a cause of cervical cancer (and others) proceeded in the literature and in conferences everywhere, an important niche opened up for a commercial application of HPV diagnostic tests. The reason was the confusion created among cytopathologists by the adoption in 1988 of the Bethesda system for reporting cervical cytological diagnoses. In an attempt to incorporate the new knowledge about the possible role of HPV in cervical carcinogenesis, the Bethesda system created a new terminology to accommodate into the original category of mild dysplasia the cytopathic effects caused by HPV infection, such as koilocytosis. It also created a new equivocal category, called atypical squamous cells of undetermined significance (ASCUS). By the early 1990s ASCUS smears became a dominant finding among cervical abnormalities in the US and some other Western countries that had adopted the Bethesda classification or modified versions thereof. Without clear guidelines and mindful of risking malpractice litigation, gynecologists faced much uncertainty on what to do with patients with ASCUS diagnoses. For the sake of patient safety, clinicians tended to refer them for immediate colposcopy and biopsy, with the attendant risks, costs, and patient anxiety that such diagnostic procedures entailed. Much of the clinical research that prevailed in the 1990s was thus geared to examining the value of HPV testing as a triage strategy to decide which women with ASCUS results needed immediate referral to colposcopy.

Luckily, the real clinical value of HPV diagnostic assays as a better technology for cervical cancer screening began to be realized by the late 1990s. Research on HPV testing[17] to help doctors decide which women with ASCUS needed to undergo colposcopy was more or less a distraction that prevented due attention to what HPV testing does well, that is, the ability to detect cervical lesions on screening with much greater sensitivity than cytology. The latter is not particularly sensitive but is very specific.[18] By 2002, screening guidelines began to incorporate HPV testing in parallel with cytology (called cotesting). It was not until 2014 that HPV testing with partial genotyping for HPVs 16/18 was approved as a stand-alone screening test, leaving cytology to play the triage

[17]Validation of commercially available HPV assays for clinical use is a complicated process. Clinically validated tests are different than the ones alluded to earlier in this chapter. Ideally, a useful HPV test that can attain good performance on screening must be calibrated to detect the levels of HPV viral load in exfoliated cervical samples that are consistent with the presence of lesions. The test must not detect minute quantities of HPV that represent incipient infections or mere deposition from recent sexual activity. A clinically useful assay must also target the 14 types of HPV that are now considered to be carcinogenic and avoid picking up infections by types that are commensal or cause at most benign or asymptomatic infections of the lower genital tract.

[18]Sensitivity and specificity are terms used in screening for disease. The former denotes the probability that a test will be positive if the disease is present, whereas the latter measures the opposite, that is, probability that the test will turn out negative if the disease is absent. Sensitivity and specificity are typically expressed as percentages. The supplements to these values (100% minus the value) are the false-negative and false-positive rates, respectively.

role. This new paradigm of primary HPV screening is gradually being adopted in many Western countries.

THE OPPORTUNITY TO PREVENT CERVICAL CANCER WITH A VACCINE

However, it was not only in improving cervical cancer screening that HPV-based technologies came to help. Following on the success of her Spain/Colombia epidemiologic study that brought strong evidence that HPV was a causal agent in cervical cancer, Nubia Muñoz attempted to use her office's influence to interest the Findley Institute in Cuba and the Oswaldo Cruz Foundation in Brazil in developing HPV vaccines. Unfortunately, her attempts were blocked by administrative and technology transfer hurdles. Unfazed, she convened a joint meeting of the IARC and Merieux Foundation in December 1994 to bring experts to agree on specific action plans towards the development of an HPV vaccine. The discussions in this meeting covered therapeutic and prophylactic vaccines, choice of endpoints, and assays for measuring immune response to HPV [51]. Foremost, the participants agreed that the tasks ahead were monumental in scope and costs, going beyond of the means of individual governmental agencies that funded health research.

The solution to this problem was not far behind. The 1995 IARC monograph that established HPVs 16 and 18 as group 1 carcinogens was the sort of endorsement by a highly respected World Health Organization agency that two giant pharmaceutical companies were waiting to have before they embarked on the costly trajectory of research and development of HPV vaccines. Both Merck and SmithKline Beecham (SKB) Biologicals were paying close attention to the evolving HPV–cervical cancer story and felt convinced that it was time to act.

Many laboratories worldwide were trying to study the immune response against HPV infection using linear epitopes, that is, peptides derived from the nucleotide sequence of selected HPV genes that could be candidates for inducing a protective, neutralizing immune response. Today's successful prophylactic vaccines against HPV finally became a reality with the development of a technology that produces HPV virus-like particles (VLP).[19] The VLP technology has its roots in research by Robert Garcea on polyomaviruses, which are closely related to HPVs [52]. Subsequent independent work between 1991 and 1993 in the laboratories of Ian Frazer in Australia's Queensland University, of Bennet Jenson and Richard Schlegel at Georgetown University in Washington, of Doug Lowy and John Schiller in Bethesda, Maryland, and of Robert Rose at University of Rochester, New York, succeeded in developing HPV VLP formulations that were suitable for large-scale production [53]. By the time of the [48] monograph Merck and SKB Biologicals[20] had already acquired the license rights to the VLP technology and began planning randomized controlled trials of their respective candidate vaccines. These trials began in the mid to late 1990s and the first results were published in 2002, 2004, and 2005 (Figs. 8.2 and 8.3). The trial results exceeded all expectations. Both initial vaccines[21] attained greater than 95% protection against

[19]This is done via expression of the major HPV capsid gene (L1) in eukaryotic cell systems (yeast or insect cells). The expressed capsid protein self-assembles as pentamers and 72 of the latter spontaneously join to form a structure that resembles an intact HPV virion. The VLP does not contain the viral DNA and thus it is not infectious. However, the ordered arrangement of epitopes in the VLP makes this formulation highly immunogenic.

[20]In 2000 Glaxo and SmithKline merged as a single company, Glaxo SmithKline (GSK).

[21]GSK's bivalent against HPVs 16 and 18 (Cervarix), and Merck's quadrivalent against HPVs 6, 11, 16, and 18 (Gardasil).

persistent infection with the target types and associated precancerous lesions. The succession of HPV vaccines (i.e., bivalent, quadrivalent, and eventually a nonavalent one) has been licensed between 2006 and 2014 and deployed after regulatory approval. The reality today is unequivocal; HPV vaccination is among the most successful public health interventions in cancer control. The first indication that HPV vaccination is beginning to reduce the incidence of cervical cancers (invasive, not precancerous lesions) has recently been published [54,55].

FROM DISCOVERY TO A NOBEL PRIZE

As the scientific evidence accumulated incriminating HPV as the cause of cervical cancer (and other cancers as well), attendance to the international papillomavirus conferences—held every 1–2 years in different countries—steadily increased. By 2000, it had been common knowledge among conference delegates that Harald zur Hausen was among those slated for receiving the Nobel Prize in Physiology or Medicine. His laboratory had discovered the molecular basis for the epidemiological association between HPV and cervical cancer and formulated a general mechanism for carcinogenesis to progress from what was apparently a common infection. The demonstration in large-scale clinical trials of HPV vaccination by Merck in 2002 and 2005, and GSK in 2004 and 2006 [56–59], that not only infection can be prevented but also the cervical lesions associated with HPV can be averted as well, provided the last and final proof that cervical cancer is caused by HPV. This argument obviously bolstered the chances that the Nobel Committee would choose zur Hausen for the coveted prize. That the chances were improving came in May 2008. He had been selected for the prestigious Canadian Gairdner award, which has the reputation of frequently preceding a Nobel Prize for the recipient. It was no exception for zur Hausen; the news eventually came in early October 2008 via a phone call from Stockholm. He was one of the 2008 awardees, with half of the prize; the other half being equally shared by Françoise Barré-Sinoussi and Luc Montagnier. In the view of many, zur Hausen was overdue for the prize. One could also argue that Barré-Sinoussi and Montagnier were also overdue to receive it. After all, their discovery of HIV as cause of AIDS had happened more or less at the same time as the finding of a new HPV DNA sequence in cervical cancers in zur Hausen's laboratory. It was a felicitous combination; the honorees were scientists who identified the causes of two diseases that kill millions of people worldwide: HIV/AIDS and cancer of the cervix.

Given the secrecy of the deliberations by the Nobel Committee at the Karolinska Institute we will never know how many times candidacies submitted on behalf of zur Hausen had been considered and eventually passed over with a judgment of being premature. We know the story behind one of the nominations. In 2006, Jørn Olsen, the President of the International Epidemiological Association (IEA) asked ELF to draft a proposal to seek a nomination of an epidemiologist for the Nobel Prize. The epidemiology contributions to the HPV–cervical cancer link were much too strong to justify some activism on the part of the epidemiology community. ELF accepted the task and proposed that there could be an opportunity to seek a joint nomination for the prize involving an epidemiologist and zur Hausen. The advantages were twofold. First, a stronger case could be made for zur Hausen by adding to his candidacy a champion for the epidemiologic evidence that made his discovery unassailable. Second, and as a self-serving reason, an epidemiologist could finally be rewarded by a Nobel Prize.[22] The IEA Council enthusiastically endorsed the idea and tasked ELF with the role of conducting the background research and drafting a nomination proposal. He led a nominating committee that included Jørn Olsen and Rodolfo Saracci.

After considerable research, interviews with members of the scientific community and going through multiple rounds of draft, the proposal was submitted on December 29, 2006 to the Nobel Committee.[23] The draft proposal became public knowledge because it was posted in the IEA website to elicit comments from the professional community. It included Harald zur Hausen and Nubia Muñoz as conominees. The latter was chosen because she epitomized the successful efforts over three decades that validated the seminal discovery by zur Hausen. With her name in the proposal, we believed that the case for the prize would have been airtight. But choosing only two conominees was not an easy task. There were so many champions in this most beautiful journey of discovery and knowledge translation in cancer prevention. We initially considered going with three names. We wished to include a third nominee representing those who earned the primacy of being the developers of the VLP technology behind todays' HPV vaccines. Our group agonized with the decision. We did not want to dilute the merit of zur Hausen, the primary nominee, but felt convinced that zur Hausen's and Muñoz's body of work required also the final proof from the vaccination efficacy studies, all of which relied on the VLP technology. Our final decision of going with only two names was reached on pragmatic reasons. In 2006, the patent dispute on the VLP technology was at its height (see above description of teams involved in the VLP development process) [53]. It would have been counterproductive to bring a controversial angle to our nomination; hence the decision to go with zur Hausen and Muñoz as only conominees. Our nomination's most cogent argument was "*the road from discovery to prevention of this important disease has its unequivocal origins in the [...] two independent research tracks* ... (NB: meaning discovery of the virus in cervical cancer and the epidemiology work that proved causation). *Eliminate any of the*[se] *two tracks and the scientific progress would not have been sufficient to permit this auspicious moment in public health history*."

The end of this story is of course known. zur Hausen was deservedly a recipient of the Nobel Prize and Muñoz went on to receive a multitude of other accolades. We will never know if the 2006 IEA nomination made a difference in nudging the Nobel committee to select zur Hausen. As this chapter's conclusion, we wish to remind those entering the field of cancer control research of the importance of multidisciplinary cooperation to make discoveries that will have an impact in improving health. Had the circumstances of scientific discourse been different, would "*the road to*

[22]The Nobel Prize has a reputation for recognizing fundamental mechanisms rather than applied scientific discoveries or their translation into health interventions. In several Nobel Prize stories, epidemiologists have made many of the observations that were *sine qua non* in validating the recipients' biological discoveries. Robert Koch was given the Nobel Prize in 1905 for his work in tuberculosis, but he could have received the prize for his isolation of the *Vibrio cholerae*. Had the Nobel Committee awarded the prize for cholera, according to their current practice they would have left out John Snow who elegantly demonstrated the environmental source of the infectious agent and how it spread, as well as showing how it could be prevented. Epidemiologists, such as Richard Doll, A. Bradford Hill, Richard Peto, Ernst Wynder, and Evarts Graham, made enormous contributions to the most important cancer prevention paradigm, the demonstration that tobacco smoking causes cancer. They received substantial recognition from the research community, but not a Nobel Prize.

[23]The right to nominate someone to the Nobel Prize is restricted to a few categories of individuals: prior recipients, those invited by the Nobel committee to submit nominations, and professors from Scandinavian universities. Luckily, Prof. Jørn Olsen, IEA President, met the third of these conditions and served as the signatory.

preventing a major human cancer" end up shorter? In an editorial so entitled, zur Hausen expressed his beliefs that the actionable science was already available in 1987 but the discomfort by epidemiologists in calling the state of the evidence as sufficient delayed the actions that could have led to preventive strategies sooner [60]. In the rebuttal to zur Hausen [61], we and others reasoned that this assertion was simplistic and ignored the state of the technologies at the time. Pharmaceutical companies would not have had the benefit of the essential VLP technology in 1987, nor the validated HPV diagnostic assays that matured only in the early 1990s. Most importantly, however, it had been the HPV epidemiology community that had become the strongest supporter of zur Hausen's body of work. It fast-tracked a monumental series of studies whose findings moved rapidly to become primary (vaccination) and secondary (screening) prevention fronts. In combination, these two fronts may eventually bring the end to a human cancer.

REFERENCES

[1] Muñoz N, Bosch X, Kaldor JM. Does human papillomavirus cause cervical cancer? The state of the epidemiological evidence. Br J Cancer 1988;57(1):1–5.
[2] Wynder EL, Cornfield J, Schroff PD, Doraiswami KR. A study of environmental factors in carcinoma of the cervix. Am J Obstet Gynecol 1954;68(4):1016–47 discussion, 1048-52.
[3] Graham S, Priore R, Graham M, Browne R, Burnett W, West D. Genital cancer in wives of penile cancer patients. Cancer 1979;44(5):1870–4.
[4] Bosch FX, Cardis E. Cancer incidence correlations: genital, urinary and some tobacco-related cancers. Int J Cancer 1990;46(2):178–84.
[5] Li JY, Li FP, Blot WJ, Miller RW, Fraumeni Jr. JF. Correlation between cancers of the uterine cervix and penis in China. J Natl Cancer Inst 1982;69(5):1063–5.
[6] Franco EL, Campos Filho N, Villa LL, Torloni H. Correlation patterns of cancer relative frequencies with some socioeconomic and demographic indicators in Brazil: an ecologic study. Int J Cancer 1988;41 (1):24–9.
[7] Doll R, Peto R. The causes of cancer: quantitative estimates of avoidable risks of cancer in the United States today. J Natl Cancer Inst 1981;66(6):1191–308.
[8] Franco EL. Viral etiology of cervical cancer: a critique of the evidence. Rev Infect Dis 1991;13 (6):1195–206.
[9] Koss LG. Cytologic and histologic manifestations of human papillomavirus infection of the uterine cervix. Cancer Detect Prev. 1990;14(4):461–4.
[10] Koss LG, Durfee GR. Unusual patterns of squamous epithelium of the uterine cervix: cytologic and pathologic study of koilocytotic atypia. Ann N Y Acad Sci. 1956;63(6):1245–61.
[11] Koss LG. Cytologic and histologic manifestations of human papillomavirus infection of the female genital tract and their clinical significance. Cancer. 1987;60(8 Suppl):1942–50.
[12] Koss LG. The Papanicolaou test for cervical cancer detection. A triumph and a tragedy. JAMA. 1989;261(5):737–43.
[13] Koss LG. Human papillomavirus--passenger, driver, or both? Hum Pathol. 1998;29(4):309–10.
[14] zur Hausen H. Intracellular surveillance of persisting viral infections: human genital cancer results from deficient cellular control of papillomavirus gene expression. Lancet 1986;2:489–91.
[15] Bedell MA, Jones KH, Laimins LA. The E6-E7 region of human papillomavirus type 18 is sufficient for transformation of NIH 3T3 and rat-1 cells. J Virol. 1987;61(11):3635–40.

[16] Matlashewski G, Schneider J, Banks L, Jones N, Murray A, Crawford L. Human papillomavirus type 16 DNA cooperates with activated ras in transforming primary cells. EMBO J. 1987;6(6):1741–6.
[17] Werness BA, Levine AJ, Howley PM. Association of human papillomavirus types 16 and 18 E6 proteins with p53. Science. 1990;248(4951):76–9.
[18] Dyson N, Howley PM, Münger K, Harlow E. The human papilloma virus-16 E7 oncoprotein is able to bind to the retinoblastoma gene product. Science. 1989;243(4893):934–7.
[19] Kessler II. Human cervical cancer as a venereal disease. Cancer Res. 1976;36(2 pt 2):783–91.
[20] Nahmias AJ, Roizman B. Infection with herpes-simplex viruses 1 and 2. 1. N Engl J Med. 1973;289 (13):667–74.
[21] Rawls WE, Tompkins WA, Figueroa ME, Melnick JL. Herpesvirus type 2: association with carcinoma of the cervix. Science. 1968;161(3847):1255–6.
[22] Vonka V, Kanka J, Hirsch I, Závadová H, Krcmár M, Suchánková A, et al. Prospective study on the relationship between cervical neoplasia and herpes simplex type-2 virus. II. Herpes simplex type-2 antibody presence in sera taken at enrollment. Int J Cancer. 1984;33(1):61–6.
[23] Meignier B, Norrild B, Thuning C, Warren J, Frenkel N, Nahmias AJ, et al. Failure to induce cervical cancer in mice by long-term frequent vaginal exposure to live or inactivated herpes simplex viruses. Int J Cancer. 1986;38(3):387–94.
[24] Kessler II. Etiological concepts in cervical carcinogenesis. Appl Pathol. 1987;5(1):57–75.
[25] Kjaer SK, Engholm G, Teisen C, Haugaard BJ, Lynge E, Christensen RB, et al. Risk factors for cervical human papillomavirus and herpes simplex virus infections in Greenland and Denmark: a population-based study. Am J Epidemiol. 1990;131(4):669–82.
[26] Reeves WC, Brinton LA, García M, Brenes MM, Herrero R, Gaitán E, et al. Human papillomavirus infection and cervical cancer in Latin America. N Engl J Med. 1989;320(22):1437–41.
[27] de Villiers EM, Wagner D, Schneider A, Wesch H, Miklaw H, Wahrendorf J, et al. Human papillomavirus infections in women with and without abnormal cervical cytology. Lancet. 1987;2(8561):703–6.
[28] Villa LL, Franco EL. Epidemiologic correlates of cervical neoplasia and risk of human papillomavirus infection in asymptomatic women in Brazil. J Natl Cancer Inst. 1989;81(5):332–40.
[29] Wagner D, Ikenberg H, Boehm N, Gissmann L. Identification of human papillomavirus in cervical swabs by deoxyribonucleic acid in situ hybridization. Obstet Gynecol. 1984;64(6):767–72.
[30] Tidy JA, Parry GC, Ward P, Coleman DV, Peto J, Malcolm AD, et al. High rate of human papillomavirus type 16 infection in cytologically normal cervices. Lancet. 1989;1(8635):434.
[31] Franco EL. The sexually transmitted disease model for cervical cancer: incoherent epidemiologic findings and the role of misclassification of human papillomavirus infection. Epidemiology. 1991;2 (2):98–106.
[32] Schiffman MH, Schatzkin A. Test reliability is critically important to molecular epidemiology: an example from studies of human papillomavirus infection and cervical neoplasia. Cancer Res. 1994;54(7 Suppl):1944s–7s.
[33] Tidy J, Farrell PJ. Retraction: human papillomavirus subtype 16b. Lancet. 1989;2(8678-8679):1535.
[34] Koutsky LA, Holmes KK, Critchlow CW, Stevens CE, Paavonen J, Beckmann AM, et al. A cohort study of the risk of cervical intraepithelial neoplasia grade 2 or 3 in relation to papillomavirus infection. N Engl J Med. 1992;327(18):1272–8.
[35] Muñoz N, Bosch FX, Shah KV, Meheus A, editors. The epidemiology of human papillomavirus and cervical cancer. Lyon: IARC Scientific Publication No. 119; 1992. ISBN-13 (Print Book).
[36] Muñoz N, Bosch FX, de Sanjosé S, Tafur L, Izarzugaza I, Gili M, et al. The causal link between human papillomavirus and invasive cervical cancer: a population-based case-control study in Colombia and Spain. Int J Cancer. 1992;52(5):743–9.

[37] Schiffman MH, Bauer HM, Hoover RN, Glass AG, Cadell DM, Rush BB, et al. Epidemiologic evidence showing that human papillomavirus infection causes most cervical intraepithelial neoplasia. J Natl Cancer Inst. 1993;85(12):958–64.
[38] Bosch FX, Lorincz A, Muñoz N, Meijer CJLM, Shah KV. The causal relation between human papillomavirus and cervical cancer. J Clin Pathol. 2002;55(4):244–65.
[39] Muñoz N, Bosch FX, de Sanjosé S, Herrero R, Castellsagué X, Shah KV, et al. Epidemiologic classification of human papillomavirus types associated with cervical cancer. N Engl J Med. 2003;348(6):518–27.
[40] Franco EL, Tota J. Invited commentary: human papillomavirus infection and risk of cervical precancer--using the right methods to answer the right questions. Am J Epidemiol. 2010;171(2):164–8.
[41] Meisels A, Fortin R. Condylomatous lesions of the cervix and vagina. I. Cytologic patterns. Acta Cytol. 1976;20(6):505–9.
[42] Purola E, Savia E. Cytology of gynecologic condyloma acuminatum. Acta Cytol. 1977;21(1):26–31.
[43] Syrjänen KJ. Two landmark studies published in 1976/1977 paved the way for the recognition of human papillomavirus as the major cause of the global cancer burden. Acta Cytol. 2017;61(4-5):316–37.
[44] Muñoz N. From causality to prevention - the example of cervical cancer: my personal contribution to this fascinating history. Public Health Genomics. 2009;12(5-6):368–71.
[45] Gissmann L, zur Hausen H. Human papilloma virus DNA: physical mapping and genetic heterogeneity. Proc Natl Acad Sci U S A. 1976;73(4):1310–13.
[46] Dürst M, Gissmann L, Ikenberg H, zur Hausen H. A papillomavirus DNA from a cervical carcinoma and its prevalence in cancer biopsy samples from different geographic regions. Proc Natl Acad Sci U S A. 1983;80(12):3812–15.
[47] Muñoz N, Bosch FX, Jensen OM, editors. Human papillomavirus and cervical cancer. Lyon: IARC Scientific Publication No. 94; 1989. ISBN-13 (Print Book).
[48] IARC. Human papillomaviruses. IARC monographs on the evaluation of carcinogenic risks to humans Volume 64. Lyon: IARC; 1995. ISBN-13 (Print Book).
[49] Walboomers JM, Jacobs MV, Manos MM, Bosch FX, Kummer JA, Shah KV, et al. Human papillomavirus is a necessary cause of invasive cervical cancer worldwide. J Pathol. 1999;189(1):12–19.
[50] Franco EL, Rohan TE, Villa LL. Epidemiologic evidence and human papillomavirus infection as a necessary cause of cervical cancer. J Natl Cancer Inst. 1999;91(6):506–11.
[51] Muñoz N, Crawford L, Coursaget P. HPV vaccines for cervical neoplasia. Lancet 1995;345:249.
[52] Garcea RL, Salunke DM, Caspar DL. Site-directed mutation affecting polyomavirus capsid self-assembly in vitro. Nature. 1987;329(6134):86–7.
[53] McNeil C. Who invented the VLP cervical cancer vaccines? J Natl Cancer Inst. 2006;98(7):433.
[54] Guo F, Cofie LE, Berenson AB. Cervical Cancer Incidence in Young U.S. Females After Human Papillomavirus Vaccine Introduction. Am J Prev Med. 2018;55(2):197–204.
[55] Luostarinen T, Apter D, Dillner J, Eriksson T, Harjula K, Natunen K, et al. Vaccination protects against invasive HPV-associated cancers. Int J Cancer. 2018;142(10):2186–7.
[56] Harper DM, Franco EL, Wheeler C, Ferris DG, Jenkins D, Schuind A, et al. Efficacy of a bivalent L1 virus-like particle vaccine in prevention of infection with human papillomavirus types 16 and 18 in young women: a randomized controlled trial. Lancet. 2004;364:1757–65.
[57] Harper DM, Franco EL, Wheeler CM, Moscicki AB, Romanowski B, Roteli-Martins CM, et al. Sustained efficacy up to 4.5 years of a bivalent L1 virus-like particle vaccine against human papillomavirus types 16 and 18: follow-up from a randomized control trial. Lancet. 2006;367(9518):1247–55.
[58] Koutsky LA, Ault KA, Wheeler CM, et al. A controlled trial of a human papillomavirus type 16 vaccine. N Engl J Med 2002;347:1645–51.

[59] Villa LL, Costa RL, Petta CA, Andrade RP, Ault KA, Giuliano AR, et al. Prophylactic quadrivalent human papillomavirus (types 6, 11, 16, and 18) L1 virus-like particle vaccine in young women: a randomised double-blind placebo-controlled multicentre phase II efficacy trial. Lancet Oncol. 2005;6:271–8.

[60] zur Hausen H. Cervical carcinoma and human papillomavirus: on the road to preventing a major human cancer. J Natl Cancer Inst. 2001;93(4):252–3.

[61] Bosch FX, Muñoz N, de Sanjosé S, Franco EL, Lowy DR, Schiffman M, et al. Re: Cervical carcinoma and human papillomavirus: on the road to preventing a major human cancer. J Natl Cancer Inst. 2001;93 (17):1349–50.

CHAPTER

9

THE NATURAL HISTORY OF HUMAN PAPILLOMAVIRUS INFECTION IN RELATION TO CERVICAL CANCER

Anna-Barbara Moscicki[1], Mark Schiffman[2] and Silva Franceschi[3]

[1]*Division of Adolescent and Young Adult Medicine, Department of Pediatrics, University of California, Los Angeles, Los Angeles, CA, United States* [2]*Division of Cancer Epidemiology and Genetics, National Cancer Institute, Bethesda, MD, United States* [3]*Aviano Cancer Center (CRO) IRCCS, Aviano, Italy*

KEY BULLET POINTS

- The majority of "new" HPV infections appear to regress as defined by the lack of detection-regardless of age at first detection.
- Persistence of high risk HPV infections, specifically HPV 16, is the key risk for the development of cervical cancer.
- The majority of CIN 2 infections regress, specifically in young women. However, this may be due to misclassification of low grade lesions.
- Cofactors thought to be additional to HPV persistence include smoking cigarettes, multiparity, and prolonged hormonal contraception. These cofactors contribute quite weak risks compared to the persistence of an oncogenic HPV.

Epidemiologic studies throughout the 20th century strongly suggested that cervical cancer (CC) was related to sexual activity. Women with CC were more likely to report multiple sexual partners and husbands with penile cancer or previous wives who also had CC. Numerous sexually transmitted diseases (as well as semen) were postulated to be the primary causal "candidates" including herpes simplex virus and *Chlamydia trachomatis*.

Human papillomavirus (HPV) was eventually implicated partly because of the histopathologic similarity of flat cervical condyloma (warts) and mild dysplasia (a recognized precursor to cancer), as noted by Meisels and Fortin [1] and Purola and Savia [2]. Around the same time in the late 1970s, zur Hausen postulated an etiologic role for HPV, confirmed in the early 1980s when DNA was found using low-stringency Southern blot by zur Hausen and his colleagues in two cases of CC [3]. Virtually immediately, epidemiologists began the search for tests for HPV that could be used in exfoliative specimens from controls. Misclassification from the earliest tests attenuated the strength of the causal association, and even masked the sexual transmission of HPV. As assays

Human Papillomavirus. DOI: https://doi.org/10.1016/B978-0-12-814457-2.00009-X

improved, most prominently due to the work of Lorincz and colleagues [4], and more HPV types were characterized, the strength of the true associations steadily increased, and the fraction of CC attributable to HPV infection rose dramatically. The discovery that virtually all CCs are caused by HPV yielded to the recognition of an unprecedented central and necessary role for HPV of approximately a dozen types [5]. With the invention and application of PCR, after initial flawed use temporarily suggested that HPV infection was ubiquitous and noncausal, it became evident that persistent detection of a limited set of HPV types explained virtually all cases of CC [5]. Moreover, it became clear that the cytologic and histologic precursors of CC represented the direct cytopathic effect of HPV infection and the secondary accumulation of somatic mutations combined with the antiapoptotic state induced by HPV [6,7].

These initial studies using HPV DNA detection resulted in confusion between prevalence and future risk of CIN 3. The high prevalence of HPV in young populations resulted in guidelines such as in the United States and Australia that recommended CC screening in young women within 1 year of the onset of sexually activity and at age 15 years in other countries [8,9]. This led to a large and unnecessary number of young women being referred to colposcopy for the benign changes of atypical squamous cells of undetermined significance (ASCUS) and low-grade squamous intraepithelial lesion (LSIL). Although later it was recognized that the causal HPV associated with cancer is likely acquired shortly after the onset of sexual activity [9], epidemiology studies demonstrated that the majority of HPV infections were transient in nature of HPV, specifically in young women [10–13] and that LSIL reflected the natural history of HPV with high rates of regression, specifically in young women [12,14]. With the understanding that only a dozen HPV types were considered carcinogenic, HPV DNA testing began to target these for management of abnormal cytology [15,16]. HPV detection was used to triage cytologic abnormalities decreasing the number of referrals to colposcopy. On the other hand, the repeated detection of multiple HPV infections in young women led guidelines to support the limited utility of HPV testing in young women [17]. Studies also began to demonstrate that multiple HPV types were often present suggesting synergistic effects. However, on closer examination within tissue samples, it was evident that when multiple infections were found, a single HPV type tended to be causal [18].

Longitudinal studies with adequate follow-up revealed the significance of persistence of high risk HPV in predicting CIN 3 development [19]. These studies also demonstrated the power of genotyping. HPV 16 persistence revealed a greatly increased risk of CIN 3 development specifically compared to other oncogenic genotypes [20–22]. Numerous studies began to demonstrate the sensitivity of HPV detection, however, the positive predictive value of a single test remains quite low since the chance of clearance of an infections is much greater than persistence [23,24]. Studies also demonstrated that clearance of HPV infections was not limited to LSIL. Once again, several observational and prospective studies showed that CIN 2 and 3 can also regress with rates exceeding 50%, specifically in young women [25,26]. In part, these high rates of regression are likely due to the subjectivity and overcall of CIN 2 diagnosis. With the recent incorporation of biomarkers such as p16 INK, this overall call could decrease if the marker is used properly [27]. Prospective studies of CIN 2 with positive p16 staining are needed to understand the role of this biomarker in predicting progression. This may be difficult in light of the tragic study of women with CIN3 who were observed and not treated demonstrating a progression rate to invasive cancer of 5% per year [28]. Persistence of HPV is thought to be necessary but not sufficient for the development of invasive cancers—although this remains controversial. Several rather weak cofactors have been identified in epidemiology studies including tobacco use, multiparity, prolonged oral contraceptive use, and chlamydia infection. Of these, tobacco has one of the most consistent associations with risk of CC and is discussed below. Our current understanding of the natural history of HPV is summarized in Figs. 9.1 and 9.2.

It has long been described that squamous CC arises from the cervical transformation zone, an epithelial area where the proximal single-layered columnar epithelium lays adjacent to the distal stratified squamous epithelium. As a part of normal continued development, uncommitted cells from the columnar epithelium are triggered to differentiate into squamous cells. The hormonal changes of puberty triggers an acceleration of this process referred to as squamous metaplasia [29]. The cervix during adolescence and young adulthood is a mixture of columnar, metaplastic, and squamous epithelia with a predominance of columnar and metaplastic compared to adult women and coincides with the age range when HPV infections are most common. Longitudinal studies have shown evidence that active metaplasia is a risk for HPV acquisition [30,31]. Given HPV's dependence on cell replication and differentiation, squamous metaplasia likely supports viral replication and infection of vulnerable reserve cells.

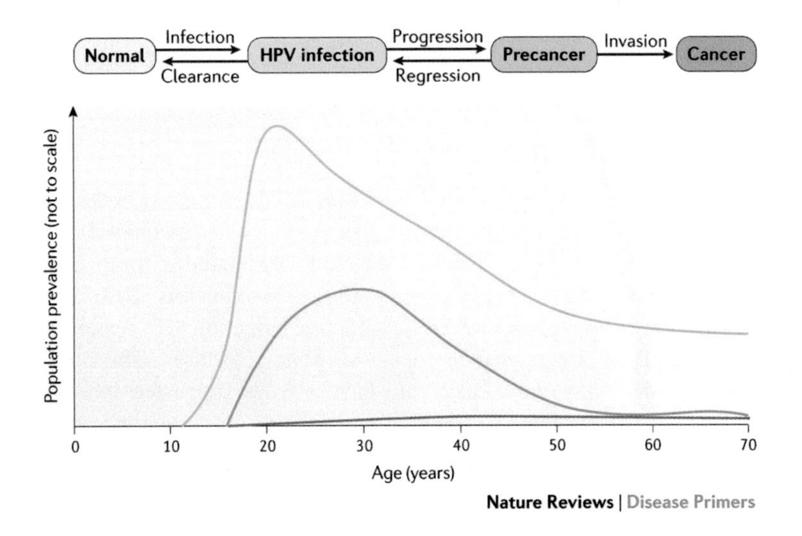

FIGURE 9.1

This figure demonstrates the time taken between a causative infection and cancer diagnosis. The natural history of HPV includes infection followed by clearance or persistence with persistence leading to precancer, which if untreated may go on to cervical cancer. Each of these stages are associated characteristically with age. The peak prevalence of HPV infection is highly associated with newly acquired HPV among youth. This peak is followed by a secondary peak of precancer occurring a number of years later depending on the intensity of screening. In turn, the rise of invasive cancer occurs decades years later.

Based on Schiffman M, Doorbar J, Wentzensen N, de Sanjose S, Fakhry C, Monk BJ, Stanley MA, Franceschi S. Carcinogenic human papillomavirus infection. Nat Rev Dis Primers 2016. doi: 10.1038/nrdp.2016.86.

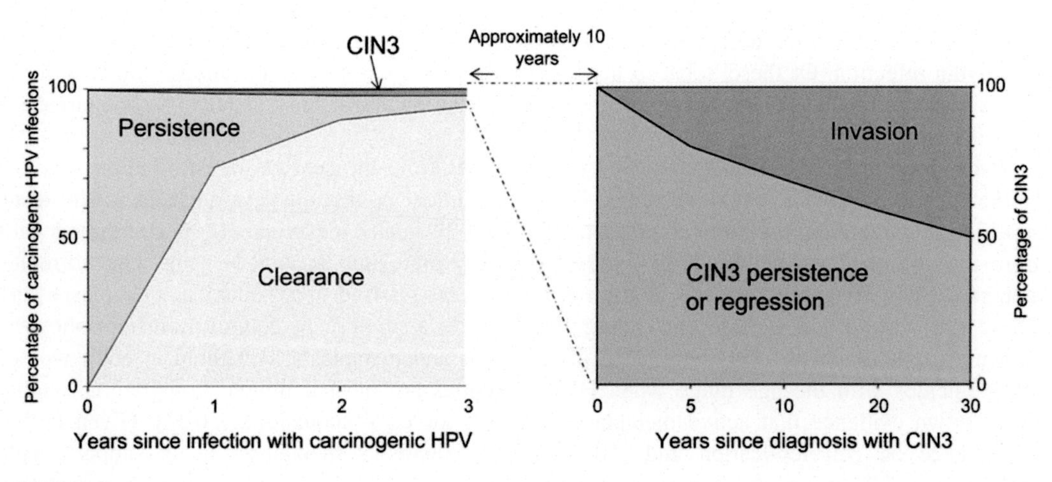

FIGURE 9.2

Left graph: Proportion of prevalent carcinogenic HPV infections that clear, persist, or progress to CIN 3 in the first 3 years after first detection. Long-term persistence without CIN3 is uncommon. *Right graph*: Proportion of untreated CIN3 lesions that will invade to cancer within 30 years following the initial diagnosis.

Based on Schiffman M, Wentzensen N, Wacholder S, Kinney W, Gage JC, Castle PE. Human papillomavirus testing in the prevention of cervical cancer. J Natl Cancer Inst 2011;103:368–83.

TRANSMISSION OF HUMAN PAPILLOMAVIRUS

HPV is the most common sexually transmitted infection (STI) in the world with a lifetime risk for cervical HPV infection of 80% [30]. This lifetime risk is somewhat incongruent with the estimated transmission rates ranging from 2% to 100%. One study estimated a transmission rate of 100% with 11 acts of sexual intercourse [31]. Consequently, approximately 20% of women with "no detectable HPV" are likely exposed but never show active infection (DNA detection). Reasons for this remain unclear but likely are related to robust immune responses. The high rates of HPV in younger women with a declining prevalence rate have always suggested that women develop an adaptive immune response to repeated exposures resulting in rapid clearance of second exposures [32]. Protection for redetection was also seen in women participating in the placebo arm of the HPV vaccine trials who were high-titer HPV16 seropositive. This group experienced lower rates of HPV 16 infection than women that were seronegative [33]. This immune protection is thought by some to explain the differences in HPV associated oral cancers by gender—with men showing dramatic increases whereas women have not experienced this same increase in incidence. Prospective studies of couples suggest that female-to-male transmission is greater than male-to-female and percent of women with antibodies to HPV are much higher than seen in men despite the fact that men have higher rates of HPV DNA detection in their genitals than women [34,35]. All of these studies support the premise that cervical infections that are cleared are linked to an adaptive memory immune response and superficial infections in men result in weaker immune protection. Some countries show a bimodal pattern for prevalence with second wave of HPV in women 50 and older or even a similar prevalence of HPV in all age groups [36]. Although latent reactivation is a possibility, certainly sexual behavior likely plays an important role [37,38]. Incident infections in older women are as likely to be controlled immunologically as younger women [39]. Although HPV DNA of genital HPV types can be found in nongenital areas including the hand, fingernails, and underwear [34,40,41], it remains doubtful that these are modes of efficient transmission. There is no good evidence that the HPV DNA detected in these nongenital sites is representative of true infections.

LACK OF STRONG EPIDEMIOLOGIC INDICATORS FOR HUMAN PAPILLOMAVIRUS PERSISTENCE (OR CLEARANCE)

Epidemiological and natural history studies have shown that the majority of anogenital HPV incident infections will clear with time, but a fraction (estimated as $<10\%$) of these infections persist and may therefore progress to cancer leading researchers to search for modifiable factors impeding persistence [42]. The longer we observed persistence; it became clear that the longer a detectable oncogenic HPV infection lasts, the higher the risk that molecular events will occur during the repeat cycles of viral replication. Three groups of factors, mainly studied in the cervix, came to light as a strong probability of affecting HPV persistence: viral, host, and behavioral variables [42].

Viral factors included genetic (and possibly epigenetic) differences between HPV types and, more recently explored, HPV variants that greatly influence the risk of precancer development. Two meta-analyses that included 41 and 86 cohort studies, respectively, gave us valuable insight into the frequency and duration of persistent infection with different HPVs [43,44]. The median duration of HPV infection was on average 9.8 months [44]. Median duration of persistency (typically defined as two or more positive HPV DNA tests 6 or 12 months apart) is influenced by the duration of follow up, the frequency of visits, the sensitivity of HPV test, and the definition of infection clearance (i.e., one or two negative tests). In general, HPV 16 is as likely to clear as other hrHPV types [42,45]. An important notion that came to light is that carcinogenicity of a HPV type is related to but not a mere function of persistence [21]. For example, it has been observed that HPV16 infection is associated with a risk of progression to CIN3 or worse that is, compared to other HPV types that are also classified as oncogenic (such as HPV51, 56, and 59), more than one order of magnitude larger whereas the risk of persistency varies less than twofold. Among control women (age 15–25 years) from the PATRICIA HPV vaccine trial, median time to clearance for HPV16 was 17 months and not significantly longer than the median (range 11–14 months) for HPV 31, 33, and 45.

With respect to host risk factors, the probability of HPV persistency (and progression) to cancer in the cervix and in other anogenital sites was shown to be strongly influenced by the immune status, notably HIV infection [46]. Of interest, the resolution of HPV infection is not necessarily accompanied by clearance of viral genome from the epithelial basal layer and immunosuppression due to HIV or transplant can lead to reactivation of HPV [42]. The influence of a woman's age has been difficult to understand. Prevalent infections are less likely to regress in older women because the likelihood they represent long-duration persistent infection and not because of women's age per se [47]. In fact, age appears not to be associated with persistency in incident HPV infections [39].

Behavioral risk factors, including multiparity, long-term use of hormonal contraceptives, and tobacco smoking, also influence the risk of HPV persistency and progression to cancer [42] (see next section). Coinfection with *C. trachomatis* has been implicated albeit inconsistently. Finally, it was suggested that diagnostic biopsies can affect the course of HPV infection, for example, a punch biopsy was shown to approximately halve the time to clearance, compared to no biopsies [48]. Other studies did not support this finding [12]. Although the identification of behavioral risk factors (often referred to as cofactors of HPV) informs etiology and pathogenesis, from a clinical perspective, their influence is relatively modest compared with viral factors.

ROLES OF PARITY, FEMALE HORMONES AND TOBACCO

Cofactors of HPV infections for cancer progression have been mainly studied in the cervix and include high parity, recent use of hormonal contraceptives, tobacco smoking, and a protective role of intrauterine device (IUD) and condom use [49].

The cervical epithelium is extremely sensitive to reproductive and menstrual events during a woman's life. The rise in ovarian activity after menarche stimulates the transformation of columnar reserve cells into squamous epithelium in the transformation zone as described above. It was observed that the CC incidence in unscreened populations, such as in India, stop rising after approximately 45 years of age, coinciding with a time when perimenopausal hormonal activity starts to decline [50]. This behavior is quite distinct from that of cancer of the cervix in well-screened populations (in whom incidence reaches a plateau at approximately age 35) and other anogenital cancers in both sexes in which the steep growth by age resemble that of the other common epithelial cancers.

It was, however, difficult to rule out the possibility that the associations of CC risk with these life-style cofactors could be a mere consequence of confounding by sexual habits and sociocultural level. The International Collaboration of Epidemiological Studies of Cervical Cancer was launched by the International Agency for Research on Cancer and Oxford University to tackle this problem gathering individual data for 16,573 women with CC and 35,509 women without CC from 26 countries worldwide, with about half of the studies being from less developed countries [51–54]. All the purported HPV cofactors for CC could be revisited in different populations and their significant role confirmed after careful adjustment for lifetime number of sexual partners, age at first intercourse, and other possible confounding factors. Selected associations between behavioral cofactors and CC risk are shown in Table 9.1. Five full-term pregnancies or more and current long-term use (>5 years) of hormonal contraceptives (including both combined oral contraceptive use and depot medroxyprogesterone) [55] were all independently associated with a doubling of CC risk whereas current smoking showed a 46% risk increase. A large study from the European Prospective Investigation in to Cancer and Nutrition showed a twofold increase in CIN 3 and CC taking into account duration and frequency [56] of smoking. Importantly no or reduced associations were found when either hormonal contraceptives or tobacco were discontinued. Only for smoking was a significant etiological heterogeneity between squamous-cell carcinoma and adenocarcinoma demonstrated. The role of other coinfections has been a consistent interest over the decades since it has always been thought that HPV persistence is necessary but not sufficient and since HPV is a

Table 9.1 Selected Associations of Behavioral Cofactors With Cervical Cancer (CC).

Risk Factor	Category	Relative Risk[a]	(95% CI)[b]
Full-term pregnancies, number (reference: nulliparae)	2	1.14	(1.06–1.23)
	5	2.11	(1.91–2.34)
Hormonal contraceptives, 5 + years use (reference: never user)	Current	1.90	(1.69–2.13)
	Former[c]	0.97	(0.90–1.94)
Tobacco smoking (reference: never smokers)			
Squamous cell CC	Current	1.46	(1.35–1.58)
	Former	1.05	(0.94–1.17)
Adenocarcinoma	Current	0.92	(0.78–1.07)
	Former	0.84	(0.68–1.03)
Body Mass Index, kg/m^2 (reference: [20–25])			
Squamous cell CC	25–30	1.05	(0.94–1.18)
	30 +	0.94	(0.78–1.14)
Adenocarcinoma	25–30	1.24	(1.02–1.50)
	30 +	1.13	(0.83–1.55)

CI, *Confidence interval.*
[a]*Stratified by study and age and adjusted for other risk factors including number of sexual partners and age at first intercourse.*
[b]*95% floated CI.*
[c]*10 + years cessation.*
Modified from International Collaboration of Epidemiological Studies of Cervical Cancer, 2006, 2007a, 2007b, and 2009.

sexually transmitted disease, a coinfection might well be a significant cofactor. Although an indicator of high risk behavior, infections with *C. trachomatis* have continued to gain interest over the years with more compelling evidence as technology develops [57,58]. Interestingly, in contrast to systemic contraceptives, epidemiology studies found IUD use to be protective with a 50% reduction in CC with a history of ever use [59].

It is worth remembering that in 2000s, when the International Collaboration was carried out, there were hopes that the comparison of CC cases (all assumed to be oncogenic HPV-positive) and HPV-positive control women only could help pinpoint the mechanisms of action of behavioral characteristics, that is, facilitator of CC development exclusively in women positive for oncogenic HPV types. The International Collaboration could not show differences in RRs from these restricted analyses and the overall results and the approach was eventually dropped. Indeed, as a necessary cause of CC, not only HPV infection cannot be considered a classic confounder but, most importantly, the stratification did not really achieve a better case-control comparison. In fact, while HPV infection in CC is necessarily a long-duration infection, among control women it can be a recent/transient one.

Full-term pregnancies and hormonal contraceptives involve exposure to high levels of both estrogens and progesterone, but which hormonal pattern is specifically associated with risk of CC is unclear—contrary to that of breast and endometrial cancer. Few studies in women [60] and HPV-carrying transgenic mice [61] suggested a predominant role of high estrogen levels. In HPV-carrying transgenic mice, CC could even be controlled by treatment with raloxifene, an estrogen-receptor antagonist [62]. Of the two proxies of unopposed estrogen exposure in women, body mass index did not show significant association with CC risk (Table 9.1) and studies of menopausal replacement therapy were hampered by the association of menopausal replacement therapy with regular CC screening. Of note, estrogen signaling in CC in humans was shown to be very different than in breast cancer—CC is associated with estrogen receptors alpha in stromal fibroblasts rather than in tumor cells as observed in breast cancer [63].

Role of progesterone only contraceptive continues to be more difficult to explain. Biologic association for smoking tobacco became evident when in vivo studies demonstrated that nicotine and its carcinogenic by-products can be measured directly in cervical mucus [64]. This may suggest either direct exposure to tobacco-associated carcinogens or be secondary to tobacco's immunosuppressive effect. The potential biologic associations with *C. trachomatis* include the induction of chronic inflammation, inhibition of apoptosis, and interference with immune response to HPV [57]. The potential mechanisms of the potential protective role of IUD remains unknown, but it has been postulated that the IUD string left at the cervix of induces a protective inflammatory state or results in a mechanical removal of small HPV-associated lesions. On the other hand, IUD users represent women who are not pregnant and not on estrogen containing hormones raising the possibility of confounding effects. Condom use has been shown to provide protection from cervical infection in the newly sexual active [65]; however; majority of studies show only partial protection from anogenital cancers—likely on account of the broad extension of HPV-infected area.

Additional behavioral risk factors for HPV-associated cancers including CC are mainly related to an increased probability of HPV infection—such as number of sexual partners of the individual and also number of partners of his or her sexual partners. Young age at first sexual intercourse is also associated with increased CC risk possibly because it is a proxy of long-duration exposure to HPV infection and hence high probability of malignant transformation [50]. On the other hand, a plausible biologic explanation has not been ruled out since it has been estimated that most of the causal HPV infections associated with cervical cancer occur in adolescents and young adulthood and the studies alluded to above demonstrated a risk of acquiring HPV during periods of active metaplasia [30,31,66]. Anogenital or orogenital intercourse is risk factors for anal cancer and oropharyngeal cancer, respectively.

In conclusion, the elucidation of the natural history of HPV has led to brand-new CC prevention strategies. It has also strongly influenced the methodology for the HPV vaccination trials as well as cervical screening trials (i.e., HPV primary testing vs cytology). Studies of transmission have been borne out with the evidence of herd immunity seen in populations with high vaccination rates [67]. HPV-associated cancers can be added to the list of good reasons to avoid smoking whereas additional studies on the possible dependence of HPV infection and CC on female hormones and their possible responsiveness to hormonal manipulation may be worth pursuing.

REFERENCES

[1] Meisels A, Fortin R. Condylomatous lesions of the cervix and vagina. I. Cytologic patterns. Acta Cytologica 1976;20(6):505–9.
[2] Purola E, Savia E. Cytology of gynecologic condyloma acuminatum. Acta Cytologica 1977;21(1):26–31.
[3] zur Hausen H. Human papillomaviruses and their possible role in squamous cell carcinomas. Curr Top Microbiol Immunol 1977;78:1–30.

[4] Lorincz AT, Temple GF, Kurman RJ, Jenson AB, Lancaster WD. Oncogenic association of specific human papillomavirus types with cervical neoplasia. J Natl Cancer Inst 1987;79(4):671–7.
[5] Munoz N, Bosch FX, de Sanjose S, Herrero R, Castellsague X, Shah KV, et al. Epidemiologic classification of human papillomavirus types associated with cervical cancer. N Engl J Med 2003;348 (6):518–27.
[6] Koss LG. Cytologic and histologic manifestations of human papillomavirus infection of the female genital tract and their clinical significance. Cancer 1987;60(8 Suppl):1942–50.
[7] Dyson JL, Walker PG, Singer A. Human papillomavirus infection of the uterine cervix: histological appearances in 28 cases identified by immunohistochemical techniques. J Clin Pathol 1984;37 (2):126–30.
[8] Moscicki AB, Hills N, Shiboski S, Powell K, Jay N, Hanson E, et al. Risks for incident human papillomavirus infection and low-grade squamous intraepithelial lesion development in young females. JAMA 2001;285(23):2995–3002.
[9] Winer RL, Feng Q, Hughes JP, O'Reilly S, Kiviat NB, Koutsky LA. Risk of female human papillomavirus acquisition associated with first male sex partner. J Infect Dis 2008;197(2):279–82.
[10] Goodman MT, Shvetsov YB, McDuffie K, Wilkens LR, Zhu X, Thompson PJ, et al. Prevalence, acquisition, and clearance of cervical human papillomavirus infection among women with normal cytology: Hawaii Human Papillomavirus Cohort Study. Cancer Res 2008;68(21):8813–24.
[11] Ho GY, Bierman R, Beardsley L, Chang CJ, Burk RD. Natural history of cervicovaginal papillomavirus infection in young women. N Engl J Med 1998;338(7):423–8.
[12] Moscicki AB, Shiboski S, Hills NK, Powell KJ, Jay N, Hanson EN, et al. Regression of low-grade squamous intra-epithelial lesions in young women. Lancet (London, England) 2004;364(9446):1678–83.
[13] Moscicki AB, Shiboski S, Broering J, Powell K, Clayton L, Jay N, et al. The natural history of human papillomavirus infection as measured by repeated DNA testing in adolescent and young women. J Pediatr 1998;132(2):277–84.
[14] Bansal N, Wright JD, Cohen CJ, Herzog TJ. Natural history of established low grade cervical intraepithelial (CIN 1) lesions. Anticancer Res 2008;28(3b):1763–6.
[15] Castle PE, Sideri M, Jeronimo J, Solomon D, Schiffman M. Risk assessment to guide the prevention of cervical cancer. J Low Genit Tract Dis 2008;12(1):1–7.
[16] Katki HA, Wacholder S, Solomon D, Castle PE, Schiffman M. Risk estimation for the next generation of prevention programmes for cervical cancer. Lancet Oncol 2009;10(11):1022–3.
[17] Moscicki AB, Ma Y, Jonte J, Miller-Benningfield S, Hanson L, Jay J, et al. The role of sexual behavior and HPV persistence in predicting repeated infections with new HPV types. Cancer Epidemiol Biomarkers Prev 2010;19(8):2055–65.
[18] van der Marel J, Berkhof J, Ordi J, Torne A, Del Pino M, van Baars R, et al. Attributing oncogenic human papillomavirus genotypes to high-grade cervical neoplasia: which type causes the lesion? Am J Surg Pathol 2015;39(4):496–504.
[19] Kjær SK, Frederiksen K, Munk C, Iftner T. Long-term absolute risk of cervical intraepithelial neoplasia grade 3 or worse following human papillomavirus infection: role of persistence. J Natl Cancer Inst 2010;102(19):1478–88.
[20] Wright Jr. TC, Stoler MH, Sharma A, Zhang G, Behrens C, Wright TL. Evaluation of HPV-16 and HPV-18 genotyping for the triage of women with high-risk HPV + cytology-negative results. Am J Clin Pathol 2011;136(4):578–86.
[21] Schiffman M, Herrero R, Desalle R, Hildesheim A, Wacholder S, Rodriguez AC, et al. The carcinogenicity of human papillomavirus types reflects viral evolution. Virology 2005;337(1):76–84.
[22] Peto J, Gilham C, Deacon J, Taylor C, Evans C, Binns W, et al. Cervical HPV infection and neoplasia in a large population-based prospective study: the Manchester cohort. Br J Cancer 2004;91(5):942–53.

[23] Schiffman M, Rodriguez AC, Chen Z, Wacholder S, Herrero R, Hildesheim A, et al. A population-based prospective study of carcinogenic human papillomavirus variant lineages, viral persistence, and cervical neoplasia. Cancer Res 2010;70(8):3159–69.
[24] Rodriguez AC, Schiffman M, Herrero R, Wacholder S, Hildesheim A, Castle PE, et al. Rapid clearance of human papillomavirus and implications for clinical focus on persistent infections. J Natl Cancer Inst 2008;100(7):513–17.
[25] Moscicki AB, Ma Y, Wibbelsman C, Darragh TM, Powers A, Farhat S, et al. Rate of and risks for regression of cervical intraepithelial neoplasia 2 in adolescents and young women. Obstet Gynecol 2010;116(6):1373–80.
[26] Fuchs K, Weitzen S, Wu L, Phipps MG, Boardman LA. Management of cervical intraepithelial neoplasia 2 in adolescent and young women. J Pediatr Adol Gynec 2007;20(5):269–74.
[27] Wentzensen N, Hampl M, Herkert M, Reichert A, Trunk MJ, Poremba C, et al. Identification of high-grade cervical dysplasia by the detection of p16INK4a in cell lysates obtained from cervical samples. Cancer 2006;107(9):2307–13.
[28] McCredie MR, Sharples KJ, Paul C, Baranyai J, Medley G, Jones RW, et al. Natural history of cervical neoplasia and risk of invasive cancer in women with cervical intraepithelial neoplasia 3: a retrospective cohort study. Lancet Oncol 2008;9(5):425–34.
[29] Moscicki AB, Singer A. The cervical epithelium during puberty and adolescence. In: Jordan JSA, Shafi M, Jones III H, editors. The cervix. Oxford, UK: Blackwell Publishing; 2006.
[30] Syrjanen K, Hakama M, Saarikoski S, Vayrynen M, Yliskoski M, Syrjanen S, et al. Prevalence, incidence, and estimated life-time risk of cervical human papillomavirus infections in a nonselected Finnish female population. Sex Transm Dis 1990;17(1):15–19.
[31] Burchell AN, Richardson H, Mahmud SM, Trottier H, Tellier PP, Hanley J, et al. Modeling the sexual transmissibility of human papillomavirus infection using stochastic computer simulation and empirical data from a cohort study of young women in Montreal, Canada. Am J Epidemiol 2006;163(6):534–43.
[32] Farhat S, Nakagawa M, Moscicki AB. Cell-mediated immune responses to human papillomavirus 16 E6 and E7 antigens as measured by interferon gamma enzyme-linked immunospot in women with cleared or persistent human papillomavirus infection. Intl J Gynecol Cancer 2009;19(4):508–12.
[33] Olsson SE, Kjaer SK, Sigurdsson K, Iversen OE, Hernandez-Avila M, Wheeler CM, et al. Evaluation of quadrivalent HPV 6/11/16/18 vaccine efficacy against cervical and anogenital disease in subjects with serological evidence of prior vaccine type HPV infection. Hum Vaccin 2009;5(10):696–704.
[34] Widdice L, Ma Y, Jonte J, Farhat S, Breland D, Shiboski S, et al. Concordance and transmission of human papillomavirus within heterosexual couples observed over short intervals. J Infect Dis 2013;207(8):1286–94.
[35] Hernandez BY, Wilkens LR, Zhu X, Thompson P, McDuffie K, Shvetsov YB, et al. Transmission of human papillomavirus in heterosexual couples. Emerg Infect Dis 2008;14(6):888–94.
[36] Bruni L, Diaz M, Castellsague X, Ferrer E, Bosch FX, de Sanjose S. Cervical human papillomavirus prevalence in 5 continents: meta-analysis of 1 million women with normal cytological findings. J Infect Dis 2010;202(12):1789–99.
[37] Moscicki AB, Ma Y, Farhat S, Darragh TM, Pawlita M, Galloway DA, et al. Redetection of cervical human papillomavirus type 16 (HPV16) in women with a history of HPV16. J Infect Dis 2013;208(3):403–12.
[38] Trottier H, Ferreira S, Thomann P, Costa MC, Sobrinho JS, Prado JC, et al. Human papillomavirus infection and reinfection in adult women: the role of sexual activity and natural immunity. Cancer Res 2010;70(21):8569–77.

[39] Rodriguez AC, Schiffman M, Herrero R, Hildesheim A, Bratti C, Sherman ME, et al. Longitudinal study of human papillomavirus persistence and cervical intraepithelial neoplasia grade 2/3: critical role of duration of infection. J Natl Cancer Inst 2010;102(5):315–24.
[40] Ferenczy A, Bergeron C, Richart RM. Human papillomavirus DNA in fomites on objects used for the management of patients with genital human papillomavirus infections. Obstet Gynecol 1989;74(6):950–4.
[41] Fu TC, Hughes JP, Feng Q, Hulbert A, Hawes SE, Xi LF, et al. Epidemiology of human papillomavirus detected in the oral cavity and fingernails of mid-adult women. Sexually Transmitted Dis 2015;42 (12):677–85.
[42] Schiffman M, Doorbar J, Wentzensen N, de Sanjose S, Fakhry C, Monk BJ, et al. Carcinogenic human papillomavirus infection. Nat Rev Dis Prim 2016;2:16086.
[43] Koshiol J, Lindsay L, Pimenta JM, Poole C, Jenkins D, Smith JS. Persistent human papillomavirus infection and cervical neoplasia: a systematic review and meta-analysis. Am J Epidemiol 2008;168(2):123–37.
[44] Rositch AF, Koshiol J, Hudgens MG, Razzaghi H, Backes DM, Pimenta JM, et al. Patterns of persistent genital human papillomavirus infection among women worldwide: a literature review and meta-analysis. Inter J Cancer 2013;133(6):1271–85.
[45] Moscicki AB, Widdice L, Ma Y, Farhat S, Miller-Benningfield S, Jonte J, et al. Comparison of natural histories of human papillomavirus (HPV) detected by clinician- and self-sampling. Int J Cancer 2010;127(8):1882–92.
[46] Monographs on the evaluation of carcinogenic risks to human volume 100B: a review of human carcinogens: biological agents. Lyon, France: International Agency for Research on Cancer. IARC; 2012.
[47] Maucort-Boulch D, Plummer M, Castle PE, Demuth F, Safaeian M, Wheeler CM, et al. Predictors of human papillomavirus persistence among women with equivocal or mildly abnormal cytology. Int J Cancer 2010;126(3):684–91.
[48] Petry KU, Horn J, Luyten A, Mikolajczyk RT. Punch biopsies shorten time to clearance of high-risk human papillomavirus infections of the uterine cervix. BMC Cancer 2018;18(1):318.
[49] Franceschi S, El-Serag HB, Forman D, Newton R, Plummer M. In: Thun M LM, Cerhan JR, Haiman CA, Schottenfeld D, editors. Cancer epidemiology and prevention. 4th ed. New York: Oxford University Press; 2018.
[50] Plummer M, Peto J, Franceschi S. Time since first sexual intercourse and the risk of cervical cancer. Int J Cancer 2012;130(11):2638–44.
[51] International Collaboration of Epidemiological Studies of Cervical Cancer. Carcinoma of the cervix and tobacco smoking: collaborative reanalysis of individual data on 13,541 women with carcinoma of the cervix and 23,017 women without carcinoma of the cervix from 23 epidemiological studies. Int J Cancer 2006;118(6):1481–95.
[52] International Collaboration of Epidemiological Studies of Cervical Cancer. Cervical cancer and hormonal contraceptives: collaborative reanalysis of individual data on 16,573 women with cervical cancer and 35,509 women without cervical cancer from 24 epidemiological studies. Lancet 2007a;370:1609–1621.
[53] International Collaboration of Epidemiological Studies of Cervical Cancer. Comparison of risk factors for invasive squamous cell carcinoma and adenocarcinoma of the cervix: Collaborative reanalysis of individual data on 8,097 women with squamous cell carcinoma and 1,374 women with adenocarcinoma from 12 epidemiological studies. Int J Cancer 2007;120:885–91.
[54] International Collaboration of Epidemiological Studies of Cervical Cancer. Cervical Carcinoma and Sexual Behavior: Collaborative Reanalysis of Individual Data on 15,461 Women with Cervical Carcinoma and 29,164 Women without Cervical Carcinoma from 21 Epidemiological Studies. Cancer Epidemiol Biomarkers Prev 2009;18(4).

[55] Thomas DB, Noonan L, Whitehead A. Breast cancer and depot-medroxyprogesterone acetate. WHO collaborative study of neoplasia and steroid contraceptives. Bull World Health Organ 1985;63(3):513–19.
[56] Roura E, Castellsague X, Pawlita M, Travier N, Waterboer T, Margall N, et al. Smoking as a major risk factor for cervical cancer and pre-cancer: results from the EPIC cohort. Int J Cancer 2014;135 (2):453–66.
[57] Karim S, Souho T, Benlemlih M, Bennani B. Cervical cancer induction enhancement potential of chlamydia trachomatis: a systematic review. Curr Microbiol 2018;75:1667–74.
[58] Smith JS, Bosetti C, Munoz N, Herrero R, Bosch FX, Eluf-Neto J, et al. Chlamydia trachomatis and invasive cervical cancer: a pooled analysis of the IARC multicentric case-control study. Int J Cancer 2004;111(3):431–9.
[59] Castellsague X, Diaz M, Vaccarella S, de Sanjose S, Munoz N, Herrero R, et al. Intrauterine device use, cervical infection with human papillomavirus, and risk of cervical cancer: a pooled analysis of 26 epidemiological studies. Lancet Oncol 2011;12(11):1023–31.
[60] Rinaldi S, Plummer M, Biessy C, Castellsagué X, Overvad K, Kruger KS, et al. Endogenous sex steroids and risk of cervical carcinoma: results from the EPIC study. Cancer Epidemiol Biomarkers Prev 2011;20 (12):2532–40.
[61] Chung SH, Franceschi S, Lambert PF. Estrogen and ERalpha: culprits in cervical cancer? Trends Endocrinol Metab 2010;21(8):504–11.
[62] Spurgeon ME, Chung SH, Lambert PF. Recurrence of cervical cancer in mice after selective estrogen receptor modulator therapy. Am J Pathol 2014;184(2):530–40.
[63] den Boon JA, Pyeon D, Wang SS, Horswill M, Schiffman M, Sherman M, et al. Molecular transitions from papillomavirus infection to cervical precancer and cancer: role of stromal estrogen receptor signaling. Proc Natl Acad Sci U S A 2015;112(25):E3255–64.
[64] Hellberg D, Nilsson S, Haley NJ, Hoffman D, Wynder E. Smoking and cervical intraepithelial neoplasia: nicotine and cotinine in serum and cervical mucus in smokers and nonsmokers. Am J Obst Gynecol 1988;158(4):910–13.
[65] Winer RL, Hughes JP, Feng Q, O'Reilly S, Kiviat NB, Holmes KK, et al. Condom use and the risk of genital human papillomavirus infection in young women. New Engl J Med 2006;354(25):2645–54.
[66] Burger EA, Kim JJ, Sy S, Castle PE. Age of acquiring causal human papillomavirus (HPV) infections: leveraging simulation models to explore the natural history of HPV-induced cervical cancer. Clin Infect Dis 2017;65(6):893–9.
[67] Baussano I LF, Ronco G, Franceschi S. Impacts of human papillomavirus vaccination for different populations: a modeling study. Int J Cancer 2018;0(0).

SECTION

HUMAN PAPILLOMAVIRUS BEYOND CERVICAL CANCER

2

INTRODUCTION

Intense, large-scale, research on the role of Human Papillomavirus (HPV) as the major carcinogen in cervical cancer has driven much of the biological and pathological development of HPV science leading to the introduction of HPV testing in cervical screening and to the initial case for HPV vaccination. Section 2, however, moves beyond the cervix to address the development of understanding of the role of HPV in cancers and other lesions in other parts of the female body and in men. This has been increasingly important in defining the whole role of HPV in cancer, showing that about 5% of all cancer globally is associated with HPV, and a disproportionate excess of HPV-driven cancer is in low-income countries: changing from 1.2% in Australia to 15% in sub-Saharan Africa and India [1].

A distinct aspect of HPV is discussed in Chapter 10, Low-risk Human Papillomavirus: Genital Warts, Cancer and Respiratory Papillomatosis. This addresses the role in disease of HPV types considered to be of low-risk for cancer, but important in genital warts and occasionally associated with distinctive, unusual types of anogenital cancer. Historically these are important: the transmissible nature of genital warts has been known at least since Roman times. The occasional involvement of low-risk HPV, mainly in rare kinds of anogenital cancer is well documented, and the clinical association of genital warts and cervical cancer has been important in setting the hypothesis that HPV might cause cervical cancer. Genital warts also are a common, visible sexually transmitted disease with an annual incidence of around 0.1%–0.2% in surveillance studies and a peak at 15–24 years of age. This made genital warts an important, accessible endpoint for demonstrating the efficacy of vaccination against HPV6 and 11. The rare, but serious nonmalignant condition of recurrent respiratory papillomatosis, causing serious respiratory distress affecting children and occasionally death has been an important, highly emotive issue in regard to prevention of HPV 6/11 infection by vaccination and the development of quadrivalent vaccination, especially in the USA.

The growth in studies of the male role in HPV infection has been very important for understanding the ease of transmission of HPV between men and women. It has led to a clearer knowledge of the natural history of HPV infection in men and its role in some male cancers, particularly anal, oropharyngeal, and penile cancer. Studies of HPV infection in men as well as women have contributed to the realization that HPV is a group of widespread, almost universal, sexually transmitted viruses that occasionally produce cancer in both sexes. This is accepted widely by the health professions and largely in the informed general population. This work has also contributed to understanding the interaction of HIV and HPV infection in women and also men, particularly men who have sex with men (MSM). The role of HIV induced and other iimmunosuppression in allowing HPV carcinogenesis, and the role of HPV in increasing risk of HIV infection are both important.

In Chapter 11, Human Papillomavirus Infection and Related Diseases Among Men, Anna Guilliano discusses how the evidence on the natural history of HPV infection in men has developed rapidly in recent years. Previously the focus on men had been mainly in relation to cervical HPV with studies of men being limited by the practical difficulties of sample taken for HPV detection. These studies had shown that small, often high-grade penile lesions were common in male consorts of women with cervical precancer but appeared to carry very little risk of progression to cancer, unlike histologically similar cervical lesions. As there was no easy, painless treatment for essentially subclinical lesions, there was little reason to investigate. Importantly, recent studies have confirmed that HPV infection is frequent and widespread in the anogenital region of men as well as women, and that HPV is easily transmitted from men to women and women to men, and men to men. This is frequent compared to other sexually transmitted diseases as a result of skin-to-skin contact and only partly prevented by condom use.

Increased understanding of the ease of transmission and the ubiquity of anogenital HPV has been important in influencing the development of gender neutral vaccination. Awareness of the ease of transmission and that HPV-driven cancers are a rare complication of HPV infection in men as well as women has helped convince all, except those with the most extreme conservative views on sexual activity in young men and women, of the value of HPV vaccination. There is further discussion in Section 4 how gender neutral HPV vaccination provides additional protection for women through herd immunity and also for young males, particularly MSM who are at high-risk and who would be difficult to identify and reach as a specific population at an early age for prophylactic vaccination.

For most of the cancers outside the cervix the range of associated HPV types is much smaller with predominance of HPV16, accounting for 90% anal cancer and HPV-positive head and neck squamous carcinoma and 70%–80% of HPV-related vulval and penile neoplasia. These cancers affect mainly the specialized skin of the anogenital regions of men as well as women, beginning historically with the relation of vulval cancer and precancer to cervical cancer, then extending to penile and anal cancer, the oropharynx, particularly the specialized tonsillar epithelium, and to male scrotal cancer, once thought to be mainly driven by chemical carcinogens in soot. The model for demonstrating the causal role of HPV in these is based on a mixture of experimental laboratory evidence and molecular epidemiological studies, including widespread use of p16 immunohistochemistry as a marker (not entirely specific) for hrHPV E7 gene activity, and thus for the presence and transforming activity of hrHPV.

Chapter 12, Anal HPV Infection and HPV-Associated Disease, is concerned with the recognition of the importance of HPV in anal cancer in both men and women and the particular issue of the high frequency of anal cancer in MSM, especially if HIV positive. Although there are important differences between the cervix and the anus, both have a transformation zone and the approach to cervical cancer prevention has provided a valuable model for preventing anal cancer in MSM. However, the frequency of multiple anal HPV infections in the squamous intraepithelial lesions of MSM and different outcomes to similar lesions of cervical intraepithelial neoplasia complicate the development of screening and treatment of anal SIL. Anal HPV infection and neoplasia remain an area for important further study.

The natural history of other cancers associated with HPV is discussed in Chapter 13, Oropharyngeal Human Papillomavirus and Head and Neck Cancer (oropharyngeal cancer) and Chapter 14, Vulvar, Penile, and Scrotal Human Papillomavirus and Non–Human Papillomavirus Cancer Pathways (vulval, penile, and scrotal cancers). These are very different from the cervix for which probably only a few true cervical cancers are truly unassociated with HPV. For the other sites between 30% and 50% of cancers are HPV-associated. The role of HPVs in head and neck cancer is a rapidly developing and changing field of research as described by Carole Fakhry and her colleagues. Although there are still many questions about the development of HPV-related cancers at this site, it is clear that there are both HPV-related and non-HPV pathways and that the natural history and response to treatment of the two types of cancer are different. For vulval, penile and scrotal cancers there are clearly defined HPV-related and unrelated histological pathways of precancer related to cancers that are different histologically. The HPV-related pathways of squamous intraepithelial lesions (AIN, VIN, and PEIN) resemble morphologically that of the cervix, and are unified under the lower anogenital squamous intraepithelial lesion terminology, but the clinical significance of lesions of similar histology at different sites is not identical, and the appropriate management of equivalent histological grades of lesions is not clear. While morphologically the hrHPV-related pathways to cancer in the vulva and penis are similar to those in the cervix and anus, non-HPV pathways are characterized by differentiated squamous hyperplasia and by the characteristic combination of squamous hyperplasia in a thin epithelium with basal lymphocytic infiltration and sub-epithelial homogenization called lichen sclerosus et atrophicus.

The study of HPV has opened up new opportunities for treating and preventing HPV-related cancers beyond those of the cervix. The natural history and outcome of invasive oropharyngeal cancer that is HPV-positive and HPV-negative differs. Those that are not related to HPV are associated with alcohol and smoking and more aggressive. This has opened up the possibility of less aggressive treatment of HPV-related oropharyngeal cancers. Screening for anal cancer in MSM modeled on cervical screening is being explored. All male and female HPV-related cancers are at least potential, accessible to primary prevention through prophylactic vaccination. The principles of prevention based on that of cervical cancer are the subject of Section 3.

REFERENCE

[1] Forman D, de Martel C, Lacey CJ, et al. Global burden of human papillomavirus and related diseases. Vaccine 2012;305:F12–23.

CHAPTER

LOW-RISK HUMAN PAPILLOMAVIRUS: GENITAL WARTS, CANCER AND RESPIRATORY PAPILLOMATOSIS

10

Charles J.N. Lacey[1], Nuria Guimera[2] and Suzanne M. Garland[3]

[1]York Biomedical Research Institute, University of York, York, United Kingdom [2]DDL Diagnostic Laboratory, Rijswijk, The Netherlands [3]Department of Obstetrics and Gynaecology, University of Melbourne, VIC, Australia

THE PROBLEM: GENITAL WARTS AND LARYNGEAL PAPILLOMATOSIS

Genital warts can also be referred to as anogenital warts and condylomata acuminata. In this chapter, we will use the term Genital Warts and the abbreviation GW for consistency. GW usually present as small papillomas/warty growths on the external anogenital skin. They also occur but less frequently on internal anogenital mucosae, that is, the cervix, vagina, anal canal, and urethral meatus. GW do not usually result in major morbidity or mortality. However, they occasionally persist for long periods and rarely such longstanding infection may progress to malignancy. Thus, ~2% of external genital cancers, namely anal, vulval, and penile cancers, are caused by HPV 6 or HPV 11 (discussed later). However, GW are usually perceived as unsightly and disfiguring by the infected person, and are associated with psychological morbidity and feelings of shame [1].

Therefore, many persons with GW will seek treatment for their lesions, and this can be in various healthcare settings depending on local services. However, treatment is by no means straightforward, as there are many therapies in use, and these are not always used in a logical and efficient way, and inadequate response and recurrence after apparent clearance are frequently seen. Indeed, some years ago we showed that in a setting of unstructured treatment practices, after 3 months of treatment and follow-up, 44% of men and 38% of women still had lesions [2]. This is not only a problem for the infected subjects but also a problem for the service providers, as large amounts of time and money are diverted to the management of GW, which in turn detracts from efforts to manage other sexually transmitted diseases (STDs). One estimate from Australia put the annual healthcare costs of GW at >A$14 million per year [3].

Laryngeal papillomatosis is a disease characterized by the growth of multiple papillomas, usually arising from the larynx. The terms laryngeal papillomatosis and recurrent respiratory papillomatosis (RRP) are both used with overlapping meanings. In this chapter, we will mainly use the term laryngeal papillomatosis for consistency and convenience. The majority of patients present with hoarseness or stridor due to the growth of papillomas on the vocal cords or in the larynx.

Human Papillomavirus. DOI: https://doi.org/10.1016/B978-0-12-814457-2.00010-6

Young age at onset (<3 years), and infection by HPV 11 is associated with increased severity of the disease, as measured by number of required surgical procedures, severity of hoarseness, or airway obstruction [4,5]. The natural history of the disease is highly variable and can either persist for many years, be interspersed with short or long periods of remission, or even result in mortality. Recurrence rates vary between patients, with the worst cases requiring surgery under general anesthesia as frequently as every 3 weeks to maintain an airway. Respiratory papillomas are primarily located in the larynx, but approximately 17% of patients have tracheal disease and 5% pulmonary involvement. The papillomas are usually benign, but malignant conversion occurs in ~2% of cases [6].

HISTORICAL ASPECTS OF GENITAL WARTS AND LARYNGEAL PAPILLOMATOSIS

GW were well known to the Greeks and the Romans, and a number of authors already considered them an STD [7]. The first clear proposal in the modern era that a transmissible agent was responsible for the development of warts was given in 1824 by Sir Astley Cooper [8]. He stated that warts "frequently secrete a matter which is able to produce a similar disease in others." He cites two examples of this, the first being the accidental injury of a surgical dresser with a scalpel that had just been used for the removal of warts, and with warts developing at the inoculation site within a short time. The second example was a case of sexual transmission of GW from husband to wife. In 1897 Cathcart provided an excellent account of GW in which he proposed that they were a specific disease independent of other genital pathology, that the lesions were benign neoplasms that were contagious, and he cites case histories of two pairs of sexual partners to support this [9].

Barrett et al. (1954) were the first to publish direct evidence supporting sexual transmission in the acquisition of GW [10]. They reported on 66 soldiers returning from the Korean War who all had sexual contact overseas and had developed GW, with 38% of the subjects having another concurrent STD. They also described 24 wives of these returning soldiers who presented with GW usually between 4 and 6 weeks after their husbands' return, and in whom 22 of the husbands had concurrent GW and 2 a recent history of such lesions. A number of authors reported attempt to demonstrate viral particles by electron microscopy of specimens of GW in the mid-1960s, but the most convincing findings were those of Oriel and Almeida (1970) in which 13 out of 25 genital wart specimens were shown to contain viral particles [11].

In 1971 Oriel's seminal work, "Natural history of genital warts," was published [12]. He reported on 332 men and women with GW. He showed that the age distribution of these subjects was virtually identical to that of gonorrhea, and that the one-third of the subjects had other concurrent sexually transmitted infections (STIs). In an assiduous exercise in contact tracing, Oriel showed that the incubation period ranged between 3 weeks and 8 months with an average of 12 weeks, and that the infectivity rate from an index case was approximately 65%. The detailed nature of his observations had a profound impact in both Europe and the United States, and brought about the unequivocal acceptance of GW as an STD.

Shortly after this in 1974, Harald zur Hausen and Lutz Gissmann began their work on human papillomavirus (HPV) as a cause of different anogenital lesions, based on anecdotal reports of the rare malignant conversion of GW into squamous cell carcinomas. In 1980 low-risk HPVs were the first genotypes isolated: HPV 6 from GW and HPV 11 from laryngeal papillomas [13,14]. HPV 6 was identified both in typical GW and also in locally invasive giant condyloma acuminata. Later, using HPV 11 as a probe, it was possible to isolate HPV 16 from cervical cancer biopsies. In 2008 Harald zur Hausen was awarded the Nobel Prize in Physiology or Medicine for this discovery and his pioneering work in this area.

Laryngeal papillomatosis was superlatively described in 1871 by the pioneering British otolaryngologist Morell Mackenzie [15]. He acknowledges probably the earliest description by Koderick in 1750, a series of 31 cases by Ehrmann in 1850, and then the first direct in vivo visualization by Czernak, the inventor of indirect laryngoscopy using a specially designed mirror. Mackenzie makes a number of startling observations, reporting 67 cases of papillomatosis, describes their histology being "of warty character" and draws parallels with GW, and reports that the lesions contain what are now referred to as koilocytes, with "dense nuclei surrounded by a halo." Following this laryngeal papillomatosis was recognized as a rare disease that was often difficult to treat. It was not until the "golden era" of HPV research that significant further advances in the understanding of RRP were made. In 1982 Phoebe Mounts, Keerti Shah, and Haskins Kashima from John Hopkins showed both HPV antigens and the presence of the recently described HPV 6 in a series of laryngeal papillomas [16]. The following year Betty Steinberg, Allan Abramson, and colleagues from the Long Island Jewish Medical Centre showed the presence of latent laryngeal HPV in RRP cases during clinical remission [17].

RISK FACTORS, NATURAL HISTORY, AND TREATMENT OF GENITAL WARTS AND LARYNGEAL PAPILLOMATOSIS

GW are quite frequently associated with other STIs. We examined the demographic and geospatial risk factors for GW and their interrelation with other STIs in an urban area in the United Kingdom [18]. Regression analysis showed that young age (15–24 years), ethnicity (Black > White > Asian), and social deprivation were independent risk factors for all STDs. There were highly significant correlations in the geospatial distribution of STD incidence rates, and GW and chlamydia had the widest geospatial distribution.

The natural history of untreated GW is poorly understood, as such studies would be unethical. Some cases of GW do undergo spontaneous resolution, and this is known to be associated with the development of a cell-mediated immune response [19]. Even with modern treatments, the average duration of an episode of GW is ~3 months but with wide variation around the mean [20].

There are a variety of medical treatments for GW and various national guidelines for their use. Some of these contain algorithms to determine treatment and treatment sequence, and patients and physicians can choose between patient-applied, multiple session physician-delivered, and single-session physician-delivered using local anesthesia options [21].

The strongest risk factor for juvenile-onset laryngeal papillomatosis is a maternal history of GW in pregnancy which is associated with a >200 times increased risk for the infant, and a "classical case" is a firstborn child delivered vaginally to a teenage mother [22]. The risk of adult-onset laryngeal papillomatosis is associated with lifetime number of sexual partners and oro-genital sex [23].

Risk factors for developing RRP among children of women with GW are poorly understood. Patients have a widespread latent HPV infection in clinically normal tissues of the larynx, trachea, and bronchi. In latency the viral DNA is present but with essentially no expression of viral RNA and no clinical or histologic evidence of disease. Recurrent papillomas are thought to be due to repeated activation of the latent infection, rather than reinfection or "seeding" of the virus during surgery, as was originally thought by many physicians. However, a significant fraction of the normal population carries latent HPV DNA in their airway but with no history of the disease, suggesting additional factors in pathogenesis.

Many therapies have been tried for RRP, with limited success and often with significant side effects. These include various surgical approaches, topical and systemic treatment with immunomodulators, antivirals, and chemotherapeutic drugs. Unfortunately, with the exception of the two large interferon studies, most reports have been either case reports or small clinical series with no controls, and at present, there are still no proven effective adjuvant therapies in addition to surgical ablation.

PATHOLOGY OF GENITAL WARTS AND LARYNGEAL PAPILLOMATOSIS

GW are usually multiple and multifocal, occurring in the vulval, penile, perianal and vaginal regions, and less commonly the cervix. HPV infects long-lived cells (probably stem cells) on the basal cell layer of the epithelium, and then undergoes its life cycle in differentiation-committed cells as they migrate upwards within the epithelium [24]. Fig. 10.1 illustrates some of the differences between low-risk and high-risk HPV lesions. Typical histological features of a genital wart include elongation of the dermal papillae (papillomatosis), hyperplasia of the stratum spinosum (acanthosis), and in the stratum granulosum, large, vacuolated cells (koilocytes) are seen (Fig. 10.2). The histology of laryngeal papillomatosis lesions is essentially the same as that of GW lesions. There may often be some features of intraepithelial neoplasia within GW lesions. Such features include a failure of orderly maturation, excessive and abnormal mitotic activity, and cells with an increased nuclear-to-cytoplasmic ratio. There are still concurrent terminologies to describe such intraepithelial neoplasia changes (i.e., IN1, IN2, IN3) and the SIL classification [low-grade squamous intraepithelial lesion (LSIL), high-grade squamous intraepithelial lesion (HSIL)].

At all anogenital sites, GW may be misdiagnosed as an LSIL. For some authors, these terms are interchangeable. If GW present moderate cytological atypia and a dysplastic growth pattern, the lesions are classified as IN 2/3 or HSIL, but usually in those cases a high-risk HPV is also present. Other differential diagnoses of GW are squamous papilloma, giant condyloma, and verrucous carcinoma. Therefore defining the role of HPVs 6 and 11 in these lesions may assist in accurate diagnosis.

A rare variant of GW is the borderline malignant tumor referred to as giant condyloma or Buschke–Lowenstein tumor. These can arise at any anogenital site and can be associated with various immunodeficiency states. Giant condyloma is induced by HPVs 6 and 11 and cause local destruction without metastatic spread. Chemoradiotherapy may be the best treatment option [25].

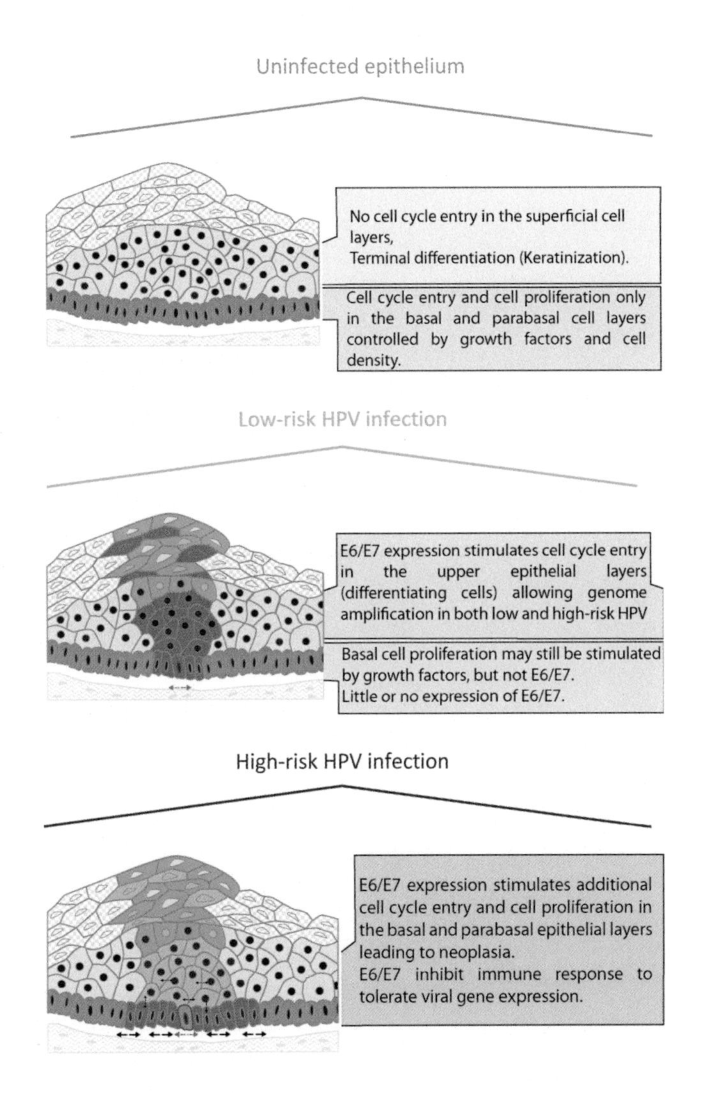

FIGURE 10.1

Explaining the different cellular differentiation states related to viral protein expression in lesions caused by low-risk and in high-risk HPV.

Adapted from Egawa N, Doorbar J. The low-risk papillomaviruses. Virus Res 2017;231:119–27.

HUMAN PAPILLOMAVIRUS IN GENITAL WARTS AND LARYNGEAL PAPILLOMATOSIS

GW and laryngeal papillomatosis are exophytic lesions arising from productive viral infection by HPV 6 or HPV 11, and in the most highly controlled studies either HPV 6 or HPV 11 are found within the lesions at rates close to 100% [26,27]. HPVs 6 and 11 are exceedingly common in

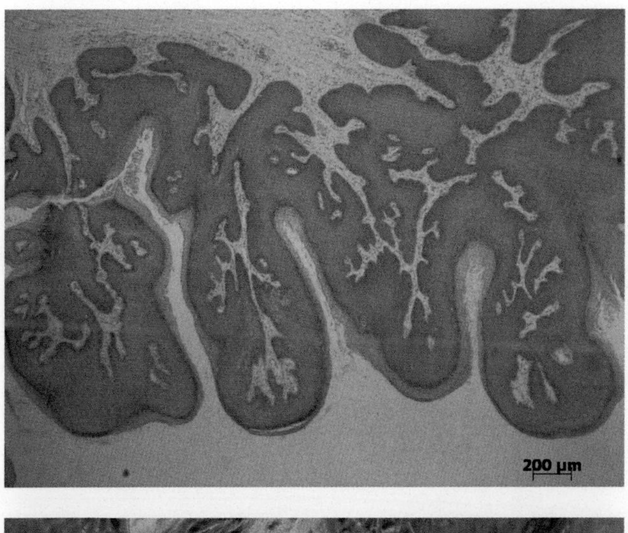

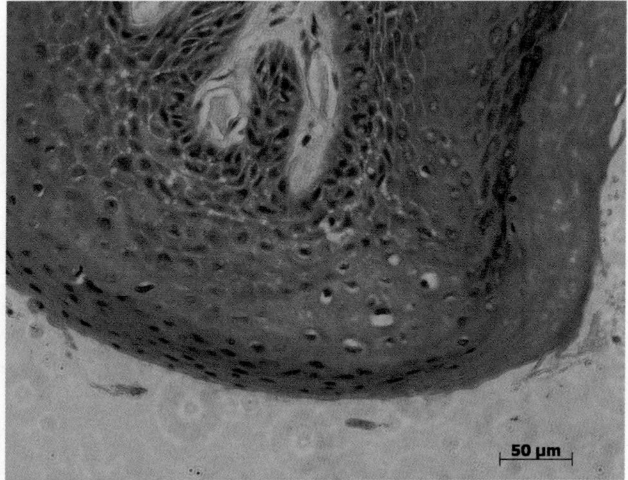

FIGURE 10.2

H&E sections of a typical GW (Lacey original).

benign HPV mucosal lesions and rarely detected in HPV-associated cancers, and therefore considered low-risk HPVs, where risk implies the potential for malignant transformation [24]. They are part of a distinct taxonomic HPV species, Alpha 10 (Fig. 10.3). The prevalence of different HPV types differs by type of lesion. HPV 6 is more prevalent than HPV 11 in GW, whereas HPVs 6 and 11 are found at similar frequencies in laryngeal papillomatosis [26–28].

HPV is usually present as a single infection and less frequently as multiple infections, which arise as separate lesional clones. The prevalence of multiple HPV infections decreases with increasing severity of the lesion, being 50%–80% in anogenital warts and 12% in cervical cancer.

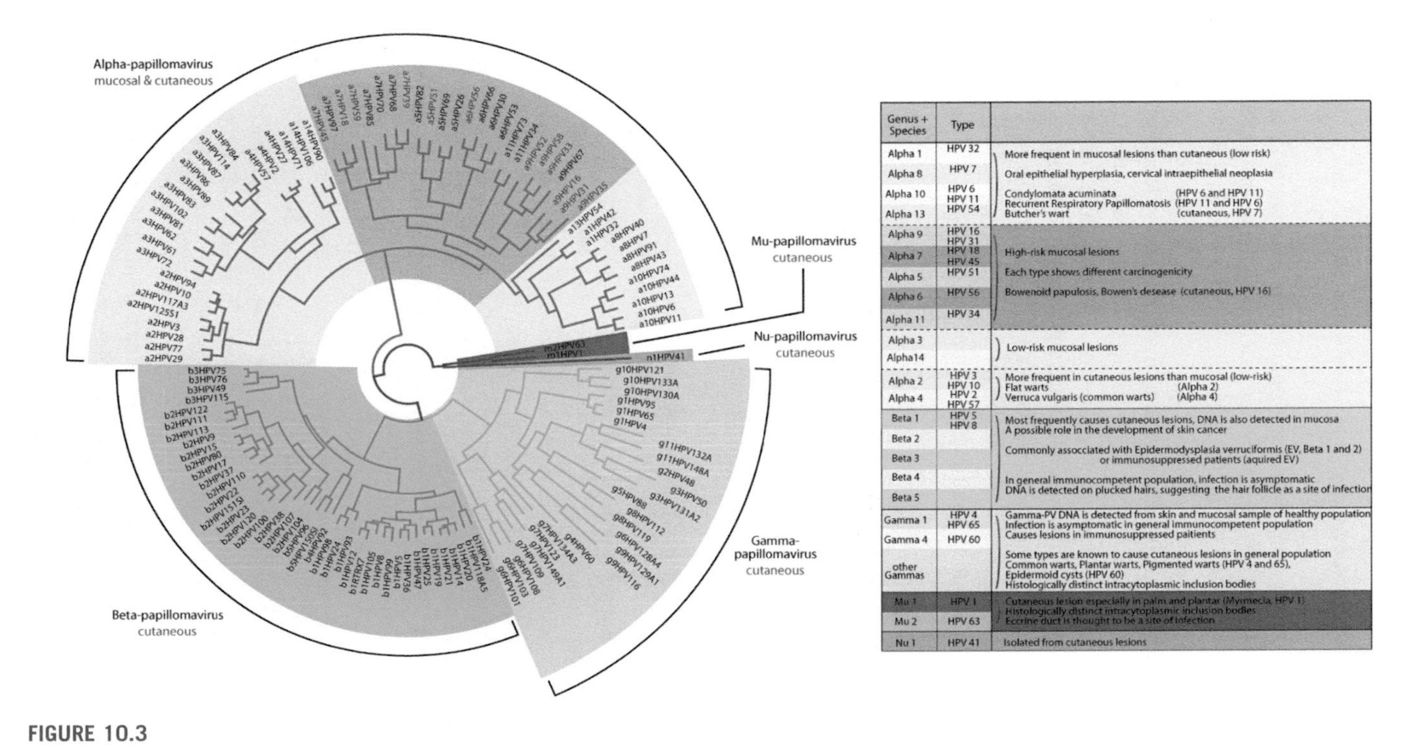

Alpha-papillomavirus
mucosal & cutaneous
Mu-papillomavirus
cutaneous
Nu-papillomavirus
cutaneous
Gamma-
papillomavirus
cutaneous
Beta-papillomavirus
cutaneous
Genus + Species
Type
Alpha 1 HPV 32 More frequent in mucosal lesions than cutaneous (low risk)
Alpha 8 HPV 7 Oral epithelial hyperplasia, cervical intraepithelial neoplasia
Alpha 10 HPV 6 HPV 11 HPV 54 Condylomata acuminata (HPV 6 and HPV 11)
Recurrent Respiratory Papillomatosis (HPV 11 and HPV 6)
Alpha 13 Butcher's wart (cutaneous, HPV 7)
Alpha 9 HPV 16 HPV 31
Alpha 7 HPV 18 HPV 45 High-risk mucosal lesions
Alpha 5 HPV 51 Each type shows different carcinogenicity
Alpha 6 HPV 56 Bowenoid papulosis, Bowen's desease (cutaneous, HPV 16)
Alpha 11 HPV 34
Alpha 3
Alpha14 Low-risk mucosal lesions
Alpha 2 HPV 3 HPV 10 More frequent in cutaneous lesions than mucosal (low-risk)
Flat warts (Alpha 2)
Alpha 4 HPV 2 HPV 57 Verruca vulgaris (common warts) (Alpha 4)
Beta 1 HPV 5 HPV 8 Most frequently causes cutaneous lesions, DNA is also detected in mucosa
A possible role in the development of skin cancer
Beta 2
Beta 3 Commonly associated with Epidermodysplasia verruciformis (EV, Beta 1 and 2) or immunosuppressed patients (aquired EV)
Beta 4
Beta 5 In general immunocompetent population, infection is asymptomatic
DNA is detected on plucked hairs, suggesting the hair follicle as a site of infection
Gamma 1 HPV 4 HPV 65 Gamma-PV DNA is detected from skin and mucosal sample of healthy population
Infection is asymptomatic in general immunocompetent population
Gamma 4 HPV 60 Causes lesions in immunosuppressed paitients
other Gammas Some types are known to cause cutaneous lesions in general population
Common warts, Plantar warts, Pigmented warts (HPV 4 and 65),
Epidermoid cysts (HPV 60)
Histologically distinct intracytoplasmic inclusion bodies
Mu 1 HPV 1 Cutaneous lesion especially in palms and plantar (Myrmecia, HPV 1)
Histologically distinct intracytoplasmic inclusion bodies
Mu 2 HPV 63 Eccrine duct is thought to be a site of infection
Nu 1 HPV 41 Isolated from cutaneous lesions

FIGURE 10.3

Showing the taxonomy and evolutionary relationships between HPVs.

Reproduced from Egawa N, Doorbar J. The low-risk papillomaviruses. Virus Res 2017;231:119–27.

Multiple infections with other HPVs include both low-risk and high-risk HPV types [26]. The prevalence of multiple infections varies with the HPV detection method used in the different studies, as not all tests have same specificity and sensitivity in the detection of the different HPV types.

Multiple infections and simultaneous lesions are common in anogenital sites prone to infection. Studies have shown how cervical HPV infections act independently in the development of preneoplastic and neoplastic lesions in the cervix and that each lesion present in a patient is the result of a specific HPV type [29,30]. Simultaneous anogenital lesions can be driven by one or more HPV types, may originate at the same or different times, and may progress independently. The high prevalence of multiple infections detected in some GW creates difficulties in the study of their natural history. In a recent study of penile lesions in men presenting with penile cancer, we found atypical GW with high-risk HPV 16 and p16 biomarker expression and in the same patient invasive carcinoma and high-grade precancers with the same high-risk HPV 16 and p16 expression pattern. This finding suggests a possible, rare precursor role of the infrequent HPV 16/p16 positive GW [31]. In laryngeal papillomatosis the presence of multiple infections is infrequent and simultaneous precancerous lesions rare [5].

LOW-RISK HUMAN PAPILLOMAVIRUS TYPES AND CANCER

HPVs are phylogenetically classified in different genera (e.g., α, β, γ, μ, ν—Fig. 10.3). There is a strong concordance between HPV phylogeny, viral natural history, and carcinogenicity. In 2003 an epidemiologic risk classification for HPV was published [32], classifying HPVs as high-risk or low-risk based on the strength of the association in a metaanalysis of each type with cervical cancer. Subsequently, the epidemiological risk classification was compared with phylogenetic relationships of DNA and amino acid sequences. Evolutionary models clustered all α-HPVs into three ancestral branches [33]. The first and third ancestral branches include low-risk HPVs (α1, α2, α3, α4, α8, α10, α13, and α14 species) and the second ancestral branch high-risk HPVs (α5, α6, α7, α9, α12, and α11 species). These findings suggested that the carcinogenic potential of α-genera HPV genotypes is closely related to their species classification.

However, contrary to their biological description as "low-risk" or "noncarcinogenic" viral genotypes, low-risk HPVs have been identified in anogenital and head and neck cancers, although with a low frequency. The prevalence of low-risk HPVs decreases with increasing severity of the lesion, reflecting the lower oncogenic potential and more frequent regression of the lesions. Low-risk HPVs 6, 11, 40, 42, and 44 have been detected in 1.7% of cervical carcinomas as single and/or multiple genotypes (0.27% and 1.4%, respectively) [34]. Low-risk HPVs 6 and 11 have a prevalence of 2% and 1.3%, respectively, in HSIL and 6.2% and 2.9%, respectively, in LSIL [35].

At the Catalan Institute of Oncology, we performed several large epidemiological studies which included 13,328 HPV-related malignant carcinomas from different anogenital sites, such as cervical, vaginal, vulvar, anal, and penile. These studies were the basis to investigate the presence of low-risk HPVs in malignant carcinomas as a large number of patients are needed to detect these rare cases. Only 57 cases showed single low-risk HPV infection [30]. Low-risk

HPV DNA was confirmed as a single infection in tumor cells in 56% (32/57) by laser capture microdissection followed by human papillomavirus polymerase chain reaction (HPV PCR) detection. All anogenital tumors with confirmed low-risk HPVs 6 and 11 showed verruco-papillary, well-differentiated, squamous, or transitional histology without p16 expression. HPVs 42 and 70 were associated with typical squamous carcinomas with p16 expression. These data support the causal involvement of low-risk HPVs in the carcinogenesis of <2% of anogenital malignancies [30].

Malignant transformation can occur in RRP but is infrequent, of the order of 2% [6,36]. Such cancer development is strongly linked with a long history, HPV 11, and involvement of the lung parenchyma. The HPV 11 genome can be integrated into such cases, suggesting deregulated viral gene expression followed by the acquisition of additional genetic and epigenetic modifications as seen for the α HR types [24].

THERAPEUTIC AND PROPHYLACTIC VACCINES AGAINST GENITAL WARTS AND LARYNGEAL PAPILLOMATOSIS

Therapeutic vaccination as a treatment for GW using poorly characterized autogenous wart preparations was first described in 1925, and similar therapy was also used in RRP [37,38]. Such approaches to treatment using autogenous vaccines were occasionally used for many years, but a double-blind controlled cross-over study versus placebo showed no differences in outcome [39]. We developed an HPV 6 L2E7 fusion protein therapeutic vaccine for GW, based on the findings of Jarrett & Campo in the BPV model, but a series of three trials showed no additional benefit over conventional therapy [40].

The foundation stone for HPV VLP vaccine development was the description by Jian Zhou and colleagues of the expression of HPV 16 L1 and its spontaneous assembly into VLPs in 1991 [41]. The subsequent development of the HPV VLP vaccines is fully covered in Chapters 15–17, but some of the key observations at this time were in HPV 6/11 systems. Rose and colleagues inoculated rabbits with HPV 11 VLPs, and using the Kreider athymic mouse model showed that vaccination-induced neutralizing antibodies against HPV 11 blocked infection [42]. Roden et al. assessed the serological relatedness of genital HPVs and showed that even in highly related HPV genotypes, such as HPVs 6 and 11, where amino acid sequence homology in L1 was >85%, there was relatively limited cross-reactivity by neutralizing antibodies, which suggested that protection induced by HPV VLPs would be predominantly type specific [43]. Lowe et al. immunized African green monkeys with HPV 11 VLPs and showed that high levels of neutralizing antibodies were induced in serum, but also that significant levels of neutralizing antibodies could be detected via transudation in cervico-vaginal secretions [44]. The development of VLP vaccines then proceeded apace and culminated in 2006 with the licensing of Gardasil by the FDA. This was followed in 2007 with a number of key publications including that of Garland et al. demonstrating 100% efficacy of the HPV 6/11/16/18 vaccine in the per protocol population against all external anogenital and vaginal lesions, and all HPV 6/11 lesions [45]. These data triggered the widespread introduction of HPV VLP vaccination across the world.

PAST, PRESENT, AND FUTURE EPIDEMIOLOGY OF GENITAL WARTS AND LARYNGEAL PAPILLOMATOSIS

GW were a significant public health problem worldwide with rates of 0.1%–0.2% in general populations for both incidence of new cases, as well as prevalence/recurrent cases, before the introduction of HPV vaccination [46]. The UK National Survey of Sexual Attitudes and Lifestyles conducted in the year 2000 was a stratified population-based survey of 11,161 men and women aged 16–44 years. This found that 3.6% of men and 4.1% of women reported ever being diagnosed with GW, which were the commonest reported STI [47].

Incidence data for laryngeal papillomatosis in the prevaccine era are sparse but show greater variation between countries than that for GW. Incidence rates of 3.6, 4.3, and 0.24 (Denmark, the United States, Canada) per 10^6 person-years (py) in children, and 3.9 and 1.8 (Denmark, the United States) per 10^6 py for adult disease have been reported [48–50]. Juvenile-onset disease usually presents between the ages of 1 and 4 years and is equally distributed between males and females. Adult-onset disease has a broad peak between ages 20 and 40 years, with a male:female ratio of ~2:1 [22].

Rapid licensing of the HPV vaccines occurred in many countries from 2007 onwards. Economic modeling of the introduction of HPV vaccines confirmed their cost-effectiveness [51]. Australia was one the first countries to introduce a national state-funded HPV vaccination program in 2007. In Australia during 2007–09, an estimated 83% of females aged 12–17 years received at least one dose of the HPV vaccine and 70% completed the three-dose HPV vaccination course [52], and this degree of coverage has been maintained since then. GW represent a short incubation period HPV disease, and so can act as a sentinel readout to assess the effectiveness of HPV vaccination at a population level. Thus Donovan et al. observed that in sexual health clinics in Australia the rate of diagnosis of GW in young female residents decreased by 59% in 2007–09, that there was also a significant decrease (28%) in GW in nonvaccinated young heterosexual men, whereas no decline was seen in female nonresidents, older women >26 years, and men who have sex with men (MSM) [53].

These early signals of HPV 6/11 disease reduction were conclusively confirmed by the meta-analysis of Drolet et al. [54]. They identified 20 eligible studies conducted in 9 high-income countries representing 140 million py of follow-up. In countries with >50% female vaccination coverage, GW declined by 61% (RR 0.39, 0.22–0.71) in girls aged 13–19 years through vaccination, and in boys <20 years by 34% (0.66, 0.47–0.91) due to herd immunity. Australia extended qHPV vaccination to boys in 2013. Recent Australian data confirm the profound on-going decline in GW, with a 74% reduction in *all* females and a 65% reduction in *all* males [55]. Other countries with lesser degrees of HPV vaccine population coverage are also reporting ongoing declines in GW (United States – [56], Sweden – [57]). Recent modeling has investigated the likely course of GW epidemiology in Australia [58]. These data predict a 95% reduction in GW cases by 2060 in all the subgroups of females, males, and MSM, and by then ~50% of cases will be occurring in non-Australian born nonvaccinated travellers.

Surveillance for juvenile-onset RRP was established in Australia in 2012, and the data until end-2016 have been reported [59]. This found that the average annual incidence rates have declined significantly from 0.16 in 2012 to 0.02 in 2016. Of the incident cases, no mothers were vaccinated prepregnancy, 20% had a maternal history of GW, and 60% were firstborn. This was the first report of a decline in laryngeal papillomatosis incidence in children following the introduction of the HPV vaccine program. A similar national-level surveillance program for pediatric RRP was established in Canada in 2007 at the time of HPV vaccine introduction, and recent data indicate that there have been a ~65% decrease in incidence since then compared to the period 1992–2016 [60].

CONCLUSION

GW and laryngeal papillomatosis are ancient historical diseases almost exclusively caused by HPVs 6 and 11. These diseases were associated with very significant physical and psychological morbidity and healthcare costs. In many countries a distinct increase in incidence and prevalence of GW was recorded in the period 1970–2000 thought to be associated with changes in sexual behavior. The astonishing HPV vaccine efficacy against HPV 6/11 observed in Phase 3 trials has led to the widespread uptake of the quadrivalent vaccine containing HPV 6/11 VLPs in many developed world countries. Such implementation from 2007 onwards can now be seen to be producing very significant reductions in GW and laryngeal papillomatosis disease incidence. This evidence strongly suggests that as HPV vaccine introduction increases worldwide, over the next 50 years, these diseases will become increasingly rare.

ACKNOWLEDGEMENTS

Thanks to Prof. Philip Quirke, University of Leeds, for histopathological advice, and Prof. John Doorbar, University of Cambridge, and Dr. Helen Ashwin, University of York, for assistance with the illustrations.

REFERENCES

[1] Maw RD, Reitano M, Roy M. An international survey of patients with genital warts: perceptions regarding treatment and impact on lifestyle. Int J STD AIDS 1998;9:571–8.
[2] Reynolds M, Murphy M, Waugh MA, Lacey CJN. An audit of treatment of genital warts: opening the feedback loop. Int J STD AIDS 1993;4:226–31.
[3] Pirotta M, Stein AN, Conway EL, Harrison C, Britt H, Garland S. Genital warts incidence and healthcare resource utilisation in Australia. Sex Transm Infect 2010;86:181–6.
[4] Wiatrak BJ, Wiatrak DW, Broker TR, Lewis L. Recurrent respiratory papillomatosis: a longitudinal study comparing severity associated with HPV types 6 and 11 and other risk factors in a large pediatric population. Laryngoscope 2004;114:1–23.
[5] Seedat RY, Thukane M, Jansen AC, Rossouw I, Goedhals D, Burt FJ. HPV types causing juvenile recurrent laryngeal papillomatosis in South Africa. Int J Pediatr Otorhinolaryngol 2010;74:255–9.
[6] Dedo HH, Yu KC. CO2 laser treatment in 244 patients with respiratory papillomas. Laryngoscope 2001;111:1639–44.
[7] Bafverstedt B. Condylomata acuminata - past and present. Acta Dermato-Venereol 1967;47:376–81.
[8] Cooper A. Surgical lectures, no 62: warts. Lancet 1824;3:336–7.
[9] Cathcart CW. Venereal warts, a contagious form of tumour. J Pathol Bacteriol 1897;4:160–72.

[10] Barrett TJ, Silbar JD, McGinley JP. Genital warts - a venereal disease. JAMA 1954;154:333–4.
[11] Oriel JD, Almeida JD. Demonstration of virus particles in human genital warts. Br J Vener Dis 1970;46:37–42.
[12] Oriel JD. Natural history of genital warts. Br J Vener Dis 1971;47:1–13.
[13] Gissmann L, de Villiers EM, zur Hausen H. Analysis of human genital warts (condylomata acuminata) and other genital tumors for human papillomavirus type 6 DNA. Int J Cancer 1982;29:143–6.
[14] Gissmann L, Diehl V, Schultz-Coulon HJ, zur Hausen H. Molecular cloning and characterization of human papilloma virus DNA derived from a laryngeal papilloma. J Virol 1982;44:393–400.
[15] Mackenzie M. Essays on growths in the larynx: with reports, and an analysis of one hundred consecutive cases. London: J &A Churchill; 1871 (Available via Google Books).
[16] Mounts P, Shah KV, Kashima H. Viral etiology of juvenile- and adult-onset squamous papilloma of the larynx. Proc Natl Acad Sci 1982;79:5425–9.
[17] Steinberg BM, Topp WC, Schneider PS, Abramson AL. Laryngeal papillomavirus infection during clinical remission. N Engl J Med 1983;308:1261–4.
[18] Monteiro EF, Lacey CJN, Merrick D. The interrelation of demographic and geospatial risk factors between four common sexually transmitted diseases. Sex Transm Inf 2005;81:41–6.
[19] Coleman N, Birley HDL, Renton AM, Hanna NF, Ryait BK, Byrne M, et al. Immunological events in regressing genital warts. Am J Clin Pathol 1994;102:768–74.
[20] Woodhall SC, Jit M, Cai C, Ramsey T, Zia S, Crouch S, et al. Cost of treatment and QALYs lost due to genital warts: data for the economic evaluation of HPV vaccines in the United Kingdom. Sex Transm Dis 2009;36:515–21.
[21] Lacey CJ, Woodhall SC, Wikstrom A, Ross J. European guideline for the management of anogenital warts. J Eur Acad Dermatol Venereol 2013;27:e263–70.
[22] Gillison ML, Alemany L, Snijders PJ, Chaturvedi A, Steinberg BM, Schwartz S, et al. Human papillomavirus and diseases of the upper airway: head and neck cancer and respiratory papillomatosis. Vaccine 2012;30(Suppl. 5):F34–54.
[23] Kashima HK, Shah F, Lyles A, Glackin R, Muhammad N, Turner L, et al. A comparison of risk factors in juvenile-onset and adult-onset recurrent respiratory papillomatosis. Laryngoscope 1992;102:9–13.
[24] Egawa N, Doorbar J. The low-risk papillomaviruses. Virus Research 2017;231:119–27.
[25] Armstrong N, Foley G, Wilson J, Finan P, Sebag-Montefiore D. Successful treatment of a large Buschke-Lowenstein tumour with chemo-radiotherapy. Int J STD AIDS 2009;20:732–4.
[26] Brown DR, Schroeder JM, Bryan JM, Stoler MH, Fife KH. Detection of multiple human papillomavirus types in condylomata acuminata lesions from otherwise healthy and immunosuppressed patients. J Clin Microbiol 1999;37:3316–22.
[27] Hawkins MG, Winder DM, Ball SL, Vaughan K, Sonnex C, Stanley MA, et al. Detection of specific HPV subtypes responsible for the pathogenesis of condylomata acuminata. Virol J 2013;10:137.
[28] Wiatrak BJ, Wiatrak DW, Broker TR, Lewis L. Recurrent respiratory papillomatosis: a longitudinal study comparing severity associated with human papilloma viral types 6 and 11 and other risk factors in a large pediatric population. Laryngoscope 2004;114(11Pt 2 Suppl. 104):1–23.
[29] Quint W, Jenkins D, Molijn A, Struijk L, van de Sandt M, Doorbar J, et al. One virus, one lesion--individual components of CIN lesions contain a specific HPV type. J Pathol 2012;227:62–71.
[30] Guimerà N, Lloveras B, Lindeman J, Alemany L, van de Sandt M, Alejo M, et al. The occasional role of low-risk human papillomaviruses 6, 11, 42, 44, and 70 in ano-genital carcinoma defined by laser capture microdissection/PCR methodology: results from a global study. Am J Surg Pathol 2013;37:1299–310.
[31] Fernández-Nestosa MJ, Guimerà N, Sanchez DF, Cañete-Portillo S, Velazquez EF, Jenkins D, et al. Human Papillomavirus (HPV) Genotypes in condylomas, intraepithelial neoplasia, and invasive carcinoma of the penis using laser capture microdissection (LCM)-PCR: a study of 191 lesions in 43 patients. Am J Surg Pathol 2017;41:820–32.

[32] Muñoz N, Bosch FX, de Sanjosé S, Herrero R, Castellsagué X, Shah KV, et al. Epidemiologic classification of human papillomavirus types associated with cervical cancer. N Engl J Med 2003;348:518–27.
[33] Schiffman M, Herrero R, Desalle R, Hildesheim A, Wacholder S, Rodriguez AC, et al. The carcinogenicity of human papillomavirus types reflects viral evolution. Virology 2005;337:76–84.
[34] De Sanjose S, Quint WG, Alemany L, Geraets DT, Klaustermeier JE, Lloveras B, et al. Human papillomavirus genotype attribution in invasive cervical cancer: a retrospective cross-sectional worldwide study. Lancet Oncol 2010;11:1048–56.
[35] HPV Information Centre, Statistics. <https://www.hpvcentre.net/datastatistics.php> [accessed 21.02.2019].
[36] Gerein V, Rastorguev E, Gerein J, Draf W, Schirren J. Incidence, age at onset, and potential reasons of malignant transformation in recurrent respiratory papillomatosis patients: 20 years experience. Otolaryngol Head Neck Surg 2005;132:392–4.
[37] Biberstein H. Versuche uber Immunotherapie der Warzen und Kondylome. Klin Wochenschr 1925;4:638–41.
[38] Stephens CB, Arnold GE, Butchko GM, Hardy CL. Autogenous vaccine treatment of juvenile laryngeal papillomatosis. Laryngoscope 1979;89:1689–96.
[39] Malison MD, Morris R, Jones LW. Autogenous vaccine therapy for condyloma acuminatum: a double-blind controlled study. Br J Vener Dis 1982;58:62–5.
[40] Vandepapeliere P, Barrasso R, Meijer CJLM, Walboomers JMM, Wettendorff M, Stanberry LR, et al. Randomised controlled trial of an adjuvanted HPV6 L2E7 vaccine: multiple HPV type infection of external anogenital warts and failure of therapeutic vaccination. J Infect Dis 2005;192:2099–107.
[41] Zhou J, Sun XY, Stenzel DJ, Frazer IH. Expression of vaccinia recombinant HPV 16 L1 and L2 ORF proteins in epithelial cells is sufficient for assembly of HPV virion-like particles. Virology 1991;1991 (185):251–7.
[42] Rose RC, Reichman RC, Bonnez W. Human papillomavirus (HPV) type 11 recombinant virus-like particles induce the formation of neutralizing antibodies and detect HPV-specific antibodies in human sera. J Gen Virol 1994;75:2075–9.
[43] Roden RBS, Hubbert NL, Kirnbauer R, Christensen ND, Lowy DR, Schiller JT. Assessment of the serological relatedness of genital human papillomaviruses by haemagglutination inhibition. J Virol 1996;70:3298–301.
[44] Lowe RS, Brown DR, Bryan JT, Cook JC, George HA, Hofmann KJ, et al. Human papillomavirus type 11 (HPV-11) neutralizing antibodies in the serum and genital mucosal secretions of African green monkeys immunized with HPV-11 virus-like particles expressed in yeast. J Infec Dis 1997;176:1141–5.
[45] Garland SM, Hernandez-Avila M, Wheeler CM, Perez G, Harper DM, Leodolter S, et al. Quadrivalent vaccine against human papillomavirus to prevent anogenital diseases. N Engl J Med 2007;356:1928–43.
[46] Forman D, de Martel C, Lacey CJ, Soerjomataram I, Lortet-Tieulent J, Bruni L, et al. Global burden of human papillomavirus and related diseases. Vaccine 2012;30(Suppl. 5):F12–21.
[47] Fenton KA, Korovessis C, Johnson AM, McCadden A, McManus S, Wellings K, et al. Sexual behaviour in Britain: reported sexually transmitted infections and prevalent genital Chlamydia trachomatis infection. Lancet 2001;358:1851–4.
[48] Lindeberg H, Elbrønd O. Laryngeal papillomas: the epidemiology in a Danish subpopulation 1965-1984. Clin Otolaryngol Allied Sci 1990;15:125–31.
[49] Derkay CS. Task force on recurrent respiratory papillomas. A preliminary report. Arch Otolaryngol Head Neck Surg 1995;121:1386–91.
[50] Campisi P, Hawkes M, Simpson K, Canadian Juvenile Onset Recurrent Respiratory Papillomatosis Working Group. The epidemiology of juvenile onset recurrent respiratory papillomatosis derived from a population level national database. Laryngoscope 2010;120:1233–45.
[51] Jit M, Choi YH, Edmunds JD. Economic evaluation of human papillomavirus vaccination in the UK. BMJ 2008;337:a769.

[52] Tabrizi SN, Brotherton JM, Kaldor JM, Skinner SR, Cummins E, Liu B, et al. Fall in human papillomavirus prevalence following a national vaccination programme. J Infec Dis 2012;206:1645–51.
[53] Donovan B, Franklin N, Guy R, Grulich AE, Regan DG, Ali H, et al. Quadrivalent human papillomavirus vaccination and trends in genital warts in Australia: analysis of national sentinel surveillance data. Lancet Infec Dis 2011;11:39–44.
[54] Drolet M, Benard E, Boily M-C, Ali H, Baandrup L, Bauer H, et al. Population-level impact and herd effects following human papillomavirus vaccination programmes: a systematic review and meta-analysis. Lancet Infec Dis 2015;15:565–80.
[55] Chow EP, Callander D, Fairley CK, Guy RJ, Regan DG, Grulich AE, et al. The continuing decline in genital warts 10 years into the national human papillomavirus vaccination programme: national sentinel surveillance data 2004-2016. IPVC 2018 Sydney.
[56] Flagg EW, Torrone EA. Declines in anogenital warts among age groups most likely to be impacted by human papillomavirus vaccination, United States 2006-2014. Am J Public Health 2018;108:112–19.
[57] Herweijer E, Ploner A, Sparen P. Substantially reduced incidence of genital warts in women and men six years after vaccine availability in Sweden. Vaccine 2018;36:1917–20.
[58] Khawar L, McGregor S, Machalek D, Regan D, Kaldor J, Donovan B, et al. When will we eliminate genital warts in Australia? IPVC 2018 Sydney; 2018.
[59] Novakovic D, Cheng ATL, Zurynski Y, Booy R, Walker PJ, Berkowitz R, et al. A prospective study of the Incidence of juvenile-onset recurrent respiratory papillomatosis after Implementation of a national HPV vaccination program. J Infect Dis 2018;217:208–12.
[60] Campisi P. Should we be screening for oral HPV-related diseases? IPVC 2018 Sydney.

CHAPTER

HUMAN PAPILLOMAVIRUS INFECTION AND RELATED DISEASES AMONG MEN

11

Anna R. Giuliano

Center for Immunization and Infection Research in Cancer (CIIRC), Moffitt Cancer Center and Research Institute, Tampa, FL, United States

WHERE THE STORY OF MALE HUMAN PAPILLOMAVIRUS INFECTION AMONG HETEROSEXUAL MEN BEGAN: THE MALE FACTOR IN CERVICAL CARCINOGENESIS

Male human papillomavirus (HPV) infection significantly contributes to infection and subsequent cervical disease in women [1–4]. Case-control studies of women with cervical cancer and their husbands have demonstrated that men's sexual behavior affects women's risk of cervical neoplasia, even when controlling for female sexual activity [1–7]. In areas with a high incidence of cervical cancer, men's sexual behavior is in itself a risk factor for cervical neoplasia [5]. Among men with a history of multiple sexual partners, male circumcision was associated with a reduced risk of penile HPV infection and a reduced risk of cervical cancer in their current female partners [6].

As male sexual behavior clearly impacts rates of HPV infection, cervical dysplasia, and invasive cervical cancer in female partners, a greater understanding of HPV infection in men is an essential component of cervical cancer prevention programs. Unfortunately, until recently, there was a paucity of studies that could shed light on HPV infection natural history among men, data needed to inform vaccine trial design in males, and interpartner transmission of HPV infection.

HUMAN PAPILLOMAVIRUS NATURAL HISTORY IN HETEROSEXUAL MEN—PAST AND CURRENT UNDERSTANDING

As recently as 2004 the information available regarding penile HPV infection was derived from primarily three sources: (1) studies where husbands of female cervical cancer cases were examined [7–11]; (2) cross-sectional studies of select populations, such as sexually transmitted infection (STI) patient and military recruits [12,13]; and (3) small prospective studies [14,15]. Our study team conducted the first HPV study among US men, comprised of an ethnically diverse population attending an STI clinic. In the early 2000s reported rates of HPV infection varied widely, in part due to the use of different analytical methods and populations studied. Early studies used aceto-whitening of the penis as a diagnostic marker for HPV infection in men, and examined men for

Human Papillomavirus. DOI: https://doi.org/10.1016/B978-0-12-814457-2.00011-8

papular, spiked, and flat whitish or grayish epithelial lesions [16]. While HPV is significantly associated with penile aceto-white lesions [17], many other benign genital conditions are also associated with these lesions [17,18], resulting in poor specificity of peniscopy for detection of HPV in male genitalia. Studies using polymerase chain reaction (PCR) detection of HPV, a method sensitive enough to detect 10–100 copies of viral DNA, or 1–2 copies per cell, have found HPV detection rates in high-risk men to be at least as high as rates in their female counterparts [19].

Depending on the population studied, the stage of the epidemic in the target population, method and anatomical site of sampling, and laboratory method, penile HPV infection has been detected in 16%–69% of healthy men. Among Finnish army recruits, [12] HPV infection was detected in 16.5% of healthy males using PCR. In a study of young Mexican men where the coronal sulcus and urethra were sampled, 42.7% of men were HPV positive [20]. Among STI clinic attendees the prevalence of HPV has varied from 30.5% among Swedish men [13], to 48% of men from Greenland, and 49% of Danish men [21]. In our US study, 31.8% of STI clinic attendees were HPV positive [22]. Among 34 men whose partners were diagnosed with cervical dysplasia, HPV was detected in 64% of men whose partner had CIN I, 16% whose partner had CIN II, and 50% among those whose partner had CIN III [23]. Finally, in a British study of 43 couples, 69% of penile biopsy samples were HPV positive [24]. Until recently, few studies determined the prevalence of specific types of HPV and few had determined the correlates of HPV infections in men [11,20,22].

The *HIM Study* was designed and implemented to fill this information gap. At the time the *HIM Study* began in 2004 the study of HPV infection in men was motivated by an interest in studying the transmission of infection to female sexual partners with the goal of informing how best to prevent cervical cancer. In 2005 IARC convened the second HPV Monograph which concluded for the first time that HPV16 was the cause of not only cancer among women but also multiple cancers among men, including, anal, penile, and oropharyngeal cancers. With this new knowledge the *HIM Study* infrastructure was rapidly expanded in 2005 to include specimen collection at each of the anatomic sites where HPV is known to cause cancer. Whereas there was a literature reporting prevalence of external genital HPV among males, there were little to no reports of anal and oral HPV among a general population of heterosexual men, let along with HPV natural history reports at these anatomic sites.

THE *HIM STUDY*: POPULATION AND PROCEDURES

The *HIM Study* was a prospective study of the natural history of HPV infections in men in three countries. Participants were healthy men, aged 18–70 years, residents of southern Florida, United States, Sao Paulo, Brazil, or Cuernavaca, Mexico; and were willing to comply with scheduled visits every 6 months for up to 7 years of follow-up. Study recruitment occurred between July 2005 and September 2009, and 4299 men provided consent.

At each study visit, questionnaire data and specimens from the genital area and anal canal, and oral rinse-and-gargle samples were obtained. Samples were analyzed for the detection of individual HPV genotypes. Also, visually distinct external genital lesions such as genital warts (condyloma) and penile intraepithelial neoplasia (PeIN) were biopsied starting in 2009. Blood samples were

Table 11.1 HPV Natural History by Anatomic Site of Infection in Men

	Any HPV Type		
	Prevalence (%)	Incidence Rate per 1000 Person-Months	Median Time to Clearance (Months)
Genital HPV	50.4	38.4	7.5
Anal HPV[a]	12.0	8.1	–
Oral HPV	4.0	5.6	6.9

[a] *Heterosexual men.*

collected at each study visit to test for herpes simplex virus type 2 (HSV-2) serostatus and to measure serum antibodies against four HPV genotypes (HPVs 6, 11, 16, and 18). Finally, men provided a first-void urine specimen for Chlamydia trachomatis (CT) infection testing. To date the *HIM Study* has led to 102 publications all of which are focused on HPV natural history among men [22,25–125].

One of the most important findings to arise from the *HIM Study* was the observation that HPV natural history differs across anatomic sites. As shown in Table 11.1, HPV prevalence is the highest at the genitals, followed by the anal canal, and the lowest at the oral cavity.

GENITAL HUMAN PAPILLOMAVIRUS INFECTION

At the external genitalia (coronal sulcus, glans penis, shaft, and scrotum) HPV prevalence was high (50.4%) and did not vary with age (Fig. 11.1). The incidence of a new genital HPV infection was correspondingly high (38.4 per 1000 person-months) with a median duration of 7.5 months for any HPV (Table 11.1) [53].

Factors independently associated with genital HPV detection were race (lower risk among Asian men), condom use (lower risk among men who always use condoms), smoking (higher risk among current smokers), CT infection and HSV-2 serostatus (higher risk among antibody positive men), and circumcision (lower risk for nononcogenic HPV types only).

The rate of progression from infection to disease differed by HPV type. Whereas high rates of genital HPV infection progression to genital warts was observed (16%–22% of men with a genital HPV6/11 infection developed an HPV6/11 condyloma), very low rates of disease progression to PeIN following genital HPV16 infection were noted [118].

ANAL HUMAN PAPILLOMAVIRUS INFECTION

Anal HPV prevalence differs by sexual behavior. Anal canal HPV prevalence was 12.2% among men who have sex with women (MSW) and 47.2% among men who have sex with men (MSM)

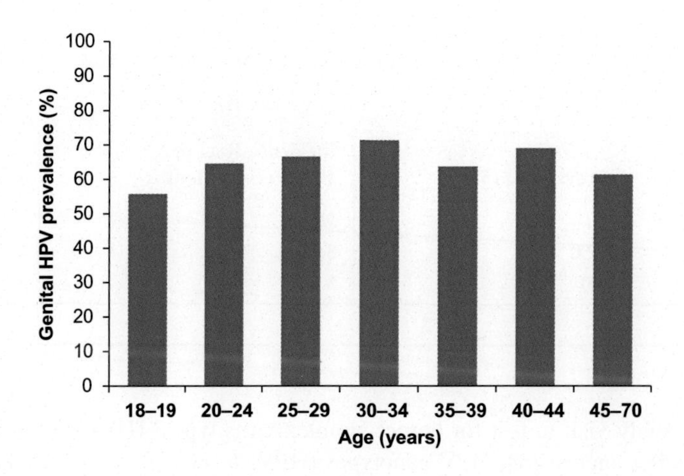

FIGURE 11.1

Age-specific prevalence of any genital HPV infection among men, showing that HPV prevalence is high and does not vary with age.

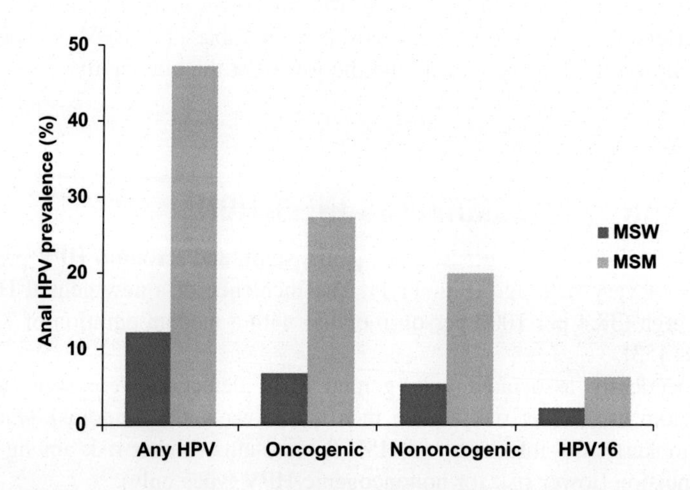

FIGURE 11.2

Anal canal HPV prevalence by sexual behavior. MSW: men who have sex with women, MSM: men who have sex with men.

(Fig. 11.2). Hence although anal HPV infection is commonly acquired by both MSW and MSM, incident events and persistence occurred more often among MSM [88].

ORAL HUMAN PAPILLOMAVIRUS INFECTION

While genital and anal HPV prevalence was relatively common in the *HIM* Study cohort, oral HPV prevalence was rare (~4%). Oral HPV prevalence was the lowest in the youngest age category

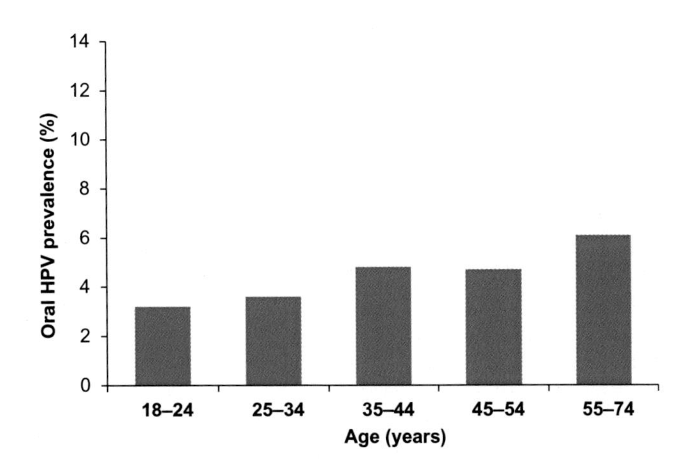

FIGURE 11.3

Age-specific prevalence of any oral HPV infection.

(18–24 year olds) and increased with increasing age, with men aged 55–74 years having the highest oral HPV prevalence (Fig. 11.3). Newly acquired oral oncogenic HPV infections in healthy men were rare. However, once acquired, oral HPV16 had a high rate of persistence [102].

HUMAN PAPILLOMAVIRUS SEROLOGY

Despite a high prevalence of genital HPV among men, the antibody response to HPV appears to be much lower than observed among women. Indeed, HPVs 6, 11, 16, and 18 seroprevalence were relatively low at 8.1%, 13.9%, 12.7%, and 10.8%, respectively. Moreover, this response appears not to confer protection against subsequent infection. HPV seropositivity following natural infection with HPVs 6, 11, and 16 was not associated with protection against subsequent type-specific genital infection, with only a possible protective effect against persistent HPV18 infection [99].

HETEROSEXUAL HUMAN PAPILLOMAVIRUS TRANSMISSION

Among 65 discordant *HIM Study* heterosexual couples (the partners were discordant for ≥1 HPV type), HPV transmission was higher from females to males (12.3 per 1000 person-months) than from males to females (7.3 per 1000 person-months) [93].

In conclusion, these findings demonstrate the susceptibility of men to HPV infection at multiple anatomic sites where HPV causes cancer and highlight the importance of HPV prevention programs, such as gender-neutral HPV vaccination.

REFERENCES

[1] Buckley JD, Harris RW, Doll R, Vessey MP, Williams PT. Case-control study of the husbands of women with dysplasia or carcinoma of the cervix uteri. Lancet 1981;2(8254):1010–15 Epub 1981/11/07. PubMed PMID: 6118477.

[2] Zunzunegui MV, King MC, Coria CF, Charlet J. Male influences on cervical cancer risk. Am J Epidemiol 1986;123(2):302–7 Epub 1986/02/01. PubMed PMID: 3753819.

[3] Agarwal SS, Sehgal A, Sardana S, Kumar A, Luthra UK. Role of male behavior in cervical carcinogenesis among women with one lifetime sexual partner. Cancer 1993;72(5):1666–9 Epub 1993/09/01. PubMed PMID: 8348498.

[4] Thomas DB, Ray RM, Pardthaisong T, Chutivongse S, Koetsawang S, Silpisornkosol S, et al. Prostitution, condom use, and invasive squamous cell cervical cancer in Thailand. Am J Epidemiol 1996;143(8):779–86 Epub 1996/04/15. PubMed PMID: 8610687.

[5] Shah KV. Human papillomaviruses and anogenital cancers. N Engl J Med 1997;337(19):1386–8. Available from: https://doi.org/10.1056/nejm199711063371911 Epub 1997/11/14. PubMed PMID: 9358136.

[6] Castellsague X, Bosch FX, Munoz N, Meijer CJ, Shah KV, de Sanjose S, et al. Male circumcision, penile human papillomavirus infection, and cervical cancer in female partners. N Engl J Med 2002;346(15):1105–12. Available from: https://doi.org/10.1056/NEJMoa011688 Epub 2002/04/12. PubMed PMID: 11948269.

[7] Bosch FX, Castellsague X, Munoz N, de Sanjose S, Ghaffari AM, Gonzalez LC, et al. Male sexual behavior and human papillomavirus DNA: key risk factors for cervical cancer in Spain. J Natl Cancer Inst 1996;88(15):1060–7 Epub 1996/08/07. PubMed PMID: 8683637.

[8] Castellsague X, Ghaffari A, Daniel RW, Bosch FX, Munoz N, Shah KV. Prevalence of penile human papillomavirus DNA in husbands of women with and without cervical neoplasia: a study in Spain and Colombia. J Infect Dis 1997;176(2):353–61 Epub 1997/08/01. PubMed PMID: 9237700.

[9] Munoz N, Castellsague X, Bosch FX, Tafur L, de Sanjose S, Aristizabal N, et al. Difficulty in elucidating the male role in cervical cancer in Colombia, a high-risk area for the disease. J Natl Cancer Inst 1996;88(15):1068–75 Epub 1996/08/07. PubMed PMID: 8683638.

[10] Thomas DB, Ray RM, Kuypers J, Kiviat N, Koetsawang A, Ashley RL, et al. Human papillomaviruses and cervical cancer in Bangkok. III. The role of husbands and commercial sex workers. Am J Epidemiol 2001;153(8):740–8 Epub 2001/04/11. PubMed PMID: 11296145.

[11] Franceschi S, Castellsague X, Dal Maso L, Smith JS, Plummer M, Ngelangel C, et al. Prevalence and determinants of human papillomavirus genital infection in men. Br J Cancer 2002;86(5):705–11. Available from: https://doi.org/10.1038/sj.bjc.6600194 Epub 2002/03/05. PubMed PMID: 11875730; PMCID: PMC2375316.

[12] Hippelainen M, Syrjanen S, Hippelainen M, Koskela H, Pulkkinen J, Saarikoski S, et al. Prevalence and risk factors of genital human papillomavirus (HPV) infections in healthy males: a study on Finnish conscripts. Sex Transm Dis 1993;20(6):321–8 Epub 1993/11/01. PubMed PMID: 8108754.

[13] Strand A, Rylander E, Evander M, Wadell G. Genital human papillomavirus infection among patients attending an STD clinic. Genitourin Med 1993;69(6):446–9 Epub 1993/12/01. PubMed PMID: 8282298; PMCID: PMC1195149.

[14] Van Doornum GJ, Prins M, Juffermans LH, Hooykaas C, van den Hoek JA, Coutinho RA, et al. Regional distribution and incidence of human papillomavirus infections among heterosexual men and women with multiple sexual partners: a prospective study. Genitourin Med 1994;70(4):240–6 Epub 1994/08/01. PubMed PMID: 7959707; PMCID: PMC1195247.

[15] Wikstrom A, Popescu C, Forslund O. Asymptomatic penile HPV infection: a prospective study. Int J STD AIDS 2000;11(2):80–4. Available from: https://doi.org/10.1177/095646240001100203 Epub 2000/03/04. PubMed PMID: 10678474.

[16] Krebs HB, Schneider V. Human papillomavirus-associated lesions of the penis: colposcopy, cytology, and histology. Obstet Gynecol 1987;70(3 Pt 1):299–304 Epub 1987/09/01. PubMed PMID: 2819795.

[17] Voog E, Ricksten A, Olofsson S, Ternesten A, Ryd W, Kjellstrom C, et al. Demonstration of Epstein-Barr virus DNA and human papillomavirus DNA in acetowhite lesions of the penile skin and the oral mucosa. Int J STD AIDS 1997;8(12):772–5. Available from: https://doi.org/10.1258/0956462971919255 Epub 1998/01/20. PubMed PMID: 9433952.

[18] Pinto PA, Mellinger BC. HPV in the male patient. Urol Clin North Am 1999;26(4):797–807 ix. Epub 1999/12/10. PubMed PMID: 10584620.

[19] Baken LA, Koutsky LA, Kuypers J, Kosorok MR, Lee SK, Kiviat NB, et al. Genital human papillomavirus infection among male and female sex partners: prevalence and type-specific concordance. J Infect Dis 1995;171(2):429–32 Epub 1995/02/01. PubMed PMID: 7844382.

[20] Lazcano-Ponce E, Herrero R, Munoz N, Hernandez-Avila M, Salmeron J, Leyva A, et al. High prevalence of human papillomavirus infection in Mexican males: comparative study of penile-urethral swabs and urine samples. Sex Transm Dis 2001;28(5):277–80 Epub 2001/05/17. PubMed PMID: 11354266.

[21] Svare EI, Kjaer SK, Nonnenmacher B, Worm AM, Moi H, Christensen RB, et al. Seroreactivity to human papillomavirus type 16 virus-like particles is lower in high-risk men than in high-risk women. J Infect Dis 1997;176(4):876–83 Epub 1997/10/23. PubMed PMID: 9333144.

[22] Baldwin SB, Wallace DR, Papenfuss MR, Abrahamsen M, Vaught LC, Kornegay JR, et al. Human papillomavirus infection in men attending a sexually transmitted disease clinic. J Infect Dis 2003;187(7):1064–70. Available from: https://doi.org/10.1086/368220 Epub 2003/03/28. PubMed PMID: 12660920.

[23] Levine RU, Crum CP, Herman E, Silvers D, Ferenczy A, Richart RM. Cervical papillomavirus infection and intraepithelial neoplasia: a study of male sexual partners. Obstet Gynecol 1984;64(1):16–20 Epub 1984/07/01. PubMed PMID: 6330630.

[24] Wickenden C, Hanna N, Taylor-Robinson D, Harris JR, Bellamy C, Carroll P, et al. Sexual transmission of human papillomaviruses in heterosexual and male homosexual couples, studied by DNA hybridisation. Genitourin Med 1988;64(1):34–8 Epub 1988/02/01. PubMed PMID: 2831137; PMCID: PMC1194144.

[25] HPV Study group in men from Brazil, USA, and Mexico. Human papillomavirus infection in men residing in Brazil, Mexico, and the USA. Salud Publica Mex 2008;50(5):408–18 Epub 2008/10/15. PubMed PMID: 18852938; PMCID: PMC3495064.

[26] Akogbe GO, Ajidahun A, Sirak B, Anic GM, Papenfuss MR, Fulp WJ, et al. Race and prevalence of human papillomavirus infection among men residing in Brazil, Mexico and the United States. Int J Cancer 2012;131(3):E282–91. Available from: https://doi.org/10.1002/ijc.27397 Epub 2011/12/14. PubMed PMID: 22161806; PMCID: PMC3458422.

[27] Albero G, Castellsague X, Giuliano AR, Bosch FX. Male circumcision and genital human papillomavirus: a systematic review and meta-analysis. Sex Transm Dis 2012;39(2):104–13. Available from: https://doi.org/10.1097/OLQ.0b013e3182387abd Epub 2012/01/18. PubMed PMID: 22249298.

[28] Albero G, Castellsague X, Lin HY, Fulp W, Villa LL, Lazcano-Ponce E, et al. Male circumcision and the incidence and clearance of genital human papillomavirus (HPV) infection in men: the HPV Infection in men (HIM) cohort study. BMC Infect Dis 2014;14:75. Available from: https://doi.org/10.1186/1471-2334-14-75 Epub 2014/02/13. PubMed PMID: 24517172; PMCID: PMC3925013.

[29] Albero G, Villa LL, Lazcano-Ponce E, Fulp W, Papenfuss MR, Nyitray AG, et al. Male circumcision and prevalence of genital human papillomavirus infection in men: a multinational study. BMC Infect

Dis 2013;13:18. Available from: https://doi.org/10.1186/1471-2334-13-18 Epub 2013/01/19. PubMed PMID: 23327450; PMCID: PMC3554597.
[30] Alberts CJ, Schim van der Loeff MF, Papenfuss MR, da Silva RJ, Villa LL, Lazcano-Ponce E, et al. Association of Chlamydia trachomatis infection and herpes simplex virus type 2 serostatus with genital human papillomavirus infection in men: the HPV in men study. Sex Transm Dis 2013;40(6):508–15. Available from: https://doi.org/10.1097/OLQ.0b013e318289c186 Epub 2013/05/18. PubMed PMID: 23680908; PMCID: PMC3904659.
[31] Anic GM, Giuliano AR. Genital HPV infection and related lesions in men. Prev Med 2011;53(Suppl. 1): S36–41. Available from: https://doi.org/10.1016/j.ypmed.2011.08.002 Epub 2011/10/14. PubMed PMID: 21962470; PMCID: PMC3495069.
[32] Anic GM, Lee JH, Stockwell H, Rollison DE, Wu Y, Papenfuss MR, et al. Incidence and human papillomavirus (HPV) type distribution of genital warts in a multinational cohort of men: the HPV in men study. J Infect Dis 2011;204(12):1886–92. Available from: https://doi.org/10.1093/infdis/jir652 Epub 2011/10/21. PubMed PMID: 22013227; PMCID: PMC3209812.
[33] Anic GM, Lee JH, Villa LL, Lazcano-Ponce E, Gage C, Jose CSR, et al. Risk factors for incident condyloma in a multinational cohort of men: the HIM study. J Infect Dis 2012;205(5):789–93. Available from: https://doi.org/10.1093/infdis/jir851 Epub 2012/01/13. PubMed PMID: 22238467; PMCID: PMC3274369.
[34] Anic GM, Messina JL, Stoler MH, Rollison DE, Stockwell H, Villa LL, et al. Concordance of human papillomavirus types detected on the surface and in the tissue of genital lesions in men. J Med Virol 2013;85(9):1561–6. Available from: https://doi.org/10.1002/jmv.23635 Epub 2013/07/16. Epub 2012/01/13. PubMed PMID: 23852680; PMCID: PMC3879682.
[35] August EM, Daley E, Kromrey J, Baldwin J, Romero-Daza N, Salmeron J, et al. Age-related variation in sexual behaviours among heterosexual men residing in Brazil, Mexico and the USA. J Fam Plann Reprod Health Care 2014;40(4):261–9. Available from: https://doi.org/10.1136/jfprhc-2012-100564 Epub 2013/10/09. PubMed PMID: 24099979; PMCID: PMC5687827.
[36] Baldwin SB, Wallace DR, Papenfuss MR, Abrahamsen M, Vaught LC, Giuliano AR. Condom use and other factors affecting penile human papillomavirus detection in men attending a sexually transmitted disease clinic. Sex Transm Dis 2004;31(10):601–7 Epub 2004/09/25. PubMed PMID: 15388997.
[37] Beachler DC, Pinto LA, Kemp TJ, Nyitray AG, Hildesheim A, Viscidi R, et al. An examination of HPV16 natural immunity in men who have sex with men (MSM) in the HPV in men (HIM) study. Cancer Epidemiol Biomarkers Prev 2018;27(4):496–502. Available from: https://doi.org/10.1158/1055-9965 Epub 2018/02/25. Epi-17-0853. PubMed PMID: 29475967; PMCID: PMC5884716.
[38] Beachler DC, Waterboer T, Pierce Campbell CM, Ingles DJ, Kuhs KA, Nyitray AG, et al. HPV16 E6 seropositivity among cancer-free men with oral, anal or genital HPV16 infection. Papillomavirus Res 2016;2:141–4. Available from: https://doi.org/10.1016/j.pvr.2016.07.003 Epub 2017/02/28. PubMed PMID: 28239675; PMCID: PMC5322843.
[39] Brancaccio RN, Robitaille A, Dutta S, Cuenin C, Santare D, Skenders G, et al. Generation of a novel next-generation sequencing-based method for the isolation of new human papillomavirus types. Virology 2018;520:1–10. Available from: https://doi.org/10.1016/j.virol.2018.04.017 Epub 2018/05/11. PubMed PMID: 29747121.
[40] Capra G, Nyitray AG, Lu B, Perino A, Marci R, Schillaci R, et al. Analysis of persistence of human papillomavirus infection in men evaluated by sampling multiple genital sites. Eur Rev Med Pharmacol Sci 2015;19(21):4153–63 Epub 2015/11/26. PubMed PMID: 26592842.
[41] da Silva RJC, Sudenga SL, Sichero L, Baggio ML, Galan L, Cintra R, et al. HPV-related external genital lesions among men residing in Brazil. Braz J Infect Dis 2017;21(4):376–85. Available from: https://doi.org/10.1016/j.bjid.2017.03.004 Epub 2017/04/12. PubMed PMID: 28399126.

[42] Daley EM, Buhi ER, Baldwin J, Lee JH, Vadaparampil S, Abrahamsen M, et al. Men's responses to HPV test results: development of a theory-based survey. Am J Health Behav 2009;33(6):728–44 Epub 2009/03/27. PubMed PMID: 19320621; PMCID: PMC3495067.

[43] Daley EM, Buhi ER, Marhefka SL, Baker EA, Kolar S, Ebbert-Syfrett J, et al. Cognitive and emotional responses to human papillomavirus test results in men. Am J Health Behav 2012;36(6):770–85. Available from: https://doi.org/10.5993/ajhb.36.6.5 Epub 2012/10/03. PubMed PMID: 23026036; PMCID: PMC3904642.

[44] Daley EM, Marhefka S, Buhi E, Hernandez ND, Chandler R, Vamos C, et al. Ethnic and racial differences in HPV knowledge and vaccine intentions among men receiving HPV test results. Vaccine 2011;29(23):4013–18. Available from: https://doi.org/10.1016/j.vaccine.2011.03.060 Epub 2011/04/05. PubMed PMID: 21459176; PMCID: PMC3092789.

[45] Daley EM, Marhefka SL, Buhi ER, Vamos CA, Hernandez ND, Giuliano AR. Human papillomavirus vaccine intentions among men participating in a human papillomavirus natural history study versus a comparison sample. Sex Transm Dis 2010;37(10):644–52 Epub 2010/09/30. PubMed PMID: 20879088; PMCID: PMC3471777.

[46] Dunne EF, Nielson CM, Hagensee ME, Papenfuss MR, Harris RB, Herrel N, et al. HPV 6/11, 16, 18 seroprevalence in men in two US cities. Sex Transm Dis 2009;36(11):671–4. Available from: https://doi.org/10.1097/OLQ.0b013e3181bc094b Epub 2009/10/08. PubMed PMID: 19809385.

[47] Flores R, Abalos AT, Nielson CM, Abrahamsen M, Harris RB, Giuliano AR. Reliability of sample collection and laboratory testing for HPV detection in men. Journal of Virological Methods 2008;149 (1):136–43. Available from: https://doi.org/10.1016/j.jviromet.2007.12.010 Epub 2008/02/19. PubMed PMID: 18279976.

[48] Flores-Diaz E, Sereday KA, Ferreira S, Sirak B, Sobrinho JS, Baggio ML, et al. HPV-11 variability, persistence and progression to genital warts in men: the HIM study. J Gen Virol 2017;98(9):2339–42. Available from: https://doi.org/10.1099/jgv.0.000896 Epub 2017/08/16. PubMed PMID: 28809141.

[49] Flores-Diaz E, Sereday KA, Ferreira S, Sirak B, Sobrinho JS, Baggio ML, et al. HPV-6 molecular variants association with the development of genital warts in men: The HIM Study. J Infect Dis 2017;215 (4):559–65. Available from: https://doi.org/10.1093/infdis/jiw600 Epub 2016/12/25. PubMed PMID: 28011919; PMCID: PMC5388291.

[50] Giuliano AR, Anic G, Nyitray AG. Epidemiology and pathology of HPV disease in males. Gynecol Oncol 2010;117(2 Suppl.):S15–19. Available from: https://doi.org/10.1016/j.ygyno.2010.01.026 Epub 2010/02/09. PubMed PMID: 20138345; PMCID: PMC4254924.

[51] Giuliano AR, Lazcano E, Villa LL, Flores R, Salmeron J, Lee JH, et al. Circumcision and sexual behavior: factors independently associated with human papillomavirus detection among men in the HIM Study. Int J Cancer 2009;124(6):1251–7. Available from: https://doi.org/10.1002/ijc.24097 Epub 2008/12/18. PubMed PMID: 19089913; PMCID: PMC3466048.

[52] Giuliano AR, Lazcano-Ponce E, Villa LL, Flores R, Salmeron J, Lee JH, et al. The human papillomavirus infection in men study: human papillomavirus prevalence and type distribution among men residing in Brazil, Mexico, and the United States. Cancer Epidemiol Biomarkers Prev 2008;17(8):2036–43. Available from: https://doi.org/10.1158/1055-9965 Epub 2008/08/19. Epi-08-0151. PubMed PMID: 18708396; PMCID: PMC3471778.

[53] Giuliano AR, Lee JH, Fulp W, Villa LL, Lazcano E, Papenfuss MR, et al. Incidence and clearance of genital human papillomavirus infection in men (HIM): a cohort study. Lancet 2011;377(9769):932–40. Available from: https://doi.org/10.1016/s0140-6736(10)62342-2 Epub 2011/03/04. PubMed PMID: 21367446; PMCID: PMC3231998.

[54] Giuliano AR, Lu B, Nielson CM, Flores R, Papenfuss MR, Lee JH, et al. Age-specific prevalence, incidence, and duration of human papillomavirus infections in a cohort of 290 US men. J Infect Dis

2008;198(6):827–35. Available from: https://doi.org/10.1086/591095 Epub 2008/07/29. PubMed PMID: 18657037.

[55] Giuliano AR, Nielson CM, Flores R, Dunne EF, Abrahamsen M, Papenfuss MR, et al. The optimal anatomic sites for sampling heterosexual men for human papillomavirus (HPV) detection: the HPV detection in men study. J Infect Dis 2007;196(8):1146–52. Available from: https://doi.org/10.1086/521629 Epub 2007/10/24. PubMed PMID: 17955432; PMCID: PMC3904649.

[56] Giuliano AR, Nyitray AG, Albero G. Male circumcision and HPV transmission to female partners. Lancet 2011;377(9761):183–4. Available from: https://doi.org/10.1016/s0140-6736(10)62273-8 Epub 2011/01/11. PubMed PMID: 21216001.

[57] Giuliano AR, Nyitray AG, Kreimer AR, Pierce Campbell CM, Goodman MT, Sudenga SL, et al. EUROGIN 2014 roadmap: differences in human papillomavirus infection natural history, transmission and human papillomavirus-related cancer incidence by gender and anatomic site of infection. Int J Cancer 2015;136(12):2752–60. Available from: https://doi.org/10.1002/ijc.29082 Epub 2014/07/22. PubMed PMID: 25043222; PMCID: PMC4297584.

[58] Giuliano AR, Sirak B, Abrahamsen M, Silva RJC, Baggio ML, Galan L, et al. Genital Wart recurrence among men residing in Brazil, Mexico, and the United States. J Infect Dis 2018. Available from: https://doi.org/10.1093/infdis/jiy533 Epub 2018/11/06. PubMed PMID: 30388232.

[59] Giuliano AR, Tortolero-Luna G, Ferrer E, Burchell AN, de Sanjose S, Kjaer SK, et al. Epidemiology of human papillomavirus infection in men, cancers other than cervical and benign conditions. Vaccine 2008;26(Suppl. 10):K17–28. Available from: https://doi.org/10.1016/j.vaccine.2008.06.021 Epub 2008/10/14. PubMed PMID: 18847554; PMCID: PMC4366004.

[60] Giuliano AR, Viscidi R, Torres BN, Ingles DJ, Sudenga SL, Villa LL, et al. Seroconversion following anal and genital HPV infection in men: the HIM Study. Papillomavirus Res 2015;1:109–15. Available from: https://doi.org/10.1016/j.pvr.2015.06.007 Epub 2015/12/22. PubMed PMID: 26688833; PMCID: PMC4680989.

[61] Hampras SS, Giuliano AR, Lin HY, Fisher KJ, Abrahamsen ME, McKay-Chopin S, et al. Natural history of polyomaviruses in men: the HPV infection in men (HIM) study. J Infect Dis 2015;211(9):1437–46. Available from: https://doi.org/10.1093/infdis/jiu626 Epub 2014/11/13. PubMed PMID: 25387582; PMCID: PMC4462655.

[62] Hampras SS, Giuliano AR, Lin HY, Fisher KJ, Abrahamsen ME, Sirak BA, et al. Natural history of cutaneous human papillomavirus (HPV) infection in men: the HIM study. PLoS One 2014;9(9):e104843. Available from: https://doi.org/10.1371/journal.pone.0104843 Epub 2014/09/10. PubMed PMID: 25198694; PMCID: PMC4157763.

[63] Ingles DJ, Lin HY, Fulp WJ, Sudenga SL, Lu B, Schabath MB, et al. An analysis of HPV infection incidence and clearance by genotype and age in men: The HPV Infection in Men (HIM) Study. Papillomavirus Res 2015;1:126–35. Available from: https://doi.org/10.1016/j.pvr.2015.09.001 Epub 2016/08/23. PubMed PMID: 27547836; PMCID: PMC4986989.

[64] Ingles DJ, Pierce Campbell CM, Messina JA, Stoler MH, Lin HY, Fulp WJ, et al. Human papillomavirus virus (HPV) genotype- and age-specific analyses of external genital lesions among men in the HPV Infection in Men (HIM) Study. J Infect Dis 2015;211(7):1060–7. Available from: https://doi.org/10.1093/infdis/jiu587 Epub 2014/10/26. PubMed PMID: 25344518; PMCID: PMC4432433.

[65] Klatser PR, de Wit MY, Fajardo TT, Cellona RV, Abalos RM, de la Cruz EC, et al. Evaluation of mycobacterium leprae antigens in the monitoring of a dapsone-based chemotherapy of previously untreated lepromatous patients in Cebu, Philippines. Leprosy Review 1989;60(3):178–86 Epub 1989/09/01. PubMed PMID: 2682104.

[66] Kreimer AR, Pierce Campbell CM, Lin HY, Fulp W, Papenfuss MR, Abrahamsen M, et al. Incidence and clearance of oral human papillomavirus infection in men: the HIM cohort study. Lancet 2013;382

(9895):877–87. Available from: https://doi.org/10.1016/s0140-6736(13)60809-0 Epub 2013/07/06. PubMed PMID: 23827089; PMCID: PMC3904652.

[67] Kreimer AR, Villa A, Nyitray AG, Abrahamsen M, Papenfuss M, Smith D, et al. The epidemiology of oral HPV infection among a multinational sample of healthy men. Cancer Epidemiol Biomarkers Prev 2011;20(1):172–82. Available from: https://doi.org/10.1158/1055-9965.Epi-10-0682 Epub 2010/12/15. PubMed PMID: 21148755; PMCID: PMC3027138.

[68] Latiff LA, Ibrahim Z, Pei CP, Rahman SA, Akhtari-Zavare M. Comparative assessment of a self-sampling device and gynecologist sampling for cytology and HPV DNA detection in a rural and low resource setting: Malaysian experience. Asian Pac J Cancer Prev 2015;16(18):8495–501 Epub 2016/01/09. PubMed PMID: 26745108.

[69] Lee JH, Han G, Fulp WJ, Giuliano AR. Analysis of overdispersed count data: application to the human papillomavirus infection in men (HIM) Study. Epidemiol Infect 2012;140(6):1087–94. Available from: https://doi.org/10.1017/s095026881100166x Epub 2011/08/31. PubMed PMID: 21875452; PMCID: PMC3471780.

[70] Liu Z, Nyitray AG, Hwang LY, Swartz MD, Abrahamsen M, Lazcano-Ponce E, et al. Human papillomavirus prevalence among 88 male virgins residing in Brazil, Mexico, and the United States. J Infect Dis 2016;214(8):1188–91. Available from: https://doi.org/10.1093/infdis/jiw353 Epub 2016/08/05. PubMed PMID: 27489299; PMCID: PMC5034958.

[71] Liu Z, Nyitray AG, Hwang LY, Swartz MD, Abrahamsen M, Lazcano-Ponce E, et al. Acquisition, persistence, and clearance of human papillomavirus infection among male virgins residing in Brazil, Mexico, and the United States. J Infect Dis 2018;217(5):767–76. Available from: https://doi.org/10.1093/infdis/jix588 Epub 2017/11/23. PubMed PMID: 29165581; PMCID: PMC5853496.

[72] Lopes R, Teixeira JA, Marchioni D, Villa LL, Giuliano AR, Luiza Baggio M, et al. Dietary intake of selected nutrients and persistence of HPV infection in men. Int J Cancer 2017;141(4):757–65. Available from: https://doi.org/10.1002/ijc.30772 Epub 2017/05/10. PubMed PMID: 28486774.

[73] Lu B, Hagensee ME, Lee JH, Wu Y, Stockwell HG, Nielson CM, et al. Epidemiologic factors associated with seropositivity to human papillomavirus type 16 and 18 virus-like particles and risk of subsequent infection in men. Cancer Epidemiol Biomarkers Prev 2010;19(2):511–16. Available from: https://doi.org/10.1158/1055-9965 Epub 2010/01/21. Epi-09-0790. PubMed PMID: 20086109.

[74] Lu B, Viscidi RP, Lee JH, Wu Y, Villa LL, Lazcano-Ponce E, et al. Human papillomavirus (HPV) 6, 11, 16, and 18 seroprevalence is associated with sexual practice and age: results from the multinational HPV Infection in Men Study (HIM Study). Cancer Epidemiol Biomarkers Prev 2011;20(5):990–1002. Available from: https://doi.org/10.1158/1055-9965.Epi-10-1160 Epub 2011/03/08. PubMed PMID: 21378268; PMCID: PMC3232028.

[75] Lu B, Viscidi RP, Wu Y, Lee JH, Nyitray AG, Villa LL, et al. Prevalent serum antibody is not a marker of immune protection against acquisition of oncogenic HPV16 in men. Cancer Res 2012;72(3):676–85. Available from: https://doi.org/10.1158/0008-5472.Can-11-0751 Epub 2011/11/30. PubMed PMID: 22123925; PMCID: PMC3474343.

[76] Lu B, Viscidi RP, Wu Y, Nyitray AG, Villa LL, Lazcano-Ponce E, et al. Seroprevalence of human papillomavirus (HPV) type 6 and 16 vary by anatomic site of HPV infection in men. Cancer Epidemiol Biomarkers Prev 2012;21(9):1542–6. Available from: https://doi.org/10.1158/1055-9965 Epub 2012/07/05. Epi-12-0483. PubMed PMID: 22761306; PMCID: PMC3466057.

[77] Lu B, Wu Y, Nielson CM, Flores R, Abrahamsen M, Papenfuss M, et al. Factors associated with acquisition and clearance of human papillomavirus infection in a cohort of US men: a prospective study. J Infect Dis 2009;199(3):362–71. Available from: https://doi.org/10.1086/596050 Epub 2009/01/13. PubMed PMID: 19133808.

[78] Marhefka SL, Daley EM, Anstey EH, Vamos CA, Buhi ER, Kolar S, et al. HPV-related information sharing and factors associated with U.S. men's disclosure of an HPV test result to their female sexual partners. Sex Transm Infect 2012;88(3):171–6. Available from: https://doi.org/10.1136/sextrans-2011-050091 Epub 2012/01/05. PubMed PMID: 22215695; PMCID: PMC3471785.
[79] Nielson CM, Flores R, Harris RB, Abrahamsen M, Papenfuss MR, Dunne EF, et al. Human papillomavirus prevalence and type distribution in male anogenital sites and semen. Cancer Epidemiol Biomarkers Prev 2007;16(6):1107–14. Available from: https://doi.org/10.1158/1055-9965 Epub 2007/06/06. Epi-06-0997. PubMed PMID: 17548671.
[80] Nielson CM, Harris RB, Dunne EF, Abrahamsen M, Papenfuss MR, Flores R, et al. Risk factors for anogenital human papillomavirus infection in men. J Infect Dis 2007;196(8):1137–45. Available from: https://doi.org/10.1086/521632 Epub 2007/10/24. PubMed PMID: 17955431; PMCID: PMC3877918.
[81] Nielson CM, Harris RB, Flores R, Abrahamsen M, Papenfuss MR, Dunne EF, et al. Multiple-type human papillomavirus infection in male anogenital sites: prevalence and associated factors. Cancer Epidemiol Biomarkers Prev 2009;18(4):1077–83. Available from: https://doi.org/10.1158/1055-9965. Epi-08-0447 Epub 2009/03/26. PubMed PMID: 19318438; PMCID: PMC5415340.
[82] Nielson CM, Harris RB, Nyitray AG, Dunne EF, Stone KM, Giuliano AR. Consistent condom use is associated with lower prevalence of human papillomavirus infection in men. J Infect Dis 2010;202 (3):445–51. Available from: https://doi.org/10.1086/653708 Epub 2010/06/24. PubMed PMID: 20569156.
[83] Nielson CM, Schiaffino MK, Dunne EF, Salemi JL, Giuliano AR. Associations between male anogenital human papillomavirus infection and circumcision by anatomic site sampled and lifetime number of female sex partners. J Infect Dis 2009;199(1):7–13. Available from: https://doi.org/10.1086/595567 Epub 2008/12/18. PubMed PMID: 19086813; PMCID: PMC5068969.
[84] Nunes EM, Lopez RVM, Sudenga SL, Gheit T, Tommasino M, Baggio ML, et al. Concordance of Beta-papillomavirus across anogenital and oral anatomic sites of men: The HIM Study. Virology 2017;510:55–9. Available from: https://doi.org/10.1016/j.virol.2017.07.006 Epub 2017/07/15. PubMed PMID: 28708973.
[85] Nunes EM, Sudenga SL, Gheit T, Tommasino M, Baggio ML, Ferreira S, et al. Diversity of beta-papillomavirus at anogenital and oral anatomic sites of men: The HIM Study. Virology 2016;495:33–41. Available from: https://doi.org/10.1016/j.virol.2016.04.031 Epub 2016/05/11. PubMed PMID: 27161202; PMCID: PMC4949595.
[86] Nyitray A, Nielson CM, Harris RB, Flores R, Abrahamsen M, Dunne EF, et al. Prevalence of and risk factors for anal human papillomavirus infection in heterosexual men. J Infect Dis 2008;197 (12):1676–84. Available from: https://doi.org/10.1086/588145 Epub 2008/04/23. PubMed PMID: 18426367.
[87] Nyitray AG, Carvalho da Silva RJ, Baggio ML, Lu B, Smith D, Abrahamsen M, et al. Age-specific prevalence of and risk factors for anal human papillomavirus (HPV) among men who have sex with women and men who have sex with men: the HPV in men (HIM) study. J Infect Dis 2011;203 (1):49–57. Available from: https://doi.org/10.1093/infdis/jiq021 Epub 2010/12/15. PubMed PMID: 21148496; PMCID: PMC3086435.
[88] Nyitray AG, Carvalho da Silva RJ, Baggio ML, Smith D, Abrahamsen M, Papenfuss M, et al. Six-month incidence, persistence, and factors associated with persistence of anal human papillomavirus in men: the HPV in men study. J Infect Dis 2011;204(11):1711–22. Available from: https://doi.org/10.1093/infdis/jir637 Epub 2011/10/04. PubMed PMID: 21964400; PMCID: PMC3203231.
[89] Nyitray AG, Carvalho da Silva RJ, Chang M, Baggio ML, Ingles DJ, Abrahamsen M, et al. Incidence, duration, persistence, and factors associated with high-risk anal human papillomavirus persistence among HIV-negative men who have sex with men: a multinational study. Clin Infect Dis 2016;62(11):1367–74.

Available from: https://doi.org/10.1093/cid/ciw140 Epub 2016/03/11. PubMed PMID: 26962079; PMCID: PMC4872291.

[90] Nyitray AG, Chang M, Villa LL, Carvalho da Silva RJ, Baggio ML, Abrahamsen M, et al. The natural history of genital human papillomavirus among HIV-negative men having sex with men and men having sex with women. J Infect Dis 2015;212(2):202–12. Available from: https://doi.org/10.1093/infdis/jiv061 Epub 2015/02/05. PubMed PMID: 25649172; PMCID: PMC4565999.

[91] Nyitray AG, da Silva RJ, Baggio ML, Lu B, Smith D, Abrahamsen M, et al. The prevalence of genital HPV and factors associated with oncogenic HPV among men having sex with men and men having sex with women and men: The HIM Study. Sex Transm Dis 2011;38(10):932–40. Available from: https://doi.org/10.1097/OLQ.0b013e31822154f9 Epub 2011/09/22. PubMed PMID: 21934568; PMCID: PMC3178038.

[92] Nyitray AG, Kim J, Hsu CH, Papenfuss M, Villa L, Lazcano-Ponce E, et al. Test-retest reliability of a sexual behavior interview for men residing in Brazil, Mexico, and the United States: the HPV in Men (HIM) Study. Am J Epidemiol 2009;170(8):965–74. Available from: https://doi.org/10.1093/aje/kwp225 Epub 2009/09/11. PubMed PMID: 19741044; PMCID: PMC2765366.

[93] Nyitray AG, Lin HY, Fulp WJ, Chang M, Menezes L, Lu B, et al. The role of monogamy and duration of heterosexual relationships in human papillomavirus transmission. J Infect Dis 2014;209(7):1007–15. Available from: https://doi.org/10.1093/infdis/jit615 Epub 2013/11/21. PubMed PMID: 24253288; PMCID: PMC3952669.

[94] Nyitray AG, Menezes L, Lu B, Lin HY, Smith D, Abrahamsen M, et al. Genital human papillomavirus (HPV) concordance in heterosexual couples. J Infect Dis 2012;206(2):202–11. Available from: https://doi.org/10.1093/infdis/jis327 Epub 2012/04/28. PubMed PMID: 22539815; PMCID: PMC3490693.

[95] Nyitray AG, Smith D, Villa L, Lazcano-Ponce E, Abrahamsen M, Papenfuss M, et al. Prevalence of and risk factors for anal human papillomavirus infection in men who have sex with women: a cross-national study. J Infect Dis 2010;201(10):1498–508. Available from: https://doi.org/10.1086/652187 Epub 2010/04/07. PubMed PMID: 20367457; PMCID: PMC2856726.

[96] Ompad DC, Bell DL, Amesty S, Nyitray AG, Papenfuss M, Lazcano-Ponce E, et al. Men who purchase sex, who are they? An interurban comparison. J Urban Health 2013;90(6):1166–80. Available from: https://doi.org/10.1007/s11524-013-9809-8 Epub 2013/05/31. PubMed PMID: 23719715; PMCID: PMC3853174.

[97] Pamnani SJ, Nyitray AG, Abrahamsen M, Rollison DE, Villa LL, Lazcano-Ponce E, et al. Sequential acquisition of anal human papillomavirus (HPV) Infection following genital infection among men who have sex with women: the HPV infection in men (HIM) study. J Infect Dis 2016;214(8):1180–7. Available from: https://doi.org/10.1093/infdis/jiw334 Epub 2016/08/05. PubMed PMID: 27489298; PMCID: PMC5034951.

[98] Pamnani SJ, Sudenga SL, Rollison DE, Ingles DJ, Abrahamsen M, Villa LL, et al. Recurrence of genital infections with 9 human papillomavirus (HPV) vaccine types (6, 11, 16, 18, 31, 33, 45, 52, and 58) among men in the HPV infection in men (him) study. J Infect Dis 2018;218(8):1219–27. Available from: https://doi.org/10.1093/infdis/jiy300 Epub 2018/05/26. PubMed PMID: 29800222.

[99] Pamnani SJ, Sudenga SL, Viscidi R, Rollison DE, Torres BN, Ingles DJ, et al. Impact of Serum antibodies to HPV serotypes 6, 11, 16, and 18 to risks of subsequent genital HPV infections in men: The HIM Study. Cancer Res 2016;76(20):6066–75. Available from: https://doi.org/10.1158/0008-5472.Can-16-0224 Epub 2016/08/19. PubMed PMID: 27535333; PMCID: PMC5065769.

[100] Pierce Campbell CM, Gheit T, Tommasino M, Lin HY, Torres BN, Messina JL, et al. Cutaneous beta human papillomaviruses and the development of male external genital lesions: A case-control study nested within the HIM Study. Virology 2016;497:314–22. Available from: https://doi.org/10.1016/j.virol.2016.08.002 Epub 2016/08/16. PubMed PMID: 27518539; PMCID: PMC5997247.

[101] Pierce Campbell CM, Giuliano AR, Torres BN, O'Keefe MT, Ingles DJ, Anderson RL, et al. Salivary secretory leukocyte protease inhibitor (SLPI) and head and neck cancer: The Cancer Prevention Study II Nutrition Cohort. Oral Oncol 2016;55:1–5. Available from: https://doi.org/10.1016/j.oraloncology.2016.02.004 Epub 2016/03/27. PubMed PMID: 27016010; PMCID: PMC4808568.

[102] Pierce Campbell CM, Kreimer AR, Lin HY, Fulp W, O'Keefe MT, Ingles DJ, et al. Long-term persistence of oral human papillomavirus type 16: the HPV Infection in Men (HIM) study. Cancer Prev Res (Phila) 2015;8(3):190–6. Available from: https://doi.org/10.1158/1940-6207.Capr-14-0296 Epub 2015/01/13. PubMed PMID: 25575501; PMCID: PMC4355174.

[103] Pierce Campbell CM, Lin HY, Fulp W, Papenfuss MR, Salmeron JJ, Quiterio MM, et al. Consistent condom use reduces the genital human papillomavirus burden among high-risk men: the HPV infection in men study. J Infect Dis 2013;208(3):373–84. Available from: https://doi.org/10.1093/infdis/jit191 Epub 2013/05/07. PubMed PMID: 23644283; PMCID: PMC3699010.

[104] Pierce Campbell CM, Messina JL, Stoler MH, Jukic DM, Tommasino M, Gheit T, et al. Cutaneous human papillomavirus types detected on the surface of male external genital lesions: a case series within the HPV infection in men study. J Clin Virol 2013;58(4):652–9. Available from: https://doi.org/10.1016/j.jcv.2013.10.011 Epub 2013/11/12. PubMed PMID: 24210970; PMCID: PMC3866698.

[105] Pierce Campbell CM, Viscidi RP, Torres BN, Lin HY, Fulp W, Abrahamsen M, et al. Human papillomavirus (HPV) L1 serum antibodies and the risk of subsequent oral HPV acquisition in men: The HIM Study. J Infect Dis 2016;214(1):45–8. Available from: https://doi.org/10.1093/infdis/jiw083 Epub 2016/03/05. PubMed PMID: 26931445; PMCID: PMC4907411.

[106] Rahman S, Giuliano AR, Rollison DE, Pawlita M, Waterboer T, Villa LL, et al. Cutaneous HPV and alpha-mucosal 9-valent HPV sero-status associations. Papillomavirus Res 2017;4:54–7. Available from: https://doi.org/10.1016/j.pvr.2017.08.001 Epub 2017/11/29. PubMed PMID: 29179870; PMCID: PMC5728424.

[107] Rahman S, Pierce Campbell CM, Rollison DE, Wang W, Waterboer T, Michel A, et al. Seroprevalence and associated factors of 9-valent human papillomavirus (HPV) types among men in the multinational HIM study. PLoS One 2016;11(11):e0167173. Available from: https://doi.org/10.1371/journal.pone.0167173 Epub 2016/12/03. PubMed PMID: 27902759; PMCID: PMC5130234.

[108] Rahman S, Pierce Campbell CM, Torres BN, O'Keefe MT, Ingles DJ, Villa LL, et al. Distribution and factors associated with salivary secretory leukocyte protease inhibitor concentrations. Oral Dis 2016;22(8):781–90. Available from: https://doi.org/10.1111/odi.12550 Epub 2016/07/30. PubMed PMID: 27470907; PMCID: PMC5056814.

[109] Rahman S, Pierce Campbell CM, Waterboer T, Rollison DE, Ingles DJ, Torres BN, et al. Seroprevalence of cutaneous human papillomaviruses (HPVs) among men in the multinational HPV Infection in Men study. J Gen Virol 2016;97(12):3291–301. Available from: https://doi.org/10.1099/jgv.0.000620 Epub 2016/12/03. PubMed PMID: 27902363; PMCID: PMC5756495.

[110] Ranjeva SL, Baskerville EB, Dukic V, Villa LL, Lazcano-Ponce E, Giuliano AR, et al. Recurring infection with ecologically distinct HPV types can explain high prevalence and diversity. Proc Natl Acad Sci USA 2017;114(51):13573–8. Available from: https://doi.org/10.1073/pnas.1714712114 Epub 2017/12/07. PubMed PMID: 29208707; PMCID: PMC5754802.

[111] Schabath MB, Thompson ZJ, Egan KM, Torres BN, Nguyen A, Papenfuss MR, et al. Alcohol consumption and prevalence of human papillomavirus (HPV) infection among US men in the HPV in men (HIM) study. Sex Transm Infect 2015;91(1):61–7. Available from: https://doi.org/10.1136/sextrans-2013-051422 Epub 2014/10/04. PubMed PMID: 25278617; PMCID: PMC4342055.

[112] Schabath MB, Villa LL, Lazcano-Ponce E, Salmeron J, Quiterio M, Giuliano AR. Smoking and human papillomavirus (HPV) infection in the HPV in Men (HIM) study. Cancer Epidemiol Biomarkers Prev

2012;21(1):102–10. Available from: https://doi.org/10.1158/1055-9965.Epi-11-0591 Epub 2011/10/22. PubMed PMID: 22016473; PMCID: PMC3253903.

[113] Schabath MB, Villa LL, Lin HY, Fulp WJ, Akogbe GO, Abrahamsen ME, et al. Racial differences in the incidence and clearance of human papilloma virus (HPV): the HPV in men (HIM) study. Cancer Epidemiol Biomarkers Prev 2013;22(10):1762–70. Available from: https://doi.org/10.1158/1055-9965.Epi-13-0303 Epub 2013/07/23. PubMed PMID: 23872745; PMCID: PMC3795913.

[114] Schabath MB, Villa LL, Lin HY, Fulp WJ, Lazcano-Ponce E, Salmeron J, et al. A prospective analysis of smoking and human papillomavirus infection among men in the HPV in Men Study. Int J Cancer 2014;134(10):2448–57. Available from: https://doi.org/10.1002/ijc.28567 Epub 2013/11/14. PubMed PMID: 24222514; PMCID: PMC3949156.

[115] Sichero L, Nyitray AG, Nunes EM, Nepal B, Ferreira S, Sobrinho JS, et al. Diversity of human papillomavirus in the anal canal of men: The HIM Study. Clin Microbiol Infect 2015;21(5):502–9. Available from: https://doi.org/10.1016/j.cmi.2014.12.023 Epub 2015/02/24. PubMed PMID: 25698660; PMCID: PMC4538945.

[116] Sichero L, Pierce Campbell CM, Ferreira S, Sobrinho JS, Luiza Baggio M, Galan L, et al. Broad HPV distribution in the genital region of men from the HPV infection in men (HIM) study. Virology 2013;443(2):214–17. Available from: https://doi.org/10.1016/j.virol.2013.04.024 Epub 2013/06/01. PubMed PMID: 23722104; PMCID: PMC3782990.

[117] Sichero L, Pierce Campbell CM, Fulp W, Ferreira S, Sobrinho JS, Baggio M, et al. High genital prevalence of cutaneous human papillomavirus DNA on male genital skin: the HPV Infection in Men Study. BMC Infect Dis 2014;14:677. Available from: https://doi.org/10.1186/s12879-014-0677-y Epub 2014/01/01. PubMed PMID: 25857319; PMCID: PMC4265346.

[118] Sudenga SL, Ingles DJ, Pierce Campbell CM, Lin HY, Fulp WJ, Messina JL, et al. Genital human papillomavirus infection progression to external genital lesions: The HIM Study. Eur Urol 2016;69(1):166–73. Available from: https://doi.org/10.1016/j.eururo.2015.05.032 Epub 2015/06/09. PubMed PMID: 26051441; PMCID: PMC4670812.

[119] Sudenga SL, Nyitray AG, Torres BN, Silva R, Villa L, Lazcano-Ponce E, et al. Comparison of anal HPV natural history among men by country of residence: Brazil, Mexico, and the United States. J Infect 2017;75(1):35–47. Available from: https://doi.org/10.1016/j.jinf.2017.03.010 Epub 2017/04/02. PubMed PMID: 28363585.

[120] Sudenga SL, Torres BN, Botha MH, Zeier M, Abrahamsen ME, Glashoff RH, et al. Cervical HPV natural history among young Western Cape, South African women: the randomized control EVRI Trial. J Infect 2016;72(1):60–9. Available from: https://doi.org/10.1016/j.jinf.2015.10.001 Epub 2015/10/18. PubMed PMID: 26476151; PMCID: PMC4698060.

[121] Sudenga SL, Torres BN, Botha MH, Zeier M, Abrahamsen ME, Glashoff RH, et al. HPV serostatus pre- and post-vaccination in a randomized phase II preparedness trial among young Western Cape, South African women: The evri trial. Papillomavirus Res 2017;3:50–6. Available from: https://doi.org/10.1016/j.pvr.2017.02.001 Epub 2017/05/10. PubMed PMID: 28480334; PMCID: PMC5417542.

[122] Sudenga SL, Torres BN, Fulp WJ, Silva R, Villa LL, Lazcano-Ponce E, et al. Country-specific HPV-related genital disease among men residing in Brazil, Mexico and The United States: The HIM study. Int J Cancer 2017;140(2):337–45. Available from: https://doi.org/10.1002/ijc.30452 Epub 2016/10/22. PubMed PMID: 27681815; PMCID: PMC5687823.

[123] Sudenga SL, Torres BN, Silva R, Villa LL, Lazcano-Ponce E, Abrahamsen M, et al. Comparison of the Natural History of Genital HPV Infection among Men by Country: Brazil, Mexico, and the United States. Cancer Epidemiol Biomarkers Prev 2017;26(7):1043–52. Available from: https://doi.org/10.1158/1055-9965 Epub 2017/04/28. Epi-17-0040. PubMed PMID: 28446543; PMCID: PMC5556383.

[124] Vaccarella S, Plummer M, Franceschi S, Gravitt P, Papenfuss M, Smith D, et al. Clustering of human papillomavirus (HPV) types in the male genital tract: the HPV in men (HIM) study. J Infect Dis 2011;204(10):1500–4. Available from: https://doi.org/10.1093/infdis/jir595 Epub 2011/09/13. PubMed PMID: 21908729; PMCID: PMC3222106.

[125] Vyas NS, Pierce Campbell CM, Mathew R, Abrahamsen M, Van der Kooi K, Jukic DM, et al. Role of histological findings and pathologic diagnosis for detection of human papillomavirus infection in men. J Med Virol 2015;87(10):1777–87. Available from: https://doi.org/10.1002/jmv.24238 Epub 2015/05/07. PubMed PMID: 25945468; PMCID: PMC4780420.

CHAPTER

ANAL HPV INFECTION AND HPV-ASSOCIATED DISEASE

12

Joel M. Palefsky

Professor of Medicine, University of California, San Francisco, CA, United States

Anal HPV infection, both precancer and cancer, share many biological similarities with their cervical counterparts. The incidence of anal cancer has been rising in both men and women in the general population for decades. However, it is less common than cervical cancer and attracted relatively little attention until the incidence was shown to be substantially elevated in at-risk groups such as men who have sex with men and HIV-infected individuals. Techniques were modified from the cervix to allow for anal HPV testing, disease assessment and treatment. Studies also show a high prevalence and incidence of anal HPV infection, anal high-grade squamous intraepithelial lesions and anal cancer in women with history of cervical and vulvar cancer, and individuals with immunocompromise due to solid organ transplant. Comparative studies between cervical and anal HPV infection and HPV-associated disease at these two anatomic sites offer valuable insight into HPV biology.

STUDIES OF ANAL HPV INFECTION: THEIR ORIGINS AND INSPIRATIONS

In general, interest in HPV infection at any given anatomic site is driven by the magnitude of the morbidity and mortality associated with the HPV-related cancer at that site. For this reason, much of the early work in the HPV field was done on cervical cancer and its precursors, along with cervical HPV infection. Likewise, much of the work done on penile HPV infection was inspired less by the risk of penile cancer in men than by the need to understand HPV as a sexually transmitted agent with its most serious consequence being cervical cancer.

Anal cancer is less common than cervical cancer in the general population, and as a result of that, anal HPV infection and HPV-related disease of the anus attracted much less attention than cervical HPV infection and its associated diseases. Several countries around the world do not collect reliable information on anal cancer incidence, but among those that do, the incidence of anal cancer is about 1.8/100,000, with the incidence slightly higher among women than men in most countries (https://seer.cancer.gov/statfacts/html/anus.html#incidencemortality.%202.%20Piketty%20C,%20Selinger-Leneman, accessed May 1, 2018). Notably, however, the incidence of anal cancer has been increasing since the 1970s in both men and women by about 2% per year (Fig. 12.1). While the reason for this is unknown, this may reflect changes in sexual behavior in the general population since the 1960s. The combination of relatively low incidence of anal cancer with societal

Human Papillomavirus. DOI: https://doi.org/10.1016/B978-0-12-814457-2.00012-X

Anal cancer incidence

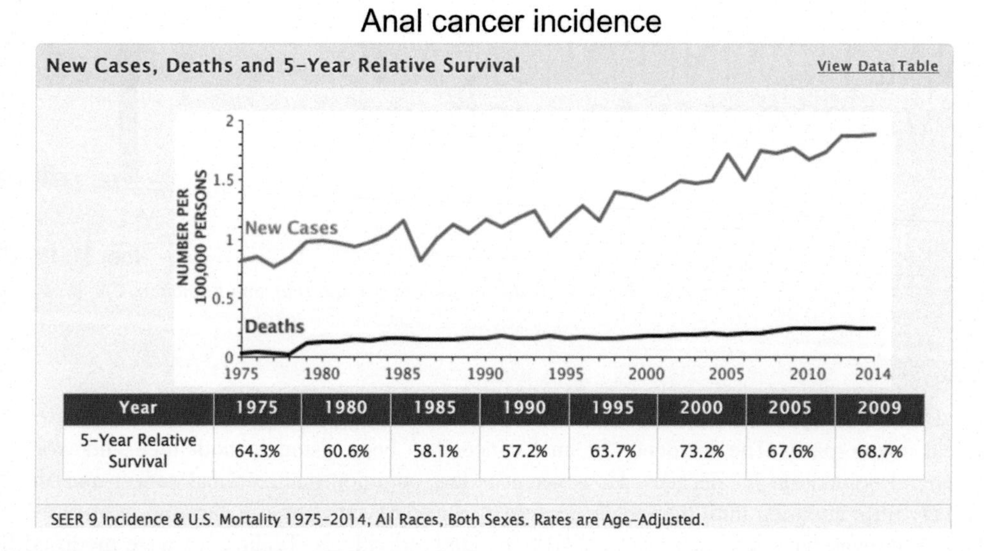

Year	1975	1980	1985	1990	1995	2000	2005	2009
5-Year Relative Survival	64.3%	60.6%	58.1%	57.2%	63.7%	73.2%	67.6%	68.7%

FIGURE 12.1 Anal cancer incidence in the population.

Data showing the increasing incidence of anal cancer in the Surveillance Epidemiology End Results (SEER) database through 2014 in the United States.

taboos about the anus and anal intercourse meant that research on anal HPV infection began in earnest much later than cervical infection.

With the increasing recognition that anal cancer was morphologically similar to cervical cancer, and that lesions morphologically similar to cervical squamous intraepithelial lesions could be found on the perianus and inside the anal canal, it was also theorized that anal cancer, like cervical cancer, was caused by HPV [1]. It was also postulated that progression from HPV infection to high-grade squamous intraepithelial lesions (HSIL) and then to cancer was also similar to that seen in the cervix.

One key difference in the epidemiology of cervical and anal cancer was the early recognition that the risk of anal cancer was largely concentrated in certain subgroups of the population. These groups are the same as those recommended for anal cytology screening, discussed later in this chapter (Table 12.1). Men who reported homosexual activity were shown to be at considerably higher risk than the general population, with an incidence estimated to be as high as 37/100000 [2]. Given the high proportion of this population that engaged in receptive anal intercourse, it was speculated that the cancer was associated with acquisition of anal HPV infection acquired in this manner.

DEVELOPMENT OF TOOLS TO STUDY ANAL HPV INFECTION AND HPV-ASSOCIATED DISEASES

Collection of anal material for HPV testing was relatively straightforward since it involved insertion of a swab or other collection device blindly into the anal canal and dislodging cells from the surface of the mucosa. The samples appeared to yield enough material for analysis and the specimens generally did not contain substances that inhibited performance of the same HPV tests that were shown to work on cervical samples. Consequently most studies of anal HPV infection used tests developed for cervical HPV detection. One of the earliest studies of anal HPV infection were therefore performed in populations of men who have sex with men (MSM) [3], and showed high rates of infection, confirmed later in multiple other studies [4,5].

The high rate of anal HPV infection and anal cancer in at-risk populations such as MSM mandated that a new set of tools be developed to assess anal squamous intraepithelial lesions (SIL) to enable further, rigorous research. Given the similarities between cervical and anal cancer, and the fact that most anal lesions are found at the transformation zone, similar to the cervix, it was believed that tools developed over decades to assess cervical SIL could be adapted for use in the anus.

Development of techniques for anal cytology and visualization of ASIL were more challenging than techniques for detection of anal HPV DNA. Following from cervical studies, moistened Dacron swabs were inserted blindly into the anal canal in an effort to obtain cells from the entire anal canal up to and including the distal rectum. Early studies showed that scraping of the surface of the anal mucosa could yield cells with HPV-associated morphologic changes similar to those seen on cervical cytology [6]. No scope was used to avoid covering some of the mucosa to be sampled. Studies of the performance of anal cytology revealed that they were more hypocellular than cervical cytology. Similar to cervical cytology, any one sampling had low sensitivity for detection of HSIL, necessitating repeat sampling at defined time intervals [7]. Similar to cervical cytology, anal cytology generally underestimated the grade of biopsy-proven disease. Efforts to try other sampling instruments such as flecked swabs or cytobrushes yielded only modest improvement, highlighting the limits of anal cytology as a screening tool for anal HSIL or cancer. Efforts to improve the screening performance of cytology by adding HPV testing and other adjunctive tests are currently underway, similar to what had been done in the cervix many years earlier.

Development of techniques to visualize and confirm anal HSIL or anal cancer on biopsy have been even more challenging. As in the cervix, determination of the grade of disease for the purposes of clinical care or research is based on histology. Visual identification of lesions and sampling them through biopsy are therefore critical. Colposcopy is used in the cervix for this purpose—since magnification is needed—and is often combined with application of 5% acetic acid or Lugol's iodine. The colposcope is also used to detect anal disease but the technique is called "high resolution anoscopy" [8]. Typically, a small plastic disposable anoscope is inserted into the anus, and after application of 5% acetic acid, visualized through the colposcope rolled up to the opening of the anoscope. Jay et al. showed that many of the changes associated with cervical lesions on colposcopy are seen with anal lesions as well, with some differences [9]. In some ways, HRA is even more challenging than anoscopy given the three-dimensional nature of the anal canal, the presence of folds, obstructing anatomic features such as hemorrhoids and hypertrophic papillae, crypts, blood and mucus, and the need for both muscular strength and excellent eye-hand coordination to perform biopsies. There is a long learning curve for HRA and the higher incidence of anal HSIL reported in more recent studies compared with the earliest studies may at least in part reflect time-based improvement in technique on the part of the clinicians as well as an evolving pattern of HPV infection and disease over time. Standards were proposed recently for learning HRA and metrics that should be achieved in determining the readiness of a clinician to perform this technique [10].

Many of the challenges that apply to HRA and measurement of anal disease apply to treatment of anal disease. In the cervix the most common treatment for HSIL is to remove it by excising relatively large portions of tissue. This can't be done in the anus, and anal treatments usually involve targeted ablation using techniques such as electrocautery or infrared coagulation. Successful treatment relies on the ability of the clinician to visually identify the lesion, which can be challenging as described above. Anal lesions—especially in the HIV-infected or other immunocompromised populations—are often large and multifocal, compounding the challenges of treatment. They have a high recurrence rate after treatment and often are found at new, nearby mucosal tissue where they were not previously seen or treated, a phenomenon known as metachronous disease. For all these reasons it is not yet clear if screening for and treating anal HSIL will reduce the incidence of anal cancer, similar to the way screening for and treating cervical HSIL has been shown to reduce the incidence of cervical cancer. In the absence of evidence that screening for anal HSIL and treatment of the lesions reduces anal cancer, there have not yet been any official guidelines (with the exception of the New York State Department of Health, US) mandating screening for anal HSIL or cancer as standard of care in high risk populations (https://www.hivguidelines.org/adult-hiv/preventive-care-screening/anal-dysplasia-cancer/, accessed May 1, 2018). To address this issue, a randomized controlled trial known as the Anal Cancer/HSIL Outcomes Research (ANCHOR) Study sponsored by the U.S. National Cancer Institute (NCI) was initiated. The trial will compare the incidence of anal cancer in over 5000 HIV-infected men and women with biopsy-proven anal HSIL in which half were or will be randomized to treatment and half to active monitoring, with follow-up for at least 5 years. In the absence of this evidence, and despite the absence of formal guidelines, many experts in the field believe that it is prudent to treat anal HSIL if the patient is not being followed as part of a research study, at least until the ANCHOR study data become available. A screening algorithm advocated by some experts in shown in Fig. 12.2. Screening for anal HSIL and cancer is only suggested for populations at high risk of cancer and are listed in Table 12.1. As with cervical screening guidelines, some experts recommend deferring initiation of anal screening among those in their teen years or early twenties, with the goal of not over-treating individuals at relatively low risk of cancer. For HIV-uninfected individuals, some recommend initiating screening at age 40 years and at age 25 years for HIV-infected individuals.

Given the challenges and time required to master HRA, and absence of standard of care guidelines, there is a shortage of clinicians trained to perform this technique. It is generally recommended that anal cytology and/or anal HPV testing *not* be done, even in high-risk populations, in case the screening clinician does not perform HRA himself or herself, or have access to someone who does. This is because cytology should not be done if there is no ability to act on an abnormal result. In contrast, even in in the absence of cytology screening, experts strongly recommend that high-risk individuals undergo an annual digital ANOrectal exam (DARE) to feel for hard lumps that may indicate the presence of cancer. Almost all clinicians are trained to perform a digital rectal exam (DRE), and can also perform a DARE, focusing on feeling for hard masses in the anal canal and perianus, in addition to the distal rectum and prostate.

ANAL CANCER AND IMMUNE SUPPRESSION

Other immunosuppressed groups in addition to those with HIV infection are known to be at increased risk for anal cancer, including men and women who were immunosuppressed to prevent

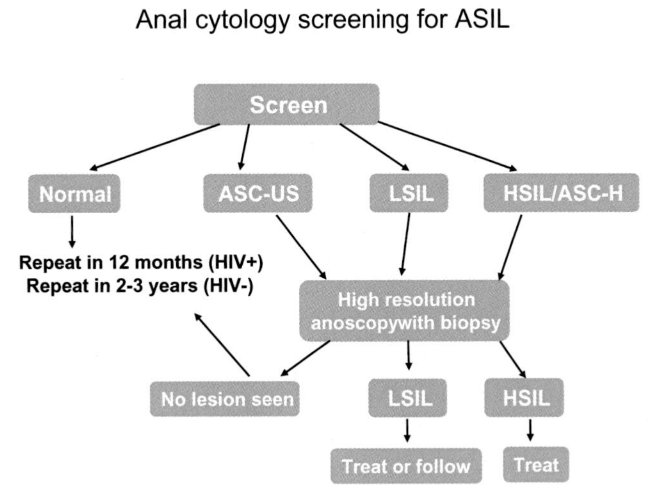

Anal cytology screening for ASIL

Screen

Normal
ASC-US
LSIL
HSIL/ASC-H

Repeat in 12 months (HIV+)
Repeat in 2-3 years (HIV-)

High resolution anoscopywith biopsy

No lesion seen
LSIL
HSIL

Treat or follow
Treat

FIGURE 12.2 Anal cytology screening algorithm.

This algorithm differs from many cervical screening guidelines in that even those with atypical squamous cells of undetermined significance are recommended to have high resolution anoscopy with biopsies of visible lesions to determine the presence of high-grade squamous intraepithelial lesions and exclude cancer. Anal cancer screening could also begin later than cervical screening given the low incidence of anal cancer under the age of 40 in immunocompetent individuals and 25 for immunocompromised individuals.

Table 12.1 Populations Recommended for Anal Cytology Screening.
• All HIV-infected men regardless of sexual orientation
• All HIV-uninfected men who have sex with men
• Women with high-grade cervical or vulvar lesions or cancer
• All HIV-infected women
• All men and women with perianal condyloma or other signs of HPV-related disease
• Solid organ transplant recipients and other forms of immunosuppression
• Over 25 years if immunosuppressed, including HIV
• Over 40 years if immunocompetent

grant rejection after organ transplant [11-13]. However, it wasn't until the advent of the HIV epidemic in the 1980s that studies of anal HPV infection, anal cancer and its precursors really began in earnest. The incidence of anal cancer was known to be elevated among HIV-infected men and women even at the earliest stages of the HIV epidemic, when many were dying of HIV-associated comorbidities before having time to progress to anal cancer. Early studies showed a high prevalence and incidence of anal HPV infection and anal HSIL in HIV-infected men, largely inversely proportional to CD4 level [8].

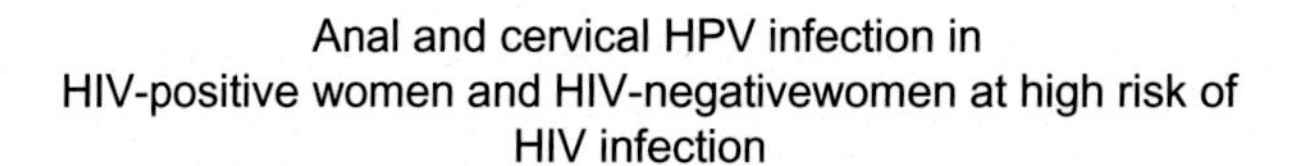

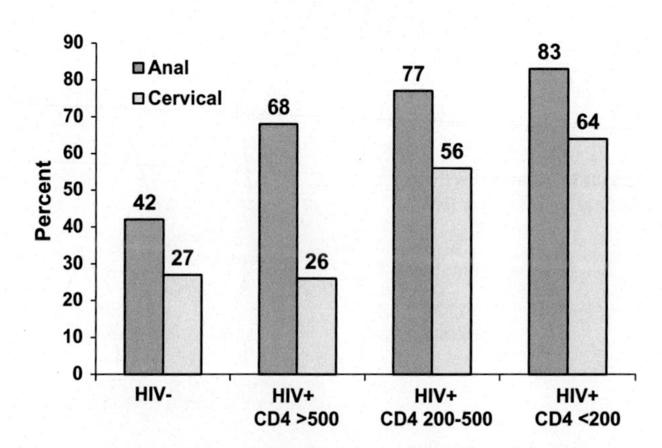

FIGURE 12.3 Anal and cervical HPV infection in HIV-infected women.

One of the largest early studies of concurrent cervical and anal HPV infection in HIV-infected women stratified by CD4 level, and HIV-uninfected women at high risk of HIV infection. Anal HPV infection was more common than cervical HPV infection in each of the groups of women.

The elevated incidence of anal cancer in the setting of HIV infection was not restricted to MSM- HIV-infected women were also at increased risk. This led to the first studies of anal HPV infection in women (Fig. 12.3), and these studies consistently showed that anal HPV was more common in these women than cervical HPV infection [14,15], with the earliest study done by Williams et al [16]. The high prevalence and incidence of anal HPV infection was confirmed in many other studies of women ranging from immunosuppressed to otherwise completely healthy and suggests that anal HPV infection is a common component of the anogenital microenvironment. Consistent with this were studies showing high rates of ASIL in women [17]. The role of anal HPV infection as a reservoir and its role in the natural history of cervical infection and disease is not known. The high prevalence of anal HPV infection in women also raised questions as to how HPV was acquired—although many women reported engaging in anal intercourse, many did not, suggesting that other modes of acquisition may be playing a role. This was also consistent with studies of HIV-infected men who have sex with women who also had a high prevalence of anal HPV infection despite reporting no history of receptive anal intercourse. It was speculated that at least some of these infections may have occurred through autoinoculation from other HPV-infected anatomic sites. A study from Tasmania provided supporting evidence for this, as women who wiped themselves from front to back after going to the toilet were more likely to have anal HPV infection than those who did not [18]. Women therefore may acquire anal HPV infection through autoinoculation from cervical or vulvar infection, or directly through anal intercourse.

ANAL CANCER IN HIV-INFECTED MEN AND WOMEN IN THE ERA OF ANTIRETROVIRAL THERAPY

The advent of antiretroviral therapy (ART) for HIV was a major breakthrough that led to immune reconstitution against HIV and dramatically lower mortality. Although this was a tremendous benefit overall, it was predicted that it would lead to increased, not decreased, incidence of anal cancer. This is because those with HSIL were now living long enough to progress to cancer, and there was (and is) no organized screening for anal cancer or HSIL, similar to the highly effective cervical cancer screening programs (Fig. 12.4). Many of the studies that were performed in the preART era were repeated in HIV-infected men and women and showed that those on ART continued to have a very high prevalence of anal HPV infection as well as anal HSIL [19,20]. Consistent with this, the incidence of anal cancer in the postART era was estimated to be as high as 131/100000 in HIV-infected MSM and 30/100000 in HIV-infected women. As we enter the next stage of the HIV epidemic in which men and women will have been on ART for many years, it appears that the incidence of anal cancer has at least levelled off, if not declined some, but is unclear whether it will drop to the same level as HIV-uninfected men and women.

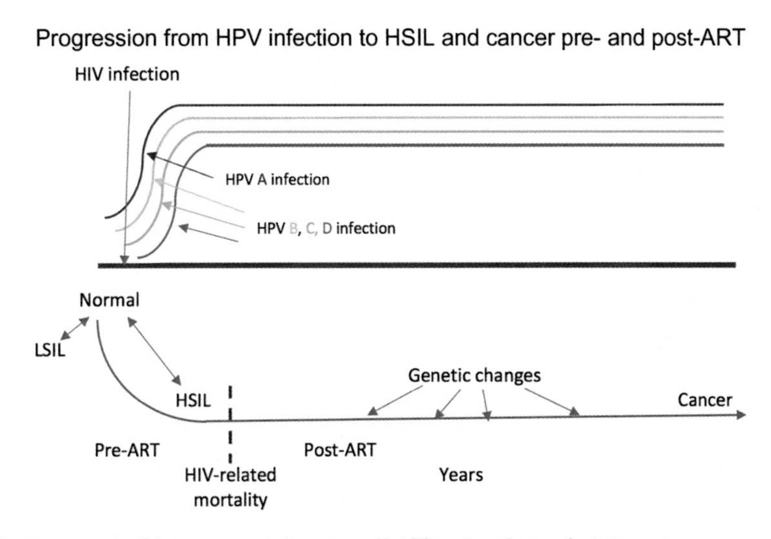

FIGURE 12.4 Anal cancer incidence pre- and post-availability of antiretroviral therapy.

Anal HPV infection is often acquired early after initiation of sexual activity. HIV infection often occurs after HPV infection and new HPV types may be acquired after HIV infection. High-grade squamous intraepithelial lesions (HSIL) develop within a few years after acquisition of HPV infection, but it may take decades for HSIL to progress to cancer. In the pre-antiretroviral (ART) era, most individuals died of HIV and HIV-associated complications before HSIL had time to progress to cancer. The incidence of anal cancer increased among HIV-infected individuals on ART in the postART era for several reasons. First, HSIL persists most of the time even with effective ART. Secondly, individuals on effective ART may now live long enough to progress to cancer, and third, most individuals with anal HSIL do not undergo screening or treatment of HSIL in an effort to reduce the risk of progression to anal cancer.

THE FUTURE: AN EMERGING FIELD, AND POTENTIAL FOR CONTRIBUTIONS TO UNDERSTANDING THE OVERALL BIOLOGY OF HPV INFECTION

The dramatic increase in anal cancer associated with HIV infection in the ART era has led to studies of anal HPV infection and ASIL in a broad spectrum of the population, and revealed a surprisingly high prevalence and incidence of anal HPV infection and HSIL in populations beyond those at highest risk for anal cancer. Compared with cervical HPV infection, the field of anal HPV infection is young, but combined with the rising incidence of anal cancer in the general population, it is clear that further research on anal HPV infection and disease in different patient populations is needed.

The anus is a highly interesting site of HPV infection from a biological and research standpoint, and offers many insights into HPV biology in general. The anus is really two organs in one, with the perianus being keratinized and closely resembling the vulva, and the anal canal being mucosal and resembling the cervix and vagina. An interesting observation is that despite the high prevalence of incidence of anal HPV infection, the incidence of anal cancer is relatively low compared with cervical cancer. Despite the fact that the same HPV types infect the cervix and the anus, including HPV 16, the incidence of cervical cancer is higher than that of anal cancer on a per-HPV infection basis. Several factors may play a role in this difference—hormonal contributions, with the anal epithelium being less responsive to estrogens and other hormones active in the cervix; the microbiome, which is very different at the two sites; difference in local immune response; different levels of exposure to environmental factors associated with sexual exposure, among others.

The anus offers some unique research opportunities that have potential to provide critical information that may be generalizable to HPV infection at all anatomic sites. Half of the participants in the ANCHOR study will have active monitoring of their anal HSIL. It is expected that some will progress to anal cancer in the course of the study. This offers a unique opportunity to understand the molecular pathogenesis of progression from HSIL to invasive cancer, a study that cannot, or should not, be done at any other anatomic site, including the cervix and oropharynx. Tissues and other specimens are being systematically collected and stored in a biobank that have the potential to reveal pathways important in progression to cancer, biomarkers of progression from HSIL to cancer, and potentially new therapeutic approaches.

The anus also offers unique opportunities to assess new therapeutic approaches for treatment of HPV-associated HSIL and cancer. Both the anal canal and perianal areas are relatively accessible for topical application of new therapeutic agents, particularly with topical application. Areas of treatment are relatively easy to visualize and to sample with biopsies. Given the paucity of high efficacy treatment alternatives, studies assessing new approaches to treating anal HSIL are generally received with enthusiasm by eligible participants. Therapeutic approaches that show promise for the treatment of anal HSIL can also be assessed for treatment of anal cancer, as well as HSIL and cancer at other anatomic sites.

Finally, the study of anal HPV infection and disease has yielded dividends critical to the overall control of HPV infection in the general population and prevention of HPV-associated cancer in both men and women. Specifically studies of anal HPV infection and disease in young MSM provided a quick pathway to approval of the quadrivalent vaccine in males. The qHPV vaccine was shown to be highly effective at preventing genital warts in males [21], but demonstration of prevention of anal HSIL and the likelihood of prevention of anal cancer [22] was pivotal in approval of routine vaccination in the U.S. of males 9–13 years of age with catch-up up to age 26 years. Gender-neutral HPV vaccination is now being adopted in a growing number of other countries worldwide.

SUMMARY

Given the lower incidence of anal cancer compared with cervical cancer, the field of HPV infection and anal cancer is relatively new. However, it has been attracting greater interest of late with the continuing rise in incidence in the general population, the onset of the HIV epidemic and the high incidence of cancer in specific at-risk populations. Development of techniques to study anal HPV infection and disease were largely adapted from those of the cervix, given the anatomic and biological similarities. These studies have been rewarding, having yielded data to support gender-neutral HPV vaccination, and insights into tissue tropism, site-specific molecular pathogenesis, the role of the immune response in controlling HPV infection, and HPV biology in general. At the same time, much remains to be done, including development and validation of techniques to predict progression of HSIL to cancer with biomarkers, and prevent anal cancer—particularly through treatment of anal HSIL; work that will likely be applicable to control of HPV at all anatomic sites.

REFERENCES

[1] Beckmann AM, Daling JR, Sherman KJ, et al. Human papillomavirus infection and anal cancer. Int J Cancer 1989;43:1042–9.

[2] Daling JR, Weiss NS, Hislop TG, Maden C, et al. Sexual practices, sexually transmitted diseases, and the incidence of anal cancer. N Engl J Med 1987;317:973–7.

[3] Kiviat N, Rompalo A, Bowden R, et al. Anal human papillomavirus infection among human immunodeficiency virus-seropositive and -seronegative men. J Infect Dis 1990;162(2):358–61.

[4] Kojic EM, Cu-Uvin S, Conley L, et al. Human papillomavirus infection and cytologic abnormalities of the anus and cervix among HIV-infected women in the study to understand the natural history of HIV/AIDS in the era of effective therapy (The SUN study). Sex Transm Dis 2011;38(4):253–9.

[5] Palefsky JM, Holly EA, Ralston ML, et al. Prevalence and risk factors for human papillomavirus infection of the anal canal in human immunodeficiency virus (HIV)-positive and HIV-negative homosexual men. J Infect Dis 1998;177:361–7.

[6] Medley G. Anal smear test to diagnose occult anorectal infection with human papillomavirus in men. Br J Ven Dis 1984;60(3):205.

[7] Palefsky JM, Holly EA, Hogeboom CJ, et al. Anal cytology as a screening tool for anal squamous intraepithelial lesions. J Acq Immun Defic Syndr 1997;14:415–22.

[8] Palefsky JM, Gonzales J, Greenblatt RM, et al. Anal intraepithelial neoplasia and anal papillomavirus infection among homosexual males with group IV HIV disease. JAMA 1990;263:2911–16.

[9] Jay N, Holly EA, Berry M, Hogeboom CJ, et al. Colposcopic correlates of anal squamous intraepithelial lesions. Dis Col Rectum 1997;40:919–28.

[10] Hillman RJ, Cuming T, Darragh T, et al. IANS International Guidelines for Practice Standards in the Detection of Anal Cancer Precursors. J Low Genit Tract Dis 2016;20(4):283–91.

[11] Adami J, Gabel H, Lindelof B, et al. Cancer risk following organ transplantation: a nationwide cohort study in Sweden. Br J Cancer 2003;89(7):1221–7.
[12] Patel HS, Silver AR, Levine T, et al. Human papillomavirus infection and anal dysplasia in renal transplant recipients. Br J Surg 2010;97:1716–21.
[13] Penn I. Cancers of the anogenital regions in renal transplant recipients. Cancer 1986;58:611–16.
[14] Melbye M, Smith E, Wohlfahrt J, et al. Anal and cervical abnormality in women--prediction by human papillomavirus tests. Int J Cancer 1996;68:559–64.
[15] Palefsky JM, Holly EA, Hogeboom CJ, et al. Virologic, immunologic, and clinical parameters in the incidence and progression of anal squamous intraepithelial lesions in HIV-positive and HIV-negative homosexual men. J Acquir Immune Defic Syndr Hum Retrovirol 1998;17:314–19.
[16] Williams AB, Darragh TM, Vranizan K, et al. Anal and cervical human papillomavirus infection and risk of anal and cervical epithelial abnormalities in human immunodeficiency virus-infected women. Obstet Gynecol 1994;83(2):205–11.
[17] Holly EA, Ralston ML, Darragh TM, et al. Prevalence and risk factors for anal squamous intraepithelial lesions in women. J Natl Cancer Inst 2001;93:843–9.
[18] Simpson Jr S, Blomfield P, Cornall A, et al. Front-to-back & dabbing wiping behaviour post-toilet associated with anal neoplasia & HR-HPV carriage in women with previous HPV-mediated gynaecological neoplasia. Cancer Epidemiol 2016;42:124–32.
[19] Palefsky JM, Holly EA, Efird JT, et al. Anal intraepithelial neoplasia in the highly active antiretroviral therapy era among HIV-positive men who have sex with men. AIDS 2005;19:1407–14.
[20] Hessol NA, Holly EA, Efird JT, Minkoff H, et al. Anal intraepithelial neoplasia in a multisite study of HIV-infected and high-risk HIV-uninfected women. AIDS 2009;23:59–70.
[21] Giuliano AR, Palefsky JM, Goldstone SE, et al. The efficacy of the quadrivalent vaccine in preventing HPV 6/11/16/18-related external genital disease and anogenital infection in young men. New Engl J Med 2011;364:401–11.
[22] Palefsky JM, Giuliano AR, Goldstone S, et al. HPV vaccine against anal HPV infection and anal intraepithelial neoplasia. N Engl J Med 2011;365:1576–85.

CHAPTER

13 OROPHARYNGEAL HUMAN PAPILLOMAVIRUS AND HEAD AND NECK CANCER

Farhoud Faraji[1] and Carole Fakhry[2,3]

[1]*Division of Otolaryngology-Head and Neck Surgery, University of California San Diego School of Medicine, San Diego, CA, United States* [2]*Department of Otolaryngology-Head and Neck Surgery, Johns Hopkins University School of Medicine, Baltimore, MD, United States* [3]*Department of Epidemiology, Johns Hopkins Bloomberg School of Public Health, Baltimore, MD, United States*

INTRODUCTION

Over three decades have passed since human papillomavirus (HPV) was first isolated from malignancies of the head and neck [1,2]. In the last 15 years, HPV has become formally recognized as an etiologic agent and prognostic factor in head and neck squamous cell carcinoma (HNSCC) [3]. Moreover, HPV-related oropharyngeal squamous cell carcinoma (HPV-OPC) is now considered a distinct subset of HNSCC, and one of only a few head and neck malignancies that continues to increase in incidence [4–6].

HNSCC encompasses a diverse group of tumors originating from the squamous epithelium of the aerodigestive tract [7]. These are clinically delineated by anatomic site into tumors arising from the oropharynx and those arising from nonoropharyngeal sites including the oral cavity, oropharynx, nasopharynx, hypopharynx, and larynx. While the overall incidence of HNSCC has declined slightly over the last 40 years, the relative contribution of each anatomic site to the overall incidence of HNSCC has shifted significantly [8]. The incidence of nonoropharyngeal tumors has declined while the incidence of OPC continues to rise, likely a result of broad societal changes in behavior that have altered exposures to HNSCC-related risk factors [9]. Patients with HPV-OPC diverge from the classical profile of HNSCC patients in that they are typically under 60 years of age and are less likely to report a history of heavy tobacco and alcohol consumption [10]. However, recent data suggest that the prevalence of HPV-OPC is growing regardless of age, sex, or race [11,12].

Two recent analyses performed on large ($n > 20,000$ HPV-OPC) nationwide multi-institutional national cancer database cohorts showed recent trends of HPV in OPC by age, race, and sex, and characterized the influence of age, race, and sex on survival for HPV-OPC [11,12]. With regard to age, from 2010 to 2014 the proportion of OPC that was HPV-positive increased from 45% to 60% for patients ≥ 70 years old showing that an increasing proportion of older patients is presenting with HPV-positive tumors. In addition, patients aged ≥ 70 years had a diminished survival benefit from HPV-positive tumor status compared to patients aged 50–59 years [11]. With respect to sex,

Human Papillomavirus. DOI: https://doi.org/10.1016/B978-0-12-814457-2.00013-1

the prevalence of HPV in OPC increased from 54% to 61% among women and from 65% to 75% among men from 2010 to 2015. With regard to race, the prevalence of HPV in OPC increased from 65% to 74% among whites, 37% to 54% among blacks, 51% to 69% among Hispanics from 2010 to 2015. Interestingly, neither sex nor race/ethnicity were independently associated with survival in patients with HPV-OPC [12].

HPV-OPC also exhibits substantially better prognosis than HPV-negative HNSCC and greater sensitivity to treatment [13,14]. These features have led to the establishment of a staging system specific to HPV-OPC [15], as well as ongoing investigations into improving treatment strategies directed toward this disease [16]. This chapter will review the epidemiological and molecular basis for distinguishing HPV-OPC as a distinct entity within HNSCC, and discuss the clinical significance of this distinction.

EPIDEMIOLOGY

Over the last four decades an unexpected epidemiological pattern has emerged in HNSCC that suggests evolving risk factor exposures may be altering the epidemiology of HNSCC (Fig. 13.1). The overall incidence of HNSCC has declined; however, the relative contribution of each anatomic site to the overall incidence of HNSCC has changed. The incidence of OPC has increased, while the incidence of tumors arising from nonoropharyngeal sites has decreased [9]. Interestingly these site-specific epidemiological trends in HNSCC have been attributed to an evolution in societal and cultural patterns of behavior that have resulted in shifts in exposure to two classes of HNSCC risk factors: chemical and infectious [18].

Chemical carcinogens induce mutations of key tumor driver genes, promoting carcinogenesis of the upper aerodigestive tract and potentially leading to HNSCC [19–22]. HPV infection may introduce molecular processes that inhibit tumor suppression (which will be discussed in more detail in the next section and also in Chapters 2, 4 and 5). The rise in HPV-OPC has been attributed to changes in societal acceptance of sexual behaviors that increase the likelihood of oncogenic HPV transmission [23]. The two major chemical risk factors in HPV-negative HNSCC are tobacco and alcohol consumption. Tobacco and alcohol exposure are each a strong, independent risk factor in HPV-negative HNSCC [24–26]. Increased public awareness of the dangers of chemical carcinogen exposure and concomitant reduction of exposure to these risk factors have contributed to declines in HPV-negative HNSCC and other carcinogen-related tumors [27,28]. Another notable independent HNSCC risk factor, primarily within parts of Asia and East Africa, is the cultural practice of chewing "paan" (areca nut with lime folded in betel leaf) [29–31].

Perhaps the most convincing evidence establishing alcohol and tobacco consumption as independent risk factors of HNSCC derives from a pooled analysis of 15 case–control studies. In never-users of alcohol, those who had ever smoked were twice as likely to develop HNSCC. A dose–response relationship between tobacco smoking and HNSCC was observed in this study, with subjects who smoked 31–40 cigarettes per day almost twice as likely to develop HNSCC than those who smoked 1–10 cigarettes per day. Among never-users of tobacco, participants who had ever used alcohol exhibited a modest, but statistically significant increase in risk. Heavy alcohol

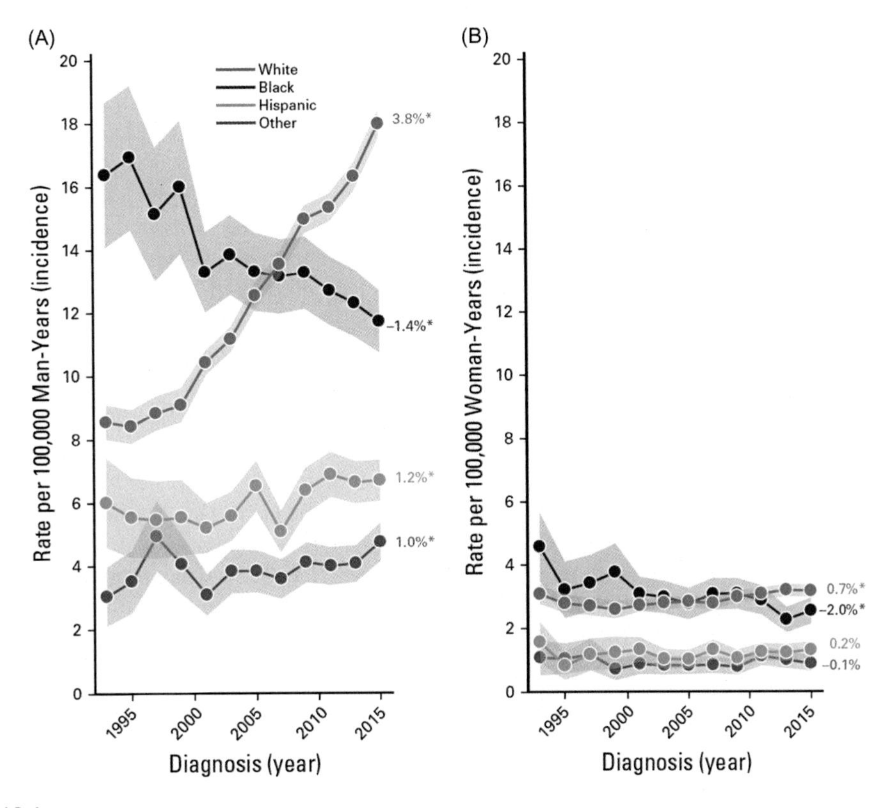

FIGURE 13.1

Incidence trends for oropharynx cancer in the United States, stratified by race and sex.
In both (A) men and (B) women, there was an increase among non-Hispanic white individuals and decline among non-Hispanic black individuals. There was also an increase among Hispanic men and men of other races (American Indian/Alaska Natives and Asian/Pacific Islanders; A), whereas incidence remained stable among Hispanic women and women of other races (B). (*) Indicates statistically significant incidence trend ($P < .05$).

Reproduced with permission from [17] Tota JE, et al. Evolution of the Oropharynx Cancer Epidemic in the United States: moderation of increasing incidence in younger individuals and shift in the burden to older individuals. J Clin Oncol 2019. JCO1900370, Available from: https://doi.org/10.1200/JCO.19.00370. © 2019 American Society of Clinical Oncology. All rights reserved.

consumption of greater than four drinks per day was associated with a nearly threefold increase in risk of HNSCC [26].

HPV is the most common sexually transmitted infection in the United States and the primary infectious cause of HNSCC [32,33]. Nearly 200 types of HPV have been identified; yet, a single type, HPV16, is found in 87%–96% of HPV-OPC and 68%–74% of HPV-positive non-OPC [34–36]. Compared to HPV-negative disease, HPV-OPC is more common in younger patients, and those with a history of high-risk sexual behaviors [6,10]. The strongest and most specific risk factor in HPV-OPC [10,37] is the number of lifetime oral sexual partners. A pooled analysis of four case–control studies found a history of four or more oral–genital sex partners most strongly increased the risk of squamous cell carcinoma within anatomic subsites of the oropharynx [38]. Younger age at sexual debut, increasing number of vaginal sexual partners, history of oral–anal sex, and a history of same-sex sexual contact among men also increased risk for HPV-OPC [38]. A large cross-sectional study conducted as part of the National Health and Nutrition Examination Survey (NHANES) extended these findings, demonstrating that individuals who ever performed oral or vaginal sex were at 2.2-fold or 5.4-fold greater risk of oral HPV infection [39], the putative precursor to HPV-OPC [40], respectively. Furthermore, oral HPV infection is associated with a 53-fold greater risk of HPV-positive HNSCC [10]. These findings support a link between oral HPV infections as a distinct risk factor for developing squamous cell carcinoma of the oropharynx.

Although tobacco and alcohol exposure are not thought to be central drivers in HPV-related oncogenesis of the upper aerodigestive tract, there is evidence that both may modify the risk of developing HPV-OPC. Compared to HPV-negative HNSCC, patients with HPV-OPC are less likely to report a history of heavy tobacco use [10]. Current tobacco users, however, are three times more likely to have oral HPV16 infection compared to never or former tobacco users [41]. Similarly, heavy alcohol consumption increases risk of oral HPV infection, with greater than 21 drinks per week resulting in a 19-fold greater risk of oral HPV infection [42]. The increased prevalence of oral HPV infection in current tobacco and heavy alcohol users indicates a potential avenue by which tobacco and alcohol use indirectly increase risk for HPV-OPC. Consistent with this notion, compared to never smokers the incidence of HPV-OPC was 2.3-fold greater in current smokers and 1.4-fold greater former smokers. In addition, the burden of HPV-OPC was also significantly higher in ever smokers than never smokers (63.3% vs 36.7%, $P < .001$) [43].

The risk factor profiles contributing to carcinogenesis in HPV-OPC and HPV-negative HNSCC are largely distinct. Sexual behaviors are the most important risk factors associated with HPV-OPC, while heavy tobacco and alcohol use are the strongest risk factors associated with HPV-negative HNSCC. Diverging incidence trends of HPV-OPC and HPV-negative HNSCC are generally attributed to shifts in societal and cultural behaviors. These changes in society, with increased education and awareness of the role of carcinogens from alcohol and tobacco, and changes in sexual practices, have driven changes in exposures to these risk factors. Whether another societal shift, HPV vaccination, will influence these epidemiologic trends remains a question under active investigation [44].

MOLECULAR PATHOPHYSIOLOGY OF HUMAN PAPILLOMAVIRUS CARCINOGENESIS

HPV infects keratinocyte progenitor cells located in the basal layer of stratified squamous epithelia and adhered to the epithelial basement membrane. It has been noted that specific structures in the head and neck are more likely to undergo HPV-related neoplastic transformation, presumably due to their proclivity to infection by HPV. Within the oropharynx, HPV disproportionately causes transformation of the palatine and lingual tonsils [45]. The palatine, lingual, tubal, and adenoid tonsils are lymphoid structures collectively known as Waldeyer's tonsillar ring. These contiguous tissues contain a specialized reticulated squamous epithelium with a fenestrated, discontinuous basement membrane thought to allow immune cells access to sample oral antigens for physiologic immune surveillance [46]. This reticulated epithelium may simultaneously allow HPV access to basal keratinocytes [47], thus facilitating viral infection of oropharyngeal structures (Fig. 13.2).

Once HPV enters a basal keratinocyte and sheds its outer protein capsid, its genetic elements are transported to the cell's nucleus and employ cellular transcriptional and translational machinery to produce viral proteins. These processes do not require incorporation of the viral genome into the host genome and, thus in many cases the HPV genome may exist and propagate as a DNA fragment separate from the host cell genome called an episome. HPV-encoded genes disturb key keratinocyte differentiation and proliferation pathways to support viral reproduction and viral life cycle completion. In high-risk HPV types, these perturbations promote neoplastic transformation.

Oncogenic potential is conferred by virulent alleles of viral proteins, E6 and E7, carried by high-risk HPV types. High-risk HPV E6 and E7 promote multiple processes leading to carcinogenesis, including genomic instability, cell cycle progression, keratinocyte immaturity, keratinocyte

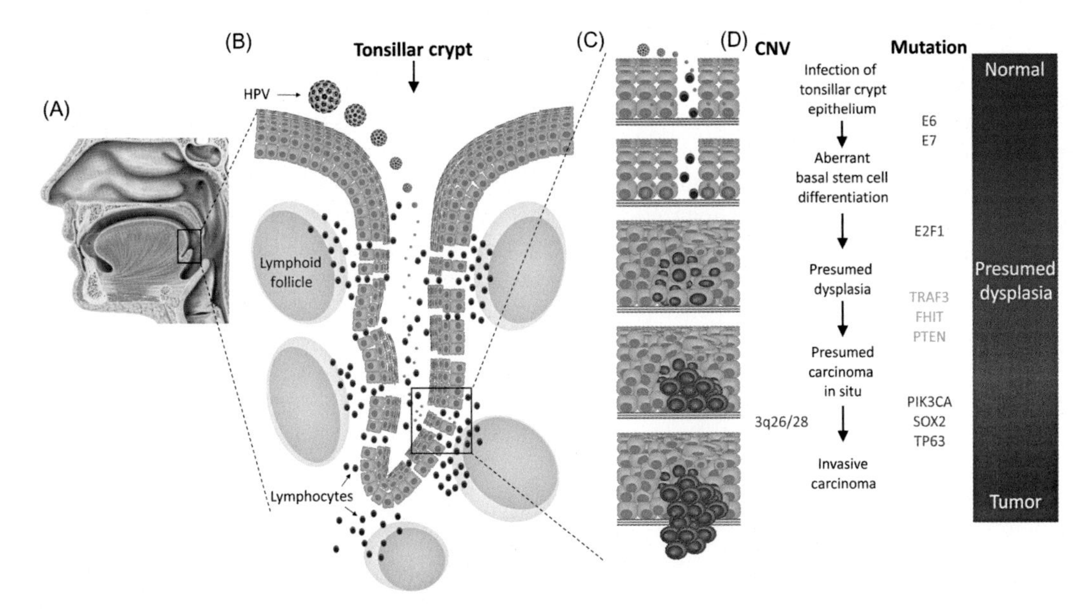

FIGURE 13.2

Tumor progression model for HPV-related carcinogenesis.

(A) In the head and neck, HPV demonstrates tropism for lymphoid-associated structure of the oropharynx, including the palatine and lingual tonsils (area in black box). (B) In the oropharynx, HPV gains access to basal keratinocyte progenitors through fenestrations in the reticulated epithelium of the tonsillar crypts. (C) Infection of the tonsillar epithelium result aberrant basal cell differentiation, presumed dysplasia, presumed carcinoma in situ, and finally invasive carcinoma. (D) Hypothetical model for somatic mutations in a multistage tumor progression model. Genes and loci in red are upregulated or activated or show increase in copy number. Genes in green undergo loss of function mutation or deletion. CNV, copy number variations.

Reproduced with permission from [48] Faraji F, Zaidi M, Fakhry C, Gaykalova DA. Molecular mechanisms of human papillomavirus-related carcinogenesis in head and neck cancer. Microbes Infect 2017;19:464–75, Available from: https://doi.org/10.1016/j.micinf.2017.06.001.

immortalization and neoplastic transformation [49]. P53 is the protein product of the *TP53* gene, a regulator of the DNA-damage response [50] as well as the G_2/M cell cycle transition [51], and the most commonly mutated tumor suppressor gene in cancer [52]. The HPV E6 oncoprotein binds to p53 and marks it for proteolytic degradation [53]. Only E6 alleles from high-risk HPV types bind p53 with high affinity, and the strength of the binding affinity of E6 alleles for p53 correlates with that HPV type's oncogenic potential. For example, HPV16, which is found in the great majority of HPV-related HNSCC compared to another high-risk type, HPV18 [54], encodes an E6 allele that binds p53 with twice the affinity of the HPV18 E6 allele [55].

Similarly, the HPV E7 oncoprotein facilitates the proteolytic destruction of Retinoblastoma (Rb)-family tumor suppressors [56]. Rb inhibits progression through the cell cycle by controlling the G1 to S-phase cell cycle transition. Break down of Rb triggered by E7 not alleviates inhibition of cell cycle progression, inducing compensatory upregulation of, $p16^{INK4A}$ (p16), a downstream tumor suppressor of the Rb pathway [57]. Indeed, diffuse and intense immunohistochemical p16-positivity is a widely accepted method for designating oropharyngeal tumors as HPV-related [58,59]. Approximately 90% p16-positive OPC are HPV-positive, and nearly 100% of HPV-OPC are p16 positive [60,61].

Early studies showed that *TP53* mutation in HNSCC was also strongly correlated with smoking and that *TP53* mutations occurred at markedly lower rates in HPV-positive tumors [62,63]. Fewer HPV-OPC tumors with mutations than HPV-negative HNSCC tumors is likely related to a lower incidence of heavy alcohol and tobacco use in this patient population [10] and that HPV-driven tumor formation may not require mutation of key tumor suppressor genes, as the HPV genome encodes viral oncogenes that inactivate host cellular tumor suppressive pathways. Tumor suppressors, like p53, represent a convergence point for two distinct carcinogenesis mechanisms leading to HNSCC. While there are overlapping points in the pathophysiology, there are many molecular pathways downstream of E6 and E7 unique to the viral infectious process. Furthermore, the consequences of this viral etiology distinguish the response to treatment overall prognosis of HPV-OPC from that of HPV-negative HNSCC.

PROGNOSIS OF HUMAN PAPILLOMAVIRUS-POSITIVE OROPHARYNGEAL SQUAMOUS CELL CARCINOMA

Perhaps the most clinically relevant feature of HPV-OPC as a unique entity is its effect on disease prognosis. Characteristics that influence survival include those related to the natural history of the disease, such as tendency to metastasize or speed of progression, and the interaction between the disease and interventions, such as response to chemotherapy or radiation. HPV-OPC is substantially more responsive to treatment than HPV-negative HNSCC and HPV-positive tumor status confers an independent and robust survival advantage [13,60].

The survival benefit of HPV-OPC was demonstrated in a prospective Eastern Cooperative Oncology Group (ECOG) 2399 Phase II trial [13]. Ninety-six patients with stage III or IV HNSCC were treated with induction chemotherapy followed by definitive chemoradiotherapy. All HPV-positive HNSCC were confined to the oropharynx, HPV-negative tumors arose from the oropharynx or larynx. Higher progression-free survival (PFS) and overall survival (OS) were observed in patients with HPV-positive disease, after a median follow-up of 39 months. Multivariate analysis found that HPV-status reduced risk of death by 64% and was a prognostic factor independent of age, tumor stage, or ECOG performance status [13]. HPV-positive status itself is an indicator of better prognosis; one important reason for this appears to be the outcome of interactions between HPV-OPC and treatment interventions compared to that with HPV-negative HNSCC.

TRENDS IN HUMAN PAPILLOMAVIRUS OROPHARYNGEAL SQUAMOUS CELL CARCINOMA STAGING

Oncologic treatment paradigms must balance therapeutic response and treatment-related morbidity. In the context of head and neck cancer, tumor proximity to structures critical for swallow, airway protection, communication, and social interaction underscores the necessity of oncologic control and anatomic function. Long-term treatment-related toxicities in head and neck cancer include dysphagia, esophageal fibrosis, xerostomia, long-term feeding tube and tracheostomy dependence, sensorineural hearing loss, and osteoradionecrosis [64]. The current standard of care for HNSCC includes combinations of surgery, radiotherapy, and chemotherapy. HPV-OPC is distinctly responsive to current standard of care therapies. This realization has led to concerns that current therapeutic regimens, which were developed primarily for laryngeal cancers and prior to the formal recognition of HPV as an etiologic and prognostic agent in HNSCC, may be more intense than necessary for treatment of HPV-positive disease [65]. As such, current clinical trials represent ongoing efforts to refine therapeutic approaches for HPV-positive disease. Clinical trials of therapeutic de-intensification protocols have thus been implemented to attempt to mitigate long-term toxicities associated with the current standard HNSCC treatment while sustaining high response rates [66]. The success of de-intensification protocols, however, depends heavily on the correct selection of patients at low risk of disease progression. Thus appropriate staging and risk stratification for HPV-OPC is essential.

Until the beginning of 2018, the 7th edition American Joint Committee on Cancer (AJCC-7th), which was developed prior to the acceptance of HPV-OPC as a distinct entity [67], comprised the staging system for all HNSCC. AJCC-7th was found to be insufficient for risk stratification of patients with HPV-OPC [68–71]. While it showed good prognostic performance in HPV-negative HNSCC, it classified 93%–95% of patients with HPV-OPC as stage III or IV, offering inadequate prognostic resolution for this disease entity. Efforts to develop and validate a prognostically effective staging system for HPV-OPC culminating in the International Collaboration on Oropharyngeal cancer Staging (ICON-S) [70] led to the revision of staging criteria in the 8th edition of the AJCC Cancer Staging Manual (AJCC-8th) [15]. The AJCC-8th now formally recognizes HPV-OPC (via its surrogate marker p16) as a distinct neoplastic entity, and defines a distinct staging system for p16-positive OPC [9,10,72]. Further, AJCC-8th delineates a surgical pathology-based staging system to classify risk in patients undergoing surgical management. Ongoing efforts to validate the pTNM system are required to determine its value in clinical decision making [73].

PREVENTION

The natural history and mechanisms of carcinogenic HPV infection in OPC remain incompletely understood. Analogous to cervical cancer [74], persistent oral HPV infection is thought to precede HPV-OPC. However, no precursor lesion has yet been identified for HPV-OPC and key aspects of oral HPV infection remain unresolved, including potential routes of oral infection and determinants of persistent infection, latency, reactivation, viral clearance, and host immune response to HPV [23]. These deficits in knowledge combined with the anatomic topography that renders the oropharynx challenging for tissue sampling have thus far hindered the development of effective screening methods for the early detection of HPV-OPC, such as pap-test equivalent strategies or serologic methods [75–78].

To date, three prophylactic HPV vaccines have been approved by the United States Food and Drug Administration (Gardasil, Gardasil 9, and Cervarix). These vaccines are based on virus-like particles (VLP) of HPV L1 capsid proteins and protect at minimum against HPV-16 and -18, which together are associated with 70% of cervical cancers [79]. A series of large, multinational clinical trials performed in young women established the efficacy of these VLP-based HPV vaccines for the prevention of cervical HPV infection and cancer [80–82]. These studies demonstrated high vaccine efficacy (98%–100%) in preventing premalignant HPV-associated cervical lesions in women without previous exposure to HPV [80–82]. Prophylactic HPV vaccination also exhibited high efficacy (90%) in preventing genital HPV-related lesions in men [83]. While no data is yet available on the efficacy of HPV vaccination in preventing HPV-OPC, vaccination has been linked to significant increases in HPV type-specific oral antibody prevalence [84]. Furthermore, oral titers were observed to correlate with serum antibody titers, suggesting that HPV vaccination may reduce oral oncogenic HPV infection (estimated efficacy for oral HPV: 93%) [84,85]. More recently, HPV vaccination has been associated with reductions in vaccine type-specific oral HPV infection in young US adults [44]. Taken together, these findings offer the first clues into the possibility that HPV vaccination reduces oral oncogenic HPV infection and may also be effective in the prevention of HPV-OPC.

CONCLUSION

Multiple lines of evidence across several fields of study have reinforced the idea that HPV-OPC is a distinct malignancy from other forms of HNSCC. HNSCC describes a diverse group of malignancies arising from the squamous epithelium of the aerodigestive tract. While these tumors have and continue to be classified primarily based on anatomic subsite, the acceptance of HPV-status underscores that molecular markers of tumor etiology can provide clinically actionable information. HPV has broadened the spectrum of patient features observed within HNSCC patient population with regard to age, sex, race, and risk factor profiles. The distinct mechanism of carcinogenesis conferred by HPV-related disease likely drives prognostic differences between HPV-OPC and HPV-negative HNSCC, an insight that will be critical for improving tailoring HPV-OPC-specific treatment strategies. In conjunction with the newly adopted AJCC-8th staging system, ongoing efforts strive to retain high therapeutic response rates while further limiting long-term side effects of treatment. Over the past two decades significant progress has been made in understanding of HPV-related carcinogenesis, epidemiology, and clinical oncology. Continued elaborations of these foundations will lead to the development of improved diagnostic and therapeutic strategies to further to reduce the burden imposed by this malignancy.

FINANCIAL DISCLOSURES

The authors have no financial disclosures.

CONFLICTS OF INTEREST

The authors have no conflicts of interest to declare.

REFERENCES

[1] de Villiers EM, Weidauer H, Otto H, zur Hausen H. Papillomavirus DNA in human tongue carcinomas. Int J Cancer 1985;36:575–8.
[2] Syrjanen K, Syrjanen S, Pyrhonen S. Human papilloma virus (HPV) antigens in lesions of laryngeal squamous cell carcinomas. ORL J Otorhinolaryngol Relat Spec 1982;44:323–34.
[3] Humans I W G o t E o C R t. Human papillomaviruses. IARC Monogr Eval Carcinog Risks Hum 2007;90:1–636.
[4] Jemal A, et al. Annual Report to the Nation on the Status of Cancer, 1975-2009, featuring the burden and trends in human papillomavirus(HPV)-associated cancers and HPV vaccination coverage levels. J Natl Cancer Inst 2013;105:175–201. Available from: https://doi.org/10.1093/jnci/djs491.
[5] Patel MA, et al. Rising population of survivors of oral squamous cell cancer in the United States. Cancer 2016;122:1380–7. Available from: https://doi.org/10.1002/cncr.29921.
[6] Gillison ML, et al. Evidence for a causal association between human papillomavirus and a subset of head and neck cancers. J Natl Cancer Inst 2000;92:709–20.
[7] Pai SI, Westra WH. Molecular pathology of head and neck cancer: implications for diagnosis, prognosis, and treatment. Annu Rev Pathol 2009;4:49–70. Available from: https://doi.org/10.1146/annurev.pathol.4.110807.092158.
[8] Carvalho AL, Nishimoto IN, Califano JA, Kowalski LP. Trends in incidence and prognosis for head and neck cancer in the United States: a site-specific analysis of the SEER database. Int J Cancer 2005;114:806–16. Available from: https://doi.org/10.1002/ijc.20740.
[9] Chaturvedi AK, et al. Human papillomavirus and rising oropharyngeal cancer incidence in the United States. J Clin Oncol 2011;29:4294–301. Available from: https://doi.org/10.1200/JCO.2011.36.4596.
[10] Gillison ML, et al. Distinct risk factor profiles for human papillomavirus type 16-positive and human papillomavirus type 16-negative head and neck cancers. J Natl Cancer Inst 2008;100:407–20. Available from: https://doi.org/10.1093/jnci/djn025.
[11] Rettig EM, et al. Oropharyngeal cancer is no longer a disease of younger patients and the prognostic advantage of Human Papillomavirus is attenuated among older patients: analysis of the National Cancer Database. Oral Oncol 2018;83:147–53. Available from: https://doi.org/10.1016/j.oraloncology.2018.06.013.
[12] Faraji F, et al. The prevalence of human papillomavirus in oropharyngeal cancer is increasing regardless of sex or race, and the influence of sex and race on survival is modified by human papillomavirus tumor status. Cancer 2018;. Available from: https://doi.org/10.1002/cncr.31841.

[13] Fakhry C, et al. Improved survival of patients with human papillomavirus-positive head and neck squamous cell carcinoma in a prospective clinical trial. J Natl Cancer Inst 2008;100:261–9. Available from: https://doi.org/10.1093/jnci/djn011.

[14] Fung N, Faraji F, Kang H, Fakhry C. The role of human papillomavirus on the prognosis and treatment of oropharyngeal carcinoma. Cancer Metastasis Rev 2017;36:449–61. Available from: https://doi.org/10.1007/s10555-017-9686-9.

[15] Lydiatt WM, et al. Head and neck cancers-major changes in the American Joint Committee on cancer eighth edition cancer staging manual. CA Cancer J Clin 2017;. Available from: https://doi.org/10.3322/caac.21389.

[16] Kelly JR, Husain ZA, Burtness B. Treatment de-intensification strategies for head and neck cancer. Eur J Cancer 2016;68:125–33. Available from: https://doi.org/10.1016/j.ejca.2016.09.006.

[17] Tota JE, et al. Evolution of the Oropharynx Cancer Epidemic in the United States: moderation of increasing incidence in younger individuals and shift in the burden to older individuals J Clin Oncol 2019; JCO1900370. Available from: https://doi.org/10.1200/JCO.19.00370.

[18] Faraji F, Eisele DW, Fakhry C. Emerging insights into recurrent and metastatic human papillomavirus-related oropharyngeal squamous cell carcinoma. Laryngoscope Investig Otolaryngol 2017;2:10–18. Available from: https://doi.org/10.1002/lio2.37.

[19] Hecht SS. Tobacco carcinogens, their biomarkers and tobacco-induced cancer. Nat Rev Cancer 2003;3:733–44. Available from: https://doi.org/10.1038/nrc1190.

[20] Applebaum KM, et al. Lack of association of alcohol and tobacco with HPV16-associated head and neck cancer. J Natl Cancer Inst 2007;99:1801–10. Available from: https://doi.org/10.1093/jnci/djm233.

[21] Boyle JO, et al. The incidence of p53 mutations increases with progression of head and neck cancer. Cancer Res 1993;53:4477–80.

[22] Brennan JA, et al. Association between cigarette smoking and mutation of the p53 gene in squamous-cell carcinoma of the head and neck. N Engl J Med 1995;332:712–17. Available from: https://doi.org/10.1056/NEJM199503163321104.

[23] Faraji F, Fakhry C. In: Durand M, Deschler D, editors. Infections of the Ears, Nose, Throat, and Sinuses. Cham: Springer; 2018. p. 349–64.

[24] Seitz HK, Stickel F. Molecular mechanisms of alcohol-mediated carcinogenesis. Nat Rev Cancer 2007;7:599–612. Available from: https://doi.org/10.1038/nrc2191.

[25] Blot WJ, et al. Smoking and drinking in relation to oral and pharyngeal cancer. Cancer Res 1988;48:3282–7.

[26] Hashibe M, et al. Alcohol drinking in never users of tobacco, cigarette smoking in never drinkers, and the risk of head and neck cancer: pooled analysis in the International Head and Neck Cancer Epidemiology Consortium. J Natl Cancer Inst 2007;99:777–89. Available from: https://doi.org/10.1093/jnci/djk179.

[27] Chaturvedi AK, et al. Worldwide trends in incidence rates for oral cavity and oropharyngeal cancers. J Clin Oncol 2013;31:4550–9. Available from: https://doi.org/10.1200/JCO.2013.50.3870.

[28] Zumsteg ZS, et al. Incidence of oropharyngeal cancer among elderly patients in the United States. JAMA Oncol 2016;2:1617–23. Available from: https://doi.org/10.1001/jamaoncol.2016.1804.

[29] Balaram P, et al. Oral cancer in southern India: the influence of smoking, drinking, paan-chewing and oral hygiene. Int J Cancer 2002;98:440–5.

[30] Mack TM. The new pan-Asian paan problem. Lancet 2001;357:1638–9.

[31] Merchant A, et al. Paan without tobacco: an independent risk factor for oral cancer. Int J Cancer 2000;86:128–31.

[32] Satterwhite CL, et al. Sexually transmitted infections among US women and men: prevalence and incidence estimates, 2008. Sex Transm Dis 2013;40:187–93. Available from: https://doi.org/10.1097/OLQ.0b013e318286bb53.

[33] Plummer M, et al. Global burden of cancers attributable to infections in 2012: a synthetic analysis. Lancet Glob Health 2016;4:e609–16. Available from: https://doi.org/10.1016/S2214-109X(16)30143-7.

[34] Kreimer AR, Clifford GM, Boyle P, Franceschi S. Human papillomavirus types in head and neck squamous cell carcinomas worldwide: a systematic review. Cancer Epidemiol Biomarkers Prev 2005;14:467–75. Available from: https://doi.org/10.1158/1055-9965.EPI-04-0551.

[35] Ndiaye C, et al. HPV DNA, E6/E7 mRNA, and p16INK4a detection in head and neck cancers: a systematic review and meta-analysis. Lancet Oncol 2014;15:1319–31. Available from: https://doi.org/10.1016/S1470-2045(14)70471-1.

[36] Mehanna H, et al. Prevalence of human papillomavirus in oropharyngeal and nonoropharyngeal head and neck cancer--systematic review and meta-analysis of trends by time and region. Head Neck 2013;35:747–55. Available from: https://doi.org/10.1002/hed.22015.

[37] D'Souza G, Agrawal Y, Halpern J, Bodison S, Gillison ML. Oral sexual behaviors associated with prevalent oral human papillomavirus infection. J Infect Dis 2009;199:1263–9. Available from: https://doi.org/10.1086/597755.

[38] Heck JE, et al. Sexual behaviours and the risk of head and neck cancers: a pooled analysis in the International Head and Neck Cancer Epidemiology (INHANCE) consortium. Int J Epidemiol 2010;39:166–81. Available from: https://doi.org/10.1093/ije/dyp350.

[39] Gillison ML, et al. Prevalence of oral HPV infection in the United States, 2009-2010. JAMA 2012;307:693–703. Available from: https://doi.org/10.1001/jama.2012.101.

[40] Suk R, et al. Trends in risks for second primary cancers associated with index human papillomavirus-associated cancers. JAMA Netw Open 2018;1:e181999. Available from: https://doi.org/10.1001/jamanetworkopen.2018.1999.

[41] Fakhry C, Gillison ML, D'Souza G. Tobacco use and oral HPV-16 infection. JAMA 2014;312:1465–7. Available from: https://doi.org/10.1001/jama.2014.13183.

[42] Smith EM, et al. Human papillomavirus in oral exfoliated cells and risk of head and neck cancer. J Natl Cancer Inst 2004;96:449–55.

[43] Chaturvedi AK, D'Souza G, Gillison ML, Katki HA. Burden of HPV-positive oropharynx cancers among ever and never smokers in the U.S. population. Oral Oncol 2016;60:61–7. Available from: https://doi.org/10.1016/j.oraloncology.2016.06.006.

[44] Chaturvedi AK, et al. Effect of Prophylactic human papillomavirus (HPV) vaccination on oral HPV infections among young adults in the United States. J Clin Oncol 2018;36:262–7. Available from: https://doi.org/10.1200/JCO.2017.75.0141.

[45] Paz IB, Cook N, Odom-Maryon T, Xie Y, Wilczynski SP. Human papillomavirus (HPV) in head and neck cancer. An association of HPV 16 with squamous cell carcinoma of Waldeyer's tonsillar ring. Cancer 1997;79:595–604.

[46] Perry ME. The specialised structure of crypt epithelium in the human palatine tonsil and its functional significance. J Anat 1994;185(Pt 1):111–27.

[47] Westra WH. The morphologic profile of HPV-related head and neck squamous carcinoma: implications for diagnosis, prognosis, and clinical management. Head Neck Pathol 2012;6(Suppl 1):S48–54. Available from: https://doi.org/10.1007/s12105-012-0371-6.

[48] Faraji F, Zaidi M, Fakhry C, Gaykalova DA. Molecular mechanisms of human papillomavirus-related carcinogenesis in head and neck cancer. Microbes Infect 2017;19:464–75. Available from: https://doi.org/10.1016/j.micinf.2017.06.001.

[49] Sherman L, et al. Inhibition of serum- and calcium-induced differentiation of human keratinocytes by HPV16 E6 oncoprotein: role of p53 inactivation. Virology 1997;237:296–306. Available from: https://doi.org/10.1006/viro.1997.8778.

[50] Bunz F, et al. Requirement for p53 and p21 to sustain G2 arrest after DNA damage. Science 1998;282:1497–501.
[51] Taylor WR, Stark GR. Regulation of the G2/M transition by p53. Oncogene 2001;20:1803–15. Available from: https://doi.org/10.1038/sj.onc.1204252.
[52] Hollstein M, Sidransky D, Vogelstein B, Harris CC. p53 mutations in human cancers. Science 1991;253:49–53.
[53] Scheffner M, Werness BA, Huibregtse JM, Levine AJ, Howley PM. The E6 oncoprotein encoded by human papillomavirus types 16 and 18 promotes the degradation of p53. Cell 1990;63:1129–36.
[54] Castellsague X, et al. HPV involvement in head and neck cancers: comprehensive assessment of biomarkers in 3680 patients. J Natl Cancer Inst 2016;108:djv403. Available from: https://doi.org/10.1093/jnci/djv403.
[55] Werness BA, Levine AJ, Howley PM. Association of human papillomavirus types 16 and 18 E6 proteins with p53. Science 1990;248:76–9.
[56] Boyer SN, Wazer DE, Band V. E7 protein of human papilloma virus-16 induces degradation of retinoblastoma protein through the ubiquitin-proteasome pathway. Cancer Res 1996;56:4620–4.
[57] Serrano M, Hannon GJ, Beach D. A new regulatory motif in cell-cycle control causing specific inhibition of cyclin D/CDK4. Nature 1993;366:704–7. Available from: https://doi.org/10.1038/366704a0.
[58] Khleif SN, et al. Inhibition of cyclin D-CDK4/CDK6 activity is associated with an E2F-mediated induction of cyclin kinase inhibitor activity. Proc Natl Acad Sci U S A 1996;93:4350–4.
[59] Riethdorf S, et al. p16INK4A expression as biomarker for HPV 16-related vulvar neoplasias. Hum Pathol 2004;35:1477–83.
[60] Ang KK, et al. Human papillomavirus and survival of patients with oropharyngeal cancer. N Engl J Med 2010;363:24–35. Available from: https://doi.org/10.1056/NEJMoa0912217.
[61] Chung CH, et al. p16 protein expression and human papillomavirus status as prognostic biomarkers of nonoropharyngeal head and neck squamous cell carcinoma. J Clin Oncol 2014;32:3930–8. Available from: https://doi.org/10.1200/JCO.2013.54.5228.
[62] Brachman DG, et al. Occurrence of p53 gene deletions and human papilloma virus infection in human head and neck cancer. Cancer Res 1992;52:4832–6.
[63] Koch WM, Lango M, Sewell D, Zahurak M, Sidransky D. Head and neck cancer in nonsmokers: a distinct clinical and molecular entity. Laryngoscope 1999;109:1544–51. Available from: https://doi.org/10.1097/00005537-199910000-00002.
[64] Epstein JB, et al. Oral complications of cancer and cancer therapy: from cancer treatment to survivorship. CA Cancer J Clin 2012;62:400–22. Available from: https://doi.org/10.3322/caac.21157.
[65] O'Sullivan B, et al. Deintensification candidate subgroups in human papillomavirus-related oropharyngeal cancer according to minimal risk of distant metastasis. J Clin Oncol 2013;31:543–50. Available from: https://doi.org/10.1200/JCO.2012.44.0164.
[66] Bhatia A, Burtness B. Human papillomavirus-associated oropharyngeal cancer: defining risk groups and clinical trials. J Clin Oncol 2015;33:3243–50. Available from: https://doi.org/10.1200/JCO.2015.61.2358.
[67] Ward MJ, et al. Staging and treatment of oropharyngeal cancer in the human papillomavirus era. Head Neck 2015;37:1002–13. Available from: https://doi.org/10.1002/hed.23697.
[68] Huang SH, et al. Refining American Joint Committee on Cancer/Union for International Cancer Control TNM stage and prognostic groups for human papillomavirus-related oropharyngeal carcinomas. J Clin Oncol 2015;33:836–45. Available from: https://doi.org/10.1200/JCO.2014.58.6412.
[69] Dahlstrom KR, Garden AS, William Jr. WN, Lim MY, Sturgis EM. Proposed staging system for patients with HPV related oropharyngeal cancer based on nasopharyngeal cancer N categories. J Clin Oncol 2016;34:1848–54. Available from: https://doi.org/10.1200/JCO.2015.64.6448.

[70] O'Sullivan B, et al. Development and validation of a staging system for HPV-related oropharyngeal cancer by the International Collaboration on Oropharyngeal cancer Network for Staging (ICON-S): a multicentre cohort study. Lancet Oncol 2016;17:440–51. Available from: https://doi.org/10.1016/S1470-2045(15)00560-4.

[71] Malm IJ, et al. Evaluation of proposed staging systems for human papillomavirus-related oropharyngeal squamous cell carcinoma. Cancer 2017;. Available from: https://doi.org/10.1002/cncr.30512.

[72] Mehanna H, Jones TM, Gregoire V, Ang KK. Oropharyngeal carcinoma related to human papillomavirus. BMJ 2010;340:c1439. Available from: https://doi.org/10.1136/bmj.c1439.

[73] Fakhry C, Zevallos JP, Eisele DW. Imbalance Between Clinical and Pathologic Staging in the Updated American Joint Commission on Cancer Staging System for Human Papillomavirus-Positive Oropharyngeal Cancer. J Clin Oncol 2018;36:217–19. Available from: https://doi.org/10.1200/JCO.2017.75.2063.

[74] Walboomers JM, et al. Human papillomavirus is a necessary cause of invasive cervical cancer worldwide. J Pathol 1999;189:12–9. Available from: https://doi.org/10.1002/(SICI)1096-9896(199909)189:1 < 12::AID-PATH431 > 3.0.CO;2-F.

[75] Fakhry C, Rosenthal BT, Clark DP, Gillison ML. Associations between oral HPV16 infection and cytopathology: evaluation of an oropharyngeal "pap-test equivalent" in high-risk populations. Cancer Prev Res (Phila) 2011;4:1378–84. Available from: https://doi.org/10.1158/1940-6207.CAPR-11-0284.

[76] Kreimer AR, et al. Evaluation of human papillomavirus antibodies and risk of subsequent head and neck cancer. J Clin Oncol 2013;31:2708–15. Available from: https://doi.org/10.1200/JCO.2012.47.2738.

[77] Lang Kuhs KA, et al. Human papillomavirus 16 E6 antibodies in individuals without diagnosed cancer: a pooled analysis. Cancer Epidemiol Biomarkers Prev 2015;24:683–9. Available from: https://doi.org/10.1158/1055-9965.EPI-14-1217.

[78] Gillison ML, Chaturvedi AK, Anderson WF, Fakhry C. Epidemiology of human papillomavirus-positive head and neck squamous cell carcinoma. J Clin Oncol 2015;33:3235–42. Available from: https://doi.org/10.1200/JCO.2015.61.6995.

[79] Lowy DR. HPV vaccination to prevent cervical cancer and other HPV-associated disease: from basic science to effective interventions. J Clin Invest 2016;126:5–11. Available from: https://doi.org/10.1172/JCI85446.

[80] Group FIS. Quadrivalent vaccine against human papillomavirus to prevent high-grade cervical lesions. N Engl J Med 2007;356:1915–27. Available from: https://doi.org/10.1056/NEJMoa061741.

[81] Lehtinen M, et al. Overall efficacy of HPV-16/18 AS04-adjuvanted vaccine against grade 3 or greater cervical intraepithelial neoplasia: 4-year end-of-study analysis of the randomised, double-blind PATRICIA trial. Lancet Oncol 2012;13:89–99. Available from: https://doi.org/10.1016/S1470-2045(11)70286-8.

[82] Joura EA, et al. A 9-valent HPV vaccine against infection and intraepithelial neoplasia in women. N Engl J Med 2015;372:711–23. Available from: https://doi.org/10.1056/NEJMoa1405044.

[83] Giuliano AR, et al. Efficacy of quadrivalent HPV vaccine against HPV Infection and disease in males. N Engl J Med 2011;364:401–11. Available from: https://doi.org/10.1056/NEJMoa0909537.

[84] Pinto LA, et al. Quadrivalent human papillomavirus (HPV) vaccine induces HPV-specific antibodies in the oral cavity: results from the mid-adult male vaccine trial. J Infect Dis 2016;214:1276–83. Available from: https://doi.org/10.1093/infdis/jiw359.

[85] Herrero R, et al. Reduced prevalence of oral human papillomavirus (HPV) 4 years after bivalent HPV vaccination in a randomized clinical trial in Costa Rica. PLoS One 2013;8:e68329. Available from: https://doi.org/10.1371/journal.pone.0068329.

CHAPTER

14 VULVAR, PENILE, AND SCROTAL HUMAN PAPILLOMAVIRUS AND NON–HUMAN PAPILLOMAVIRUS CANCER PATHWAYS

Elmar Joura[1], David Jenkins[2] and Nuria Guimera[3]

[1]Department of Gynecology and Obstetrics, Medical University of Vienna-AKHComprehensive Cancer Center Vienna, Vienna, Austria [2]Emeritus Professor of Pathology, University of Nottingham, United Kingdom [3]DDL Diagnostic Laboratory, Rijswijk, The Netherlands

HISTORY OF VULVAR, PENILE, AND SCROTAL CANCER REFLECTS BOTH HUMAN PAPILLOMAVIRUS–RELATED AND NON–HUMAN PAPILLOMAVIRUS–RELATED ETIOLOGY

Vulvar, penile, and scrotal cancer are uncommon tumors of external genitalia. Despite being cancers affecting male and female there are many similarities in their relation to human papillomavirus (HPV) and non-HPV pathways, possibly related to comparable embryological development of their respective organs and epithelia [1]. Their low incidence makes their etiology and pathogenesis difficult to study. However, international research collaborations have led to a better estimate of the etiological role of HPV in these tumors, as well as to more accurate pathological classification linked to etiology and epidemiological patterns of disease.

Vulvar cancer was initially described as a rare disease of older women in their eighties. An increasing number of cases of invasive vulvar cancer has been seen among younger women, primarily related to an increasing incidence of vulvar intraepithelial neoplasia (VIN) following HPV infections in the latter part of the 20th century [2,3].

Penile cancer is rare in western countries but is more common in some other countries. Penile cancer can account for up to 10% of cancers among men in some parts of Africa, South America, and Asia [4]. Penile cancer has been associated with several environmental risk factors. As with the vulva and scrotum, penile neoplasias are thought to develop through two different carcinogenic pathways: a non-HPV-related pathway and an HPV-related pathway. The two pathways of penile carcinogenesis show different histological features as later described this chapter. The separation of penile neoplasia into HPV-related and non-HPV-related has led to reclassification of neoplasia reflected in the new WHO histologic classification [5].

Scrotal cancer is a very rare malignancy and was the first cancer associated with occupational carcinogens including soot, lubricating, and cutting oils [6,7]. Since the 1970s occupational-related scrotal cancer incidence has decreased in the United Kingdom as working conditions have

Human Papillomavirus. DOI: https://doi.org/10.1016/B978-0-12-814457-2.00014-3

improved [7]. More recent reports, however, indicate a steady incidence in the Netherlands and increasing incidence in the United States [8,9]. This increasing incidence suggested other nonoccupational factors were involved in the carcinogenesis of scrotal cancer. In 2017, the biological evidence of an etiological relationship between HPV16 and a small number of scrotal cancers was provided [10]. The dual pathways of oncogenesis HPV and non-HPV related and relation between histological type of scrotal cancer and HPV were similar to that in penile cancers.

The relation of an important fraction of vulvar, penile, and scrotal cancers to HPV, reinforces the potential benefit of HPV vaccines in the reduction of HPV-related external genitalia malignancies in both men and women.

BURDEN OF VULVAR, PENILE, AND SCROTAL CANCER

Vulvar cancer is a rare entity with incidence rates ranging from 0.5 to 1.5 per 100,000 women and represents 4% of all gynecological cancers [11]. However, recent reports indicate an increase in the incidence of the disease especially among young women [12]. Penile cancer is also a rare disease with an estimated 22,000 new cases reported worldwide [11], but the incidence varies widely from country to country. The highest rates are in Africa, South America, and Asia, where penile cancer can account for up to 10% of cancers among men [4]. Scrotal cancer is an extremely rare skin malignancy related largely to occupational carcinogens. Knowledge of occupational risk factors led to improvements in working conditions and to a decreased of related scrotal carcinoma in the United Kingdom [7]. It was expected that scrotal cancer would disappear but, new reports indicate stabilized incidence in the Netherlands or even an increase in United States [8,9] maybe due to the newly detected role of HPV in scrotal cancer.

PATHOLOGY OF VULVAR, PENILE, AND SCROTAL NEOPLASIA

Several studies have shown the relation of HPV presence to certain common histological patterns of vulva, penile cancers, and precancers and there has been one study showing the same for scrotal cancer. In all of these the most common type of cancer is squamous cell carcinoma (SCC), although other histological types of cancer have been described and SCC is subclassified histologically. Vulvar neoplastic lesions most often affect the inner edges of the labia majora or the labia minora (Fig. 14.1). Penile neoplastic lesions usually arise from the epithelium of the inner prepuce or the glans. The scrotum consists of multiple layers, including skin, fascia, and muscle and scrotal cancer originates from the scrotal epidermis.

Vulvar SCC accounts for more than 90% of vulvar cancer. SCC subtypes described includes keratinizing, nonkeratinizing, basaloid, warty, warty–basaloid, and verrucous carcinoma [13]. Basaloid and warty variants represent about 33% of cases and are commoner in younger women. As shown in Table 14.1, they are often associated with HPV DNA detection, while keratinizing types are usually HPV negative [14].

Preneoplastic lesions classified by the LAST criteria [17] as Squamous Intraepithelial Lesions (SIL) or previously as VIN are also important and increasing in number. Low-grade VIN is caused

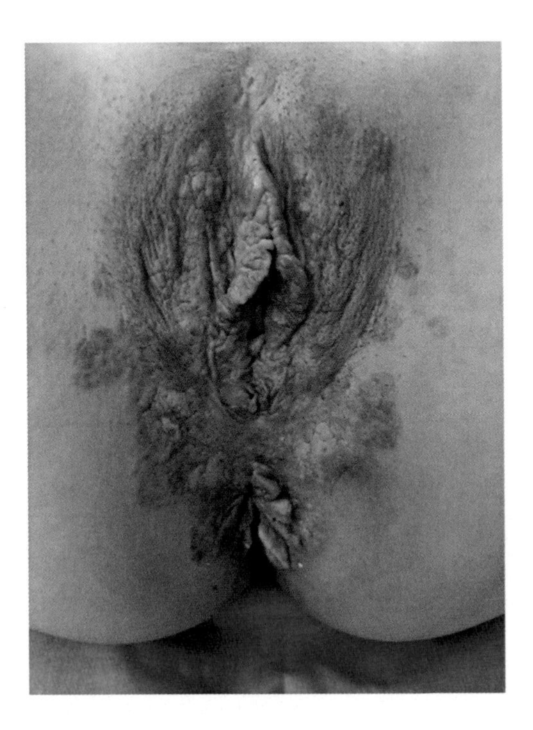

FIGURE 14.1

Extensive HSIL of the vulva and the anus.

by the same HPV types as genital warts, predominately HPV6, but coinfections with oncogenic HPV types are not uncommon [18,19]. These lesions have a low potential for progression, but the rare verrucous carcinoma of the vulva can be caused by HPV6. Differentiated VIN are non-HPV-related lesions (Table 14.1), in some of these lesion HPV may be detected [14] as the methods usually used for HPV detection do not precisely determine HPV localization in the specimen. As observed in penile preneoplastic lesions [20], the presence of other HPV-related lesions in the same specimen may overestimate the prevalence of HPV in non-HPV-related lesions such us differentiated VIN. Laser capture microdissection (LCM)-PCR is a technic that allows the accurate histological localization of the HPV detected and may help refine the prevalence of HPV in genital lesions.

As in vulvar carcinoma the majority of penile tumors are SCCs. The well differentiated and keratinizing SCCs are usually non-HPV related (Fig. 14.2). On the contrary, SCC subtypes basaloid and warty (condylomatous) are HPV related (Figs. 14.3 and 14.4) [21]. The frequency of finding HPV by histological types has been described [15] and shown in Table 14.1. In about 62% of cases more than one lesion (premalignant and/or malignant) is present per specimen [22], making precise determination of the role of HPV types difficult to estimate in penile neoplasia. WHO updated the histological classification of penile carcinoma reflecting the two pathways of penile carcinogenesis (HPV related and non-HPV related). In this histological classification penile tumors and preneoplastic lesions are basically subclassified in SCC or PeIN differentiated, and SCC or PeIN warty or

Table 14.1 Intraepithelial Neoplasia and Invasive Vulva, Penile, and Scrotal Carcinomas by Histology and HPV Prevalence.

Histology	% HPV Prevalence
Vulvar Squamous Intraepithelial Lesions	
Warty/basaloid	90
Differentiated	2
Overall HPV positive (%)	87
Invasive Vulvar Cancer	
SCC warty/basaloid	94
SCC keratinizing	69
SCC mixed	71
Other	78
Overall HPV positive (%)	29
Penile Intraepithelial Neoplasia	
Warty/basaloid	97
Differentiated	31
Both	100
Overall HPV positive (%)	87
Invasive Penile Cancer	
SCC warty/basaloid	75
SCC non-warty/basaloid	15
SCC mixed	45
Other	52
Overall HPV positive (%)	33
Invasive Scrotal Cancer	
SCC warty/basaloid	100 (3/3)
Usual SCC	0 (3/3)
Overall HPV positive (%)	50 (3/6)

SCC, squamous cell carcinoma; *HPV*, human papillomavirus.

Adapted from Guimera N, Alemany L, Halec G, Pawlita M, Wain GV, Vailen JSS, et al. Human papillomavirus 16 is an aetiological factor of scrotal cancer. Br J Cancer. 2017;116(9):1218–22 [10], de Sanjose S, Alemany L, Ordi J, Tous S, Alejo M, Bigby SM, et al. Worldwide human papillomavirus genotype attribution in over 2000 cases of intraepithelial and invasive lesions of the vulva. Eur J Cancer. 2013;49(16):3450–61 [14], Alemany L, Cubilla A, Halec G, Kasamatsu E, Quiros B, Masferrer E, et al. Role of human papillomavirus in penile carcinomas worldwide. Eur Urol. 2016;69(5):953–961 [15], and Hoang LN, Park KJ, Soslow RA, Murali R. Squamous precursor lesions of the vulva: current classification and diagnostic challenges. Pathology. 2016;48(4):291–302 [16].

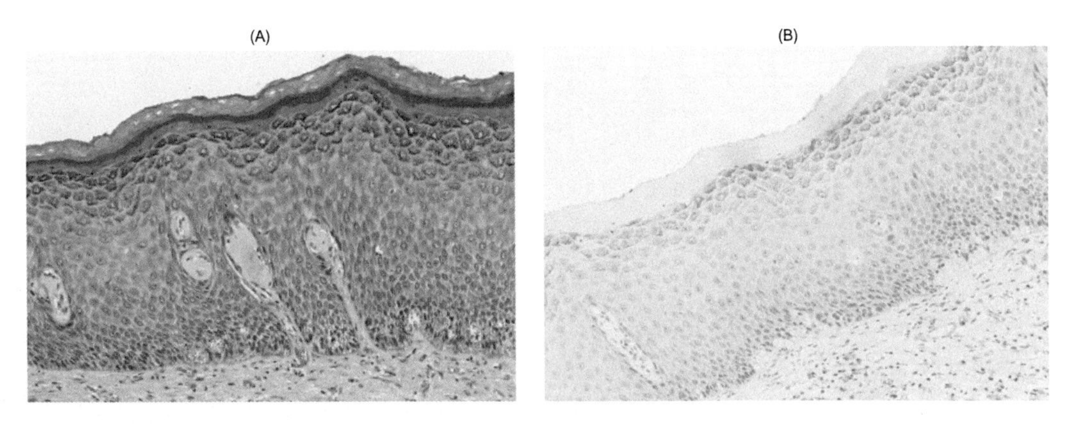

FIGURE 14.2

Differentiated PeIN. (A) Keratinizing lesion with hyper and parakeratosis, acanthosis, and lower third basal atypias; this lesion shows lichen sclerosus. (B) p16 staining is negative. *PeIN*, penile intraepithelial neoplasia.

Adapted from Fernandez-Nestosa MJ, Guimera N, Sanchez DF, Canete-Portillo S, Velazquez EF, Jenkins D, et al. Human papillomavirus (HPV) genotypes in condylomas, intraepithelial neoplasia, and invasive carcinoma of the penis using laser capture microdissection (LCM)-PCR: a study of 191 lesions in 43 patients. Am J Surg Pathol. 2017;41(6):820–32 [20].

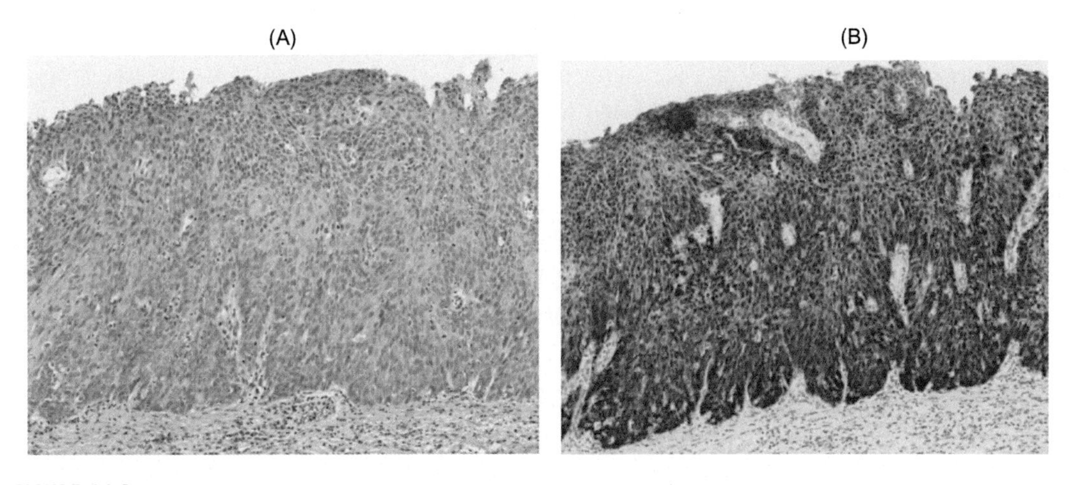

FIGURE 14.3

HPV16-positive Basaloid PeIN. (A) Epithelium full thickness is replaced by small basophilic anaplastic cells. (B) Diffuse overexpression of p16 staining is present. *PeIN*, penile intraepithelial neoplasia.

Adapted from Fernandez-Nestosa MJ, Guimera N, Sanchez DF, Canete-Portillo S, Velazquez EF, Jenkins D, et al. Human papillomavirus (HPV) genotypes in condylomas, intraepithelial neoplasia, and invasive carcinoma of the penis using laser capture microdissection (LCM)-PCR: a study of 191 lesions in 43 patients. Am J Surg Pathol. 2017;41(6):820–32 [20].

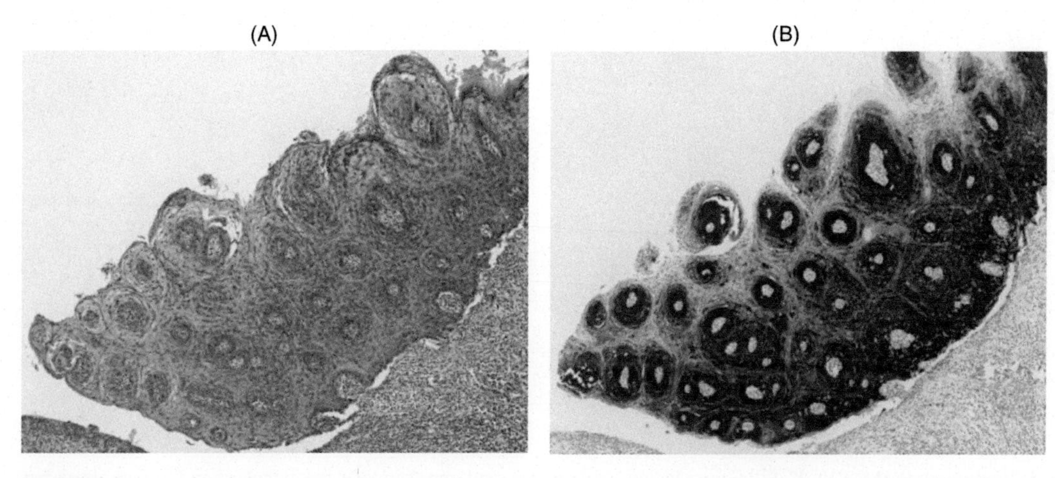

FIGURE 14.4

HPV56-positive Warty PeIN. (A) Papillomatous hyperpara keratotic lesions with pleomorphic cells and focal koilocytosis. (B) Diffuse overexpression of p16 staining is present.

Adapted from Fernandez-Nestosa MJ, Guimera N, Sanchez DF, Canete-Portillo S, Velazquez EF, Jenkins D, et al. Human papillomavirus (HPV) genotypes in condylomas, intraepithelial neoplasia, and invasive carcinoma of the penis using laser capture microdissection (LCM)-PCR: a study of 191 lesions in 43 patients. Am J Surg Pathol. 2017;41(6):820–32 [20].

basaloid morphology. Other subtypes such as papillary, sarcomatoid, and verrucous are also recognized [5].

Scrotal SCC is classified histologically as usual, warty or basaloid like vulvar and penile carcinomas. The presence of HPV has been also linked to warty/basaloid tumors and absent in usual keratinizing SCC (Table 14.1, Figs. 14.5 and 14.6) [10]. As Scrotal malignancies are very rare, and there is no published data describing its preneoplastic status.

CANCER PATHWAYS OF VULVAR, PENILE, AND SCROTAL CANCER: HUMAN PAPILLOMAVIRUS AND NON–HUMAN PAPILLOMAVIRUS RELATED

The HPV etiological contribution differs in each anatomical location reflecting differences in the natural history and viral tissue tropism. In HPV-related cancers such as cervical, vaginal, and anal; HPV is the main etiological factor with an overall HPV prevalence of >90%, 70%, and 85%, respectively. However, vulvar, penile, and scrotal cancers show two major different carcinogenic pathways: one HPV related and another one non-HPV related (Table 14.1).

HPV is present in 29% of cases of vulvar invasive carcinoma (Table 14.1) [14], and the rest is associated with chronic vulvar dermatoses such as lichen sclerosus, squamous cell hyperplasia, and differentiated VIN that are not linked to HPV [23]. These cancers tend to occur in older women. Although the dermatoses are associated with increased epithelial cell turnover and lichen sclerosus

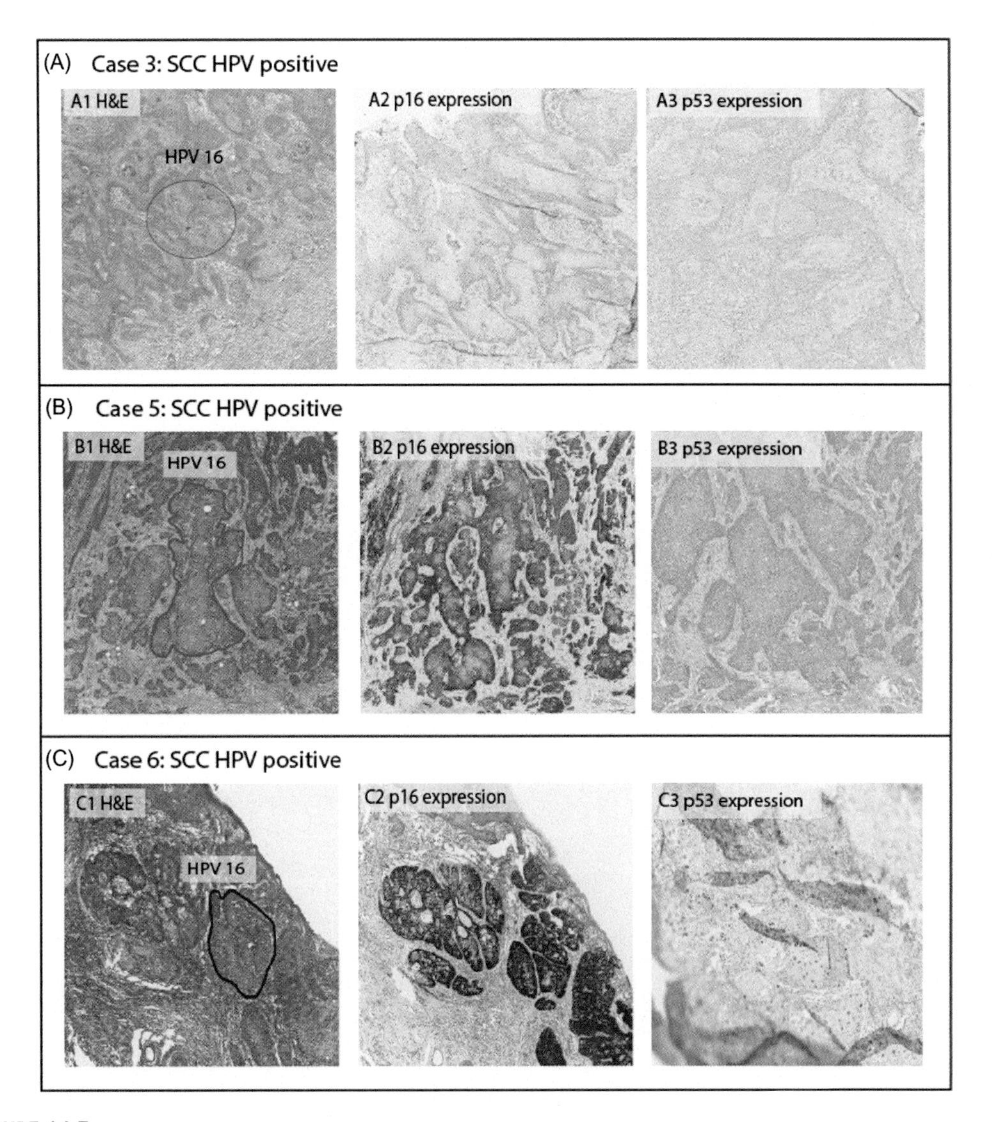

FIGURE 14.5

H&E histological images, p16^{INK4a} and p53 expression pattern of the three HPV16-associated scrotal cancers. The quality of Case C was very poor and the expression of p53 was difficult to analyze. *SCC*, squamous cell carcinoma; *HPV*, human papillomavirus

Adapted from Guimera N, Alemany L, Halec G, Pawlita M, Wain GV, Vailen JSS, et al. Human papillomavirus 16 is an aetiological factor of scrotal cancer. Br J Cancer. 2017;116(9):1218–22 [10].

is an inflammatory condition the molecular pathway of non-HPV-related vulvar cancers remains unknown. The large worldwide epidemiological study established the HPV contribution in VIN as well as invasive vulvar cancer. In this study, more than 2000 (pre) invasive vulvar cases were

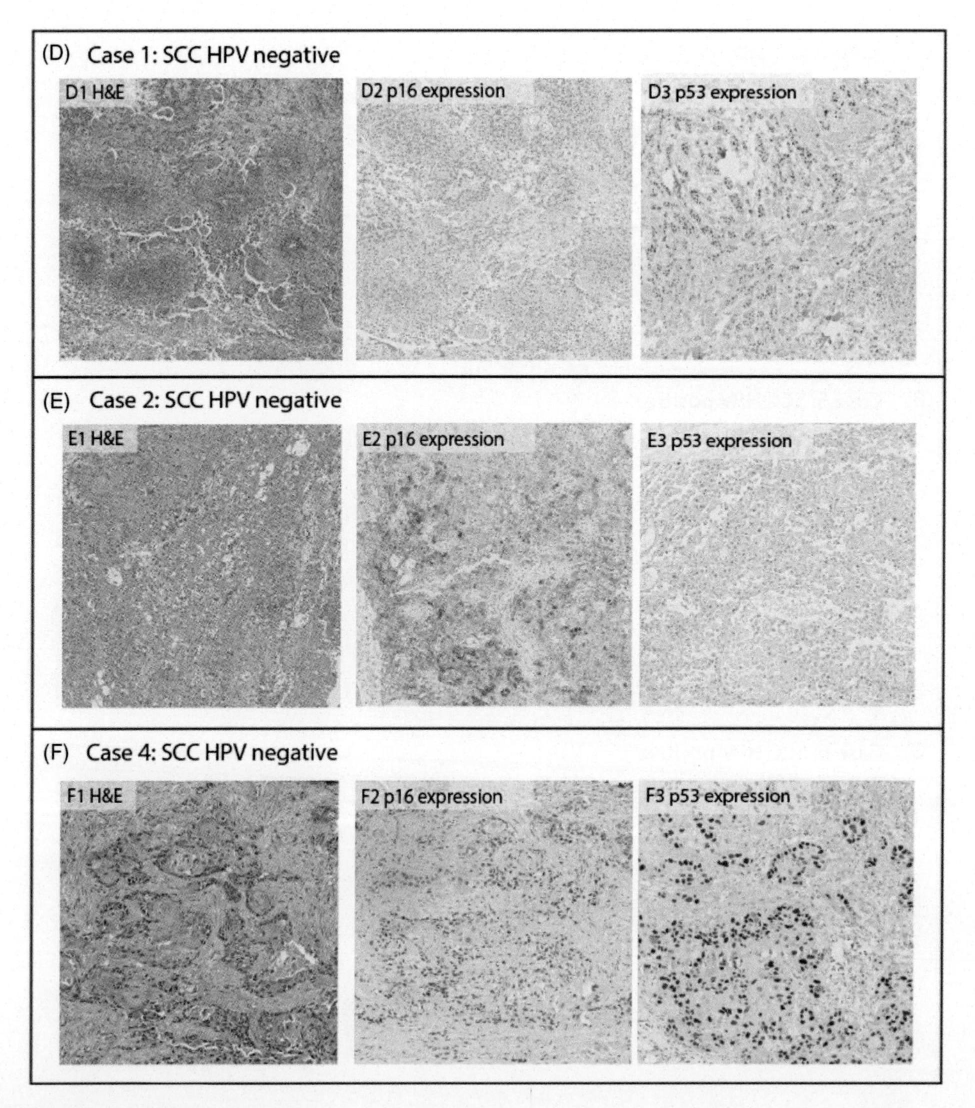

FIGURE 14.6

H&E histological images, p16INK4a and p53 expression pattern of the three HPV-negative scrotal cancers. *SCC*, squamous cell carcinoma; *HPV*, human papillomavirus.

Adapted from Guimera N, Alemany L, Halec G, Pawlita M, Wain GV, Vailen JSS, et al. Human papillomavirus 16 is an aetiological factor of scrotal cancer. Br J Cancer. 2017;116(9):1218–22 [10].

analyzed by HPV PCR and P16^{INK4a} as a biomarker of HPV E7 oncoprotein expression. Among the 29% of invasive vulvar cancer that were HPV related, HPV16 was the commonest type (72.5%) followed by HPV33 (6.5%) and HPV18 (4.6%). A history of cervical High grade squamous intraepithelial lesion (HSIL) increased the risk of invasive vulvar cancer five times [23]. Histologically, warty or basaloid SCCs are more likely to be HPV positive compared to keratinizing SCCs. VIN cases from this study were mainly warty/basaloid (91%), and mainly HPV related. HPV16 was the most common genotype. HPV was mostly detected as a single genotype and multiple HPV infections were uncommon (<10%).

Penile cancer has been associated with several risk factors and associated conditions, such as phimosis with chronic inflammation including lichen sclerosus at atrophicus as in, poor hygiene, older age, smoking, and HPV infection. Some of the factors are similar to those in vulvar cancer. HPV accounts for 33% of penile carcinomas and 87% of high-grade squamous intraepithelial neoplasia or penile intraepithelial neoplasia (PeIN), as established by a large worldwide epidemiological study [15]. HPV16 was the most frequently detected genotype in HPV-positive invasive or high-grade PeIN, 69% and 80%, respectively. Interestingly, low-risk HPV6 was the second most frequent genotype in invasive carcinomas (4% of HPV-positive cases). The relation of HPV6 and invasive penile cancer was further investigated. In 75% of the cases (six out of eight), HPV6 DNA was localized in tumor cells by LCM-PCR confirming the association of this low-risk HPV and penile invasive carcinomas. Other HPV genotypes were also detected, such as HPV35, HPV33, or low-risk HPV11. HPV prevalence was 87% in high-grade PeIN and HPV16 was as well the most common genotype detected. HPV multiple infections were present in 9% in invasive penile carcinomas and 18% in preinvasive lesions. In another study, 43 cases were studied from Paraguay where the incidence of penile cancer related to HPV is 2.6% (https://www.hpvcentre.net/). PeIN lesions studied were multifocal and histologically heterogeneous. In some of these cases, a wide range of HPV genotypes was identified, usually one genotype per lesion explaining the presence of multiple HPV genotypes in PeIN. In some cases, both pathways HPV and non-HPV related were present at the same patient. HPV16 and expression of $p16^{INK4a}$ was detected in penile condylomas suggesting that in some cases condylomas harboring HPV16 may participate in the carcinogenic process and should be followed up [20]. As in vulvar cancer, little is known regarding the molecular pathway of non-HPV-related penile carcinomas.

Scrotal cancer was long thought to be only related to occupational carcinogens including soot, and lubricating or cutting oils [6,7]. A recent study has confirmed the etiological role of HPV in three out of six cases of scrotal cancer. Indicating a dual pathway of oncogenesis, HPV and non-HPV related. Although limited by the small number of cases, this study provided biological evidence of an etiological relationship between scrotal cancer and HPV16, the only genotype identified. Larger epidemiological studies are needed to determine if different HPV genotypes could be involved, and the relative role of HPV compared to other environmental and genetic factors.

COMMON CHARACTERISTICS BETWEEN VULVA, PENILE, AND SCROTAL CARCINOMA

Vulvar, penile, and scrotal carcinomas share important similarities, they are uncommon diseases, they involve external sexual organs and HPV, mainly HPV16 is an important etiological agent with morphologically similar, but distinctive pathways of HPV-driven precancerous lesions leading to histologically characteristic cancers.

External vulva, penile, and scrotal cancers arises in an epithelium that is morphologically genital skin. External genitals share similar embryologic differentiation and may have specific cells vulnerable to HPV16 infection [1]. HPV16 has been occasionally linked to skin tumors outside the genital region such as ungual and periungual sites showing a broader epithelial tropism of HPV16 consistent with its strong carcinogenicity in anogenital sites.

Non-HPV-related pathways are less clear. Epithelial hyperplasia seen histologically and Inflammation such as seen in lichen sclerosus et atrophicus increases epithelial turnover may partly explain the origin of non-HPV-related carcinomas in these areas.

CLINICAL MANAGEMENT OF VULVAR, PENILE, AND SCROTAL NEOPLASIA

The principles of clinical management of these neoplasias are similar, involving excision which may be local or more radical including sentinel lymph node biopsy and lymph node dissection. Radiotherapy may also be used. Standard treatment of HSIL is surgery. Excision is preferable to locally destructive treatment like LASER, since invasive areas might be missed. The same caveat applies to the topical treatment with imiquimod, an immune modulator which is helpful in the treatment of extensive vulvar lesions [24]. Not surprisingly, in clinical trials with imiquimod cases of invasive cancers occurred. These nonexcisional treatments should be restricted to specialized units.

Vulvar HSIL was the first precancer outside the female cervix which was shown to be prevented by an HPV vaccine [25]. In patients with a history of cervical disease the vulva (and the anus) should be inspected at every gynecologic examination. In the cases where symptoms are noted, or abnormal areas observed, a biopsy should be performed.

The HPV-positive vulvar cancers found in younger women appear to have a better prognosis [26,27]. The standard treatment for these cancers is surgery. Changing from radical vulvectomy with en-bloc resection of the groins to local excision and the application of the Sentinel lymph node mapping has dramatically reduced the morbidity of these treatments [28,29].

With the broad implementation of HPV vaccination more than 90% of the HPV-positive vulvar cancers may be prevented in the future [30,31] and with increasing male vaccination and extension to the countries with a high incidence of penile cancer in Africa, Asia, and South America the same may be anticipated to apply to penile and scrotal cancers.

CONCLUSION

Vulvar, penile, and scrotal neoplasms are (very) rare diseases. They have a similar embryological background and for this reason the diseases show many similarities similar diseases. We can find in each of these entities an HPV driven and a non-HPV-related pathway to cancer. The non-HPV-related pathway is related to chronic inflammation on the basis of skin dermatoses such as lichen sclerosus. In the HPV-related tumors, HPV16 is by far the most common strain, but in both, the vulva and the penis HPV6 can also cause cancer, although this is considered to be a low-risk HPV type. In HPV-related vulvar cancer HPV33 and HPV18 are the second and third most common types. Infections with all these clinical relevant HPV types can be prevented in the future by vaccination.

REFERENCES

[1] Moore Keith L, Persaud TVN, G.Torchia Mark. The developing human: clinically oriented embryology. 9th ed Philadelphia, PA, USA: Elsevier; 2011. 560 p.

[2] Jones RW, Baranyai J, Stables S. Trends in squamous cell carcinoma of the vulva: the influence of vulvar intraepithelial neoplasia. Obstet Gynecol 1997;90(3):448–52.

[3] Joura EA, Losch A, Haider-Angeler MG, Breitenecker G, Leodolter S. Trends in vulvar neoplasia. Increasing incidence of vulvar intraepithelial neoplasia and squamous cell carcinoma of the vulva in young women. J Reprod Med 2000;45(8):613–15.

[4] Bleeker MC, Heideman DA, Snijders PJ, Horenblas S, Dillner J, Meijer CJ. Penile cancer: epidemiology, pathogenesis and prevention. World J Urol 2009;27(2):141–50.

[5] Moch H, Cubilla AL, Humphrey PA, Reuter VE, Ulbright TM. The 2016 WHO classification of tumours of the urinary system and male genital organs-part a: renal, penile, and testicular tumours. Eur Urol 2016;70(1):93–105.

[6] Melicow MM. Percivall Pott (1713-1788): 200th anniversary of first report of occupation-induced cancer scrotum in chimmey sweepers (1775). Urology 1975;6(6):745–9.

[7] Sorahan T, Cooke MA, Wilson S. Incidence of cancer of the scrotum, 1971-84. Br J Ind Med 1989;46 (6):430–1.

[8] Wright JL, Morgan TM, Lin DW. Primary scrotal cancer: disease characteristics and increasing incidence. Urology 2008;72(5):1139–43.

[9] Verhoeven RH, Louwman WJ, Koldewijn EL, Demeyere TB, Coebergh JW. Scrotal cancer: incidence, survival and second primary tumours in the Netherlands since 1989. Br J Cancer 2010;103 (9):1462–6.

[10] Guimera N, Alemany L, Halec G, Pawlita M, Wain GV, Vailen JSS, et al. Human papillomavirus 16 is an aetiological factor of scrotal cancer. Br J Cancer 2017;116(9):1218–22.

[11] Forman D, de Martel C, Lacey CJ, Soerjomataram I, Lortet-Tieulent J, Bruni L, et al. Global burden of human papillomavirus and related diseases. Vaccine 2012;30(Suppl. 5):F12–23.

[12] Baandrup L, Varbo A, Munk C, Johansen C, Frisch M, Kjaer SK. In situ and invasive squamous cell carcinoma of the vulva in Denmark 1978-2007-a nationwide population-based study. Gynecol Oncol 2011;122(1):45–9.

[13] Tavassoéli F, Devilee P. Pathology and genetics of tumours of the breast and female genital organs. Lyon: IARC Press; 2003.

[14] de Sanjose S, Alemany L, Ordi J, Tous S, Alejo M, Bigby SM, et al. Worldwide human papillomavirus genotype attribution in over 2000 cases of intraepithelial and invasive lesions of the vulva. Eur J Cancer 2013;49(16):3450–61.

[15] Alemany L, Cubilla A, Halec G, Kasamatsu E, Quiros B, Masferrer E, et al. Role of human papillomavirus in penile carcinomas worldwide. Eur Urol 2016;69(5):953–61.

[16] Hoang LN, Park KJ, Soslow RA, Murali R. Squamous precursor lesions of the vulva: current classification and diagnostic challenges. Pathology 2016;48(4):291–302.

[17] Darragh TM, Colgan TJ, Cox JT, Heller DS, Henry MR, Luff RD, et al. The lower anogenital squamous terminology standardization project for HPV-associated lesions: background and consensus recommendations from the College of American Pathologists and the American Society for Colposcopy and Cervical Pathology. J Low Genit Tract Dis 2012;16(3):205–42.

[18] De Vuyst H, Clifford GM, Nascimento MC, Madeleine MM, Franceschi S. Prevalence and type distribution of human papillomavirus in carcinoma and intraepithelial neoplasia of the vulva, vagina and anus: a meta-analysis. Int J Cancer 2009;124(7):1626–36.

[19] Garland SM, Steben M, Sings HL, James M, Lu S, Railkar R, et al. Natural history of genital warts: analysis of the placebo arm of 2 randomized phase III trials of a quadrivalent human papillomavirus (types 6, 11, 16, and 18) vaccine. J Infect Dis. 2009;199(6):805–14.

[20] Fernandez-Nestosa MJ, Guimera N, Sanchez DF, Canete-Portillo S, Velazquez EF, Jenkins D, et al. Human papillomavirus (HPV) genotypes in condylomas, intraepithelial neoplasia, and invasive carcinoma of the penis using laser capture microdissection (LCM)-PCR: a study of 191 lesions in 43 patients. Am J Surg Pathol 2017;41(6):820–32.

[21] Chaux A, Velazquez EF, Amin A, Soskin A, Pfannl R, Rodriguez IM, et al. Distribution and characterization of subtypes of penile intraepithelial neoplasia and their association with invasive carcinomas: a pathological study of 139 lesions in 121 patients. Hum Pathol 2012;43(7):1020–7.

[22] Cubilla AL, Velazquez EF, Young RH. Epithelial lesions associated with invasive penile squamous cell carcinoma: a pathologic study of 288 cases. Int J Surg Pathol 2004;12(4):351–64.

[23] Cohen PA, Anderson L, Eva L, Scurry J. Clinical and molecular classification of vulvar squamous precancers. Int J Gynecol Cancer 2019;29(4):821–8.

[24] van Seters M, van Beurden M, ten Kate FJ, Beckmann I, Ewing PC, Eijkemans MJ, et al. Treatment of vulvar intraepithelial neoplasia with topical imiquimod. N Engl J Med 2008;358(14):1465–73.

[25] Joura EA, Leodolter S, Hernandez-Avila M, Wheeler CM, Perez G, Koutsky LA, et al. Efficacy of a quadrivalent prophylactic human papillomavirus (types 6, 11, 16, and 18) L1 virus-like-particle vaccine against high-grade vulval and vaginal lesions: a combined analysis of three randomised clinical trials. Lancet 2007;369(9574):1693–702.

[26] Wakeham K, Kavanagh K, Cuschieri K, Millan D, Pollock KG, Bell S, et al. HPV status and favourable outcome in vulvar squamous cancer. Int J Cancer 2017;140(5):1134–46.

[27] Hinten F, Molijn A, Eckhardt L, Massuger L, Quint W, Bult P, et al. Vulvar cancer: two pathways with different localization and prognosis. Gynecol Oncol 2018;149(2):310–17.

[28] Hacker NF. Radical resection of vulvar malignancies: a paradigm shift in surgical approaches. Curr Opin Obstet Gynecol 1999;11(1):61–4.

[29] Te Grootenhuis NC, van der Zee AG, van Doorn HC, van der Velden J, Vergote I, Zanagnolo V, et al. Sentinel nodes in vulvar cancer: long-term follow-up of the GROningen INternational Study on Sentinel nodes in Vulvar cancer (GROINSS-V) I. Gynecol Oncol 2016;140(1):8–14.

[30] Garland SM, Joura EA, Ault KA, Bosch FX, Brown DR, Castellsague X, et al. Human papillomavirus genotypes from vaginal and vulvar intraepithelial neoplasia in females 15-26 years of age. Obstet Gynecol 2018;132(2):261–70.

[31] Huh WK, Joura EA, Giuliano AR, Iversen OE, de Andrade RP, Ault KA, et al. Final efficacy, immunogenicity, and safety analyses of a nine-valent human papillomavirus vaccine in women aged 16-26 years: a randomised, double-blind trial. Lancet 2017;390(10108):2143–59.

SECTION

3 USING HUMAN PAPILLOMAVIRUS KNOWLEDGE TO PREVENT CERVICAL AND OTHER CANCERS

INTRODUCTION

From the earliest studies of HPV16 and 18 DNA in cervical cancer and cervical squamous intraepithelial lesions it became evident that cervical testing for Human Papillomavirus (HPV) could provide a useful additional test in cervical cancer prevention for use in cervical screening and the management of screen-detected abnormalities. With only a small range of HPV types that could be detected by the tests available in the early 1980s, interest focused on the minimal possibility of showing that low-grade cytological or histological lesions associated with HPV6 and 11 were unlikely to progress to a high-grade precancer or cancer requiring treatment, whereas those associated with HPV16 or 18 needed careful follow-up or treatment [1].

Comparing HPV6/11 with HPV16/18 provided an initial basis for the concept of high- and low-risk HPVs, and for the Bethesda classification of both cervical cytology and histopathology of cervical intraepithelial neoplasia into high- and low-grade lesions. There was, however, the important and necessary addition to the Bethesda classification of cytology of the category atypical squamous cells of uncertain significance. This category recognized the difficulty and resulting uncertainties of distinguishing on cytology the wide range of minor microscopic changes associated with HPV and other causes, and identifying among women with these abnormalities the small but important proportion with high-grade precancer or occasionally a cancer requiring treatment.

The response in the USA was powerful as reported in Chapter 15, Triage of Women With ASCUS and LSIL Abnormal Cytology: The ALTS Experience and Beyond, by Attila Lorincz and Marc Arbyn, and led to the ALTS trial, an important collaboration between government and the diagnostics industry and to many subsequent studies across the world of the performance of different tests for DNA and RNA of different panels of HPVs. It is important, however, not to ignore the specific reasons why such a study was needed in the USA. In the USA, cervical screening was conducted by individual gynecologists as part of an annual gynecological check-up available to women of a wide age range. There was no national program and coverage was very variable. Most women with abnormal smears were investigated directly by colposcopy and biopsy of any lesions found. This was an expensive system with over-investigation of some women with transient abnormal smears but poor coverage for other women. Reducing unnecessary colposcopy was an important aim of the ALTS trial, and new HPV technology was the likely answer.

The organization of cytological cervical screening was very different in other countries. In Northern Europe there were national and regional public health programs and issues such as screening coverage and population effectiveness were considered more important. Also in some countries (especially the UK) follow-up of abnormal smears was done by cytologists requesting repeat cytology through the woman's general practitioner and there were also concerns about the psychological and other effects of unnecessary colposcopy and treatment. Cytologists, having been made aware of the variable and frequently poor diagnostic performance of cytology, wished to improve this through better quality control of cytology rather than a change in technology. The local practices and factors affecting these led to a very variable initial response to the ALTS trial results in different settings.

The very high sensitivity for HSIL of testing cytological samples for a panel of hrHPV types also led to investigation of the possibility of primary screening using HPV testing. In Chapter 16, Primary Screening by Human Papillomavirus Testing: Development, Implementation, and Perspectives, Chris Meijer and Jack Cuzick described the development of HPV testing in primary screening leading to the decision in 2011 to introduce primary HPV screening in the Netherlands.

Primary HPV screening has also brought new issues related to the management of screen-detected HPV positivity. The use of cytology in triage of HPV-positive women is part of the Dutch HPV screening program as it was introduced, but as discussed in Chapter 16 and specifically in Chapter 17, additional specific molecular testing, such as testing for methylation of somatic tumour suppressor genes would offer substantial advantages in convenient, acceptable and automated testing and in more accurate selection for treatment. In particular it could open the possibility of extending self-testing to the triage of hrHPV-positive women, in addition to the use of self-testing for women who do not respond to calls for attending routine cytology screening. This could lead to a much simpler less invasive approach to cervical screening and the identification of women who really need treatment.

Chapter 17, Infection to Cancer—Finding Useful Biomarkers for Predicting Risk of Progression to Cancer, explores the role of specific molecular markers in screening and in improving the triage of hrHPV-positive women or those with abnormal cytology. The problem with both cervical cytology and HPV DNA testing for cervical or anal screening or triage is that they lack specificity, particularly for predicting precisely the stage of neoplastic progression of the underlying lesion. Even when a biopsy is taken the need for treatment is not predicted precisely and overtreatment is a major part of current practice. Terms such as "high-grade" (HSIL) may improve reproducibility and avoid undertreatment but do not improve the predicitive ability of the diagnosis. In the case of CIN2 in young women regression may reach 70%, and in many studies is around 30%–50% [2] with rates of 20%–30% for CIN3.

The first biomarker identified was the L1 protein of BPV which was expressed in superficial differentiated cells in productive HPV infections. This has been superseded by HPV E4 protein which is a more widely expressed indicator of completion of the HPV life cycle, but its clinical use is still under investigation. HPV E6 and E7 transforming gene activity has been difficult to study directly and Chapter 17, Infection to Cancer—Finding Useful Biomarkers for Predicting Risk of Progression to Cancer, includes a detailed discussion of the development and use of the important surrogate marker p16 from the work in Heidelberg led by Magnus vonKnebel Doeberitz. The chapter also includes a discussion of the potential of inactivating methylation of both human tumor suppressor genes and viral genes as future markers.

Chapter 18, Immune Responses to Human Papillomavirus and the Development of Human Papillomavirus Vaccines, and Chapter 19, Clinical Trials of Human Papillomaviruses Vaccines, discuss the role of the natural immune responses to HPV and the development and clinical trials of prophylactic HPV vaccination. Early studies of immune cell populations in cervical HPV infection and CIN of different grades using immunohistochemistry of biopsies established that HPV infection, particularly hrHPV infection with evidence of transformation was associated with a decrease in local cell populations of antigen presenting cells and T cells, and that this decrease was particularly marked in the presence of the cofactor smoking [3]. This suggested that one mechanism for HPV survival was escaping as much as possible the attention of the immune system. In Chapter 18, Immune Responses to Human Papillomavirus and the Development of Human Papillomavirus Vaccines, Margaret Stanley explores the role of the natural immune response including the local immune system and the antibody response to viral capsid antigens that led to the development of viral like particles (VLPs) based on the L1 capsid antigen. These VLPs then provided the basis for the experimental and then clinical development of prophylactic HPV vaccines.

Chapter 19, Clinical Trials of Human Papillomaviruses Vaccines, is about the clinical development of prophylactic HPV vaccination. The development of these was undertaken by two large pharmaceutical companies who had the global resources for very large studies, billion dollar undertakings involving many countries and hospitals, initially in the study of vaccination in women under 25 years but then extended up and down in age. The strategies of the companies were different. Merck, based in the USA, after the initial successful trials of a HPV16 vaccine, decided that, despite initial concerns about the strength of the immune response induced by the different VLPs adjuvanted with a conventional adjuvant, they would develop a vaccine that protected against both the two most important hrHPV types, HPV16 and 18, and the two important lrHPV types HPV6/11 which were important in genital warts and the rare, but very emotive, juvenile recurrent laryngeal papillomatosis. The protection against genital warts as well as cervical cancer and precancer proved to be an easily achievable endpoint and an effective marketing strategy with strong public health interest, although contributing almost nothing to cancer prevention.

GlaxoSmithKline vaccine division was based in the conservative, Walloon part of Belgium and the management took a very different strategic view. There was concern to optimize the strength of the antibody response to the VLPs and introduced a stronger adjuvant, aiming to ensure long-lasting protective immunity. Also there was concern about marketing a vaccine that was explicitly directed against a clearly sexually transmitted disease, and favored an anti-cancer vaccine based on a strong response to HPV16 and 18. For both vaccines a three-dose schedule was chosen for phase 2 and large phase 3 clinical trials although the possibility of a simpler schedule was certainly an issue from the start.

As detailed in Chapter 19, Clinical Trials of Human Papillomaviruses Vaccines, the different trials showed that both vaccines were effective against HPV16 and 18, but that the bivalent vaccine provided additional cross-protection against other hrHPV types. The response of Merck was to produce a nine-valent vaccine to provide wide protection against a range of high-risk HPVs as well as HPV 6/11. Despite completion of clinical trials and the potential value of the GSK bivalent vaccine as a single-dose vaccine for countries where delivery of a complex vaccination schedule is difficult, GSK stopped promoting its vaccine.

The further development of approaches to prevention of HPV-associated cancer, the political, safety, and other issues raised, and the use of screening and vaccination in practice in the 12 years since registration of an HPV preventive vaccine are discussed in Section 4.

REFERENCES

[1] Campion MJ, McCance DJ, Cuzick J, et al. Progressive potential of mild cervical atypia. Lancet 1986; ii:237–40.

[2] Moscicki A-B, Schiffman M, Burchell A et al. Updating the natural history of human papillomavirus and anogenital cancers. Vaccine 305 (2012) F24–F33.

[3] Barton SE, Jenkins D, Cuzick J, et al. Effect of cigarette smoking on cervical epithelial immunity. Lancet 1988;ii:652–4.

CHAPTER

TRIAGE OF WOMEN WITH ASCUS AND LSIL ABNORMAL CYTOLOGY: THE ALTS EXPERIENCE AND BEYOND

15

Attila Lorincz[1], Cosette Marie Wheeler[2] and Marc Arbyn[3]

[1] *Wolfson Institute of Preventive Medicine, Queen Mary University of London, London, United Kingdom*

[2] *University of New Mexico Comprehensive Cancer Center, Albuquerque, NM, United States*

[3] *Unit of Cancer Epidemiology, Belgian Cancer Centre, Sciensano, Brussels, Belgium*

EVIDENCE THAT HUMAN PAPILLOMAVIRUS TESTING MAY BE USEFUL FOR TRIAGE OF SOME ABNORMAL CYTOLOGY

Low-to-moderate risk abnormal cytology of potential clinical importance has been given various names over the decades including borderline, mild and moderate dysplasia, class II/III Papanicolaou smear, condylomatous atypia, and many other variations. There was little disagreement that women with one or more of the moderate versions of cytological abnormality should be sent to colposcopy, however, there was constant angst regarding the much more numerous borderline and mild cytological results.

The term ASCUS (more recently called ASC-US) is an abbreviation for atypical squamous cells of undetermined significance. This diagnostic category was the first to formally recognize that cytopathology is a subjective art dependent on the experience and intuition of the reader. Prior to this time most people, with the exception of some experts, had accepted cytology results as a kind of gold standard. ASCUS was created by The Bethesda System (TBS) in 1988 and then updated to include subcategories of ASCUS in TBS 2001 [1,2] . ASCUS arose from a desire by morphologists to have a simpler, more consistent, and more realistic classification system for cytology. At the time there was a preponderance of diverse and confusing terminologies existing around the world, even within countries some institutions had their own bespoke terminologies that were difficult to compare with other nomenclatures or to translate to a common morphological language. Much of the world has now come to accept the term ASCUS/ASC-US to mean an uncertain or borderline cytology result.

Prior to the formal classification of ASCUS there was little incentive to seek triage because cytology definitions were regarded as accurate. Suggestions for the addition of HPV DNA testing were met with apathy and sometimes hostility from many in the medical and research communities, being viewed as unnecessary and a poorly justified attempt for financial gain at the expense of an overburdened cytology system. Colposcopy with or without repeat cytology was seen as the triage

Human Papillomavirus. DOI: https://doi.org/10.1016/B978-0-12-814457-2.00015-5

test and in some European countries, such as Germany and much of Latin America screening colposcopy was vigorously employed. There was a lack of appreciation of the costs of colposcopy in terms of clinician hours and missed lesions due to rushing the procedure in understaffed and inexperienced clinics [3]. Even expert colposcopy that relies on visual judgement alone has quite poor sensitivity, multiple biopsies are needed to compensate for inadequate performance but this leads to a large burden of over-treatment.

ASCUS terminology changed everything. Here was a terminology that begged for a method of triage, if the true significance of the cellular abnormalities were unknown then there should be a way to get clear answers on the risks of a malignant outcome. The power of ASCUS terminology was the door through which validated HPV DNA screening finally entered the fortress of medical practice. Of course, there were many earlier studies that looked at HPV screening in combination with Pap smear, HPV testing on women with mild dysplasia, early HPV tests for ASCUS etc. but none of the outcomes were persuasive. ASCUS was the first application of HPV DNA testing that won US FDA approval, which was then and remains today, the world's strictest and most referenced medical regulatory system; FDA approval was the key for access and was absolutely required for the HPV test to be widely reimbursable in the United States [4,5].

FDA-approval allowed early-adopters to introduce HPV triage into their routine clinics to generate large-scale field data from which they could give lectures on their results to colleagues. This eventually persuaded late adopters ($\sim$90% of users) into a movement toward the routine use of HPV DNA testing. The slow HPV translational process is still ongoing decades after the original definitive research studies were published. In Europe, reflex HPV testing to triage women with ASCUS was recommend in the 2008 edition of the EU guidelines for Quality Assurance for Cervical Cancer Screening [6].

ORIGIN AND DEVELOPMENT OF THE ALTS STUDY

Not long after the introduction of ASCUS terminology some clinicians began to complain about the uncertainties of this cytology result category, seeing it as a source of possible added legal liability [7]. In effect the cytopathologists had moved some of the risk to clinicians who were faced with a new uncertainty on what to do with women who had ASCUS Papanicolau (Pap) smears. Although the reality was that this new uncertainty was a correct but uncomfortable realization of morphological uncertainty, the medical community expressed reservations for a long time. In contrast to the ASCUS category, clinicians were quite comfortable with another product of TBS, the low-grade squamous intraepithelial lesion (LSIL), which is a lesion similar to a mild dysplasia. The path for most women with LSIL was fairly clear, with a majority of women in the United States being sent for colposcopy and possible cervical transformation zone excision or ablation, while in Europe, 6-monthly repeat cytology was recommended to determine if the LSIL was persistent before colposcopy and treatment [6]. It was ASCUS that spurred the search for a triage test because it produced a heavy workload for clinicians, representing about 2 million or approximately 5%–6% of all cytology results in the United States. The ASCUS LSIL Triage Study (ALTS) sought to assess HPV testing as a tool for triage of ASCUS, while LSIL was included as a convenient addition to the trial [7,8].

ALTS was the brainchild of Mark Schiffman who spent many years in the early 1990s canvassing the medical community, attending meetings to propose a definitive study to find the best way to deal with ASCUS and certain other challenges around TBS. It was his persistence in the face of strong opposition from some medical groups that paid off. At the end, Schiffman successfully persuaded the US National Cancer Institute to allocate 20 million dollars to the ALTS study. Finally, ALTS could get going in 1995, the same year that Cox et al. published an ASCUS study using the Hybrid Capture I HPV DNA (HC1) test; which was the first study to show that HPV triage of ASCUS may have clinical utility [4].

ALTS planned to use HC1 as the standard HPV test, partly because HC1 had received FDA-approval for ASCUS triage that same year. However, the proposal to use HC1 met with strong opposition from some clinicians and cytopathologists because it was seen to be favouring one particular commercial interest, Digene Inc. of Gaithersburg, MD, USA. Others felt that an HPV PCR test could be more suitable. For a while there were some different HPV tests under consideration for ALTS but these were eventually dropped due to not having any validation or proof of routine high-quality manufacturability. In addition, there was an important need for a test with FDA-approval. In the United States, there is a disadvantage in using a non–FDA-approved test for a large expensive clinical validation study because that test cannot be legally sold across the United States, meaning that much of the efforts of ALTS would have been delayed while awaiting approvals. Many observers protested about conflicts of interest, but at the end ALTS, went ahead to a good level of success, albeit with the HC2 test [9], which was FDA-approved in 1999, [https://www.accessdata.fda.gov/scripts/cdrh/cfdocs/cfpma/pma.cfm?id = P890064S006] 1 year before publication of the first paper on ALTS [8]. ALTS is a good example, relatively unique until today, of a governmental–commercial collaboration in diagnostics that came to benefit almost everyone.

MAIN RESULTS OF THE ALTS STUDY

Prior to and in parallel to ALTS, liquid-based cytology (LBC) methods based on monolayer cell preparations were being developed to improve the ease of reviewing Pap smears; these efforts were accompanied by the hope that more accurate detection of abnormal cells could be realized. ThinPrep LBC (Cytyc Corporation, subsequently acquired by Hologic, Marlborough, MA, USA) was approved by the FDA as an alternative to conventional Pap smears in 1996 [https://www.accessdata.fda.gov/scripts/cdrh/cfdocs/cfpma/pma.cfm?id = P950039] just as the ALTS trial design was being finalized. ThinPrep LBC cervical specimen collection was selected by the ALTS Steering Committee as the single cervical specimen source for both cytology and the HC2 high-risk HPV (hrHPV) DNA assay. As the enrollment of the ALTS trial approached, the HC2 assay was in its final stages of refinement and as noted earlier it had not yet been approved by the FDA when the ALTS trial began enrollment in November 1996. The ALTS trial put in place several Quality Control Groups including the HPV Quality Control Group, a partnership of academic researchers and Digene investigators. The group designed strategies to independently characterize all test methods including HC2 and to monitor the HC2 test throughout the trial enrollment and follow-up.

The randomized ALTS trial compared three alternative clinical management strategies in women with baseline ASCUS and LSIL Pap test results including (1) immediate colposcopy, (2) cytologic follow-up, and (3) triage by HPV DNA testing [8]. Four US clinical centers recruited female patients for ALTS (University of Washington, Seattle; University of Alabama, Birmingham; University of Pittsburgh, Pittsburgh; and University of Oklahoma, Oklahoma City). Women were invited to participate if they were not pregnant, aged 18 years and older with an entry ASCUS or LSIL cytology result and had no prior hysterectomy or cervical ablation. The study included 3488 and 1572 eligible and consenting participants with baseline ASCUS or LSIL cytology, respectively.

Early in the trial, the HPV triage arm for women referred with LSIL was closed by the ALTS Steering Committee as a result of an interim analysis that showed that 83% of women with baseline LSIL were positive for hrHPV DNA by HC2, which was too high for effective clinical triage [10]. In contrast to the interim results in women with LSIL, 56% of women referred with entry ASCUS cytology were positive for hrHPV DNA and more importantly, HC2 identified 96% of women with histologically confirmed cervical intraepithelial neoplasia (CIN) grade 3, and cancer (CIN3 +). A single repeat cytology at a triage threshold of high-grade squamous intraepithelial lesions (HSIL) identified 44% of women with histologically confirmed CIN3 + , while referring 7% of women to colposcopy. Repeat cytology at the ASCUS triage threshold had higher sensitivity, identifying 85% of women with CIN3 + . The colposcopy referral rate at the ASCUS threshold was 58.6%, slightly higher than the referral rate of HPV DNA testing. It was on the basis of these initial ALTS results that HPV testing moved into the lead position as a triage test due to a clearly higher sensitivity and equal or better specificity than cytology [11].

Though the question of HPV detection as a screening tool, either adjunctively or as a replacement for cytology, had been ongoing for quite some time, the ALTS trial and additional studies provided essential data showing that testing for hrHPV DNA was clearly useful and offered a management option for women with ASCUS screening results with the added value of enabling reflexing to the HC2 test using the remaining residual LBC sample. HPV testing had significantly greater sensitivity for detection of CIN3 + and equal specificity compared to a single repeat ASCUS or worse cytology [11]. Multiple repeat cytology evaluations could increase sensitivity, but would be accompanied by expected loss to follow-up, costs of additional clinical visits, and increased referral to colposcopy. Follow-up data from ALTS at 2 years also showed that among women with ASCUS, if colposcopy failed to detect CIN2 or CIN3 + all women who were initially diagnosed with less than CIN2 (negative and CIN1) should be managed similarly based on an observed equivalent 10%–13% risk for subsequent CIN3 + [12].

Women with ASCUS who were negative for hrHPV DNA could be saved from repeat cytology testing and/or referral to colposcopy and use of HPV testing would reduce the number of women needing colposcopy by almost half. Cost-effectiveness of cervical screening would be realized through better management of ASCUS cytology and improved clinical resource utilization. Given that HPV had risen in general awareness and became accepted as a necessary cause of cervical cancer through the mid to late 1990s, a negative test for hrHPV DNA was seen as a means to reduce a woman's fears and associated anxieties. The negative HPV result could provide strong assurance that she was highly unlikely to be at-risk for cervical cancer in the next 10 years. ALTS and the study of Manos et al. were particularly helpful in changing cervical screening algorithms in the United States and worldwide and supporting an FDA-approved indication for routine use of hrHPV DNA testing in cervical screening [5,11]. An example of core data from the ALTS trial are shown in Table 15.1.

Table 15.1 Triage Performance of HPV DNA Testing (HC2) and Repeat Cytology Read at Different Thresholds for Detecting CIN2 + and CIN3 + in the ALTS Trial; Featuring The Sensitivity of the Tests and Associated Percentages of Women Referred To Colposcopy, With Positive Predictive Values and Negative Predictive Values.

Disease Triage method	% Sensitivity	% Colposcopy	% PPV	% NPV
CIN3 +				
HPV	96.3	56.1	10.0	99.5
HSIL	44.1	6.9	37.5	96.5
LSIL	64.0	26.2	14.3	97.1
ASCUS	85.3	58.6	8.5	97.9
CIN2 +				
HPV	95.9	56.1	19.6	98.9
HSIL	34.8	6.9	58.1	92.0
LSIL	59.2	26.2	25.9	93.6
ASCUS	85.0	58.6	16.7	95.8

Adapted from Solomon D, Schiffman M, Tarone B, et al. Comparison of three management strategies for patients with Atypical Squamous Cells of Undetermined Significance (ASCUS): baseline results from a randomized trial. J Natl Cancer Inst 2001;93:293–9.

IMPACT OF THE ALTS STUDY IN CHANGING MEDICAL PRACTICE IN THE UNITED STATES AND WORLDWIDE

In 2001 based on ALTS and the triage study of Manos et al., major organizations including the American Society for Colposcopy and Cervical Pathology (ASCCP) met to standardize the management of women with cytological and histological abnormalities. Prior to the 2001 ASCCP Guidelines which emerged from this meeting, women with ASCUS or LSIL cytology in the United States were often followed by repeat Pap cytology every 4–6 months with the requirement of 3–4 normal Pap tests before being returned to routine annual screening. The 2001 ASCCP Guidelines recommended molecular hrHPV DNA testing as the preferred management strategy and ASCUS results from an LBC were to be reflexed to hrHPV testing [13]. Women with HPV-positive triage results were referred to colposcopy and HPV negative women were to be returned to routine annual screening straight away. Women with LSIL were recommended to be triaged to immediate colposcopy. In ALTS, the overall cumulative risk for CIN 2 + of hrHPV-positive ASCUS and LSIL (regardless of HPV results) were essentially the same (26.7% and 27.6%, respectively) and it was therefore recommended that women referred to colposcopy for these two cytology interpretations should be managed similarly.

Global interest and evidence supporting a variety of utilities for hrHPV testing and its role in cervical screening applications, including use in cotesting and primary screening continued to grow in parallel with and subsequent to ALTS. A variety of HPV tests have been approved by the US

FDA or clinically validated according to international guidelines for ASCUS triage, as well as for other uses such as test-of-cure and screening, either as a cotest with cytology or as a standalone HPV test. Cumulatively all these data on HPV DNA testing have helped to secure hrHPV testing of an initial ASCUS cytology result as the preferred triage option in countries that still use cytology screening and that can afford HPV DNA testing. The evolution, validation, and approval of hrHPV tests, including tests based on DNA, mRNA and type-specific hrHPV detection are described thoroughly in Chapter 17 of this book.

METAANALYSES ON THE CLINICAL ACCURACY OF HUMAN PAPILLOMAVIRUS DNA TESTING TO TRIAGE WOMEN WITH ASCUS

A metaanalysis of the diagnostic test accuracy of HC2 to detect cervical precancer in women with ASCUS confirmed the findings of the ALTS study and concluded that hrHPV testing was more sensitive and similarly specific compared to repeat cytology [14,15]. This metaanalysis formed the basis to also recommend HPV-based ASCUS triage in the European Union. An update to the first metaanalysis, in the format of a Cochrane review, corroborated the consistency of the initial conclusion [16] (Table 15.2). However, in triage of women with LSIL, hrHPV DNA testing showed very low specificity making triage hardly useful in clinical practice. Indeed the posttest risk of CIN2 + or CIN3 + in LSIL hrHPV positive women was very similar to the pretest risk [16]. hrHPV DNA triage of ASCUS with other hrHPV DNA tests shows, in general, similar results as with HC2 (Table 15.2). HPV genotyping for the most carcinogenic types, HPV16 or HPV18, is substantially more specific than hrHPV testing in triage of ASC-US but is not sufficiently sensitive to allow a hrHPV positive but HPV16/18 negative woman to be released to routine screening [17]. HPV16/18 genotyping could be a useful second triage test in hrHPV-positive women to identify those women with ASCUS who need immediate management. ASCUS cases with HPV DNA of other viral types still needs further follow-up but eventually might be followed more conservatively.

Identifying transcripts of the E6/E7 of 14 hrHPV types with the APTIMA assay (Hologic, San Diego, CA, USA) is as sensitive to detect CIN2 + and CIN3 + but significantly more specific compared to HC2, making this test particularly interesting for ASCUS triage [18]. mRNA testing for five HPV types (HPV16, HPV18, HPV31, HPV33, and HPV45) with the Pretect HPV Proofer is substantially more specific in triage of ASCUS and even of LSIL than HC2 but it is significantly less sensitive compared to hrHPV DNA testing [19]. A positive Pretect HPV Proofer assay allows for immediate referral for colposcopy and biopsy with high positive predictive value but a negative Pretect HPV Proofer does not justify to release a woman with ASCUS to routine screening. Overexpression of cell-cycle regulating proteins, p16 and p16/Ki67, identified through cyto-immunochemistry, is very specific to exclude high-grade cervical precancer and appears to be as sensitive as hrHPV DNA testing [20].

FUTURE OF CYTOLOGY AND HUMAN PAPILLOMAVIRUS TRIAGE

The future of cytology appears to lie in the realm of artificial intelligence (AI). There have been amazing strides in AI showing that the intelligent machine can beat the human in highly complex games. AI

Table 15.2 Clinical Absolute and Relative Accuracy for CIN2 + or CIN3 + of Assays Used to Triage Women With ASC-US Derived From Published Metaanalyses.

	Absolute Accuracy		Absolute	Comparator	Relative Estimate	
Triage Test	**Parameter**	**Outcome**	**Estimate**	**Test**	**(95%CI)**	**Reference**
HC2	Sensitivity	CIN2 +	90.9%	**Repeat**	1.27 (1.16–1.39)	[16]
	Sensitivity	CIN3 +	94.8%	**Cytology**	1.14 (1.06–1.22)	
	Specificity	≤CIN1	60.7%		0.99 (0.97–1.03)	
Other hrHPV DNA Assays						
Amplicor	Sensitivity	CIN2 +	90.4%	HC2	0.98 (0.92-1.05)	[20]
	Specificity	≤CIN1	58.3%		0.87 (0.80-0.95)	
Abbott	Sensitivity	CIN2 +	94.7%	HC2	0.97 (0.90-1.04)	
RT PCR	Specificity	≤CIN1	39.9%		1.18 (0.99-1.41)	
Linear Array	Sensitivity	CIN2 +	93.8%	HC2	1.02 (0.99-1.06)	
	Specificity	≤CIN1	46.2%		0.90 (0.79-1.03)	
Cervista	Sensitivity	CIN2 +	95.9%	HC2	0.98 (0.95-1.07)	
	Specificity	≤CIN1	49.6%		1.15 (1.06-1.24)	
HPV16/18 DNA Genotyping						
HPV16/18	Sensitivity	CIN2 +	56.8%	hrHPV DNA	0.59 (0.54–0.65)	[17]
	Sensitivity	CIN3 +	70.7%		0.75 (0.68–0.83)	
	Specificity	≤CIN1	82.9%		1.70 (1.51–1.90)	
HPV mRNA Testing						
APTIMA	Sensitivity	CIN2 +	95.7%	HC2	1.01 (0.97–1.06)	[18]
	Sensitivity	CIN3 +	96.2%		1.01 (0.96–1.06)	
	Specificity	≤CIN1	56.4%		1.19 (1.08–1.31)	
PreTect	Sensitivity	CIN2 +	75.4%	HC2	0.80 (0.73–0.87)	[19]
Proofer	Sensitivity	CIN3 +	86.1%		0.89 (0.80–0.99)	
	Specificity	≤CIN1	77.9%		1.98 (1.70–2.30)	

Adapted from Arbyn M, Roelens J, Simoens C, Buntinx F, Paraskevaidis E, Martin-Hirsch PPL, et al. Human papillomavirus testing versus repeat cytology for triage of minor cytological cervical lesions. Cochrane Database Syst Rev 2013a;3:1–201; Arbyn M, Roelens J, Cuschieri K, Cuzick J, Szarewski A, Ratnam S, et al. The APTIMA HPV assay versus the Hybrid Capture II test in triage of women with ASC-US or LSIL cervical cytology: a meta-analysis of the diagnostic accuracy. Int J Cancer 2013b;132:101–8.

machines can be taught to learn virtually anything logical that can be learned by humans and they can then do these jobs tirelessly and at lower costs. Already, automated cytology readers such as the ThinPrep Imager Duo Imaging Station can do a very good job at the initial screening. Many such machines have replaced human routine readers [https://www.hologic.com/sites/default/files/package-insert/MAN-03133-001_004_02.pdf; https://www.accessdata.fda.gov/cdrh_docs/pdf/P950009S008b.pdf]

The role of many cervical cytopathologists nowadays in developed countries is to supervise the machines and to check on discrepancies and possible errors. Even that job is likely to disappear over the coming years as AI machines learn to pick out the most minute of occult morphological abnormalities and provide an overall validated result, as accurate as any that the morphological approach is able to muster. These calls will be complemented with the ability of the instruments to retrieve millions of archived images to serve as reference material to compare the new and the most similar archived cases. Given these instrumented capabilities, cytology may well stay with us for many decades to come, perhaps as a triage test for HPV screening positives and perhaps for other uses such as detection of rare HPV negative precancerous lesions. The challenge will be to show whether morphology can add anything cost-effectively valuable to molecular triage tests that may be based on automated DNA methylation and other approaches [21,22]. In low- and middle-income countries where labor is relatively inexpensive it will take much longer for the cytology workforce to be reassigned to other jobs. Political considerations will slow the pace of progress as has already happened for HPV DNA testing. In fact, strong opposition of vested parties, most particularly entrenched commercial interests including laboratories themselves, combined with concerned staff in the medical system may well slow the rate of progress towards improved screening and triage.

HPV DNA testing is also going to morph into something totally different than we have today. In just a few decades there will be a revolution in HPV testing, most probably going into the Next Generation Sequencing realm. Such advances in DNA and RNA testing are covered in more detail in Chapter 17, Infection to Cancer—Finding Useful Biomarkers for Predicting Risk of Progression to Cancer, on the Development of HPV DNA tests.

REFERENCES

[1] Solomon D. The 1988 Bethesda System for reporting cervical/vaginal cytologic diagnoses: developed and approved at the National Cancer Institute Workshop in Bethesda, Maryland, December 12-13, 1988. Hum Pathol 1990;21(7):704–8 Review.

[2] Solomon D, Davey D, Kurman R, Moriarty A, O'Connor D, Prey M, et al. The Bethesda System 2001: terminology for reporting the results of cervical cytology. JAMA 2002;287:2114–19.

[3] Lazcano E, Lorincz AT, Torres L, Salmeron J, Cruz A, Rojas R, et al. Specimen self-collection and HPV DNA screening in a pilot study of 100,242 women. Int J Cancer 2013. Available from: https://doi.org/10.1002/ijc.28639 7 Dec. [Epub ahead of print].

[4] Cox JT, Lorincz AT, Schiffman MH, Sherman ME, Cullen A, Kurman RJ. Human papillomavirus testing by hybrid capture appears to be useful in triaging women with a cytologic diagnosis of atypical squamous cells of undetermined significance. Am J Obstet Gynecol 1995;172(3):946–54.

[5] Manos MM, Kinney WK, Hurley LB, et al. Identifying women with cervical neoplasia: using human papillomavirus DNA testing for equivocal Papanicolaou results. JAMA 1999;281:1605–10.

[6] Arbyn M, Anttila A, Jordan J, Ronco G, Schenck U, Segnan N, et al. European guidelines for quality assurance in cervical cancer screening. Second edition - summary document. Ann Oncol 2010;21:448–58.

[7] McNeil C. Getting a handle on ASCUS: a new clinical trial could show how. JNCI 1995;87(11):787–9. Available from: https://doi.org/10.1093/jnci/87.11.787 Published: 07 June 1995.

[8] Schiffman M, Adrianza ME. ASCUS-LSIL triage study. Design, methods and characteristics of trial participants. Acta Cytol 2000;44(5):726–42 PMID:11015972.

[9] Lörincz AT. Hybrid Capture™ method for detection of human papillomavirus DNA in clinical specimens: a tool for clinical management of equivocal Pap smears and for population screening. J Obstet Gynaecol Res 1996;22(6):629–36.

[10] ALTS group. Human papillomavirus testing for triage of women with cytologic evidence of low-grade squamous intraepithelial lesions: baseline data from a randomized trial. J Natl Cancer Inst 2000;92:397–402.
[11] Solomon D, Schiffman M, Tarone B, et al. Comparison of three management strategies for patients with Atypical Squamous Cells of Undetermined Significance (ASCUS): baseline results from a randomized trial. J Natl Cancer Inst 2001;93:293–9.
[12] Walker JL, Wang SS, Schiffman M, Solomon D, ASCUS LSIL Triage Study Group. Predicting absolute risk of CIN3 during post-colposcopic follow-up: results from the ASCUS-LSIL triage study (ALTS). Am J Obstet Gynecol 2006;195(2):341–8.
[13] Ferris DG. The 2001 ASCCP management guidelines for cervical cytology. Am Fam Physician 2004;70 (10):1866–8.
[14] Arbyn M, Buntinx F, Van Ranst M, Paraskevaidis E, Martin-Hirsch P, Dillner J. Virologic versus cytologic triage of women with equivocal Pap smears: a meta-analysis of the accuracy to detect high-grade intraepithelial neoplasia. J Natl Cancer Inst 2004;96:280–93.
[15] Arbyn M, Dillner J, Van Ranst M, Buntinx F, Martin-Hirsch P, Paraskevaidis E. Re: Have we resolved how to triage equivocal cervical cytology? J Natl Cancer Inst 2004;96:1401–2.
[16] Arbyn M, Roelens J, Simoens C, Buntinx F, Paraskevaidis E, Martin-Hirsch PPL, et al. Human papillomavirus testing versus repeat cytology for triage of minor cytological cervical lesions. Cochrane Database Syst Rev 2013;3:1–201.
[17] Arbyn M, Xu L, Verdoodt F, Cuzick J, Szarewski A, Belinson J, et al. Genotyping for human papillomavirus types 16 and 18 in women with minor cervical lesions: a systematic review and meta-analysis. Ann Intern Med 2017;166:118–27.
[18] Arbyn M, Roelens J, Cuschieri K, Cuzick J, Szarewski A, Ratnam S, et al. The APTIMA HPV assay versus the Hybrid Capture II test in triage of women with ASC-US or LSIL cervical cytology: a meta-analysis of the diagnostic accuracy. Int J Cancer 2013;132:101–8.
[19] Verdoodt F, Szarewski A, Halfon P, Cuschieri K, Arbyn M. Triage of women with minor abnormal cervical cytology: meta-analysis of the accuracy of an assay targeting messenger ribonucleic acid of 5 high-risk human papillomavirus types. Cancer Cytopathol 2013;121:675–87.
[20] Roelens J, Reuschenbach M, von Knebel-Doeberitz M, Wentzensen N, Bergeron C, Arbyn M. p16INK4a immunocytochemistry versus HPV testing for triage of women with minor cytological abnormalities: a systematic review and meta-analysis. Cancer 2012;120:294–307.
[21] Lorincz A. The virtues and weaknesses of DNA methylation as a test for cervical cancer prevention. Acta Cytol 2016;60:501–12. Available from: https://doi.org/(DOI:10.1159/000450595).
[22] Cook DA, Krajden M, Brentnall AR, Gondara L, Chan T, Law JH, et al. Evaluation of a validated methylation triage signature for human papillomavirus positive women in the HPV FOCAL cervical cancer screening trial. Int J Cancer 2018. Available from: https://doi.org/10.1002/ijc.31976 2018 Nov 9.[Epub ahead of print].

FURTHER READING

Arbyn M, Snijders PJ, Meijer CJLM, Berkhof H, Cuschieri K, Kocjan BJ, et al. Which high-risk HPV assays fulfil criteria for use in primary cervical cancer screening? Clin Microbiol Infect 2015;21:817–26.
Meijer CJLM, Castle PE, Hesselink AT, Franco EL, Ronco G, Arbyn M, et al. Guidelines for human papillomavirus DNA test requirements for primary cervical cancer screening in women 30 years and older. Int J Cancer 2009;124:516–20.

CHAPTER

PRIMARY SCREENING BY HUMAN PAPILLOMAVIRUS TESTING: DEVELOPMENT, IMPLEMENTATION, AND PERSPECTIVES

16

Chris J.L.M. Meijer[1], J. Cuzick[2], W.W. Kremer[1], D.A.M. Heideman[1] and G. Ronco[3]

[1]*Amsterdam UMC, Vrije Universiteit Amsterdam, Pathology, Cancer Center Amsterdam, Amsterdam, Netherlands*

[2]*Wolfson Institute of Preventive Medicine, Queen Mary University of London, London, United Kingdom*

[3]*International Agency for Research on Cancer, Lyon, France*

DEVELOPMENT OF HPV-BASED SCREENING

Activity to exploit Professor Harald zur Hausen's discovery that the human papillomavirus (HPV) was the primary cause of cervical cancer [1], which earned him the Nobel Prize in medicine, began shortly thereafter. Early studies focused on cancers, biopsies, and screening samples taken in women with disease. In the beginning, HPV testing on cervical tissues was mainly performed by type-specific DNA or RNA in situ hybridization techniques and Southern blot hybridization, but techniques were laborious and sensitivity was low. Lorincz and colleagues performed the first comprehensive study [2], in which low-stringency Southern blot hybridization assays were used to identify most of the HPV types associated with cervical cancer in a predominantly normal population, augmented by high-grade cervical intraepithelial neoplasia (CIN), and invasive cancer.

An early test suitable for a screening sample was Digene's ViraPap test, which was followed by hybrid capture 1 (HC1), but the sensitivities of these tests for CIN lesions were limited. Then hybrid capture 2 (HC2) was developed, which was the workhorse for many of the early clinical studies [3]. Development of more sensitive and type-specific PCRs increased sensitivity for HPV detection considerably. With the development of PCR tests based on consensus primers (general primer PCR) and subsequent genotyping of the amplimers by type-specific oligo probes [4,5] much higher values for overall HPV prevalence in the general population were obtained, in addition to HPV genotype information which was until then only obtained with separate tests.

The development of the GP5/6 general primer PCR opened the way for a PCR-based HPV test suitable for use in clinical practice, since only two primers were needed and pooled detection by an enzyme immunoassay (EIA) format was relatively simple. Important results were obtained with this test. First, analysis of cervical smears from symptom-free women with normal cytology from a screening population revealed an overall HPV prevalence of 3.5%, whereas in women with normal cytology from a gynecological referral population an overall HPV prevalence of 14% was found [6]. It appeared that these women with normal cytology from the referral population often had a

Human Papillomavirus. DOI: https://doi.org/10.1016/B978-0-12-814457-2.00016-7

history of CIN. In addition, Van den Brule and co-workers showed that the overall prevalence of HPV in mild dysplasia (CIN1), moderate dysplasia (CIN2), carcinoma in situ (CIN3), and invasive carcinoma increased (70%, 84%, 100%, and 100%, respectively) [5]. These findings resulted in the hypothesis that HPV testing with the GP5/6 general primer PCR might be used for cervical screening [7]. Additional work from the Netherlands showed that HPV prevalence in women was age-dependent with peak prevalence around 20 years [8].

In Europe, the elongated version of the GP5/6 PCR, that is, the GP5 + /6 + PCR-EIA developed by Walboomers and his Dutch colleagues [9,10], was widely used and provided genotyping information. Initially with the GP5/6 general primer PCR and later with the elongated GP5 + /6 + primers, many cervical cancer case-control studies initiated by Nubia Muñoz and Xavier Bosch, and later furthered by Silvia Franceschi and Gary Clifford of IARC, were performed. These studies showed HPV prevalence in cervical cancers ranging from 86% to 95% [11].

Subsequent work of Nobbenhuis et al. in an observational cohort showed that a persistent infection with HPV was necessary for the progression of CIN lesions (Fig. 16.1) [12]; absence or clearance of HPV was associated with regression of CIN lesions [13]. Following work focused on the role of HPV triage in the management of women with low-grade cytological abnormalities [atypical squamous cells of unknown significance (ASC-US) or low-grade squamous intraepithelial lesion (LSIL) in the Bethesda classification, or borderline or mild dysplasia (BMD) in the CISOE-A classification] [4,14]. The most notable activity in this area was the ALTS trial in the United States [15],

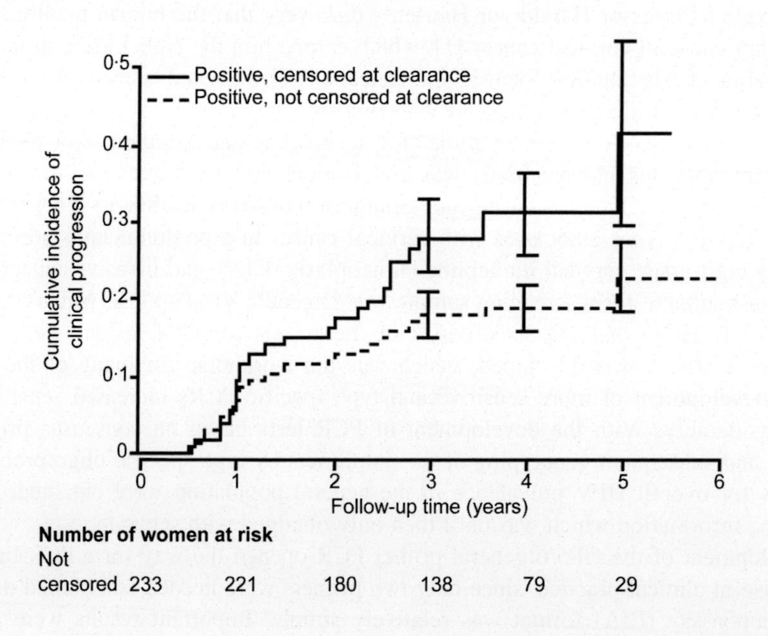

FIGURE 16.1

Cumulative incidence of clinical progression among women positive for high-risk human papillomavirus at baseline.

From Nobbenhuis MAE, Walboomers JMM, Helmerhorst TJM, Rozendaal L, Remmink AJ, Risse EKJ, et al. Relation of human papilloma virus status to cervical lesions and consequences for cervical-cancer screening: a prospective study. Lancet 1999;354(9172), 20–5, with permission.

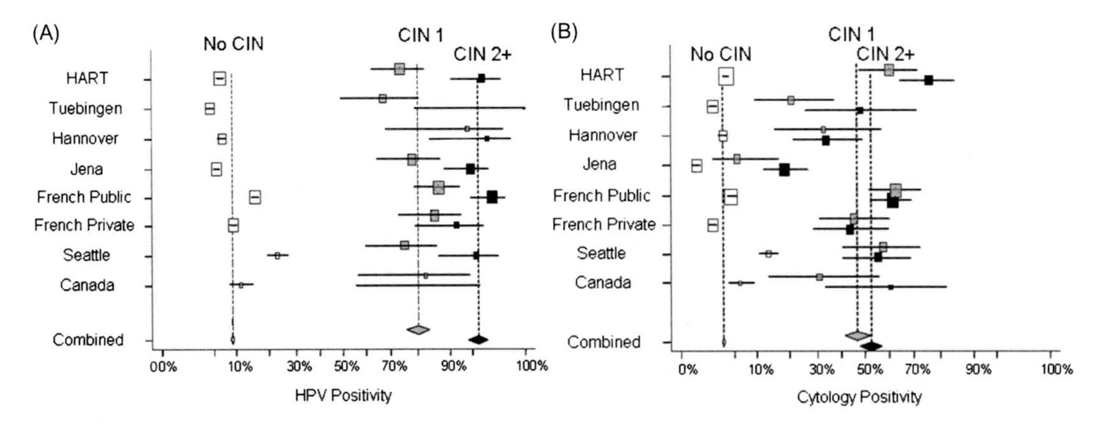

FIGURE 16.2

Relative detection rates of CIN2 + for HPV testing (A) and cytology (B) in 8 dual testing crosss-sectional studies.

From Cuzick J, Clavel C, Petry KU, Meijer CJ, Hoyer H, Ratnam S, et al. Overview of the European and North American studies on HPV testing in primary cervical cancer screening. Int J Cancer 2006;119(5):1095–1101, with permission

in which 3488 women with a referral diagnosis of ASC-US were randomized to three management strategies: (1) immediate colposcopy (considered to be the reference standard); (2) triage to colposcopy based on HPV results from HC2 and thin-layer cytology results; or (3) triage based on cytology results alone. At the same time in the United Kingdom, evaluation of HPV testing as the primary screening tool in 1985 women was undertaken in the Margaret Pyke study [16]. This was the first of several studies in which women provided a sample for both cytology and HPV testing. A summary of the studies is provided in Fig. 16.2 based on an overview by Cuzick et al. [17]. This design is the most efficient for comparing the sensitivity and specificity of different tests, but is not suitable for long-term follow-up because interventions based on one sample can confound the long-term evaluation of the other test. Nevertheless the studies provided clear evidence of a very high sensitivity of HPV testing [96% for CIN2 or worse (CIN2 +)] and highlighted the poorer than anticipated sensitivity of cytology (only 53% in this overview, and never better than 70%). The downside of HPV testing was a poorer specificity (92% for HPV vs 97% for cytology), emphasizing the need for good triage tests when HPV testing is used as the primary screen, as described in the "triage of HPV positive women section" below.

HPV TESTING VERSUS CYTOLOGY

These studies were followed by randomized trials comparing screening with cytology alone versus HPV testing, either alone or with cytology. One of the first of these was a large cluster randomized trial of 130,000 women in rural India [18] which compared HPV testing to cytology, visual inspection with acetic acid (VIA), and a control group. Over a follow-up period of 8 years, HPV testing reduced deaths from cervical cancer by 48%, which was highly significant, whereas the reductions associated with cytology or VIA were 11% and 14%, respectively, and were not significant. Concurrently, four major randomized trials were conducted in Europe. They were conducted in the Netherlands (POBASCAM) [19], Sweden (Swedescreen) [20], United Kingdom (ARTISTIC) [21], and Italy (NTCC) [22], and involved 176,464 women aged 20–64 years. These studies compared HPV and cytology co-testing in England, Sweden, the Netherlands, and in the first phase of the Italian study, and HPV as stand-alone test in the second phase of the Italian study, against cytology alone over at least two screening rounds. All these studies reported strongly increased detection of high-grade CIN in the HPV-arm compared to the cytology-arm in the first screening round, followed by, with the exception of the English study, the opposite (lower detection of high-grade CIN in the HPV-arm than in the cytology-arm at the second screening round). These studies proved HPV testing allowed earlier detection than cytology of persistent, clinically relevant CIN, which is an obvious premise for increased protection against invasive cervical cancer. In addition the data supported the use of HPV as a primary screening test every 5 years from the age of 30. Yet, for women under the age of 30 there is still uncertainty about the best screening method or if screening is needed at all. A randomized trial conducted in Finland [23] comparing stand-alone HPV testing to cytology also documented higher detection of CIN2 + at the entry round, but no data on the second round have yet been published.

A pooled analysis of the above four trials [24] showed a more than 60% greater reduction in invasive cancer in women randomized to HPV testing (Fig. 16.3). Shortly after this publication, both EU and WHO recommended HPV screening [25,26] and several countries, including Sweden, the Netherlands, Australia, and several regions of Italy, New Zealand, and Turkey have recently switched or are considering switching (United Kingdom and Belgium) from cytological screening to primary HPV screening with extended intervals.

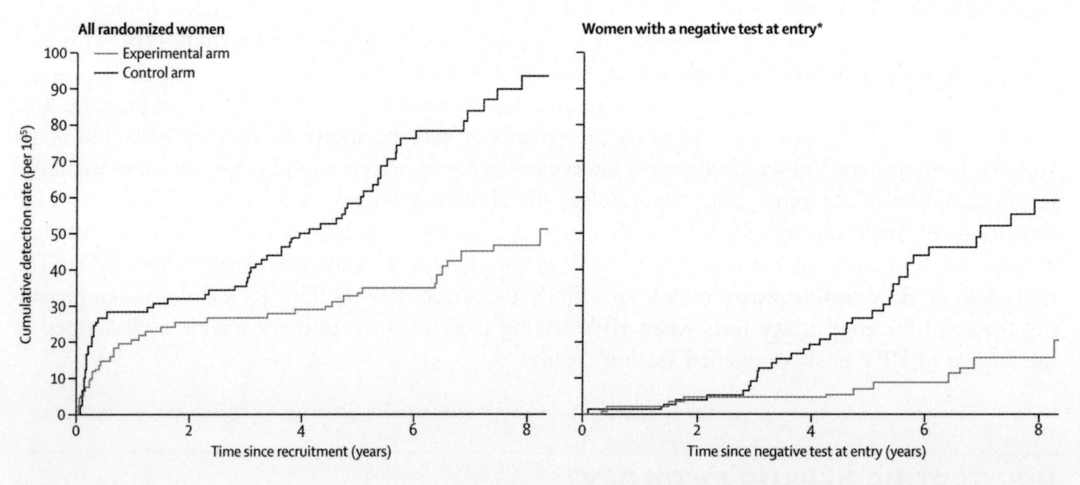

FIGURE 16.3

Cumulative detection of invasive cervical carcinoma. *Observations are censored 2.5 years after CIN2 or CIN3 detection, if any.

From Ronco G, Dillner J, Elfstrom KM, Tunesi S, Snijders PJ, Arbyn M, et al. Efficacy of HPV-based screening for prevention of invasive cervical cancer: follow-up of four European randomised controlled trials. Lancet 2014;383 (9916):524–32, with permission.

SOLE HPV TESTING OR HPV TESTING COMBINED WITH CYTOLOGY?

In the United States, the move toward primary HPV testing has been partial in that co-testing with cytology is currently most widely used, whereas in Europe and elsewhere, most countries are moving directly to HPV alone with cytology reserved as part of triage in HPV positive women. A meta-analysis clearly showed that the gain in sensitivity for CIN3 or worse (CIN3 +) of co-testing by cytology (ASC-US cut-off) and HPV (HC2) over stand-alone HPV is very small (relative CIN3 + ratio HPV/cytology co-testing versus only cytology 1.04 [95% confidence interval 0.92–1.17)] [27]. Dillner et al. [28] showed that adding cytology to HPV adds little to sensitivity, but reduces specificity by also identifying low-grade cytology in women who are HPV negative, but in whom little high-grade disease is found histologically in the subsequent biopsy. However, Katki et al. [29,30] did find a very small increase in CIN3 + risk in HPV negative women with low-grade cytology compared to women who are negative for both tests (Table 16.1). A recent update of this cohort concluded that "added sensitivity of co-testing versus HPV alone for detection of treatable cancer affected extremely few women" [31] and very recently, Horn et al. confirmed these findings in a German trial in which no CIN3 + lesions were found in HPV negative women with abnormal cytology [32].

Table 16.1 CIN3 + Risks After 5 Years of Follow-Up Stratified by Baseline HPV and Cytology Cotesting Result

5-Year CIN3 + Risk by Baseline Cotest Result		
HPV/Pap Result	**Frequency in Women Aged 30–64 years (%)**	**CIN3 + Risk (%)**
HPV + /HSIL	0.20	49
HPV + /AGC	0.05	33
HPV–/HSIL	0.01	30
HPV + /ASC-H	0.12	25
HPV–/ASC-H	0.05	3.5
HPV–/AGC	0.16	0.9
HPV + /ASC-US	1.1	6.8
HPV + /LSIL	0.81	6.1
HPV + /Pap–	3.6	4.5
HPV–/LSIL	0.19	2.0
HPV–/ASC-US	1.8	0.43
HPV–/Pap–	92.0	0.08

Data from Kaiser Permanente Northern California cohort: Katki HA, Schiffman M, Castle PE, Fetterman B, Poitras NE, Lorey T, et al. Five-year risks of CIN 2 + and CIN 3 + among women with HPV-positive and HPV-negative LSIL Pap results. J Low Genit Tract Dis 2013;17(5 Suppl. 1):S43–9.

SCREENING INTERVALS, FOLLOW-UP OF HPV POSITIVE WOMEN, AND AGE TO START SCREENING

The pooled analysis by Ronco et al. also showed that the risk of cancer within 5 years after a negative HPV test is about half of the risk of cancer within 3 years after a normal cytology test [24], and Dillner et al. showed a fourfold reduction in CIN3 + at 6 years [28]. In addition, Dijkstra et al. showed the cumulative incidence of cervical cancer (0.09%) and CIN3 + (0.56%) among HPV negative women in the intervention group after three rounds of screening were similar to the cumulative incidence of cervical cancer and CIN3 + among women with negative cytology in the control group after two rounds (0.09% and 0.69%, respectively) [33]. Thus, with primary HPV screening prolonged intervals of at least 5 years are safe and avoid unneeded costs for health services and testing of women.

Long-term follow-up studies of women screened with HPV and cytology show that HPV negative women have a very low risk for cervical cancer and high-grade CIN lesions (Table 16.2). In contrast, long-term CIN3 + risk in HPV positive women with a negative triage test (repeat cytology or cytology combined with HPV16/18 genotyping) have a 5 year CIN3 + risk of ~4% and ~7%, respectively. Therefore only in HPV screen-negative women the screening interval can be extended over 5 years (Table 16.2).

The pooled analysis also clearly showed the need for triage of HPV positive women. In fact, ARTISTIC, POBASCAM, and Swedescreen used triage (with some differences, but all based on triage by immediate cytology and HPV testing in women with normal cytology), while NTCC directly referred all HPV positive women for colposcopy. There was no statistically significant heterogeneity between studies as for efficacy, but there was for the biopsy rate. In NTCC the biopsy rate was doubled in the HPV compared to the cytology-arm while the biopsy rate was equal in the

Table 16.2 Long-Term Cancer, CIN3 + and CIN2 + Risks in HPV-Negative Women

		HPV-Negative Women		
Cohort	**Follow-Up Period**	**Cancer Risk**	**CIN3 + Risk**	**CIN2 + Risk**
Dutch cohorts				
VUSA-Screen [34]	5 years	–	0.09%	0.21%
POBASCAM [33]	14 years	0.03%	0.56%	–
International cohorts				
ARTISTIC [35]	6 years	–	0.28%	0.87%
Swedescreen [36]	13 years	–	0.84%	1.74%
Kaiser Permanente [37]	18 years	–	0.90%	1.85%

From Polman NJ, de Haan Y, Veldhuijzen NJ, Heideman DAM, de Vet HCW, Meijer C, et al. Experience with HPV self-sampling and clinician based sampling in women attending routine cervical screening in the Netherlands. Prev Med 2019;125:5–11.

two arms in the trials using cytology triage. Although the best triage procedures still have to be defined (see "Triage of HPV positive women" section), it is clear that protocols based on short-term HPV re-testing and referral of those still positive entail a high number of colposcopies. The alternatives are either just repeating cytology, as done in the Netherlands (see "The new HPV based programme in the Netherlands" section), or repeating HPV at longer intervals, when the large majority of infections will have been cleared.

The best available data [38] suggest that only about 1/3 of CIN3 progress to invasive cancer in 30 years. Therefore even with cytology-based screening a relevant proportion of CIN treatments do not result in a benefit for the woman. Moreover, excisional treatments are associated with an increased risk of pregnancy-related adverse events [39–42], which are obviously more relevant for CIN detected in women of child-bearing age. Data from NTCC, in which all HPV positive women were referred for colposcopy, showed a much higher cumulative detection of CIN2 and CIN3 in the HPV-arm than in the cytology-arm over the first two screening rounds in women below 35 and even more in women below 30 years of age [22]. This could have been the result of increased over-diagnosis of nonprogressive lesions in young women with HPV, or a very large gain in lead time with HPV in these women. Ongoing analysis of the updated long-term follow-up data of NTCC will clarify this point. Data from POBASCAM, wherein women were invited from age 30 and triaged by repeated cytology do not show this effect of overdiagnosis of nonprogressive CIN [19]. In addition, in the pooled analysis of European trials, the largest increase in protection with HPV over cytology was observed at age 30–35, which supports starting screening at age 30 [24]. On the other hand, starting screening at age 35 or above is plausibly advisable in vaccinated cohorts given their strongly reduced cancer risk.

TRIAGE OF HPV POSITIVE WOMEN

MICROSCOPY-BASED TRIAGE STRATEGIES

Cytology

At present cytology is the advocated triage strategy for HPV positive women. However, cytology is subjective and quality varies considerably among countries. In large, nation-wide screening studies (ATHENA $n = 4275$, VUSA-Screen $n = 1303$, POBASCAM $n = 1100$), triage of cytology by a single cytology test had CIN3 + sensitivities varying between 52.8% and 75.4% with corresponding specificities of 78.0% and 85.6% [43–45]. These results were obtained by cytologists unknown of the HPV status of the women. Providing the HPV status to the cytologists increased sensitivity (85.6% in HPV positive women [Ref. 46]) but significantly decreased specificity [46–48].

Currently, HPV positive women with abnormal cytology (although at different cut-off) are directly referred to colposcopy in all countries that recommend HPV-based screening.

However, CIN3 + risk after a single negative cytology results in HPV positive women is too high (5.2% in the VUSA-screen study [49]) to dismiss these women from follow-up during the screening interval. Similar results have been shown by others [44,45]. Therefore an additional cytology test at 6 or 12 months has been introduced in the current HPV-based screening program in the Netherlands (Table 16.3). However this extra follow-up step comes with a loss to follow-up.

Table 16.3 Short-Term CIN3 + Risk Estimates of Various Triage Strategies Within the VUSA-Screen and POBASCAM Cohorts

Triage Strategy	Triage Result	CIN3 + Risk (95% CI)	Management
Cytology with repeat cytology	Negative	0.7% (0.2–1.9)[a]	Regular screening interval maintained
		1.5% (0.7–3.2)[b]	Regular screening interval maintained
	Positive	37.5% (32.6–42.6)[a]	Direct referral for colposcopy
		34.0% (29.4–39.0)[b]	Direct referral for colposcopy
Cytology with HPV16/18 genotyping	Negative	2.9% (1.6–5.1)[a]	Follow-up required
		1.2% (0.5–3.0)[b]	Regular screening interval maintained
	Positive	26.0% (22.1–30.4)[a]	Direct referral for colposcopy
		28.5% (24.4–32.8)[b]	Direct referral for colposcopy
Cytology, HPV16/18 genotyping and repeat cytology	Negative	0.3% (0.1–1.6)[a]	Regular screening interval maintained
		0.4% (0.1–2.0)[b]	Regular screening interval maintained
	Positive	25.6% (21.9–29.7)[a]	Direct referral for colposcopy
		25.6% (22.0–29.6)[b]	Direct referral for colposcopy
Cytology with repeat HPV testing	Negative	0.1% (0.1–2.4)[a]	Regular screening interval maintained
		0.0% (0.0–1.8)[b]	Regular screening interval maintained
	Positive	19.5% (16.5–22.9)[a]	Direct referral for colposcopy
		22.5 (19.3–26.3)[b]	Direct referral for colposcopy

Management based on CIN3 + risk thresholds: CIN3 + risk <2%, regular screening interval; CIN3 + risk 2%–20%, follow-up required; CIN3 + risk >20%, direct referral for colposcopy.
[a]VUSA-Screen cohort.
[b]POBASCAM cohort.
Adapted from Polman NJ, de Haan Y, Veldhuijzen NJ, Heideman DAM, de Vet HCW, Meijer C, et al. Experience with HPV self-sampling and clinician-based sampling in women attending routine cervical screening in the Netherlands. Prev Med 2019;125:5–11.

For this reason only cytology is repeated after 6 months in the Netherlands (see "The new HPV based programme in the Netherlands" section). In Sweden longer persistence of HPV infection is required for referral to colposcopy (repeat HPV testing after 3 years). A mixed strategy has been applied in the English pilot, where women who initially tested HPV positive/cytology negative were referred to colposcopy if they developed cytological abnormalities after 1 year, and in some laboratories the remaining women were recalled after 24 months and referred if persistently HPV positive (regardless of cytology result) [50].

P16/Ki-67 Dual Staining

To improve cytology triage testing p16 immunostaining was introduced [51,52], later improved by adding Ki-67 staining. Expression of the anti-proliferative p16 protein, induced by HPV E7, and the proliferative marker Ki-67 in one cell is indicative of cell dysregulation by a transforming HPV infection. Using this double stain (CINtec PLUS, Roche) provides less subjective reading of cytology resulting in higher reproducibility and speed of reporting. However the cut-off of a positive p16/Ki-67 test is based on one positive cell making accurate positive reading difficult in case of low cellularity or insufficient intact abnormal cells. Sensitivities for CIN3 of p16/Ki-67 staining is not much higher than that of cytology, but the specificity for p16/Ki-67 is consistently higher [53–56]. A very low risk of CIN3 + following an HPV positive, p16/Ki-67 negative test is maintained for at least 2–3 years [57,58].

MOLECULAR TRIAGE STRATEGIES

HPV Testing With HPV Genotyping

Twelve HPV genotypes have been widely recognized as indicating a higher risk of a high-grade precancerous lesion or cancer [high-risk HPV (hrHPV)]. According to the WHO, these 12 types are carcinogenic (class 1), that is, 16, 18, 31, 33, 35, 39, 45, 51, 52, 56, 58, and 59. The HPV types that have been classified as *carcinogenic to humans* can differ by an order of magnitude in risk for cervical cancer. HPV68 is probably carcinogenic (class 2A), and HPV 26, 53, 66, 67, 70, 73, and 82 are classified as possibly carcinogenic (class 2B) [59]. HPV tests used in clinical settings detect these hrHPV types combined with two or more of the most prevalent HPV types from classes 2A and 2B.

HPV16 and 18 test more frequently positive in invasive carcinoma than in CIN3 [60]. From the other hrHPV types only HPV45 is marginally more frequently positive in invasive carcinoma than CIN3. This has been translated in a triage algorithm for HPV positive women in which HPV16/18 positive women are directly send for colposcopy and non-HPV16/18 positive women are referred to colposcopy when they test positive for cytology (≥ASC-US or ≥BMD). The advantage over repeat cytology triage is immediate triage without loss of women to follow-up. Despite the limited sensitivity of genotyping for HPV16/18 alone, overall evaluation of combined genotyping and cytology in several international studies, including both gynecological referral populations and screening populations, showed good overall sensitivity for CIN3 +, but limited positive predictive values (PPVs), resulting in high colposcopy referral rates [45,61]. Adding repeat cytology increases PPV, but then the advantage of immediate triage at baseline does no longer apply [43,44].

Direct referral to colposcopy of women positive for HPV16 and 18 is recommended in the United States [62] and Australia [63]. The remaining non-16/18 HPV positive women are re-invited for repeat cytology and HPV testing at short intervals.

It should be kept in mind that, given the rarity of invasive cancers, existing data are only based on the risk of high-grade CIN. In fact this is only an imprecise proxy as the probability of progression to invasion is surely different for different CIN. Incidentally, triage markers able to exclude the nonprogressive ones would strongly reduce overtreatment; methylation markers seem very promising in this sense [64] (see "Viral or host methylation markers" section). Plausibly conclusions will be possible only by observing cancer occurrence during the nation-wide application of triage protocols. In choosing triage strategies it should, however, be considered that an invasive cancer will occur if a high-grade CIN missed by an initial triage test has the time to progress to invasion before being detected by new testing. Thus, the cancer risk will increase with increasing screening interval. Long intervals need high sensitivity for CIN, but gains in sensitivity are irrelevant if the interval is short. An interesting question is "what is the discriminative power for detecting CIN2/3 lesions by HPV16/18 genotyping in HPV positive women?". Although at the first screening round HPV16/18 genotyping detects the majority of CIN2, CIN3, and cervical cancers, in the subsequent screening round (3–5 years later), HPV16/18 genotyping will mainly detect incident HPV infections without cervical lesions. Therefore over the course of a whole screening lifetime, HPV genotyping is probably not the best triage tool to include in nation-wide cervical screening programs [65].

Another question is "should extensive HPV genotyping be incorporated in cervical screening?". HPV16 stands out as the most important type both in terms of PPV and prevalence [66]. HPV18 is the second commonest genotype in cervical cancer, but has a lower PPV for CIN2 + in a screening context, possibly because it is more associated with endocervical lesions which are often not apparent on colposcopy. HPV31 and 33 emerge as types more associated with high-grade lesions than HPV18 or the other high-risk types [67]. A hierarchy of types was developed by Cuzick et al. [68] and is shown in Fig. 16.4.

At present, none of the widely used HPV tests provide a read-out for specific genotypes beyond HPV16 or 18, so that triage using other genotypes is not currently feasible. Use of vaccines only containing virus like particles (VLPs) for HPV16 and 18 as high-risk types may exacerbate this problem, as the relative importance of HPV31 and 33 would then be greater in those vaccinated, but widespread use of the nonavalent vaccine also containing VLPs for HPV31, 33, 45, 52, and 58 would make this less of an issue.

Viral or Host Methylation Markers

DNA methylation analysis of viral and/or host cell genes has been suggested as a method to identify women with CIN and cervical cancer. Viral methylation silences viral genes and is one of the many mechanisms by which the host protects itself against viral invasion. Specific combinations of CpG site methylation within the L1 and L2 genes of HPV16 and HPV18 have been shown to be associated with increasing CIN grade and cervical cancer (reviewed by Refs. [69,70]).

Also, methylation levels of promotor regions of host cell genes involved in cervical carcinogenesis increase with increasing CIN grade and are extremely high in cervical cancer (so-called cancer-like methylation-high profile). Investigated genes include *CADM1*, *EPB41L3*, *FAM19A4*, *MAL*, *miR124-2*, *PAX1*, and *SOX1*. The high methylation levels (cancer-like methylation-high profile) are particularly found in CIN3 and part of CIN2 lesions that are associated with a long duration of HPV infection (≥5 years) and show many copy number aberrations alike cervical cancer, suggesting that these lesions are "advanced lesions" with a high short term risk for progression to cervical cancer [71]. A recent systematic analysis of 12 methylated genes showed that methylation levels in clinical specimens increased proportional to disease severity, with the cancer-like methylation-high profile observed in 72% of CIN3, and 55% of CIN2 [72].

In vitro studies on HPV transformed cell lines in the same paper revealed that methylation of these genes typically started at the pre-tumorigenic stage (early or late immortal stage) [73]. Knowing that CIN2 lesions consist partly of productive and partly of transforming lesions, van

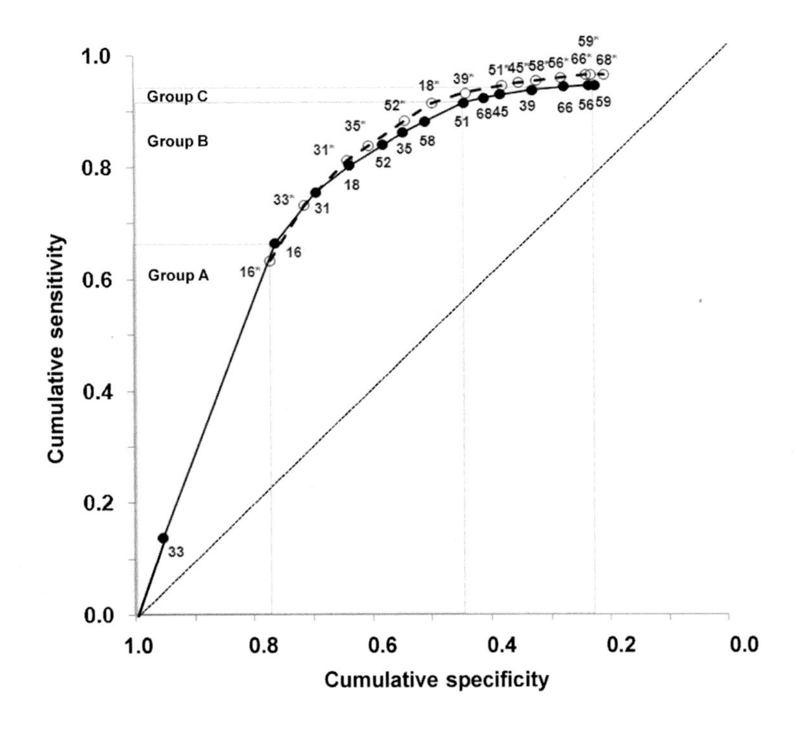

FIGURE 16.4

ROC curve of cumulative sensitivity and specificity for CIN2 + and CIN3 + . Genotypes are ordered according to PPV.

From Cuzick J, Ho L, Terry G, Kleeman M, Giddings M, Austin J, et al. Individual detection of 14 high risk human papilloma virus genotypes by the PapType test for the prediction of high grade cervical lesions. J Clin Virol 2014;60(1):44–9, with permission.

Zummeren et al. have reclassified CIN2 lesions based on the presence of p16/Ki-67 (indicative of transforming infection) and HPV-E4 (indicative of productive infection) and investigated the relation to *CADM1*, *MAL*, and *miR124-2* methylation [64]. Extensive E4 expression decreased with increasing CIN grade and was inversely related with increased methylation levels. Extensive E4 expression, as observed in some high-grade lesions, was associated with a negative methylation pattern. These data appear to confirm that the presence of host cell gene hypermethylation in smears of HPV positive women is indicative of advanced lesions in need of treatment.

It is out of our scope to discuss here the results of different research studies on triage of HPV positive women by methylation marker tests, and we will limit ourselves to assays that have been validated in clinical studies and/or are commercially available. The QIASure Methylation Test (Qiagen, Hilden, Germany) includes two genes, *FAM19A4* and *miR124-2*, and is validated in several clinical settings [74–78]. The S5 classifier is based on methylation of the human gene *EPB41L3* together with HPV16L1, HPV16L2, HPV18L2, HPV31L1, and HPV33L2 [79]. The GynTect assay (Oncgnostics, Jena, Germany) is a combination of six host cell genes, *ASTN1*, *DLX1*, *ITGA4*, *RXFP3*, *SOX17*, and *ZNF671* [80,81]. The assays have a CIN3 + sensitivity of

70%–90% with a specificity of 70%–85%. Post hoc analyses within the POBASCAM study show that the long-term cervical cancer risk is lower following a negative *FAM19A4/miR124-2* methylation test compared to a negative cytology test [76], and that the long-term CIN3 + risk of HPV positive women following a negative *FAM19A4/miR124-2* methylation test is comparable to the CIN3 + risk following a negative cytology test [77].

VALIDATION OF HPV TESTS FOR CERVICAL SCREENING

The HPV-cytology comparison studies increased interest in the use of HPV as a primary screening test enormously, and led to a number of commercially available, mostly PCR-based, new assays. These HPV tests were studied widely and in the Predictors series of studies many of them were compared to each other on samples from the same women in both the referral and screening contexts [82–85]. In addition, international guidelines for clinical validation of HPV tests were developed [86]. The guidelines are based on data derived from large clinical studies in which HPV testing by HC2 and GP5 + /6 + -PCR-EIA performed considerably better in the detection of CIN2/3 than cytology, and are widely used to clinically validate new commercial HPV tests on the basis of equivalence. Study frameworks for test comparison and validation of HPV assays, such as VALGENT, have been initiated to evaluate the clinical performance of various HPV tests relative to that of a validated and accepted comparator test in a formalized and uniform manner [87]. Additional large commercially supported trials were also conducted to provide data for licensing of specific tests for both ASC-US triage and primary screening by the FDA in the United States. The first such test to be approved was the DIGENE (now Qiagen) HC2 assay, which was to be used as a co-test in combination with cytology. Some years later Roche developed the COBAS test, which was based on PCR methodology, and it was validated in the ATHENA trial, which has led to FDA approval [88]. Subsequent approvals have been granted for Hologic's APTIMA RNA based test [89] and Becton Dickinson's Onclarity test [90], which provides a wider readout of positivity for different HPV types. Several other tests have been widely used and achieved various degrees of approval, particularly in European markets. These have been summarized by Arbyn, Poljak, Poljak et al. [91–93].

SELF-SAMPLING

With the availability of HPV testing, several efforts were made to evaluate whether HPV testing on self-samples was feasible. It soon became clear that HPV testing on cervico-vaginal self-samples was at least as sensitive as cytology on clinician-collected samples (reviewed by [94,95]). Moreover Arbyn et al. showed in a meta-analysis that some PCR-based HPV tests have similar sensitivity on both self-samples and clinician-based samples [95,96]. Recent studies have also evaluated the performance of urine for HPV detection [97,98].

HPV testing on self-collected cervico-vaginal samples (HPV self-sampling) was first mainly used to increase attendance of women who did not attend cervical screening. In the Netherlands this percentage is ~30%. Thirty percent of the non-attendees in the Netherlands used the HPV self-sampling kit which was sent to them by mail, whereas only 6%−17% responded to a re-invitation letter [99−101]. This increase in participation in the cervical screening program has been confirmed in several studies (reviewed by [102−104]) and has stimulated HPV testing on self-samples. It appeared that the way self-sampling is offered to women is important for the attendance rate. Sending women a self-collection kit by mail (opt-out) yields higher responses than an opt-in version in which women should ask for the kit [105,106].

When considering implementation of HPV self-sampling in a nation-wide screening program it is important to know whether the accuracy for detecting CIN2 + is similar to HPV testing on clinician taken samples when offered to women regularly responding to screening. This question has recently been evaluated in a randomized paired screen positive noninferiority trial within a primary screening setting in the Netherlands. HPV testing done with a clinically validated PCR-based assay (GP5 + /6 + PCR-EIA) on self-collected cervico-vaginal samples was noninferior to clinician-collected samples in terms of the detection of CIN2 + or CIN3 + lesions [107]. These results strongly support the use of HPV self-sampling as a primary screening method in routine screening. When combined with triage by a methylation assay HPV self-sampling may be an interesting option for primary screening because it improves poorer specificity due to passenger HPV and leads to full molecular self-screening.

THE NEW HPV BASED PROGRAM IN THE NETHERLANDS

From the countries that have implemented HPV-based screening, that is, Sweden, the Netherlands, Italy, Denmark, Australia and New Zealand, the Netherlands was one of the first countries to decide to convert from nation-wide cytology screening to HPV-based screening [108]. Cytology-based cervical screening of women aged 35−55 years with a screening interval of 3 years started in The Netherlands regionally in the 70s and spread over the country in the 80s. Nation-wide screening was introduced in 1994 [34]. Women of 30−60 years were targeted with a screening interval of 5 years. Although cytology has moderate sensitivity for CIN2 + , the long lead time for cervical cancer and the short screening intervals made cervical screening to a success in countries where attendance was over 70%. The effect of cervical screening by cytology resulted in a decrease of the incidence and mortality of cervical cancer, but since 2006 these parameters seem to level off [109,110]. Therefore a change in the primary screening tool, that is, cytology, was needed. When the results of HPV-cytology comparison trials showed that HPV provided much better protection against cervical cancer (60%) and CIN3 lesions (45%−50%) than cytology, and clinically validated HPV tests came available, the Ministry of Health decided in 2011 that cytology as primary screening tool should be replaced by HPV testing. However the factual implementation took place only in January 2017 because the Institute of Public Health and the Environment had to perform a feasibility study and introduced major changes in organization of cervical screening including strong reduction of number of screening labs, changes in logistics, and improvements in the follow-up chain of women with abnormal test results. At the same time significant resistance of cytologists against the implementation of automated HPV testing was noted.

COLPOSCOPY TRIAGE OF HPV POSITIVE WOMEN

Based on already mentioned data from population-based screening studies in the setting of the Dutch nation-wide screening program [i.e., the POBASCAM study (intake 1999–2002) and the VUSA-SCREEN cohort study (intake 2003–2005)] several triage strategies were compared [43,49] and repeat cytology at baseline and after 6 months was chosen in the new Dutch program.

This approach complies with the accepted short-term CIN3 + risk or, equivalently, the PPV of a triage test, requiring clinical management in the Netherlands that are based on the previous cytology-based program. A CIN3 + risk of <2% means maintain regular screening intervals, a CIN3 + risk of ~20% means direct referral for colposcopy, whereas a CIN3 + risk of 2%–20% means follow-up within 6–18 months. In the Dutch trials repeat cytology testing at 6–12 months reduced the risk from 5.2% to 1.6%, which is sufficiently low for maintaining regular screening intervals [49]. This strategy gives the lowest referral percentage (Table 16.3) and has a low cross-sectional CIN3 + risk for HPV positive women with a negative triage test (i.e., repeat cytology) but has the disadvantage that HPV positive women with normal cytology at baseline have to return after 6 months for a repeat cytology smear, which may result in loss to follow-up.

However as the performance of the triage strategies and the accepted balance between safety and screening-related burden varies considerably between countries, different triage strategies might be more suitable in other countries. For instance in the United States, a PPV of >10% is accepted for immediate referral, and triage of HPV16/18 negative women by cytology is an accepted strategy.

SCREENING INTERVALS

In the Dutch screening program, the screening interval has been extended up to 10 years for HPV negative women above 40 years of age. The decision to extend the interval was based on cost-effectiveness assessments commissioned by the Dutch Health Council [107]. The interval extension is supported by recent long-term follow-up data of the POBASCAM trial, showing that the CIN3 + and cancer risks after three screening rounds following a negative HPV test are similar to the risks after two rounds following a negative cytology result. Furthermore post hoc analyses of the Dutch trial showed that for HPV positive women with a negative cytology triage test, the screening interval cannot be extended over 5 years because these women continue to have a non-negligible cumulative 5 year CIN3 + risk of ~4% [33,111]). These results are in accordance with the Dutch screening recommendations as HPV positive women of 40, 50, or 60 years old with negative triage will have an extra screening round after 5 years.

At present no consensus exists at what age and after which test result women can exit the screening program. In the new HPV based program women of 60 years old who are HPV screen-negative are exiting the program. They have a lower CIN3 + risk than in the previous cytology program in which women with a negative cytology test after the age of 60 did not receive an invitation for screening. Moreover in the new program HPV positive, triage negative women will receive an extra screening invitation at 65 years thereby minimizing the CIN3 + risk of 60-year-old women.

SELF-SAMPLING

In the new program, women who do not respond to the screening invitation obtain a re-invitation within a month, together with a letter stating that if they prefer collecting cervico-vaginal material at home over a smear taken by their physician, they can return the letter with a prefilled, post-free envelope to the screening organization to obtain a self-sampling device. Recently it was shown that HPV testing on self-samples is as sensitive for the detection of CIN2 + and CIN3 + as HPV testing on a physician taken smear [112]. Questionnaires send to women participating in this trial revealed that women prefer the use of HPV self-sampling as a primary screening instrument. Self-sampling was associated with less discomfort, pain, and nervousness and more privacy. However although the accuracy of HPV testing for CIN3 + on self-collected cervico-vaginal specimen was noninferior to a physician taken smear, women were more uncertain about the result of an HPV self-sampling test [112]. More and detailed information about HPV self-sampling should be provided by screening organizations and the National Institute of Public Health and the Environment to make self-sampling as primary screening instrument a success.

HPV NEGATIVE AND LOW-RISK HPV POSITIVE CIN3 AND CARCINOMAS

As with every screening test, HPV testing does not detect all cervical carcinomas. True HPV negative carcinomas account for a small proportion of all cervical cancers [66,113]. According to the Cancer Genome research network, 5% of all cervical cancers are HPV negative [114]. Reasons for a negative hrHPV test result are (1) poor quality of the specimens with low DNA content due to necrosis and low cellularity; (2) use of insensitive or not clinically validated HPV tests covering not all hrHPV types associated with CIN3 + ; and (3) true negative CIN3 and cervical cancers. Squamous cell carcinomas (70%–90% of all cervical carcinomas) are nearly always HPV positive [115], whereas most HPV negative carcinomas are found among cervical adenocarcinomas. Cervical adenocarcinoma subtypes with a high prevalence of HPV include the usual type, and intestinal villo-glandular, signet-ring cell and endometrioid adenocarcinomas which originate from the squamo-columnar junction and account for around 90% of all adenocarcinomas. True HPV negative carcinomas include endocervical adenocarcinoma of the gastric type, minimal deviation adenocarcinoma, clear cell carcinoma, and mesonephric adenocarcinomas [116,117]. A recent study of 777 cervical carcinomas tissue from the United States showed that in nearly 60% of HPV negative cases no histological distinction could be made between cervical adenocarcinoma and primary endometrial carcinoma [118]. Diagnosis of HPV negative cervical adenocarcinomas should be supported by immunostainings including estrogen receptor (ER), progesteron receptor (PR), vimentin, CEA, p16, CD10, and CD34 [115].

From these findings it cannot be concluded that HPV testing cannot be used as primary cervical screening tool. The PPVs of hrHPV testing depend on the prevalence of disease in the general population. Moreover retrospective studies without surgical staging overestimate the proportion of HPV negative cervical cancers [119]. Data about the accuracy of cytology for adenocarcinomas are limited and false-negative results are common [120]. Moreover implementing quality assurance in cytology labs still remains a point of concern especially in low- and middle-income countries. Given the results of the prospective screening trials as described in 1.2, in which the reassurance against CIN3 + and cancer of a negative HPV test is much higher compared to cytology, HPV testing is at present the best test to offer to women.

Only a few publications mention the presence of non-hrHPV in cervical cancer giving estimates of 1%–2% of cervical cancers [119,121]. Some cases of HPV6 associated cervical carcinomas have been described [122]. Screening with clinically validated hrHPV tests will miss these cancers.

FUTURE PERSPECTIVES FOR CERVICAL SCREENING

The use of prophylactic vaccination will have a major decreasing effect on the incidence of cervical cancer and its precursor lesions. In properly vaccinated young women the screening intervals can be considerably extended.

However for the next decennia, a cohort of nonvaccinated women will still exist due to the limited uptake of prophylactic vaccination making cervical screening still relevant. At present HPV testing appears to be the best screening tool to offer to women.

In the follow-up algorithm of HPV positive women, triage by cytology can be made less subjective by adding the p16/Ki-67 staining resulting in more reproducible reading of cytology.

With increasing use of self-sampling, molecular triage will also gain interest because women do not have to visit their physician to have a cytology triage test. This way cervical screening will become more cost-effective.

DNA methylation assays may play an increasing role in molecular triage, especially on self-collected cervico-vaginal samples and will allow full molecular self-screening by primary HPV testing and methylation-based triage.

On the long run we foresee that with increased education, women may obtain their invitation and their previous screen results via a mobile application or e-mail, with advice on test frequency and visits to the physician. Tailoring screening to individuals may then improve screening efficiency and eventually provide optimal prevention for all women. This will require linkage between screening registers and vaccination registers.

Such approaches via mobile applications or e-mail or can also increase access of women in low- and middle-income countries to prophylactic vaccination and cervical screening and are important to prevent cervical cancer globally.

CONFLICTS OF INTEREST

(1) CJLMM and DAMH are minority shareholders of Self-screen B.V., a spin-off company of VU University Medical Center; (2) Self-screen B.V. holds patents related to the work (i.e., high-risk HPV test and methylation markers for cervical screening) and has developed and manufactured IVD assays, which are licensed to QIAGEN (QIAscreen HPV PCR Test and QIAsure Methylation Test); (3) CJLMM is part-time CEO of Self-screen B.V.; (4) CJLMM has received speakers fee from GSK, QIAGEN, and SPMSD/Merck, and served occasionally on the scientific advisory board (expert meeting) of GSK, QIAGEN, and SPMSD/Merck; (5) CJLMM has a very small number of shares of MDxHealth and QIAGEN; (6) JC has received research

support from Qiagen, Becton Dickinson, Genera, and Aventis Pharma. He has received honoraria from Merck, Roche, and Qiagen, and has participated in the speakers bureau of Becton Dickinson and Hologic; (7) DAMH has been on the speakers bureau of QIAGEN and serves occasionally on the scientific advisory boards of Pfizer and Bristol-Myers Squibb; (8) WWK and GR declare no conflicts of interest.

DISCLAIMER

The views expressed in this chapter are the sole responsibility of the authors and do not necessarily represent the views, decisions or policies of the institutions with which they are affiliated.

REFERENCES

[1] Durst M, Gissmann L, Ikenberg H, zur Hausen H. A papillomavirus DNA from a cervical carcinoma and its prevalence in cancer biopsy samples from different geographic regions. Proc Natl Acad Sci U S A 1983;80(12):3812–15.

[2] Lorincz AT, Reid R, Jenson AB, Greenberg MD, Lancaster W, Kurman RJ. Human papillomavirus infection of the cervix: relative risk associations of 15 common anogenital types. Obstet Gynecol 1992;79(3):328–37.

[3] Cuzick J, Arbyn M, Sankaranarayanan R, Tsu V, Ronco G, Mayrand MH, et al. Overview of human papillomavirus-based and other novel options for cervical cancer screening in developed and developing countries. Vaccine 2008;26(Suppl. 10):K29–41.

[4] Manos MM, Kinney WK, Hurley LB, Sherman ME, Shieh-Ngai J, Kurman RJ, et al. Identifying women with cervical neoplasia: using human papillomavirus DNA testing for equivocal Papanicolaou results. JAMA 1999;281(17):1605–10.

[5] van den Brule AJ, Snijders PJ, Gordijn RL, Bleker OP, Meijer CJ, Walboomers JM. General primer-mediated polymerase chain reaction permits the detection of sequenced and still unsequenced human papillomavirus genotypes in cervical scrapes and carcinomas. Int J Cancer 1990;45(4):644–9.

[6] Van Den Brule AJ, Walboomers JM, Du Maine M, Kenemans P, Meijer CJ. Difference in prevalence of human papillomavirus genotypes in cytomorphologically normal cervical smears is associated with a history of cervical intraepithelial neoplasia. Int J Cancer 1991;48(3):404–8.

[7] Meijer CJ, van den Brule AJ, Snijders PJ, Helmerhorst T, Kenemans P, Walboomers JM. Detection of human papillomavirus in cervical scrapes by the polymerase chain reaction in relation to cytology: possible implications for cervical cancer screening. IARC Sci Publ 1992;119:271–81.

[8] Melkert PW, Hopman E, van den Brule AJ, Risse EK, van Diest PJ, Bleker OP, et al. Prevalence of HPV in cytomorphologically normal cervical smears, as determined by the polymerase chain reaction, is age-dependent. Int J Cancer 1993;53(6):919–23.

[9] Husman AMDR, Walboomers JMM, Vandenbrule AJC, Meijer CJLM, Snijders PJF. The use of general primers Gp5 and Gp6 elongated at their 3' ends with adjacent highly conserved sequences improves human papillomavirus detection by Pcr. J Gen Virol 1995;76:1057–62.

[10] Jacobs MV, Snijders PJ, van den Brule AJ, Helmerhorst TJ, Meijer CJ, Walboomers JM. A general primer GP5 + /GP6(+)-mediated PCR-enzyme immunoassay method for rapid detection of 14 high-risk and 6 low-risk human papillomavirus genotypes in cervical scrapings. J Clin Microbiol 1997;35(3):791–5.

[11] Clifford GM, Smith JS, Aguado T, Franceschi S. Comparison of HPV type distribution in high-grade cervical lesions and cervical cancer: a meta-analysis. Br J Cancer 2003;89(1):101–5.
[12] Nobbenhuis MAE, Walboomers JMM, Helmerhorst TJM, Rozendaal L, Remmink AJ, Risse EKJ, et al. Relation of human papilloma virus status to cervical lesions and consequences for cervical-cancer screening: a prospective study. Lancet 1999;354(9172):20–5.
[13] Nobbenhuis MA, Helmerhorst TJ, van den Brule AJ, Rozendaal L, Voorhorst FJ, Bezemer PD, et al. Cytological regression and clearance of high-risk human papillomavirus in women with an abnormal cervical smear. Lancet 2001;358(9295):1782–3.
[14] Cox JT, Lorincz AT, Schiffman MH, Sherman ME, Cullen A, Kurman RJ. Human papillomavirus testing by hybrid capture appears to be useful in triaging women with a cytologic diagnosis of atypical squamous cells of undetermined significance. Am J Obstet Gynecol 1995;172(3):946–54.
[15] Solomon D, Schiffman M, Tarone R, group AS. Comparison of three management strategies for patients with atypical squamous cells of undetermined significance: baseline results from a randomized trial. J Natl Cancer Inst 2001;93(4):293–9.
[16] Cuzick J, Szarewski A, Terry G, Ho L, Hanby A, Maddox P, et al. Human papillomavirus testing in primary cervical screening. Lancet 1995;345(8964):1533–6.
[17] Cuzick J, Clavel C, Petry KU, Meijer CJ, Hoyer H, Ratnam S, et al. Overview of the European and North American studies on HPV testing in primary cervical cancer screening. Int J Cancer 2006;119(5):1095–101.
[18] Sankaranarayanan R, Nene BM, Shastri SS, Jayant K, Muwonge R, Budukh AM, et al. HPV screening for cervical cancer in rural India. N Engl J Med 2009;360(14):1385–94.
[19] Rijkaart DC, Berkhof J, Rozendaal L, van Kemenade FJ, Bulkmans NW, Heideman DA, et al. Human papillomavirus testing for the detection of high-grade cervical intraepithelial neoplasia and cancer: final results of the POBASCAM randomised controlled trial. Lancet Oncol 2012;13(1):78–88.
[20] Naucler P, Ryd W, Tornberg S, Strand A, Wadell G, Elfgren K, et al. Human papillomavirus and Papanicolaou tests to screen for cervical cancer. N Engl J Med 2007;357(16):1589–97.
[21] Kitchener HC, Almonte M, Thomson C, Wheeler P, Sargent A, Stoykova B, et al. HPV testing in combination with liquid-based cytology in primary cervical screening (ARTISTIC): a randomised controlled trial. Lancet Oncol 2009;10(7):672–82.
[22] Ronco G, Giorgi-Rossi P, Carozzi F, Confortini M, Dalla Palma P, Del Mistro A, et al. Efficacy of human papillomavirus testing for the detection of invasive cervical cancers and cervical intraepithelial neoplasia: a randomised controlled trial. Lancet Oncol 2010;11:249–57.
[23] Anttila A, Kotaniemi-Talonen L, Leinonen M, Hakama M, Laurila P, Tarkkanen J, et al. Rate of cervical cancer, severe intraepithelial neoplasia, and adenocarcinoma in situ in primary HPV DNA screening with cytology triage: randomised study within organised screening programme. BMJ 2010;340:c1804.
[24] Ronco G, Dillner J, Elfstrom KM, Tunesi S, Snijders PJ, Arbyn M, et al. Efficacy of HPV-based screening for prevention of invasive cervical cancer: follow-up of four European randomised controlled trials. Lancet 2014;383(9916):524–32.
[25] von Karsa L, Arbyn M, De Vuyst H, Dillner J, Dillner L, Franceschi S, et al. European guidelines for quality assurance in cervical cancer screening. Summary of the supplements on HPV screening and vaccination. Papillomavirus Res. 2015;1:22–31.
[26] Comprehensive cervical cancer control: a guide to essential practice. Second ed. Geneva, Switzerland: World Health Organization; 2014.
[27] Arbyn M, Ronco G, Anttila A, Meijer CJ, Poljak M, Ogilvie G, et al. Evidence regarding human papillomavirus testing in secondary prevention of cervical cancer. Vaccine 2012;30(Suppl. 5):F88–99.
[28] Dillner J, Rebolj M, Birembaut P, Petry KU, Szarewski A, Munk C, et al. Long term predictive values of cytology and human papillomavirus testing in cervical cancer screening: joint European cohort study. BMJ 2008;337:a1754.

[29] Katki HA, Schiffman M, Castle PE, Fetterman B, Poitras NE, Lorey T, et al. Five-year risks of CIN 2 + and CIN 3 + among women with HPV-positive and HPV-negative LSIL Pap results. J Low Genit Tract Dis 2013;17(5 Suppl. 1):S43–9.
[30] Katki HA, Schiffman M, Castle PE, Fetterman B, Poitras NE, Lorey T, et al. Five-year risks of CIN 3 + and cervical cancer among women with HPV testing of ASC-US Pap results. J Low Genit Tract Dis 2013;17(5 Suppl. 1):S36–42.
[31] Schiffman M, Kinney WK, Cheung LC, Gage JC, Fetterman B, Poitras NE, et al. Relative performance of HPV and cytology components of cotesting in cervical screening. J Natl Cancer Inst 2018;110(5):501–8.
[32] Horn J, Denecke A, Luyten A, Rothe B, Reinecke-Luthge A, Mikolajczyk R, et al. Reduction of cervical cancer incidence within a primary HPV screening pilot project (WOLPHSCREEN) in Wolfsburg, Germany. Br J Cancer 2019;120(10):1015–22.
[33] Dijkstra MG, van Zummeren M, Rozendaal L, van Kemenade FJ, Helmerhorst TJ, Snijders PJ, et al. Safety of extending screening intervals beyond five years in cervical screening programmes with testing for high risk human papillomavirus: 14 year follow-up of population based randomised cohort in the Netherlands. BMJ 2016;355:i4924.
[34] Hanselaar AG. The population-based screening for cervical cancer: a uniform framework for cytopathological diagnosis. Medisch Contact 1995;50:1590–2.
[35] Kitchener HC, Gilham C, Sargent A, Bailey A, Albrow R, Roberts C, et al. A comparison of HPV DNA testing and liquid based cytology over three rounds of primary cervical screening: extended follow up in the ARTISTIC trial. Eur J Cancer 2011;47(6):864–71.
[36] Elfström KM, Smelov V, Johansson ALV, Eklund C, Nauclér P, Arnheim-Dahlströ L, et al. Long term duration of protective effect for HPV negative women: follow-up of primary HPV screening randomised controlled trial. BMJ 2014;348:g130.
[37] Castle PE, Glass AG, Rush BB, Scott DR, Wentzensen N, Gage JC, et al. Clinical human papillomavirus detection forecasts cervical cancer risk in women over 18 years of follow-up. J Clin Oncol 2012;30 (25):3044–50.
[38] McCredie MRE, Sharples KJ, Paul C, Baranyai J, Medley G, Jones RW, et al. Natural history of cervical neoplasia and risk of invasive cancer in women with cervical intraepithelial neoplasia 3: a retrospective cohort study. Lancet Oncol 2008;9(5):425–34.
[39] Kyrgiou M, Mitra A, Arbyn M, Stasinou SM, Martin-Hirsch P, Bennett P, et al. Fertility and early pregnancy outcomes after treatment for cervical intraepithelial neoplasia: systematic review and meta-analysis. BMJ 2014;349:g6192.
[40] Kyrgiou M, Athanasiou A, Paraskevaidi M, Mitra A, Kalliala I, Martin-Hirsch P, et al. Adverse obstetric outcomes after local treatment for cervical preinvasive and early invasive disease according to cone depth: systematic review and meta-analysis. BMJ 2016;354:i3633.
[41] Kyrgiou M, Koliopoulos G, Martin-Hirsch P, Arbyn M, Prendiville W, Paraskevaidis E. Obstetric outcomes after conservative treatment for intraepithelial or early invasive cervical lesions: systematic review and meta-analysis. Lancet 2006;367(9509):489–98.
[42] Arbyn M, Kyrgiou M, Simoens C, Raifu AO, Koliopoulos G, Martin-Hirsch P, et al. Perinatal mortality and other severe adverse pregnancy outcomes associated with treatment of cervical intraepithelial neoplasia: meta-analysis. BMJ 2008;337:a1284.
[43] Rijkaart DC, Berkhof J, van Kemenade FJ, Coupe VM, Hesselink AT, Rozendaal L, et al. Evaluation of 14 triage strategies for HPV DNA-positive women in population-based cervical screening. Int J Cancer 2012;130(3):602–10.
[44] Dijkstra MG, van Niekerk D, Rijkaart DC, van Kemenade FJ, Heideman DA, Snijders PJ, et al. Primary hrHPV DNA testing in cervical cancer screening: how to manage screen-positive women? A POBASCAM trial substudy. Cancer Epidem Biomar Prev 2014;23(1):55–63.

[45] Castle PE, Stoler MH, Wright Jr. TC, Sharma A, Wright TL, et al. Performance of carcinogenic human papillomavirus (HPV) testing and HPV16 or HPV18 genotyping for cervical cancer screening of women aged 25 years and older: a subanalysis of the ATHENA study. Lancet Oncol 2011;12(9):880–90.
[46] Bergeron C, Giorgi-Rossi P, Cas F, Schiboni ML, Ghiringhello B, Dalla Palma P, et al. Informed cytology for triaging HPV-positive women: substudy nested in the NTCC randomized controlled trial. J Natl Cancer Inst 2015;107(2).
[47] Moriarty AT, Nayar R, Arnold T, Gearries L, Renshaw A, Thomas N, et al. The Tahoe Study: bias in the interpretation of Papanicolaou test results when human papillomavirus status is known. Arch Pathol Lab Med 2014;138(9):1182–5.
[48] Richardson LA, El-Zein M, Ramanakumar AV, Ratnam S, Sangwa-Lugoma G, Longatto-Filho A, et al. HPV DNA testing with cytology triage in cervical cancer screening: influence of revealing HPV infection status. Cancer Cytopathol 2015;123(12):745–54.
[49] Rijkaart DC, Berkhof J, van Kemenade FJ, Coupe VM, Rozendaal L, Heideman DA, et al. HPV DNA testing in population-based cervical screening (VUSA-Screen study): results and implications. Br J Cancer 2012;106(5):975–81.
[50] Rebolj M, Rimmer J, Denton K, Tidy J, Mathews C, Ellis K, et al. Primary cervical screening with high risk human papillomavirus testing: observational study. BMJ 2019;364:l240.
[51] Carozzi F, Confortini M, Dalla Palma P, Del Mistro A, Gillio-Tos A, De Marco L, et al. Use of p16-INK4A overexpression to increase the specificity of human papillomavirus testing: a nested substudy of the NTCC randomised controlled trial. Lancet Oncol 2008;9(10):937–45.
[52] Carozzi F, Gillio-Tos A, Confortini M, Del Mistro A, Sani C, De Marco L, et al. Risk of high-grade cervical intraepithelial neoplasia during follow-up in HPV-positive women according to baseline p16-INK4A results: a prospective analysis of a nested substudy of the NTCC randomised controlled trial. Lancet Oncol 2013;14(2):168–76.
[53] Wentzensen N, Fetterman B, Castle PE, Schiffman M, Wood SN, Stiemerling E, et al. p16/Ki-67 dual stain cytology for detection of cervical precancer in HPV-positive women. J Natl Cancer Inst 2015;107 (12):djv257.
[54] Gustinucci D, Rossi PG, Cesarini E, Broccolini M, Bulletti S, Carlani A, et al. Use of cytology, E6/E7 mRNA andp16(INK4a)-Ki-67 to define the management of human papillomavirus (HPV)-positive women in cervical cancer screening. Am J Clin Pathol 2016;145(1):35–45.
[55] Luttmer R, Dijkstra MG, Snijders PJ, Berkhof J, van Kemenade FJ, Rozendaal L, et al. p16/Ki-67 dual-stained cytology for detecting cervical (pre)cancer in a HPV-positive gynecologic outpatient population. Mod Pathol 2016;29(8):870–8.
[56] Wentzensen N, Clarke MA, Bremer R, Poitras N, Tokugawa D, Goldhoff PE, et al. Clinical evaluation of human papillomavirus screening with p16/ki-67 dual stain triage in a large organized cervical cancer screening program. JAMA Intern Med 2019.
[57] Qian QP, Zhang X, Ding B, Jiang SW, Li ZM, Ren ML, et al. Performance of P16/Ki67 dual staining in triaging hr-HPV-positive population during cervical Cancer screening in the younger women. Clin Chim Acta 2018;483:281–5.
[58] Uijterwaal MH, Polman NJ, Witte BI, van Kemenade FJ, Rijkaart D, Berkhof J, et al. Triaging HPV-positive women with normal cytology by p16/Ki-67 dual-stained cytology testing: baseline and longitudinal data. Int J Cancer 2015;136(10):2361–8.
[59] International Agency for Research on Cancer. Biological Agents. IARC Monogr Eval Carcinog Risks Hum 2012;100B:1–475.
[60] Guan P, Howell-Jones R, Li N, Bruni L, de Sanjose S, Franceschi S, et al. Human papillomavirus types in 115,789 HPV-positive women: a meta-analysis from cervical infection to cancer. Int J Cancer 2012;131(10):2349–59.

[61] Wright Jr. TC, Stoler MH, Sharma A, Zhang G, Behrens C, Wright TL, et al. Evaluation of HPV-16 and HPV-18 genotyping for the triage of women with high-risk HPV + cytology-negative results. Am J Clin Pathol 2011;136(4):578–86.

[62] Huh WK, Ault KA, Chelmow D, Davey DD, Goulart RA, Garcia FA, et al. Use of primary high-risk human papillomavirus testing for cervical cancer screening: interim clinical guidance. J Low Genit Tract Di 2015;19(2):91–6.

[63] Simms KT, Hall M, Smith MA, Lew JB, Hughes S, Yuill S, et al. Optimal management strategies for primary HPV testing for cervical screening: cost-effectiveness evaluation for the national cervical screening program in Australia. PLoS One 2017;12(1):e0163509.

[64] Zummeren MV, Kremer WW, Leeman A, Bleeker MCG, Jenkins D, Sandt MV, et al. HPV E4 expression and DNA hypermethylation of CADM1, MAL, and miR124-2 genes in cervical cancer and precursor lesions. Mod Pathol 2018;31(12):1842–50.

[65] Veldhuijzen NJ, Polman NJ, Snijders PJF, Meijer C, Berkhof J. Stratifying HPV-positive women for CIN3 + risk after one and two rounds of HPV-based screening. Int J Cancer 2017;141(8):1551–60.

[66] Li N, Franceschi S, Howell-Jones R, Snijders PJ, Clifford GM. Human papillomavirus type distribution in 30,848 invasive cervical cancers worldwide: variation by geographical region, histological type and year of publication. Int J Cancer 2011;128(4):927–35.

[67] Cuzick J, Wheeler C. Need for expanded HPV genotyping for cervical screening. Papillomavirus Res 2016;2:112–15.

[68] Cuzick J, Ho L, Terry G, Kleeman M, Giddings M, Austin J, et al. Individual detection of 14 high risk human papilloma virus genotypes by the PapType test for the prediction of high grade cervical lesions. J Clin Virol 2014;60(1):44–9.

[69] Clarke MA, Wentzensen N, Mirabello L, Ghosh A, Wacholder S, Harari A, et al. Human papillomavirus DNA methylation as a potential biomarker for cervical cancer. Cancer Epidemiol Biomarkers Prev 2012;21(12):2125–37.

[70] Lorincz AT. Virtues and weaknesses of DNA methylation as a test for cervical cancer prevention. Acta Cytol 2016;60(6):501–12.

[71] Steenbergen RD, Snijders PJ, Heideman DA, Meijer CJ. Clinical implications of (epi)genetic changes in HPV-induced cervical precancerous lesions. Nat Rev Cancer 2014;14(6):395–405.

[72] Verlaat W, Snoek BC, Heideman DAM, Wilting SM, Snijders PJF, Novianti PW, et al. Identification and validation of a 3-gene methylation classifier for HPV-based cervical screening on self-samples. Clin Cancer Res 2018;.

[73] Verlaat W, Van Leeuwen RW, Novianti PW, Schuuring E, Meijer C, Van Der Zee AGJ, et al. Host-cell DNA methylation patterns during high-risk HPV-induced carcinogenesis reveal a heterogeneous nature of cervical pre-cancer. Epigenetics 2018;13(7):769–78.

[74] De Strooper LM, Verhoef VM, Berkhof J, Hesselink AT, de Bruin HM, van Kemenade FJ, et al. Validation of the FAM19A4/mir124-2 DNA methylation test for both lavage- and brush-based self-samples to detect cervical (pre)cancer in HPV-positive women. Gynecol Oncol 2016;141 (2):341–7.

[75] Luttmer R, De Strooper LM, Berkhof J, Snijders PJ, Dijkstra MG, Uijterwaal MH, et al. Comparing the performance of FAM19A4 methylation analysis, cytology and HPV16/18 genotyping for the detection of cervical (pre)cancer in high-risk HPV-positive women of a gynecologic outpatient population (COMETH study). Int J Cancer 2016;138(4):992–1002.

[76] De Strooper LMA, Berkhof J, Steenbergen RDM, Lissenberg-Witte BI, Snijders PJF, Meijer C, et al. Cervical cancer risk in HPV-positive women after a negative FAM19A4/mir124-2 methylation test: a post hoc analysis in the POBASCAM trial with 14 year follow-up. Int J Cancer 2018;.

[77] Dick S, Kremer WW, De Strooper LMA, Lissenberg-Witte BI, Steenbergen RDM. Meijer CJLM, et al. Long-term CIN3 + risk of HPV positive women after triage with FAM19A4/miR124-2 methylation analysis. Gynecol Oncol. 2019;154(2):368–73.
[78] Vink FJ, Meijer CJLM, Clifford GM, Poljak M, Oštrbenk A, Petry KU, et al. FAM19A4/miR124-2 methylation in invasive cervical cancer: A retrospective cross-sectional worldwide study. Int J Cancer 2019 Aug 7. Available from: https://doi.org/10.1002/ijc.32614 [Epub ahead of print].
[79] Cook DA, Krajden M, Brentnall AR, Gondara L, Chan T, Law JH, et al. Evaluation of a validated methylation triage signature for human papillomavirus positive women in the HPV FOCAL cervical cancer screening trial. Int J Cancer 2019;144(10):2587–95.
[80] Schmitz M, Wunsch K, Hoyer H, Scheungraber C, Runnebaum IB, Hansel A, et al. Performance of a methylation specific real-time PCR assay as a triage test for HPV-positive women. Clin Epigenetics 2017;9:118.
[81] Schmitz M, Eichelkraut K, Schmidt D, Zeiser I, Hilal Z, Tettenborn Z, et al. Performance of a DNA methylation marker panel using liquid-based cervical scrapes to detect cervical cancer and its precancerous stages. BMC Cancer 2018;18(1):1197.
[82] Szarewski A, Ambroisine L, Cadman L, Austin J, Ho L, Terry G, et al. Comparison of predictors for high-grade cervical intraepithelial neoplasia in women with abnormal smears. Cancer Epidemiol Biomarkers Prev 2008;17(11):3033–42.
[83] Szarewski A, Mesher D, Cadman L, Austin J, Ashdown-Barr L, Ho L, et al. Comparison of seven tests for high-grade cervical intraepithelial neoplasia in women with abnormal smears: the Predictors 2 study. J Clin Microbiol 2012;50(6):1867–73.
[84] Cuzick J, Cadman L, Mesher D, Austin J, Ashdown-Barr L, Ho L, et al. Comparing the performance of six human papillomavirus tests in a screening population. Br J Cancer 2013;108(4):908–13.
[85] Cuzick J, Ahmad AS, Austin J, Cadman L, Ho L, Terry G, et al. A comparison of different human papillomavirus tests in PreservCyt versus SurePath in a referral population-PREDICTORS 4. J Clin Virol 2016;82:145–51.
[86] Meijer CJ, Berkhof J, Castle PE, Hesselink AT, Franco EL, Ronco G, et al. Guidelines for human papillomavirus DNA test requirements for primary cervical cancer screening in women 30 years and older. Int J Cancer 2009;124(3):516–20.
[87] Arbyn M, Depuydt C, Benoy I, Bogers J, Cuschieri K, Schmitt M, et al. VALGENT: A protocol for clinical validation of human papillomavirus assays. J Clin Virol 2016;76(Suppl. 1):S14–21.
[88] Wright TC, Stoler MH, Behrens CM, Sharma A, Zhang G, Wright TL. Primary cervical cancer screening with human papillomavirus: end of study results from the ATHENA study using HPV as the first-line screening test. Gynecol Oncol 2015;136(2):189–97.
[89] Reid JL, Wright Jr. TC, Stoler MH, Cuzick J, Castle PE, et al. Human papillomavirus oncogenic mRNA testing for cervical cancer screening: baseline and longitudinal results from the CLEAR study. Am J Clin Pathol 2015;144(3):473–83.
[90] Wright Jr. TC, Stoler MH, Parvu V, Yanson K, Eckert K, et al. Detection of cervical neoplasia by human papillomavirus testing in an atypical squamous cells-undetermined significance population: results of the Becton Dickinson Onclarity Trial. Am J Clin Pathol 2019;151(1):53–62.
[91] Arbyn M, Snijders PJ, Meijer CJ, Berkhof J, Cuschieri K, Kocjan BJ, et al. Which high-risk HPV assays fulfil criteria for use in primary cervical cancer screening? Clin Microbiol Infect 2015;21(9):817–26.
[92] Poljak M. Towards cervical cancer eradication: joint force of HPV vaccination and HPV-based cervical cancer screening. Clin Microbiol Infect 2015;21(9):806–7.
[93] Poljak M, Kocjan BJ, Ostrbenk A, Seme K. Commercially available molecular tests for human papillomaviruses (HPV): 2015update. J Clin Virol 2016;76(Suppl. 1):S3–13.

[94] Snijders PJ, Verhoef VM, Arbyn M, Ogilvie G, Minozzi S, Banzi R, et al. High-risk HPV testing on self-sampled versus clinician-collected specimens: a review on the clinical accuracy and impact on population attendance in cervical cancer screening. Int J Cancer 2013;132(10):2223–36.
[95] Arbyn M, Verdoodt F, Snijders PJ, Verhoef VM, Suonio E, Dillner L, et al. Accuracy of human papillomavirus testing on self-collected versus clinician-collected samples: a meta-analysis. Lancet Oncol 2014;15(2):172–83.
[96] Arbyn M, Smith SB, Temin S, Sultana F, Castle P. Collaboration on S-S, et al. Detecting cervical precancer and reaching underscreened women by using HPV testing on self samples: updated meta-analyses. BMJ 2018;363:k4823.
[97] Leeman A, Del Pino M, Molijn A, Rodriguez A, Torne A, de Koning M, et al. HPV testing in first-void urine provides sensitivity for CIN2 + detection comparable with a smear taken by a clinician or a brush-based self-sample: cross-sectional data from a triage population. BJOG 2017;124(9):1356–63.
[98] Cuzick J, Cadman L, Ahmad AS, Ho L, Terry G, Kleeman M, et al. Performance and diagnostic accuracy of a urine-based human papillomavirus assay in a referral population. Cancer Epidemiol Biomarkers Prev 2017;26(7):1053–9.
[99] Gok M, Heideman DA, van Kemenade FJ, Berkhof J, Rozendaal L, Spruyt JW, et al. HPV testing on self collected cervicovaginal lavage specimens as screening method for women who do not attend cervical screening: cohort study. BMJ 2010;340:c1040.
[100] Gok M, Heideman DA, van Kemenade FJ, de Vries AL, Berkhof J, Rozendaal L, et al. Offering self-sampling for human papillomavirus testing to non-attendees of the cervical screening programme: characteristics of the responders. Eur J Cancer 2012;48(12):1799–808.
[101] Gok M, van Kemenade FJ, Heideman DA, Berkhof J, Rozendaal L, Spruyt JW, et al. Experience with high-risk human papillomavirus testing on vaginal brush-based self-samples of non-attendees of the cervical screening program. Int J Cancer 2012;130(5):1128–35.
[102] Verdoodt F, Jentschke M, Hillemanns P, Racey CS, Snijders PJ, Arbyn M. Reaching women who do not participate in the regular cervical cancer screening programme by offering self-sampling kits: a systematic review and meta-analysis of randomised trials. Eur J Cancer 2015;51(16):2375–85.
[103] Sultana F, English DR, Simpson JA, Drennan KT, Mullins R, Brotherton JM, et al. Home-based HPV self-sampling improves participation by never-screened and under-screened women: Results from a large randomized trial (iPap) in Australia. Int J Cancer 2016;139(2):281–90.
[104] Nelson EJ, Maynard BR, Loux T, Fatla J, Gordon R, Arnold LD. The acceptability of self-sampled screening for HPV DNA: a systematic review and meta-analysis. Sex Transm Infect 2017;93(1):56–61.
[105] Del Mistro A, Frayle H, Ferro A, Fantin G, Altobelli E, Giorgi Rossi P. Efficacy of self-sampling in promoting participation to cervical cancer screening also in subsequent round. Prev Med Rep 2017;5:166–8.
[106] Ma'som M, Bhoo-Pathy N, Nasir NH, Bellinson J, Subramaniam S, Ma Y, et al. Attitudes and factors affecting acceptability of self-administered cervicovaginal sampling for human papillomavirus (HPV) genotyping as an alternative to Pap testing among multiethnic Malaysian women. BMJ Open 2016;6(8):e011022.
[107] Polman NJ, Ebisch RMF, Heideman DAM, Melchers WJG, Bekkers RLM, Molijn AC, et al. Performance of human papillomavirus testing on self-collected versus clinician-collected samples for the detection of cervical intraepithelial neoplasia of grade 2 or worse: a randomised, paired screen-positive, non-inferiority trial. Lancet Oncol 2019;20(2):229–38.
[108] Dutch Health Council. Cervical Screening (In Dutch). 2011.
[109] de Kok IM, van der Aa MA, van Ballegooijen M, Siesling S, Karim-Kos HE, van Kemenade FJ, et al. Trends in cervical cancer in the Netherlands until 2007: has the bottom been reached? Int J Cancer 2011;128(9):2174–81.

[110] National Monitoring Population0based Cervical Screening 2017 (In Dutch). Available at: https://www.rivm.nl/documenten/landelijke-evaluatie-van-bevolkingsonderzoek-baarmoederhalskanker-leba-tm-2017.
[111] Uijterwaal MH, Polman NJ, Van Kemenade FJ, Van Den Haselkamp S, Witte BI, Rijkaart D, et al. Five-year cervical (pre)cancer risk of women screened by HPV and cytology testing. Cancer Prev Res (Phila) 2015;8(6):502–8.
[112] Polman NJ, de Haan Y, Veldhuijzen NJ, Heideman DAM, de Vet HCW, Meijer C, et al. Experience with HPV self-sampling and clinician-based sampling in women attending routine cervical screening in the Netherlands. Prev Med 2019;125:5–11.
[113] Liebrich C, Brummer O, Von Wasielewski R, Wegener G, Meijer C, Iftner T, et al. Primary cervical cancer truly negative for high-risk human papillomavirus is a rare but distinct entity that can affect virgins and young adolescents. Eur J Gynaecol Oncol 2009;30(1):45–8.
[114] Cancer Genome Atlas Research N, Albert Einstein College of M, Analytical Biological S, Barretos Cancer H, Baylor College of M, Beckman Research Institute of City of H, et al. Integrated genomic and molecular characterization of cervical cancer. Nature 2017;543(7645):378–84.
[115] Pirog EC. Cervical adenocarcinoma: diagnosis of human papillomavirus-positive and human papillomavirus-negative tumors. Arch Pathol Lab Med 2017;141(12):1653–67.
[116] Pirog EC, Lloveras B, Molijn A, Tous S, Guimera N, Alejo M, et al. HPV prevalence and genotypes in different histological subtypes of cervical adenocarcinoma, a worldwide analysis of 760 cases. Mod Pathol 2014;27(12):1559–67.
[117] Holl K, Nowakowski AM, Powell N, McCluggage WG, Pirog EC, Collas De Souza S, et al. Human papillomavirus prevalence and type-distribution in cervical glandular neoplasias: Results from a European multinational epidemiological study. Int J Cancer 2015;137(12):2858–68.
[118] Hopenhayn C, Christian A, Christian WJ, Watson M, Unger ER, Lynch CF, et al. Prevalence of human papillomavirus types in invasive cervical cancers from 7 US cancer registries before vaccine introduction. J Low Genit Tract Dis 2014;18(2):182–9.
[119] Petry KU, Liebrich C, Luyten A, Zander M, Iftner T. Surgical staging identified false HPV-negative cases in a large series of invasive cervical cancers. Papillomavirus Res 2017;4:85–9.
[120] Ault KA, Joura EA, Kjaer SK, Iversen OE, Wheeler CM, Perez G, et al. Adenocarcinoma in situ and associated human papillomavirus type distribution observed in two clinical trials of a quadrivalent human papillomavirus vaccine. Int J Cancer 2011;128(6):1344–53.
[121] Zappacosta R, Lattanzio G, Viola P, Ianieri MM, Gatta DM, Rosini S. A very rare case of HPV-53-related cervical cancer, in a 79-year-old woman with a previous history of negative Pap cytology. Clin Interv Aging 2014;9:683–8.
[122] de Sanjose S, Quint WGV, Alemany L, Geraets DT, Klaustermeier JE, Lloveras B, et al. Human papillomavirus genotype attribution in invasive cervical cancer: a retrospective cross-sectional worldwide study. Lancet Oncol 2010;11(11):1048–56.

CHAPTER

17 INFECTION TO CANCER—FINDING USEFUL BIOMARKERS FOR PREDICTING RISK OF PROGRESSION TO CANCER

Magnus von Knebel Doeberitz[1], Chris J.L.M. Meijer[2], Attila Lorincz[3], John Doorbar[4] and Annemiek Leeman[5]

[1]*Department of Applied Tumor Biology, Institute of Pathology, Heidelberg University Hospital, Heidelberg, Germany* [2]*Amsterdam UMC, Vrije Universiteit Amsterdam, Pathology, Cancer Center Amsterdam, Amsterdam, The Netherlands* [3]*Wolfson Institute of Preventive Medicine, Queen Mary University of London, London, United Kingdom* [4]*Department of Pathology, University of Cambridge, Cambridge, United Kingdom* [5]*DDL laboratories, Rijswijk, the Netherlands*

INTRODUCTION

With a lifetime risk of high-risk human papillomavirus (hrHPV) infection around 80% and less than 1% of those infections leading to cervical cancer, the development of a cervical carcinoma is a rare complication of hrHPV infection. How this happens is discussed in detail in Section 1 of this book and in Chapter 19, Clinical Trials of Human Papillomavirus Vaccines. Most infections are cleared by the host immune system within 2 years and do not always produce a definite dysplastic lesion. Several determinants have been found to influence the progression of oncogenic HPV infection, such as smoking, number of sexual partners, contraceptive use, and an effective immune response. Cervical cancer develops through a series of stages and takes years. The large majority of HPV lesions, however, are productive lesions which produce and shed viral particles without signs of cellular transformation. The morphology of HPV-induced lesions is such that it is not possible to reliably distinguish which lesions are transforming to cancer from those with a high chance of spontaneous regression. The transforming process is driven by viral oncogenes E6 and E7 overexpression but these genes are also relevant to productive infection, leading to formation and release of new virus particles. Which hrHPV infections lead to productive infections which will never progress and which ones lead to progressive transforming lesions and warrant treatment is not clear. The site of infection is the basis of one hypothesis that has been formulated. Squamo-columnar junction cells, which are localized ectocervical squamous epithelium and endocervical glandular epithelium, are thought to be highly susceptible to transforming HPV infection while productive infections are thought to arise from squamous epithelium from the ectocervix.

Dysplastic lesions are currently classified by consensus cytologic and histomorphological criteria based on observable cellular changes. The classification describes the stage of lesion

Human Papillomavirus. DOI: https://doi.org/10.1016/B978-0-12-814457-2.00017-9

development on the biological continuum through which a hrHPV infection develops into cervical cancer according to the original concept: hrHPV persistence, hrHPV-mediated epithelial transformation, development of precancerous lesions, and finally invasive cervical cancer. For histology, two schemes are used to describe severity of a precancerous lesion: cervical intraepithelial neoplasia (CIN) grades 1–3, and squamous intraepithelial lesion (SIL), low-grade or high-grade. Lesions that have similar histologic features as CIN 1 are called low-grade SIL (LSIL), and lesions with similar histologic features as CIN2 and CIN3 are called high-grade SIL (HSIL), which was done since histologic distinction between CIN2 and CIN3 is poorly reproducible and all CIN2 + lesions are treated. However, neither of these schemes reflects the nature and extent of transformation at the cellular level and the schemes do not provide for a distinction between productive and transforming lesions. The limitations of these classifications and their limited reliability suggest that histology as a sole diagnostic method does not provide adequate information to categorize lesions according to similar biologic behavior. Therefore, molecular markers which would make diagnoses more accurate, reflect cellular transformation, and are more reproducible among pathologists would have significant impact on management of women with CIN/SIL, in selection for surgical excision or as targets for new specific therapies (Fig. 17.1).

Ideally these markers would identify hrHPV infections at an early stage which will persist and eventually cause transforming lesions if not treated. Overexpression of HPV E6 and E7, seen in transforming infections, leads to chromosomal instability and increased susceptibility to accumulation of alterations in cancer genes of the host cell increasing cancer risk. However direct detection of HPV E6

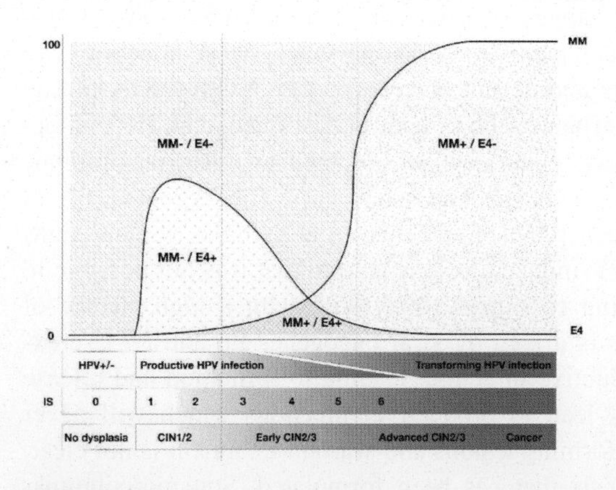
Molecular progression of –IN: loss of E4 and relation to other molecular markers
100
MM
MM- / E4-
MM+ / E4-
MM- / E4+
MM+ / E4+
0
E4
HPV+/-
Productive HPV infection
Transforming HPV infection
IS
0
1
2
3
4
5
6
No dysplasia
CIN1/2
Early CIN2/3
Advanced CIN2/3
Cancer
Van Zummeren et al, Modern Pathol 2018 (N=93 CIN lesions):
• E4 expression was inversely related to an increasing hypermethylation.
• Extensive E4 expression was mostly associated with a negative methylation marker status.
• There is a gradual transition of productive high-grade CIN (reflected by extensive p16 and E4 expression), to advanced transforming CIN (reflected by negative E4 and extensive hypermethylation of CADM1, MAL and miR124-2) and cancer.

FIGURE 17.1

Illustrates the inverse relation between methylation markers CADM1, MAL, and miR124-4 and HPV E4 on biopsy. In addition, the immunoscore (IS) which is a composite of p16 score (0–3) and Ki-67 (0–3) is shown. It demonstrates that E4 expression, identifying productive infection, is inversely related to increasing hypermethylation levels as seen in advanced transforming infections. The IS increases together with hypermethylation. Figure by van Zummeren et al. [13].

and E7 overexpression remains difficult, and other prognostic biomarkers to distinguish between cells undergoing malignant transformation by hrHPV and cells in which HPV will remain present extrachromosomally and will not cause dysplasia have been investigated for their value in practice.

Molecular markers offer a number of opportunities to improve the screening process for cervical cancer, and the development of screening for anal and oropharyngeal cancer in high-risk populations. These include the possibility of more specific diagnostic markers for use in triage of HPV-positive women or those with borderline or low-grade (ASC-US/LSIL) cytology; the possibility of distinguishing between productive and transforming lesions more specifically and reliably than morphology, and also identifying how far transforming lesions have progressed, thus avoiding the overtreatment implicit in most screening programs, particularly those where all lesions classified as HSIL are treated surgically, as these include a substantial proportion of lesions that would spontaneously regress. A wide range of biomarkers have been studied over the last 20 years but few have shown clear continuing, established clinico-pathological use. Wentzenen and von Knebel Doeberitz in 2007 reviewed a large number of biomarkers that had been explored and not shown clear clinical use, and acknowledged the importance of designing suitable studies to identify and compare different biomarkers. The present discussion of biomarkers begins with the most established biomarker: immunocytochemical staining for the cellular protein p16 as a convenient and reliable surrogate marker for hrHPV E7 gene expression that has been used in both histology and cytology. It is the only biomarker included in the United States' most recent guidelines (The Lower Anogenital Squamous Terminology, LAST) to support the diagnosis of HSIL/CIN2. Other biomarkers which are being investigated are then discussed and are either focussed around the detection of viral genes involved in productive infection, or the identification of viral or cellular changes required for the progression of transforming infection to cancer.

BIOMARKERS FOR TRANSFORMING HUMAN PAPILLOMAVIRUS -INFECTIONS: P16^{INK4A} AND KI-67

p16 is a cyclin-dependent kinase inhibitor and an integral component of normal cell cycle control. It inhibits phosphorylation of the cyclin D–dependent kinases 4 and 6 complex, which hyperphosphorylates the pRB gene product, resulting in the inactivation of pRB and the release of bound E2F transcription factors. Increased free levels of E2F lead to activation of S-phase progression genes which pushes the cell into the S-phase of the cell cycle. Increased levels of p16 result in cell cycle arrest and induction of senescence which protects the cell against genomic damage as a result of oncogenic stress such as oncogene activation or aging of the cell. HPV E6 and E7 proteins are the universal oncogenic drivers of HPV-associated cancers and persistent E6/E7 expression is necessary for maintenance of the transformed state. HPV E6 and E7 function by associating with cellular proteins and reprogramming cellular signal transduction pathways. E7 inactivates the essential pRB-mediated control functions of the cell cycle, leading to continued proliferative activity despite high levels of p16 expression. This makes p16 a highly valuable biomarker for clinical detection of high-grade precursor lesions, used as a surrogate marker for the overexpression of hrHPV E7. The importance of p16 expression also creates a potential target for therapy [1].

The finding that p16 was expressed immunocytochemically in cervical biopsy samples and smears in 2001 [2] was followed rapidly by the demonstration that p16 immunohistochemistry improves interobserver agreement compared to conventional hematoxylin and eosin-stained specimens and was restricted to cervical cancer, CIN2/3 or CIN1 associated with hrHPV [3,4]. This has been followed by numerous studies demonstrating the reproducibility of p16 and other immunocytochemical reading by microscopy and the potential value of p16 in identifying transforming HPV driven lesions in anal, penile, and oropharyngeal neoplasia, as discussed in other chapters of this book as well as in cervical histopathology and cytology [5–7].

p16 has become an important tool in diagnosing histological HSIL, which has shown a strong correlation with diffuse strong p16-positive staining in the lower third of the epithelium or more ([5,6,8]. Even though scoring p16 staining has an excellent interobserver reproducibility, there is currently insufficient evidence to support or discourage its use as a standalone marker to prospectively determine high-grade versus low-grade disease. In particular the outcome in terms of progression and regression of hrHPV-positive lesions showing CIN1 morphology and diffuse expression of p16 in the lower third of the epithelium is unclear [9,10]. The current place of p16 in diagnostic medicine was described in the LAST recommendations, which recommends its use when the H&E morphologic differential diagnosis is between unspecified intraepithelial neoplasia (-IN 2 or -IN 3) and a mimic of precancer. Strong and diffuse block-positive p16 results support a categorization of precancer (HSIL), but any identified p16 positive area must meet H&E morphologic criteria for a high-grade lesion to be reinterpreted as such. At the same time, recommendations against the use of p16 IHC as a routine adjunct to histologic assessment of biopsy specimens with morphologic interpretations of negative, -IN1 or -IN3, were made (Darragh et al., 2013). The natural history of p16-positive -IN1 and p16-negative -IN3 are uncertain, and so the use of p16 to upgrade an unequivocal -IN1 or downgrade an unequivocal -IN3 is not recommended. In addition without interpretation of morphology, p16 alone cannot differentiate between CIN2 and CIN3 and cannot separate precisely transforming from productive infections, which leads to the lack of important information that holds implications for progression risk. p16 is therefore an important marker in daily practice for diagnosing HSIL, but there is still a need for objective, reproducible markers that show distinguishing expression patterns in productive lesions with a higher chance of spontaneous regression and transforming lesions which might progress to cancer and should be treated as discussed in relation to HPV E4 and L1 expression and patterns of viral and human genomic methylation. When found, the possibilities of a scoring system that is solely based on IHC marker scoring to provide an alternative for conventional histological assessment of H&E morphology could be explored. Scoring of immunomarkers has been found to have a better inter and intraobserver agreement [3,4,11] and might be of great value in standardizing clinical practice between centers and allowing better comparison of different approaches to diagnosis, treatment, and follow-up in clinical research studies and clinical trials. Previous work on a scoring system based on the combination of p16 and Ki-67 IHC on cervical biopsies showed improved agreement between pathologists and reliable identification of CIN3 in a cross-sectional setting [12–14].

In cytology, p16 is used in combination with proliferation marker Ki-67 to identify women at risk of CIN2 + . With the architecture of tissue completely lost in cytology samples, it is difficult to distinguish metaplastic or aging cells from the upper cell layers, which can be p16 positive, from p16 positive HPV-transformed cells from the lower cell layers. Morphological features could help discriminate between the two, first screening for p16-positive cells and in a second step scoring cells on nuclear abnormalities. Wentzensen and von Knebel Doeberitz [15] categorized their observations into a binary score which could be incorporated in the cytology algorithm. The combination of p16 positivity and nuclear abnormalities in ASC-US ad LSIL cases resulted in a sensitivity approaching that of HPV testing but with a much higher specificity. However, interpretation of morphological features remains subjective.

The addition of a second immunocytochemical stain, which in normal cervical epithelial cells should never be expressed concordantly with p16, could help identify dysplastic cells in an objective way. Ki-67 is a proliferation marker which is expressed in basal cells and cells in the lower cell layers of the epithelium by cells in cycle. Metaplastic or aging p16-positive cells should be in cell cycle arrest due to activation of p16 and pRB and therefore should not also express Ki-67. Concomitant expressions of p16 and Ki-67 indicate inactivation of pRB as a result of HPV E6 and E7 overexpression and are only seen in HPV-transformed dysplastic cells. In various studies, p16/Ki-67 staining has been used as a combined stain in triage after a Pap/HPV negative co-testing screening result or after an ASC-US/LSIL cytology screening result, and the results show that this combination has a good sensitivity and specificity for CIN2 + in both settings, implying that p16/Ki-67 dual staining might gain an important role as a triage marker in HPV-based screening [5,6]. Studies that compared the use of p16/Ki-67 dual staining as a primary triage marker to cytology-based and HPV-based screening found that the dual stain was substantially more sensitive than cytology without a difference in specificity, and slightly less sensitive than HPV testing but significantly more specific for the detection of CIN2 + . These benefits compared to HPV testing were most evident in women 30 years or younger, where HPV testing has its limitations. Therefore, p16/Ki-67 dual staining could be an interesting, objective marker for primary screening of women younger and older than 30.

BIOMARKERS FOR PRODUCTIVE HUMAN PAPILLOMAVIRUS -INFECTIONS: HPV E4, L1

L1 immunohistochemistry was the first immunohistochemical biomarker for identifying productive HPV infection and showed that this was a feature of much CIN1 and of some CIN2 associated with hrHPV 16 as well as low-risk HPV infections, but was localized to only a few superficial cells [16,17]. HPV E4 is an accessory protein which is highly expressed in productive HPV infections and marks the initiation of the late stage of the virus life cycle. L1 is the major capsid protein and shows more limited expression in cells containing virus particles. The biology of HPV infection and the role of E4 have been discussed in detail in Chapter 6, The Pathology of Cervical Precancer and Cancer and its Importance in Clinical Practice.

HPV is thought usually to infect the basal layer through a micro laceration: it first maintains itself episomally at a low copy number. As epithelial cells differentiate, mature, and migrate toward the surface of the epithelium, high-risk-HPV expresses E6 and E7 proteins in order to stimulate cell proliferation, and to ensure that differentiating cells are retained in the cell cycle. Overexpression of E6 and E7 stimulates the synthesis of cellular proteins that are necessary for S-phase entry, allowing the replication of viral episomes and initiation of the productive viral phase. In LSIL/-IN 1 cases, overexpression of E6 and E7 is restricted to the lower layers of the epithelium, allowing completion of a productive virus life cycle in the upper epithelial layers. In the superficial layers of the epithelium, the hrHPV infected cells elevate viral replication by an increase in late promotor activity and accessory proteins (E1, E2, E4), after which the infectious virions are assembled and shed. Late proteins (L1 and L2, but also E4 which is more abundant) of the virus can only be expressed in terminally differentiated squamous epithelial cells which are capable of replicating the HPV-particles. Therefore, the detection of the late gene products E4 and L1 has been suggested as a marker for continuing productive hrHPV-infections even in the presence of p16 expression [18] (Fig. 17.2).

When E6 and E7 overexpression is not restricted to the lower parts of the epithelium, but extends into the upper two-third of the epithelium or even in the full thickness, as is seen in persisting HPV infections resulting in HSIL/-IN2 + lesions, this transforms the phenotype of the cell which then no longer allows for initiation of the productive phase of the HPV life cycle. p16 is a protein which is activated in cells overexpressing viral oncogenes E6 and E7 and is therefore useful as a marker of transforming infection when diffusely present. Although most HSIL expresses

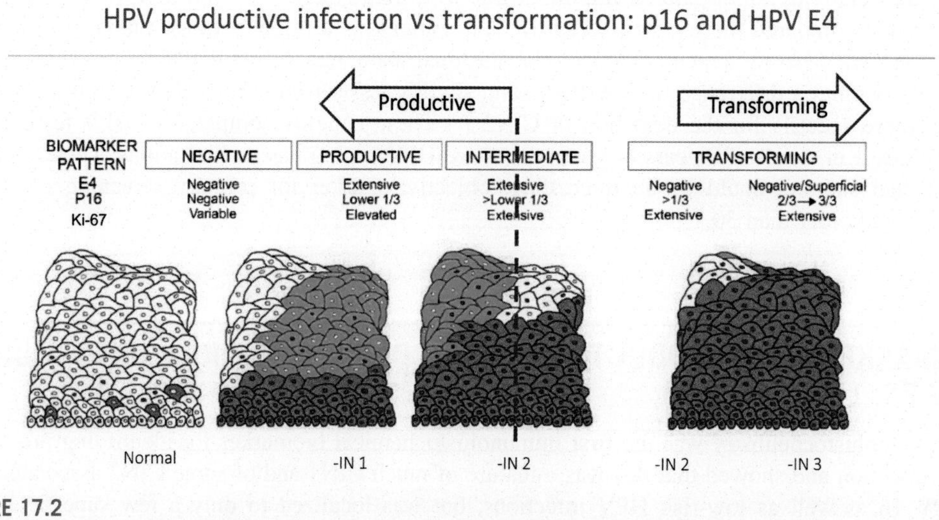

FIGURE 17.2

Biomarker expression of markers HPV E4, p16, and Ki-67 in different grades of cervical and anal intraepithelial neoplasia (-IN). HPV E4 is available is an immunohistochemical marker that stains the matured epithelial cells that express E4 in productive infections. When combined with widely used immunomarker p16, a surrogate marker for E7 overexpression and indication of malignant cellular transformation, and Ki-67, a marker of cell proliferation, different biomarker expression patterns can be identified. p16 expression is shown in brown, Ki-67 positive nuclei are red and E4 positive cells are displayed in green. In normal biopsies, all three markers are negative implying no virus driven changes. In productive lesions, most likely CIN1 lesions, p16 expression is restricted to the lower 1/3 of the epithelium. A pattern of increased cell proliferation is also seen in the lower 1/3 of the epithelium and in the upper layers of the epithelium there is extensive E4 expression. In intermediate, CIN2 lesions, p16 and Ki-67 increase up to 2/3 of the epithelium and E4 expression is still seen in the adjacent matured epithelial cells. Also, a group of E4 negative CIN2 lesions can be found. These lesions are more like CIN3 lesions, which show extensive transforming features such as p16 and Ki-67 staining up to full thickness of the epithelium.

diffuse p16 positivity, only a minority of HSIL/CIN2+ lesions progress to cancer and by setting the treatment threshold at HSIL/CIN2, most likely many self-limiting productive infections are overtreated [19]. By using a set of complementary markers which can both identify the transforming aspect of the lesion (p16), but can also show viral production in the upper layers of the epithelium (HPV E4), it might be possible to distinguish persisting transforming infections with a chance of progression to cancer from productive HPV infections in which spontaneous regression is more likely.

Multiple studies have shown that by using p16 and HPV E4 as complementary markers in diagnosing cervical or anal intraepithelial neoplasia, a number of biomarker expression patterns can be identified. Incipient, non-HPV related lesions are p16 and E4 negative while about half of all -IN1 and -IN2 lesions are E4 positive with p16 positivity increasing with the severity of the lesion [11,18,20]. Over 90% of all -IN3 lesions show diffuse p16 positivity in at least 2/3 of the epithelium and are E4 negative. The use of p16 and E4 might help further subdivide the ambiguous category of HSIL especially those with -IN2 histology. The potential for this approach is supported by the demonstration that the absence of E4 in cervical HSIL was associated with increasing methylation of somatic tumor suppressor genes associated with longer duration of CIN3 [12–14] (Figs. 17.3–17.5).

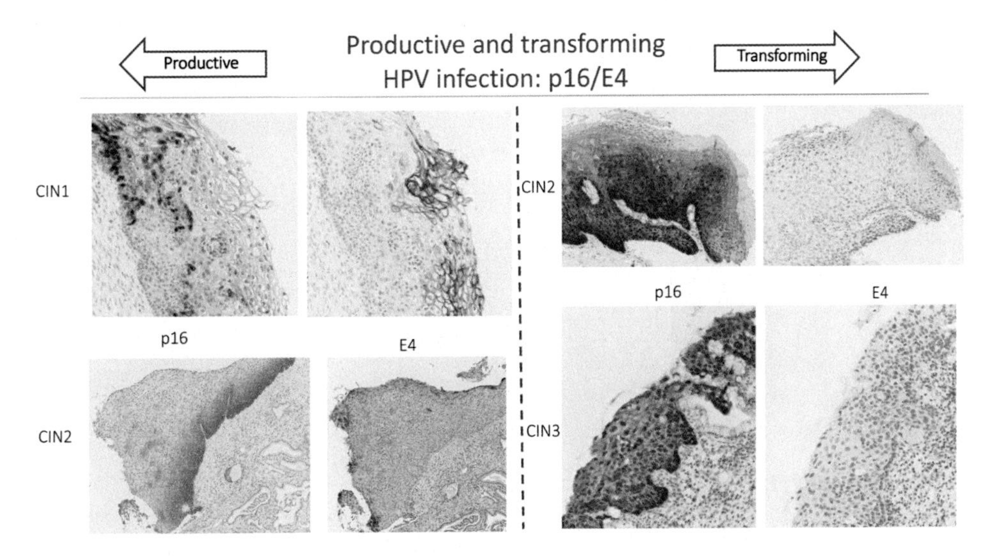

FIGURE 17.3

Two examples of p16 and E4 staining that show resemblance in E4 expression; the upper lesion is an -IN1 lesion that shows extensive E4 expression in the upper layers of the epithelium. The p16 staining shows patchy staining. The other lesion is an -IN2 lesion that also shows extensive E4 expression in the upper cell layers, but shows diffuse block staining in the lower 1/3 of the epithelium. Both lesions are productive lesions as reflected by the extensive expression of E4. The other two examples show transforming lesions as reflected by the extensive block staining of p16 and the absence of E4 expression: one -IN2 lesion with extensive p16 expression in at least the lower 2/3 of the epithelium, but no E4 expression. The -IN3 lesion shows full thickness p16 expression and again no E4 expression.

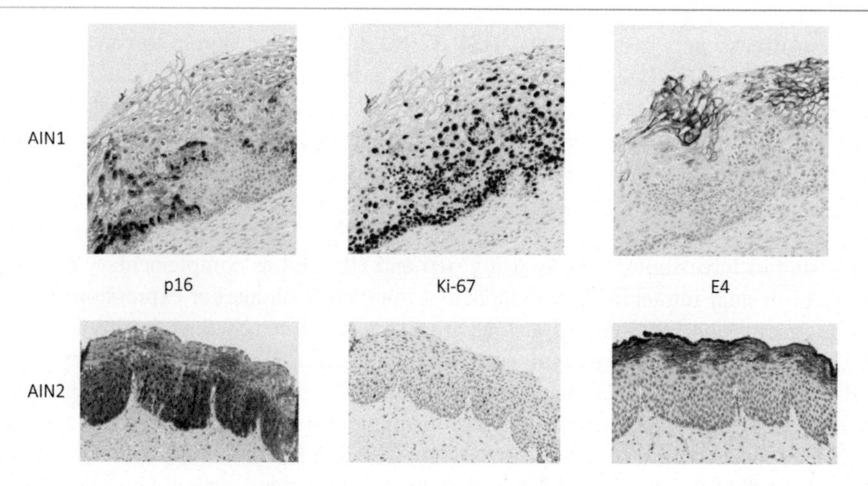

FIGURE 17.4

Two examples of p16, Ki-67, and E4 staining that show resemblance in E4 expression; the upper lesion is an AIN1 lesion that shows extensive E4 expression in the upper layers of the epithelium. The p16 staining shows patchy staining and Ki-67 staining shows proliferating cells in the lower 1/3 of the epithelium. The other lesion is an AIN2 lesion that also shows extensive E4 expression in the upper cell layers, but shows diffuse block staining in the lower 2/3 of the epithelium and also the Ki-67 shows more positive, proliferating cells in the higher layers of the epithelium.

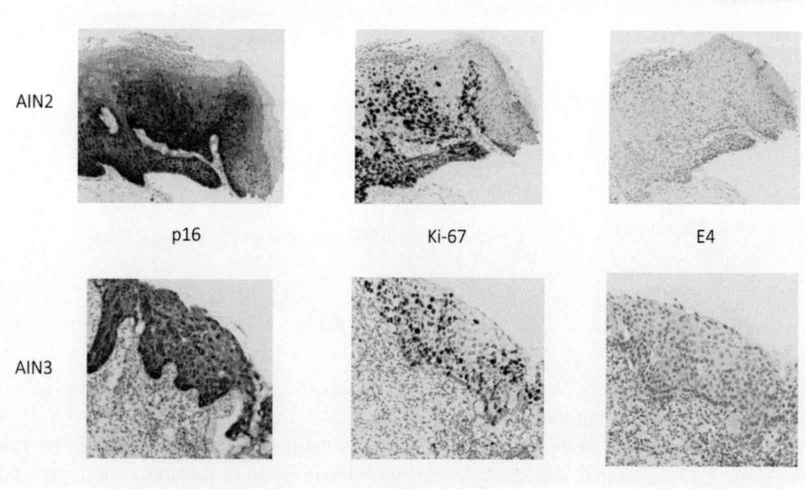

FIGURE 17.5

An example of a transforming AIN2 lesion. p16 and Ki-67 show positivity in the lower 2/3 of the epithelium, but this lesion is negative for E4 and so there is no elevated viral replication and no productive infection.
This AIN2 lesion shows more resemblance to AIN3 lesion: extensive, full thickness p16 and Ki-67 staining and without any E4 expression.

BIOMARKERS FOR TRANSFORMING HUMAN PAPILLOMAVIRUS -INFECTIONS: METHYLATION OF CELLULAR GENES

Where E6 and E7 initially endorse replication of viral episomes and initiation of the productive viral phase, long duration of their overexpression can lead to oncogenic events that drive the progression from transforming precancer to cancer through human chromosome instability [21]. This instability allows for accumulation of aberrations which can result in loss of function of human tumor suppressor genes or activation of human oncogenes [22]. Human DNA methylation is among these aberrations and involves the addition of a methyl group (CH_3) on specific cytosine nucleotides located 5′ of a guanine to generate a 5-methylcytosine. These so-called CG dyads are connected by a phosphodiester bond (p), forming a CpG site. Methylation of CpG sites often occurs in CpG islands which are regions with multiple CpG's allocated in the gene body close to the gene promotor. The DNA methyltransferases responsible for CpG methylation can be activated by hrHPV E6 via p53 and directly by E7 [23]. The addition of multiple methyl groups on a given CpG island will make the DNA sequence less accessible for proteins or even entirely transcriptionally inert, leading to temporary or permanent changes in expression of genes and in addition, DNA instability.

Over the last years, many studies have focussed on DNA methylation of cellular genes as a molecular marker of cervical cancer and precancer. These mostly involved methylation of human tumor suppressor genes which were identified as markers of other human cancers. Hypermethylation patterns of these genes were studied in cervical cancer cell lines, cervical tissue samples, and cervical scrapes. To date, over 100 human genes have been proposed as possible cervical cancer markers and the list is getting longer. Markers are often presented as a panel of multiple tumor suppressor genes. This clustering is done in the search for a highly sensitive and specific marker for disease detection, aiming to identify all cervical cancer and all precancer that the screenings programs target for (CIN2 + /HSIL).

The most frequently used technique to detect hypermethylation is through bisulphite conversion of DNA during which unmethylated cytosines only are conversed to uracil, followed by quantitative methylation-specific polymerase chain reaction. Hypermethylation of multiple genes can be detected in one test using a multiplex PCR, and such a test can be performed on tissue samples as well as liquid-based samples such as physician-taken cervical scrapes and self-collected cervicovaginal samples. Several marker combinations have been studied in triage populations, mostly consisting of hrHPV-positive women, extensively: JAM3/EPB41L3/TERT/C13ORF18, JAM3/C13ORF18/ANKRD18CP [24–27] and several CADM1, MAL, and miR124-2 and FAM19A4 combinations [28–37]. Asian studies more often focussed on SOX1 and PAX1 in several combinations [38,39], and single markers among which POU4F3 have shown promising results in smaller studies but have not yet been confirmed by large prospective follow-up studies [40–42].

The combination of JAM3/C13ORF18/ANKRD18CP showed a sensitivity of 74% for ≥CIN2 in a population of cytology screen-positive women, which was comparable to the performance of hrHPV detection as a triage test (79%), with a higher specificity (76% vs 42%) [25]. The combination of JAM3/EPB41L3/TERT/C13ORF18 has shown to have a detection rate of ≥CIN3 of 82% in a hrHPV-positive screening population, with performance comparable to that of cytology triage [26]. This combination of markers has also been tested on self-sampled cervicovaginal cells, showing good correlation to its performance on physician-taken samples [24]. The CADM1/MAL marker panel showed a sensitivity of 70% and a specificity of 71% for ≥CIN3 in a population of hrHPV-positive women [43], and the combination of MAL/miR124-2 showed a sensitivity of 70% and a specificity of 76% for ≥CIN3 when performed on brush-based self-samples [30]. This same combination was found to be at least as sensitive as cytology for the detection of ≥CIN2 in a population of hrHPV-positive former non-responders [37]. Finally, tested on self-samples, the combination of FAM19A4/miR124-2 was found to have a sensitivity of 69.4%–70.5% and a specificity of 67.8%–76.4% for the detection of ≥CIN3 in a hrHPV population. In combination with HPV16/18 genotyping, FAM19A4/miR124-2 performed well, with a higher sensitivity of 84.7% and a decreased specificity of 54.9% for the detection of ≥CIN3 [30]. To improve the performance of methylation marker panels, combinations with not only hrHPV genotyping but also detection of viral hypermethylation show potential, with studies exploring the combination of EPB41L3 with L1 and L2 regions of HPV16/18/31/33 as an example [44].

Hypermethylation levels and performance of most of the identified genes vary between studies which may not only have to do with storage and test characteristics, but may also represent differences between populations and heterogeneity of cervical cancer and precancer development or other unknown factors. New techniques and platforms for discovery of new marker genes are arising, allowing for agnostic profiling studies that specifically focus on cervical carcinogenesis. Genome-wide methylation profiling using next-generation sequencing of methyl-binding domain-enriched DNA studies are now being conducted to identify novel markers which can be used for screening or triage purposes [45].

Another use of methylation markers may be found in the identification of precursor lesions with a higher risk of progression to cancer. Increasing levels of hypermethylation have not only been correlated to increasing grade of cervical intraepithelial neoplasia, but also to duration of hrHPV infection (>5 years) [46]. Early productive viral infections with CIN1 or CIN2 lesions, as a result, have a low risk of progression and can be distinguished from more advanced transforming CIN (CIN2-3/HSIL) by methylation levels [47]. Advanced transforming CIN has a higher chance of progression to cancer and detection of increased methylation of tumor suppressor genes could be used as an indication for treatment while absence of hypermethylation could allow for close follow-up, preventing unnecessary treatment of women of reproductive age. In women who underwent treatment of CIN2+/HSIL by loop electrosurgical excision procedure, methylation testing could be performed on a cervical scrape as a test-of-cure [48]. Preceding hrHPV infections result in higher methylation levels, allowing for the detection of residual disease with a higher chance of progression to cancer.

BIOMARKERS FOR TRANSFORMING HPV-INFECTIONS: METHYLATION OF VIRAL GENES

Differential changes in carcinogenic HPV DNA methylation in various grades of cervical lesions were first reported for HPV16 in 2003 [49]. In HPV16 and other hrHPVs CpGs are relatively sparse and generally do form any areas that fit the strict definition of CpG islands. However carcinogenic HPV types do have extensive regions containing differentially methylated CpG sites [50]. Changes in methylation of CpG sites can be found throughout the HPV genome but the regions with the largest differences are the LI, L2, and E2 open reading frames. While some CpG sites in these regions are not methylated at all, there are others that show progressively increased methylation changes that are highly significant with large effect-size associations to cervical carcinogenesis [51,52]. Similarly to methylation patterns seen more recently in other genital hrHPV types HPV16 genomes in CIN3 and cancer have quite high levels of methylation of the viral late regions but relatively low levels in the upstream regulatory and early region [50]. Expanding the HPV16 methylation classifier to a small panel of additional hrHPVs (HPV18, HPV31, and HPV33) has proven to be diagnostically worthwhile [53]. Accurate measurement of individual CpG DNA methylation on HPV genomes appears to be useful in both epidemiological studies and in the triage or follow-up HPV infected women. Several papers have shown that CpG methylation in HPV16 increases with time of persistence at a rate of almost 1% per year and this effect can be used to predict increasing risk of cervical cancer [44,50,54]. It is possible that this feature may be used in the context of deep sequencing to track infections in individual patients for determining if a given HPV infection is old or new. DNA methylation may potentially be useful in populations over time to look at patterns of transmission although we do not yet know if HPV methylation is reset to low levels with every transmission event. An interesting question is to see if some patients have a faster methylation rate than others and if so whether this is linked to a higher risk of cancer. A recent study has shown that measurement of HPV DNA methylation can be useful for predicting which CIN2 progress versus regress [55]. HPV DNA methylation also appears useful for detecting anal and oropharyngeal cancers from exfoliated cell specimens [56,57].

REFERENCES

[1] Munger K, Gwin TK, McLaughlin-Drubin. P16 in HPV-associated cancer. Oncotarget 2013;4:1864–5.

[2] Klaes R, Friedrich T, Spitovsky D, et al. Overexpression ofp16(INK4a) as a specific marker for dysplastic and neoplastic epithelial cells of the cervix uteri. Int J Cancer 2001;92:276–84.

[3] Klaes R, Benner A, Friedrich T, Ridder R, Herrington S, Jenkins D, et al. p16INK4a immunohistochemistry improves interobserver agreement in the diagnosis of cervical intraepithelial neoplasia. Am J Surg Pathol 2002;26:1389–99.

[4] Klaes R, Benner A, Friedrich T, Ridder R, Herrington S, Jenkins D, et al. p16INK4a immunohistochemistry improves interobserver agreement in the diagnosis of cervical intraepithelial neoplasia. Am J Surg Pathol 2002;26:1389–99.

[5] Bergeron C, Ronco G, Reusenbach M. The clinical impact of using p16ink4a immunohistochemistry in cervical histopathology and cytology: an update of recent developments. Int J Cancer 2015;136:2741–51.

[6] Bergeron C, et al. Prospective evaluation of p16/Ki67 dual-stained cytology for manageing women with abnormal Papanicolaou cytology: PALMS study results. Cancer Cytopathol 2015;123:373–81.

[7] Bergeron C, Ordi J, Schmidt D, et al. Conjunctive p16INK4a testing significantly increases accuracy in diagnosing high-grade cervical intraepithelial neoplasia. Am J Clin Pathol 2010;133:395–406.

[8] Darragh Tm, Colgan TJ, Cox JT, Heller DS, Henry MR et al The Lower Anogenital Squamous Terminology Stanadardisation project for HPV-associated lesions. Arch pathol Lab Med 2012; 136: 1266–1297.

[9] del Pino M, Garcia S, Fuste V, Alonso I, Fuste P, Torne A, et al. Value ofp16(INK4a) as a marker of progression/regression in cervical intraepithelial neoplasia grade 1. Am J Obstet Gynecol 2009;201 488e1-7.

[10] Sagasta A, Castillo P, Saco A, Torne A, Esteve R, Marimon L, et al. p16 staining has limited value in predicting the outcome of histological low-grade squamous intraepithelial lesions of the cervix. Mod Pathol 2016;29:51–9.

[11] van Baars R, Griffin H, Wu Z, Soneji YJ, van de Sandt MM, Arora R, et al. Investigating diagnostic problems of CIN1 and CIN2 associated with high-risk HPV by combining the novel molecular biomarker PanHPVE4 with P16INK4a. Am J Surg Pathol 2015;39:1518–28.

[12] Van Zummeren M, Leeman A, Kremer W, et al. A three-tiered score for Ki67 and p16INK4a improves accuracy and reproducibility of grading CIN lesions. J Clin Pathol 2018;205271.

[13] Van Zummeren M, Kremer W, Leeman A, et al. HPV-E4 expression and DNA hypermethylation of CADM1, MAL and MiR124-2 genes in cervical cancer and precursor lesions. Mod Pathol 2018.

[14] van Zummeren M, Leeman A, Kremer WW, Bleeker MCG, Jenkins D, van de Sandt M, et al. Three-tiered score for Ki-67 andp16(ink4a) improves accuracy and reproducibility of grading CIN lesions. J Clin Pathol 2018;71:981–8.

[15] Wentzensen N, von Knebel Doeberitz M. Biomarkers in cervical cancer screening. Dis Markers 2007;23:315–30.

[16] Kurman RJ, Jenson VB, Lancaster WD. Papillomavirus infection of the cervix II. Relationship to intraepithelial neoplasia based on the presence of specific viral structural proteins. Am J Surg Pathol 1983;7:39–52.

[17] Jenkins D, Tay SK, McCance DJ, et al. Histological and immunocytochemical study of CIN associated with HPV 6 and HPV 16 infections. J Clin Pathol 1986;39:1177–80.

[18] Griffin H, Soneji Y, Van Baars R, Arora R, Jenkins D, van de Sandt M, et al. Stratification of HPV-induced cervical pathology using the virally encoded molecular marker E4 in combination with p16 or MCM. Mod Pathol 2015;28:977–93.

[19] Moscicki A-B, Schiffman M, Burchell A, et al. Updating the natural history of human papillomavirus and anogenital cancer. Vaccine 2012;30S:F24–33.

[20] Leeman A, Pirog EC, Doorbar J, van de Sandt MM, van Kemenade FJ, Jenkins D, et al. Presence or absence of significant HPVE4 expression in high-grade anal intraepithelial neoplasia with p16/Ki-67 positivity indicates distinct patterns of neoplasia: a study combining immunohistochemistry and laser capture microdissection PCR. Am J Surg Pathol 2018;42:463–71.

[21] Wilting SM, Steenbergen RDM. Molecular events leading to HPV-induced high grade neoplasia. Papillomavirus Res 2016;2:85–8.

[22] Steenbergen RD, Snijders PJ, Heideman DA, Meijer CJ. Clinical implications of (epi)genetic changes in HPV-induced cervical precancerous lesions. Nat Rev Cancer 2014;14:395–405.

[23] Au Yeung CL, Tsang WP, Tsang TY, Co NN, Yau PL, Kwok TT. HPV-16 E6 upregulation of DNMT1 through repression of tumor suppressor p53. Oncol Rep 2010;24:1599–604.

[24] Boers A, Bosgraaf RP, van Leeuwen RW, Schuuring E, Heideman DA, Massuger LF, et al. DNA methylation analysis in self-sampled brush material as a triage test in hrHPV-positive women. Br J Cancer 2014;111:1095–101.

[25] Boers A, Wang R, van Leeuwen RW, Klip HG, de Bock GH, Hollema H, et al. Discovery of new methylation markers to improve screening for cervical intraepithelial neoplasia grade 2/3. Clin Epigenetics 2016;8:29.

[26] Eijsink JJ, Lendvai A, Deregowski V, Klip HG, Verpooten G, Dehaspe L, et al. 'A four-gene methylation marker panel as triage test in high-risk human papillomavirus positive patients'. Int J Cancer 2012;130:1861–9.

[27] van Leeuwen RW, Ostrbenk A, Poljak M, van der Zee AGJ, Schuuring E, Wisman GBA. DNA methylation markers as a triage test for identification of cervical lesions in a high risk human papillomavirus positive screening cohort. Int J Cancer 2019;144:746–54.

[28] Bu Q, Wang S, Ma J, Zhou X, Hu G, Deng H, et al. The clinical significance of FAM19A4 methylation in high-risk HPV-positive cervical samples for the detection of cervical (pre)cancer in Chinese women. BMC Cancer 2018;18:1182.

[29] De Strooper LMA, Berkhof J, Steenbergen RDM, Lissenberg-Witte BI, Snijders PJF, Meijer C, et al. Cervical cancer risk in HPV-positive women after a negative FAM19A4/mir124-2 methylation test: a post hoc analysis in the POBASCAM trial with 14 year follow-up. Int J Cancer 2018;143:1541–8.
[30] De Strooper LMA, Verhoef VMJ, Berkhof J, Hesselink AT, de Bruin HME, van Kemenade FJ, et al. Validation of the FAM19A4/mir124-2 DNA methylation test for both lavage- and brush-based self-samples to detect cervical (pre)cancer in HPV-positive women. Gynecol Oncol 2016;141:341–7.
[31] De Strooper LM, Meijer CJ, Berkhof J, Hesselink AT, Snijders PJ, Steenbergen RD, et al. Methylation analysis of the FAM19A4 gene in cervical scrapes is highly efficient in detecting cervical carcinomas and advanced CIN2/3 lesions. Cancer Prev Res (Phila) 2014;7:1251–7.
[32] De Strooper LM, van Zummeren M, Steenbergen RD, Bleeker MC, Hesselink AT, Wisman GB, et al. CADM1, MAL and miR124-2 methylation analysis in cervical scrapes to detect cervical and endometrial cancer. J Clin Pathol 2014;67:1067–71.
[33] Floore A, Hesselink A, Ostrbenk A, Alcaniz E, Rothe B, Pedersen H, et al. Intra- and inter-laboratory agreement of the FAM19A4/mir124-2 methylation test: results from an international study. J Clin Lab Anal 2019;4:e22854.
[34] Luttmer R, De Strooper LM, Berkhof J, Snijders PJ, Dijkstra MG, Uijterwaal MH, et al. Comparing the performance of FAM19A4 methylation analysis, cytology and HPV16/18 genotyping for the detection of cervical (pre)cancer in high-risk HPV-positive women of a gynecologic outpatient population (COMETH study). Int J Cancer 2016;138:992–1002.
[35] Luttmer R, De Strooper LM, Dijkstra MG, Berkhof J, Snijders PJ, Steenbergen RD, et al. FAM19A4 methylation analysis in self-samples compared with cervical scrapes for detecting cervical (pre)cancer in HPV-positive women. Br J Cancer 2016;115:579–87.
[36] van Baars R, van der Marel J, Snijders PJ, Rodriquez-Manfredi A, ter Harmsel B, van den Munckhof HA, et al. CADM1 and MAL methylation status in cervical scrapes is representative of the most severe underlying lesion in women with multiple cervical biopsies. Int J Cancer 2016;138:463–71.
[37] Verhoef VM, Bosgraaf RP, van Kemenade FJ, Rozendaal L, Heideman DA, Hesselink AT, et al. Triage by methylation-marker testing versus cytology in women who test HPV-positive on self-collected cervicovaginal specimens (PROHTECT-3): a randomised controlled non-inferiority trial. Lancet Oncol 2014;15:315–22.
[38] Chen Y, Cui Z, Xiao Z, Hu M, Jiang C, Lin Y, et al. PAX1 and SOX1 methylation as an initial screening method for cervical cancer: a meta-analysis of individual studies in Asians. Ann Transl Med 2016;4:365.
[39] Lai HC, Ou YC, Chen TC, Huang HJ, Cheng YM, Chen CH, et al. PAX1/SOX1 DNA methylation and cervical neoplasia detection: a Taiwanese Gynecologic Oncology Group (TGOG) study. Cancer Med 2014;3:1062–74.
[40] Chang CC, Ou YC, Wang KL, Chang TC, Cheng YM, Chen CH, et al. Triage of atypical glandular cell by SOX1 and POU4F3 methylation: a Taiwanese Gynecologic Oncology Group (TGOG) Study. PLoS One 2015;10:e0128705.
[41] Kocsis A, Takacs T, Jeney C, Schaff Z, Koiss R, Jaray B, et al. Performance of a new HPV and biomarker assay in the management of hrHPV positive women: subanalysis of the ongoing multicenter TRACE clinical trial (n > 6,000) to evaluate POU4F3 methylation as a potential biomarker of cervical precancer and cancer. Int J Cancer 2017;140:1119–33.
[42] Pun PB, Liao YP, Su PH, Wang HC, Chen YC, Hsu YW, et al. Triage of high-risk human papillomavirus-positive women by methylated POU4F3. Clin Epigenetics 2015;7:85.
[43] Verhoef VM, van Kemenade FJ, Rozendaal L, Heideman DA, Bosgraaf RP, Hesselink AT, et al. Follow-up of high-risk HPV positive women by combined cytology and bi-marker CADM1/MAL methylation analysis on cervical scrapes. Gynecol Oncol 2015;137:55–9.

[44] Cook DA, Krajden M, Brentnall AR, Gondara L, Chan T, Law JH, et al. Evaluation of a validated methylation triage signature for human papillomavirus positive women in the HPV FOCAL cervical cancer screening trial. Int J Cancer 2019;144:2587–95.

[45] Boers R, Boers J, de Hoon B, Kockx C, Ozgur Z, Molijn A, et al. Genome-wide DNA methylation profiling using the methylation-dependent restriction enzyme LpnPI. Genome Res 2018;28:88–99.

[46] Bierkens M, Hesselink AT, Meijer CJ, Heideman DA, Wisman GB, van der Zee AG, et al. CADM1 and MAL promoter methylation levels in hrHPV-positive cervical scrapes increase proportional to degree and duration of underlying cervical disease. Int J Cancer 2013;133:1293–9.

[47] Verlaat W, Van Leeuwen RW, Novianti PW, Schuuring E, Meijer C, Van Der Zee AGJ, et al. Host-cell DNA methylation patterns during high-risk HPV-induced carcinogenesis reveal a heterogeneous nature of cervical pre-cancer. Epigenetics 2018;13:769–78.

[48] Uijterwaal MH, van Zummeren M, Kocken M, Luttmer R, Berkhof J, Witte BI, et al. Performance of CADM1/MAL-methylation analysis for monitoring of women treated for high-grade CIN. Gynecol Oncol 2016;143:135–42.

[49] Badal V, Chuang LSH, Tan EH-H, Badal S, Villa LL, Wheeler CM, et al. CpG methylation of human papillomavirus type 16 DNA in cervical cancer cell lines and in clinical specimens: genomic hypomethylation correlates with carcinogenic progression. J Virol 2003;77(11):6227–34. Available from: https://doi.org/10.1128/JVI.77.11.6227-6234 0022-538X/03/$08.000.

[50] Mirabello L, Schiffman M, Ghosh A, Rodriguez AC, Vasiljevic N, Wentzensen N, et al. Elevated methylation of HPV16 DNA is associated with the development of high grade cervical intraepithelial neoplasia. Int J Cancer 2013;132:1412–22.

[51] Louvanto K, Franco EL, Ramanakumar AV, Vasiljević N, Scibior-Bentkowska D, Koushik A, et al. Methylation of viral and host genes and severity of cervical lesions associated with human papillomavirus type 16. Int J Cancer. 2015;136(6):E638–45.

[52] Lorincz AT, Brentnall AR, Scibior-Bentkowska D, Reuter C, Banwait R, Cadman L, et al. Validation of a DNA methylation HPV triage classifier in a screening sample. Int J Cancer. 2016. Available from: https://doi.org/10.1002/ijc.30008 [Epub ahead of print] PMID: 26790008.

[53] Brentnall AR, Vasiljevic N, Scibior-Bentkowska D, Cadman L, Austin J, Cuzick J, et al. HPV33 DNA methylation measurement improves cervical pre-cancer risk estimation of an HPV16, HPV18, HPV31 and *EPB41L3* methylation classifier. Cancer Biomark 2015;15:669–75. Available from: https://doi.org/10.3233/CBM-150507 PMID:26406956.

[54] Lorincz AT, Brentnall AR, Vasiljević N, Scibior-Bentkowska D, Castanon A, Fiander A, et al. HPV16 L1 and L2 DNA methylation predicts high grade cervical intraepithelial neoplasia in women with mildly abnormal cervical cytology. Int J Cancer 2013;133:637–44.

[55] K. Louvanto, K. Aro, B. Nedjai, R. Bützow, M. Jakobsson, I. Kalliala, et al., Methylation in predicting progression of untreated high-grade cervical intraepithelial neoplasia. Clin Infect Dis. 2019 Jul 25. pii: ciz677. Available from: https://doi.org/10.1093/cid/ciz677. [Epub ahead of print].

[56] Lorincz AT, Nathan M, Reuter C, Warman R, Thaha M, Sheaff M, et al. Methylation of HPV and a tumor suppressor gene reveals anal cancer and precursor lesions. Oncotarget 2017. Available from: https://doi.org/10.18632/oncotarget.17984 May 18. [Epub ahead of print] PMID: 28591708.

[57] A.R. Giuliano, B. Nedjai, A.T. Lorincz, M.J. Schell, S. Rahman, R. Banwait, et al., Methylation Of Hpv 16 And *Epb41l3* In Oral Gargles: Associations With Oropharyngeal Cancer Detection And Tumor Characteristics. Intl J Cancer (In Press). Available from: https://doi.org/10.1002/ijc.32570.

CHAPTER

18 IMMUNE RESPONSES TO HUMAN PAPILLOMAVIRUS AND THE DEVELOPMENT OF HUMAN PAPILLOMAVIRUS VACCINES

Margaret Stanley

Department of Pathology, University of Cambridge, Tennis Court Road, Cambridge, United Kingdom

INTRODUCTION

The world, from the perspective of infectious disease, is a dangerous place in which *Homo sapiens* is continuously assaulted and challenged by the massed regiments of viruses, bacteria and eukaryotic parasites against which sophisticated, flexible and lethal host defenses (immunity) must be engaged constantly. These defenses are a partnership between innate immunity (phagocytes, innate lymphocytes, soluble proteins, e.g., cytokines, complement and epithelial barriers) together with adaptive immunity (antibody, cytotoxic effector cells). In simple terms the innate immune system detects the pathogen and acts as first line defense clearing (it is estimated) up to 90% of microbial assaults alone. Innate immunity has no specific memory but crucially activates the appropriate adaptive immune response that generates lethal effector responses of exquisite specificity for, and long lived-cells with, memory of the insult, and the capacity to respond again to it. Thus for human papillomavirus (HPV), as for most viral infections, the adaptive responses of antibody-mediated humoral immunity clear free virus particles from body fluids and can prevent reinfection by virus, those of cell-mediated immune (CMI) responses are essential for the clearance of virus-infected cells and the generation of immune memory.

INNATE IMMUNITY

Innate immunity is alerted and activated by cell injury or cell death and manifested by inflammation (the local vascular response to injury). In the inflammatory process, soluble and cellular innate immune effectors are recruited and local parenchymal cells and phagocytes (both recruited and local) are activated to secrete inflammatory cytokines and other defense molecules. Crucially dendritic cells [the only antigen presenting cells (APC) that can activate naïve T lymphocytes] after taking up antigen are activated to trek from the mucosa to the local draining lymph node and kick start the adaptive immune response making the APC a critical player in that process. The APC, in turn is activated to move, process and present antigen by signals received from the receptor ligand

Human Papillomavirus. DOI: https://doi.org/10.1016/B978-0-12-814457-2.00018-0

interactions between the APC and the pathogen and also by the cytokines released by the APC and other cells in the immediate vicinity. The APC and the local proinflammatory cytokines released are the bridge between innate and adaptive immunity.

ADAPTIVE IMMUNITY

T lymphocytes are the generals of the adaptive immune system with central roles in both cell mediated and humoral immunity (Fig. 18.1). In the T cell/APC interaction, the T cell receptor (TCR) recognizes antigen processed into short peptides bound to the major histocompatibility complex (MHC) proteins and presented as a membrane bound receptor complex on the cell surface of the APC. These responses are exquisitely specific; the TCR of the naïve T cell can only recognize and be activated by binding to a unique peptide MHC complex geometry generating T cell populations with the specificity for only that complex. There are two major subsets of T cells identified by the surface markers CD4 or CD8. CD4 + T cells recognize antigen presented by

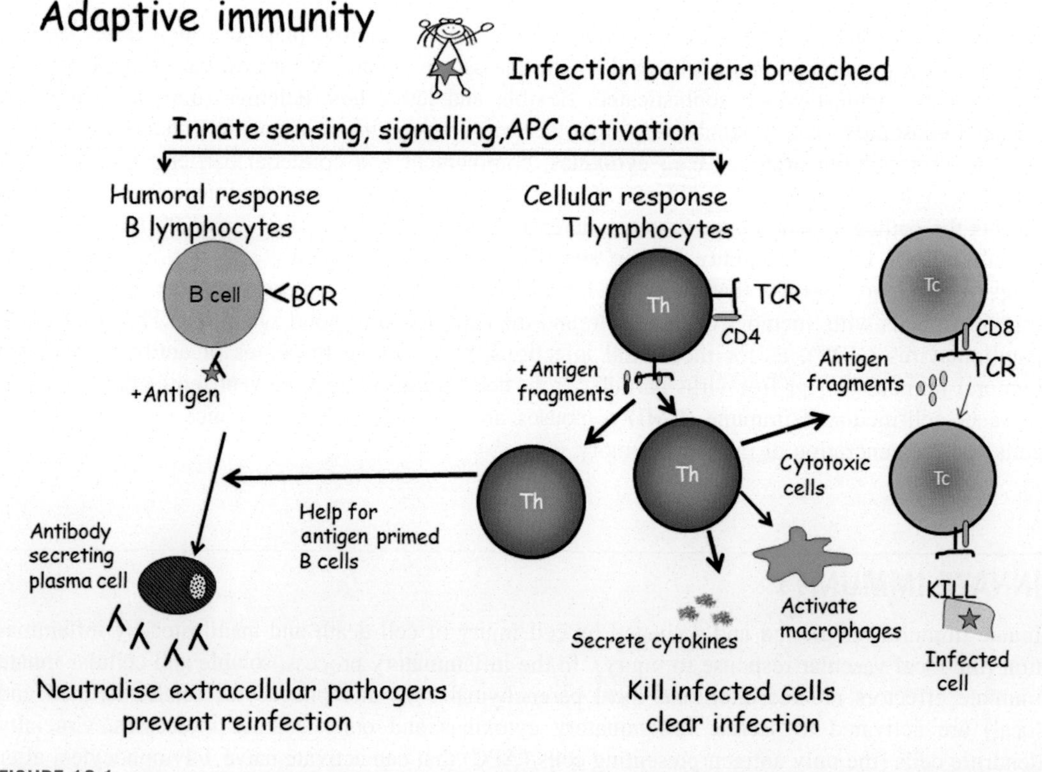

FIGURE 18.1

T lymphocytes are the generals of the adaptive immune system with central roles in both cell mediated and humoral immunity.

class II MHC; CD8 + T cells recognize antigen presented by class I MHC. CD4 + T cell activation results in the secretion of a repertoire of small proteins or cytokines that help and regulate other cells in defense responses and regulation of these responses. CD8 + T cell activation generates lethal cytotoxic cells targeted to seeking out and killing any cell expressing the unique peptide MHCI complex that initially activated the naïve T cell (Fig. 18.1).

Unlike T lymphocytes, B lymphocytes can recognize antigen in the natural conformation but only a few native antigens can directly activate B cells and generate plasma cells. In most cases antigen binding by the B cell receptor primes the B cell that then requires cognate help from the T cell in the form of receptor ligand interactions and cytokines to go through the differentiation program that results in antibody secreting plasma cells and memory B cells. T cell help is crucial for the process of class switching, that is, the generation of the different antibody classes and isotypes and the development of antigen specific memory B cells.

IMMUNITY TO HUMAN PAPILLOMAVIRUS: A SUCCESSFUL INFECTIOUS AGENT

HPVs have two central biological properties; they have a tightly restricted host range and exquisite tissue tropism—productive virus growth is confined to the fully differentiating squamous epithelia of humans. More than 200 HPV genotypes are known. They are numbered sequentially by the time of isolation and divided into five genera α, β, γ, μ, ν and, in general, have a predilection either for cutaneous or mucosal surfaces [1]. HPVs are very successful infectious agents. The majority of the β, μ, and ν HPV species reside anonymously in their epidermal niches at very low copy number resulting in no overt disease until, and unless, host immunity is compromised [2], implying immune control in these scenarios. The α-HPVs, which include most of the types that cause clinical disease, induce chronic infections that have virtually no systemic sequelae, rarely kill the host and, over weeks and months, periodically shed large amounts of infectious virus for transmission to naive individuals. To achieve these evolutionarily successful lifestyles, HPVs must avoid host defense systems and the key to understanding how this is achieved is the virus replication cycle which is in itself an immune evasion mechanism that inhibits and delays the host immune response to HPV infection. The most intensively studied over the past 3–4 decades, have been the oncogenic HR αHPVs particularly HPV 16 and unless specifically stated the following discussion reviews data on the host response to HRHPVs infecting keratinocytes at anogenital mucosae.

HUMAN PAPILLOMAVIRUS INFECTIOUS CYCLE

HPVs are exclusively intraepithelial pathogens and infection and vegetative virus growth are absolutely dependent upon the expression of the complete program of keratinocyte (KC) differentiation [3]. After epithelial micro-abrasion, virus infects primitive basal keratinocytes, probably wound keratinocytes that assume the stem cell phenotype during the wounding process. However, high-level viral gene expression, viral protein production, and virus assembly occur only in the upper differentiated layers of the stratum spinosum and granulosum of squamous epithelia.

Thus the infectious cycle of HPVs is tailored to the differentiation program of its target cell the keratinocyte and it raises several important issues with respect to immune recognition. First, infection and vegetative growth are completely dependent upon the program of keratinocyte differentiation, from basal cell to terminally differentiated superficial squames. Second, the time from infection to disease is highly variable. Clinical studies of infection with LRHPVs show that the period between infection and the appearance of lesions can range from weeks to months to years [4]. Third, there is no cytolysis or cytopathic death as a consequence of virus replication and assembly. These key events for the virus occur in the fully differentiating keratinocyte, a cell destined for apoptotic death and desquamation far from the sites of immune activity: there is no viral cytolytic death (a potent stimulator of inflammation [5] and no inflammation). Thus for most of the duration of the HPV infectious cycle, there appears to be little or no release into the local milieu of proinflammatory cytokines, important for recruitment, activation and migration of APCs and the recruitment of adaptive immune effector cells to the infected site. HPV is an exclusively intraepithelial pathogen, there is as far as is known no blood born or viremic phase of the lifecycle and only minimal amounts of virus are exposed to immune defenses. In effect the virus is practically invisible to the host defenses which remain ignorant of the presence of the pathogen for long periods of time [6].

HUMAN PAPILLOMAVIRUS EVADES INNATE IMMUNE DEFENSES

Central to this achievement of immune ignorance is the ability of HPV, particularly the high risk HPV types (HRHPVs), to compromise the role of keratinocytes as innate immune sentinels. Keratinocytes can respond to cell injury and cell stress and can sense pathogens, thus mediating immune responses [7]. Eukaryotic cells express germ line encoded receptors of the innate immune system, pathogen recognition receptors (PRRs) that recognize invariant molecular motifs known as pathogen-associated molecular patterns (PAMPs). There are four groups of PRR: (1) nucleotide-binding oligomerization domain-like receptors (NLRs), (2) C-type lectin receptors (CLRs), (3) retinoic acid inducible gene 1 (RIG I)-like receptor family (RLRs), and (4) toll-like receptors (TLRs). The four PRR families usually differ in their ligand recognition, signal transduction and subcellular localization but upon activation, they induce cellular responses that ultimately will result in the elimination of the pathogen. Ligand activated PRRs bind adaptor proteins and recruit protein kinases initiating signal transduction cascades that activate transcription factors. These translocate to the nucleus, stimulate antiviral gene transcription generating interferons (IFNs) and inflammatory cytokines. The upregulation of type I IFN expression, release of interferon and ligation of IFN receptors in turn stimulates transcription of hundreds of interferon stimulated genes (ISGs) that commit neighboring cells to an antiviral state.

TLRs are expressed by both epidermal and genital keratinocytes on the plasma membrane (TLR1, TLR2, TLR4, TLR5, and TLR6) and the endosomal membrane (TLR3, TLR9) [8]. Activation of TLRs and RLRs trigger a cascade of signals leading to the activation of nuclear factor kappa B (NF-κB) dependent expression of proinflammatory cytokines such as interleukin 6 (IL-6) and tumor necrosis factor alpha (TNF-α) and to the expression of type I interferon genes via activation of the transcription factors interferon regulatory factor 3 (IRF3) and/or IRF7 [9]. The secretion of proinflammatory cytokines, chemokines, and their receptors by keratinocytes as well as antigen-processing and -presenting molecules is central to the activation of tissue resident APC and macrophages and the recruitment of effector T cells all of which kick start adaptive immune responses to the local injury or infection. HRHPVs dampen these crucial responses almost from the start of the infectious cycle [10] Fig. 18.2.

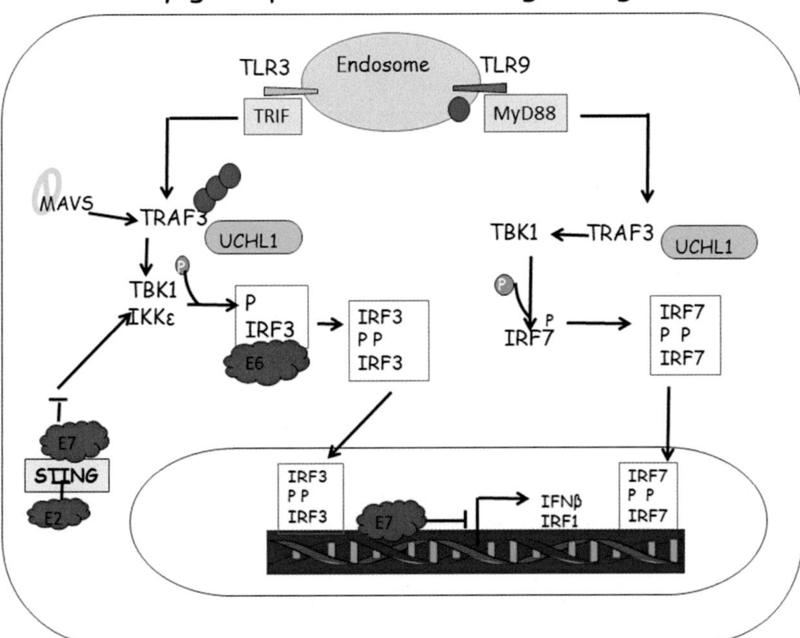

FIGURE 18.2

All TLRs except TLR3 signal via the MyD88 adaptor protein initiating a cascade of protein–protein interactions, and the end result of which is the phosphorylation and activation of IRF7. TLR3 and TLR4 signals via TRIF, the cytosolic RNA sensors signal via MAVS, and the cytosolic DNA sensors via the adaptor protein STING, and the end result of which is the phosphorylation and activation of IRF3. Activated IRF3 and IRF7 homodimerise translocate to the nucleus inducing expression of type I interferons. The HRHPV early proteins, E6, E7, and E2 directly interact with components of these signaling cascades at multiple points, inhibiting signaling via PRR ligation.

HRHPVs DOWNREGULATE INTERFERON RESPONSES

The type 1 interferons, principally IFN-α (13 subtypes) and IFN-β but also IFNs ε, τ, κ, ω, δ, and ζ are key players in antiviral defenses, inducing an antiviral state, interfering with viral transcription and activating innate and adaptive immunity [11]. Evasion of type 1 interferon response is a characteristic of many viruses and the papillomaviruses are no exception. IFNs interact with their specific cell surface receptors activating the canonical Janus Kinase (JAK)—signal transducer and activator of transcription (STAT) pathway leading to the transcription of ISGs. Inhibition of interferon stimulating genes (ISGs) is common to HPV 16, 18, and 31 [12–14]. The HRHPV early proteins, E6, E7, and E2 [15,16] directly interact with components of these signaling cascades at multiple points (Fig. 18.2) inhibiting signaling via PRR ligation, disrupting the IFN receptor-activated pathways, and the canonical NFκB pathway downstream of immune receptors, probably via the inhibition of the constitutive transcription of IFN-κ.

HUMAN PAPILLOMAVIRUS INTERACTIONS WITH DENDRITIC CELLS PROFESSIONAL ANTIGEN PRESENTING CELLS

As far as is known, HPV infections are exclusively intraepithelial so theoretically HPV antigens should be processed and presented by the professional APC of squamous epithelium the Langerhans cell (LC). Viral capsid entry is usually an activating signal for dendritic cells but the evidence is that LC are not activated but rather are tolerized by the uptake of HPV capsids [17]. In contrast stromal or dermal DCs are activated by virus-like particle (VLP) uptake and initiate HPV specific T cell responses to L1 [18]. In recurrent respiratory papillomatosis (RRP), HPV 6/11 keratinocytes secrete low levels of the proinflammatory cytokine IL-36γ compromising maturation and activation of LC in the lesions [19]. LC numbers are significantly reduced in HPV 16 cervical [20,21] and vulval [22] intraepithelial neoplastic lesions (CIN and VIN) and there are viral mediated mechanisms that could contribute to this phenomenon. HPV 16 E6 and E7 downregulate both MIP1α/CCL20 [23], a chemokine necessary for LC recruitment to the epidermis and E cadherin [20], the adhesion molecule that binds LC to the keratinocyte and is important in LC retention in the epidermis. Dermal DCs and monocyte macrophages recruited to HPV-infected epithelium in inflammatory situations may be key players in the recognition of HPV antigens and the induction of effector responses a notion supported by studies on cutaneous warts [24].

THE IMMUNE RESPONSE TO HUMAN PAPILLOMAVIRUS IN NATURAL INFECTIONS

Despite the best efforts of the virus to evade host defenses at least 80%–90% of genital HPV infections will resolve with time [25]. Anogenital warts and CIN1 lesions regress as a result of a successful CMI response directed against early viral proteins specifically E2 and E6 [26–28]. Immunohistochemical studies clearly show that regression of oral warts in natural papillomavirus infections in dogs [29,30] and anogenital and cutaneous warts in humans [31] is accompanied by a massive infiltration into the lesion of mononuclear cells (CD4+, CD8+, CD56+, and macrophages) and expression of Th1 cytokines [32]. However, despite this intense local response, systemic antigen specific T cell responses are weak and often transient [30].

The cellular effectors in these responses are still not unequivocally identified. However, regression of histologically confirmed CIN1 lesions strongly correlated with the presence of intralesional granzyme B+ve CD8+ and CD56+ cells [33]. Immunohistochemical studies have shown CD8+T cells expressing the α4/β7 integrin to be present in CIN1/koilocytic cervical lesions but absent or present in reduced numbers in CIN3 [34]. Virtually all intraepithelial cervical lymphocytes express α4/β7, the mucosal homing receptor for lymphocytes and CIN regression has been shown to correlate with the presence of intraepithelial CD8+ T cells and by the expression of the ligand for α4/β7, the mucosal addressin cell adhesion molecule 1 (MadCAM1), on the vascular endothelium in dysplastic lesions [35].

IMMUNE RESPONSES ARE DEREGULATED DURING HUMAN PAPILLOMAVIRUS-ASSOCIATED NEOPLASTIC PROGRESSION

Natural history studies suggest that about 10%–20% of individuals develop persistent cervical HPV infection; it is this group that is at high risk for progression for CIN2/3 [25]. In these persistent HPV infections, the absence of cell death means that the inflammatory signals that would activate intraepithelial APC such as LC, recruit stromal DC, and macrophages to the epithelium and plasmacytoid DC to the infected focus are absent. Furthermore, HPVs downregulate innate sensing signaling pathways in the infected keratinocyte [10], proinflammatory cytokines, particularly the type 1 interferons are not released and again the signals for LC activation and migration and the recruitment of stromal dendritic cells and macrophages are either not present or inadequate. In this scenario, there are long periods of uninterrupted virus replication in the epithelium during which the host is ignorant of virus. This is a high-risk strategy for the host when the infection is with an oncogenic genital HPV as it increases the risk of "accidents" in virus replication that result in the deregulated expression of viral E6 and E7 oncoproteins, the bypassing of cell cycle checkpoints and neoplastic transformation. With neoplastic transformation and genomic instability the expression of key cytokines, adhesion molecules, chemokines and chemokine receptors on the infected epithelium and on the underlying microvascular endothelium of the stroma are deregulated [36,37], resulting in the downregulation of key receptors essential for the ingress of antigen specific T cells and other cytotoxic effectors into the epithelium [35].

Central to this discussion of local mucosal immunity is the role of T cell memory. A key subset of memory T cells is tissue-resident memory T cells (Trm). These reside in epithelial tissues including the cervix and vagina and can respond rapidly to pathogen challenge without recruitment of T cells from the blood mediating the rapid protective response that is the hallmark of adaptive immunity. If this is inadequate for pathogen clearance, naïve and antigen experienced T cells can be recruited from the blood and the pool of Trm maintained. In murine genital herpes simplex virus infections CD4 T cells must first enter the mucosa to provide the recruiting signals for the entry of CD8 T cells to the focus [38], a scenario reminiscent of the events of regression of HPV-infected genital warts. Therapeutic HPV vaccines generate Trm in neoplastic cervical lesions and these are protective against reinfection [39]. The molecular program that facilitates the antigen-specific Trm phenotype in the tissue locales is dictated by the cytokine and chemokine milieu of the infected focus and it can be speculated that these determine effector or suppressive responses of the Trm leading to either protective immunity or persistent infection and the increased risk of neoplastic progression.

THE MUCOSAL ENVIRONMENT

The impact of the mucosal environment on HPV infection and the modulation of immune responses have received relatively little attention but potentially is of central importance in the natural history of HPV infection in the different tissue locales—anus, cervix vagina, and oropharynx. HPV is a field infection in the female genital tract but high-grade CIN arise at the squamocolumnar junction (SCJ) imply that the SCJ is vulnerable both to HPV infection and the development of effective innate responses to this. It is well established that sex hormones modulate immune responses [40] and that the SCJ and transformation zone exhibit high expression of both estrogen and progesterone receptors [41]. While estradiol most often promotes the production of IFNs and proinflammatory cytokines, innate responses can be enhanced or dampened by ER signaling; for review see [42]. Recently, it has been shown that IFN-ε in contrast to other IFNs is not induced after PRR ligation but is constitutively expressed by the cervico-vaginal epithelium, the expression is regulated by sex hormones and mice deficient in IFN-ε were susceptible to common STIs such as chlamydia and herpes simplex [43]. Finally, it is now recognized that the microbiome profoundly influences innate and adaptive responses at mucosal and cutaneous surfaces and this will differ at the different target surfaces for HPV infection. The mucosal environment at the different locales is complex but viral infection altering the commensal and hormonal balance and the innate responses modulated by these are likely culprits in association with the virus-mediated downregulation of KC-immune functions in promoting an immunosuppressive environment and neoplastic progression.

HUMORAL IMMUNITY TO HUMAN PAPILLOMAVIRUS AND THE DEVELOPMENT OF HUMAN PAPILLOMAVIRUS VACCINES

Prophylactic vaccines that generate virus-specific neutralizing antibody are the most effective means to control viral diseases, and so an early question, in the field, in the late 1980s and early 1990s was, would vaccines generating neutralizing antibody to HPV protect against HPV infection and disease? In the exclusively intraepithelial life cycle of HPVs there is no or little viremia and infectious virions are shed from mucosal surfaces with poor access to blood, lymph, and draining lymph nodes. As a result, systemic antibody responses to the genital HPVs are slow and weak (the average sero-conversion time for HPV 16 is 8–9 months after the first detection of HPV DNA) and circulating antibody levels are low. All of this would not make one optimistic that serum virus neutralizing antibody would be effective in preventing infection of epithelial surfaces. However, early studies by Shope with the cottontail rabbit papillomavirus (CRPV) showed that rabbits immunized systemically by intramuscular (i.m.) injection with infectious virus did not develop visible lesions or papillomas but it developed neutralizing antibodies and were protected against a cutaneous challenge with CRPV [44].

WHAT ANIMAL PAPILLOMAVIRUS INFECTIONS TOLD US

Subsequent serological studies exploiting natural infections in animals (dog, rabbit, and cow) showed clearly that there were serum responses to viral capsid proteins in individuals who were or had been infected (for review see [45]). In the animal models, sero-positive individuals were resistant to subsequent high-dose viral challenge. In early immunization experiments in the rabbit and cow, bacterially expressed L1 and L2 capsid proteins were used as antigens and both were protective in the rabbit immunization with either L1 or L2 protected against CRPV challenge but L1 was more effective and generated high neutralizing antibody concentrations. Importantly in these experiments, neutralizing antibody was generated only by full length native L1 protein.

In animal infections such as those in the cow, rabbit, and dog where infectious virus could be obtained in large amounts from warts, studies showed clearly that there were serum responses to viral capsid proteins in animals who were or had been infected and importantly sero-positive animals were resistant to subsequent viral challenge. However studies on humoral immunity to HPV, particularly to the high-risk genital HPVs, were seriously hampered by the lack of suitable antigenic targets for serological assays since neither clinical lesions nor in vitro culture systems were practical sources of virus. It was clear from studies on HPV1, a virus that could be harvested in good amounts from plantar warts [46] and HPV 11 [47] where virions could be generated using an athymic nude mouse xenograft system of HPV 11 infected genital wart tissue, that antibody responses to capsid proteins occurred in infected individuals, and that these were to both conformational and linear epitopes. However, the dominant immune response, which was type specific, was to conformational determinants on the intact virus particle [46], and therefore antigen targets in sero-assays and any prophylactic vaccine candidates had to include correctly folded native proteins.

GENERATION OF VIRUS-LIKE PARTICLES

Salunke et al. in 1986 had showed that purified VP1 capsid protein of polyoma virus after expression via *Escherichia coli* self-assembled in the absence of VP2 and VP3 into capsid-like structures or VLPs [48]. However, efforts to express correctly folded HPV L1 proteins via prokaryotic expression vectors were disappointing. Bacterially expressed fusion proteins of HPV L1 were insoluble and did not assume the native conformation but the challenge of generating native correctly folded HPV coat protein was met in a series of seminal studies in the early 1990s using eukaryotic expression systems. First, at the University of Queensland (using the vaccinia virus expression system he had worked with in the Cambridge laboratory of Lionel Crawford), Zhou et al. coexpressed both the coat proteins of HPV 16 L1 and L2 and showed that these proteins self-assembled to form a capsid-like structure or VLP [49] although the yield was low. In a technical development in this report and in contrast to previously published data, Zhou used the second rather than the first of the two potential start codons in the HPV 16 L1 open reading frame. Then it was shown that the HPV 16 L1 capsid protein alone could self-assemble into VLPs [50]. In these experiments, the yield of VLPs were low but the HPV 16 L1 gene used to generate the recombinant vectors was that of the prototype HPV 16 DNA isolated from a cervical carcinoma. However, when an L1 cloned from HPV 16 infecting a low-grade lesion was used, VLP yield was increased threefold and it was shown that the prototype L1 gene compared to wild type L1 had a mutation that affected the efficiency with which VLPs could be assembled [51]. Successful expression and formation of HPV 11 L1 VLPs using recombinant baculovirus that recognized antibodies in the sera of infected individuals was then demonstrated [52,53] and it became clear that this was a technology that could be successfully used to generate VLPs of any papillomavirus including the common genital HPVs. L1 VLPs lacked the minor capsid protein L2 and viral DNA but were morphologically similar to virions and closely approximated the antigenic characteristics of wild type virions [54].

DID THE ANTIBODIES GENERATED BY L1 VIRUS-LIKE PARTICLES PROTECT—PRECLINICAL ANIMAL STUDIES

VLPs were clearly candidate vaccine immunogens but were the neutralizing antibodies generated protective? In the dog, cow, and rabbit immunization with species specific L1 VLPs induced circulating, neutralizing antibody to the L1 capsid protein. Immunized animals were completely resistant to challenge with high virus inocula [55–57] and low levels (nanogram doses) were protective. Importantly, in the rabbit model relatively long-term protection was induced by VLP immunization. The data from the dog was particularly relevant for genital HPVs since the canine oral papillomavirus (COPV) is a mucosatrophic virus infecting the oral cavity. Conformational epitopes on intact VLPs appeared critical for the induction of neutralizing IgG and for successful vaccination since denatured L1 protein failed to generate neutralizing antibody or protect against virus challenge but formaldehyde fixed COPV VLPs retained their immunogenicity and protected. In these animal models, successful immunization was species specific and passive transfer experiments in rabbits [57] and dogs [58] showed that serum antibody alone was protective.

HUMAN PAPILLOMAVIRUS VIRUS-LIKE PARTICLE VACCINES

These animal studies that showed that prophylactic intervention against HPV infection could be successful constituted the scientific case that was the crucial factor in stimulating commercial interest in HPV vaccine development and the commitment by "big pharma" to a massive financial investment [59]. This was high-risk since although the animal studies were clear that generating a successful humoral response protected it was not at all evident at that time in the early 1990s that this would necessarily apply to human genital infections and proof of principle trials with HPV VLPs were crucial.

A number of randomized double blind placebo controlled Phase I studies with recombinant HPV L1 VLP vaccines were undertaken to show immunogenicity and safety [60–62]. These showed that all VLP-immunized subjects, but no subjects in the placebo arms, sero-converted and made anti-VLP antibody responses substantially greater than that identified in natural infections. The dominant antibody response was of the IgG1 subclass and was shown to be neutralizing by an HPV 16 pseudovirion neutralizing assay. Overall the data from published Phase I studies using HPV VLP vaccines including HPV 16 or18 and/or HPV 6/11 L1 VLPs showed that the HPV VLP vaccines were safe, well tolerated and highly immunogenic inducing high levels of both binding and neutralizing type antibodies. The results from these proofs of principle studies laid the foundation for the large randomized control clinical trials enrolling more than 60,000 women that would show the impressive efficacy and safety of HPV VLP vaccines and result in the licensure of these vaccines.

A VIGNETTE: THE STORY OF JIAN ZHOU

1. From Dr Xiao Yi Sun widow of Jian Zhou one of the inventors of HPV VLPs;

Jian and Xiaoyi had a habit of going for a walk after their son had gone to sleep. Jian would often suddenly come up with an idea, and sometime Xiaoyi would note it down in her hand at the time and then experiment in the laboratory later. One day at the end of 1990, they went for a walk as usual. Jian suddenly said to Xiaoyi: "We have now L1 & L2 (the major ingredients that constitute the HPV late protein and virus coat) well presented and purified, why don't we put them in a test tube under certain conditions and see if anything happens?" Xiaoyi said: "I laughed at him at the moment, how could that be possible, just simply put two things together? If so, we wouldn't be doing this here now because people would have seen the virus particle already."

Two weeks later, Jian asked Xiaoyi if she had done the experiment. Xiaoyi replied: "I noted it down at the time, but thought you were only joking." Jian again asked her to do the experiment. Following his idea, Xiaoyi "put the two existing HPV late proteins into a test tube, adding a little bit of this and then a little bit of that, as if children in the kindergarten were playing games, it was that simple".

Approximately 2 weeks later, they observed their experiment under the electron microscope. Both of them were shocked as soon as they saw the result. A virus particle had been constructed! They did see a virus produced outside human body! That was a really exciting moment!

Xiaoyi said: "It was very lucky indeed. We immediately told Ian about it and he couldn't stop smiling at the news. It was an exciting moment we could not forget for a lifetime, a break through with least expectation."

REFERENCES

[1] Egawa N, Doorbar J. The low-risk papillomaviruses. Virus Res 2017;231:119–27.

[2] Weissenborn S, Neale RE, Waterboer T, Abeni D, Bavinck JN, Green AC, et al. Beta-papillomavirus DNA loads in hair follicles of immunocompetent people and organ transplant recipients. Med Microbiol Immunol 2012;201(2):117–25.

[3] Doorbar J, Quint W, Banks L, Bravo IG, Stoler M, Broker TR, et al. The biology and life-cycle of human papillomaviruses. Vaccine 2012;30(Suppl 5):F55–70.

[4] Oriel JD. Natural history of genital warts. Br J Vener Dis 1971;47(1):1–13.

[5] Pasparakis M, Haase I, Nestle FO. Mechanisms regulating skin immunity and inflammation. Nat Rev Immunol 2014;14(5):289–301.

[6] Stanley MA, Sterling JC. Host responses to infection with human papillomavirus. Curr Problems Dermatol 2014;45:58–74.

[7] Nestle FO, Di Meglio P, Qin JZ, Nickoloff BJ. Skin immune sentinels in health and disease. Nat Rev Immunol 2009;9(10):679–91.

[8] Nasu K, Narahara H. Pattern recognition via the toll-like receptor system in the human female genital tract. Mediators Inflamm 2010;2010:976024.

[9] Taylor KE, Mossman KL. Recent advances in understanding viral evasion of type I interferon. Immunology 2013;138(3):190–7.

[10] Karim R, Meyers C, Backendorf C, Ludigs K, Offringa R, van Ommen GJ, et al. Human papillomavirus deregulates the response of a cellular network comprising of chemotactic and proinflammatory genes. PLoS One 2011;6(3):e17848.

[11] Diamond MS, Farzan M. The broad-spectrum antiviral functions of IFIT and IFITM proteins. Nat Rev Immunol 2013;13(1):46–57.

[12] Nees M, Geoghegan JM, Hyman T, Frank S, Miller L, Woodworth CD. Papillomavirus type 16 oncogenes downregulate expression of interferon-responsive genes and upregulate proliferation-associated and NF-kappaB-responsive genes in cervical keratinocytes. J Virol 2001;75(9):4283–96.

[13] Karstensen B, Poppelreuther S, Bonin M, Walter M, Iftner T, Stubenrauch F. Gene expression profiles reveal an upregulation of E2F and downregulation of interferon targets by HPV18 but no changes between keratinocytes with integrated or episomal viral genomes. Virology 2006;353(1):200–9.

[14] Chang YE, Laimins LA. Microarray analysis identifies interferon-inducible genes and Stat-1 as major transcriptional targets of human papillomavirus type 31. J Virol 2000;74(9):4174–82.

[15] Tummers B, Burg SH. High-risk human papillomavirus targets crossroads in immune signaling. Viruses 2015;7(5):2485–506.

[16] Sunthamala N, Pientong C, Ohno T, Zhang C, Bhingare A, Kondo Y, et al. HPV16 E2 protein promotes innate immunity by modulating immunosuppressive status. Biochem Biophys Res Commun 2014;446:977–82.

[17] Fausch SC, Da Silva DM, Rudolf MP, Kast WM. Human papillomavirus virus-like particles do not activate Langerhans cells: a possible immune escape mechanism used by human papillomaviruses. J Immunol 2002;169(6):3242–9.

[18] Da Silva DM, Fausch SC, Verbeek JS, Kast WM. Uptake of human papillomavirus virus-like particles by dendritic cells is mediated by Fcgamma receptors and contributes to acquisition of T cell immunity. J Immunol 2007;178(12):7587–97.

[19] Devoti J, Hatam L, Lucs A, Afzal A, Abramson A, Steinberg B, et al. Decreased Langerhans cell responses to IL-36gamma: altered innate immunity in patients with recurrent respiratory papillomatosis. Mol Med 2014;.

[20] Matthews K, Leong CM, Baxter L, Inglis E, Yun K, Backstrom BT, et al. Depletion of Langerhans cells in human papillomavirus type 16-infected skin is associated with E6-mediated down regulation of E-cadherin. J Virol 2003;77(15):8378–85.

[21] Leong CM, Doorbar J, Nindl I, Yoon HS, Hibma MH. Deregulation of E-cadherin by human papillomavirus is not confined to high-risk, cancer-causing types. Br J Dermatol 2010;.

[22] van Seters M, Beckmann I, Heijmans-Antonissen C, van Beurden M, Ewing PC, Zijlstra FJ, et al. Disturbed patterns of immunocompetent cells in usual-type vulvar intraepithelial neoplasia. Cancer Res 2008;68(16):6617–22.

[23] Guess JC, McCance DJ. Decreased migration of Langerhans precursor-like cells in response to human keratinocytes expressing human papillomavirus type 16 E6/E7 is related to reduced macrophage inflammatory protein-3alpha production. J Virol 2005;79(23):14852–62.

[24] Nakayama Y, Asagoe K, Yamauchi A, Yamamoto T, Shirafuji Y, Morizane S, et al. Dendritic cell subsets and immunological milieu in inflammatory human papilloma virus-related skin lesions. J Dermatol Sci 2011;63(3):173–83.

[25] Moscicki AB, Schiffman M, Kjaer S, Villa LL. Chapter 5: Updating the natural history of HPV and anogenital cancer. Vaccine 2006;24(Suppl 3):S3/42–3/451.

[26] de Jong A, van der Burg SH, Kwappenberg KM, van der Hulst JM, Franken KL, Geluk A, et al. Frequent detection of human papillomavirus 16 E2-specific T-helper immunity in healthy subjects. Cancer Res 2002;62(2):472–9.

[27] Welters MJ, de Jong A, van den Eeden SJ, van der Hulst JM, Kwappenberg KM, Hassane S, et al. Frequent display of human papillomavirus type 16 E6-specific memory t-Helper cells in the healthy population as witness of previous viral encounter. Cancer Res 2003;63(3):636–41.

[28] Woo YL, van den Hende M, Sterling JC, Coleman N, Crawford RA, Kwappenberg KM, et al. A prospective study on the natural course of low-grade squamous intraepithelial lesions and the presence of HPV16 E2-, E6- and E7-specific T-cell responses. Int J Cancer 2010;126(1):133–41.

[29] Nicholls PK, Moore PF, Anderson DM, Moore RA, Parry NR, Gough GW, et al. Regression of canine oral papillomas is associated with infiltration of CD4+ and CD8+ lymphocytes. Virology 2001;283 (Apr 25. 1):31–9.

[30] Jain S, Moore RA, Anderson DM, Gough GW, Stanley MA. Cell-mediated immune responses to COPV early proteins. Virology 2006;356(1-2):23–34.

[31] Coleman N, Birley HD, Renton AM, Hanna NF, Ryait BK, Byrne M, et al. Immunological events in regressing genital warts. Am J Clin Pathol 1994;102(6 Dec):768–74.

[32] Stanley MA. Imiquimod and the imidazoquinolones: mechanism of action and therapeutic potential. Clin Exp Dermatol 2002;27(7):571–7.

[33] Woo Y, Sterling J, Damay I, Coleman N, Crawford R, van der Burg Sh, et al. Characterising the local immune responses in cervical intraepithelial neoplasia: a cross-sectional and longitudinal analysis. BJOG 2008;115(13):1616–22.

[34] McKenzie J, King A, Hare J, Fulford T, Wilson B, Stanley M. Immunocytochemical characterization of large granular lymphocytes in normal cervix and HPV associated disease. J Pathol 1991;165(1):75–80.

[35] Trimble CL, Clark RA, Thoburn C, Hanson NC, Tassello J, Frosina D, et al. Human papillomavirus 16-associated cervical intraepithelial neoplasia in humans excludes CD8 T cells from dysplastic epithelium. J Immunol 2010;185(11):7107–14.

[36] Coleman N, Greenfield IM, Hare J, Kruger-Gray H, Chain BM, Stanley MA. Characterization and functional analysis of the expression of intercellular adhesion molecule-1 in human papillomavirus-related disease of cervical keratinocytes. Am J Pathol 1993;143(2):355–67.

[37] Coleman N, Stanley MA. Characterization and functional analysis of the expression of vascular adhesion molecules in human papillomavirus related disease of the cervix. Cancer 1994;74(3 Aug 1):884–92.

[38] Nakanishi Y, Lu B, Gerard C, Iwasaki A. CD8(+) T lymphocyte mobilisation to virus-infected tissue requires CD4(+) T cell help. Nature 2009;462:510–13.

[39] Maldonado L, Teague JE, Morrow MP, Jotova I, Wu TC, Wang C, et al. Intramuscular therapeutic vaccination targeting HPV16 induces T cell responses that localize in mucosal lesions. Science Transl Med 2014;6(221):221ra13.

[40] Fish EN. The X-files in immunity: sex-based differences predispose immune responses. Nat Rev Immunol 2008;8(9):737–44.

[41] Remoue F, Jacobs N, Miot V, Boniver J, Delvenne P. High intraepithelial expression of estrogen and progesterone receptors in the transformation zone of the uterine cervix. Am J Obstet Gynecol 2003;189 (6):1660–5.
[42] Kovats S. Estrogen receptors regulate innate immune cells and signaling pathways. Cell Immunol 2015;294(2):63–9.
[43] Fung KY, Mangan NE, Cumming H, Horvat JC, Mayall JR, Stifter SA, et al. Interferon-epsilon protects the female reproductive tract from viral and bacterial infection. Science 2013;339(6123):1088–92.
[44] Shope RE. Immunization of rabbits to infectious papillomatosis. J Exp Med 1937;65:607–24.
[45] Stanley M. The immunology of genital human papilloma virus infection. Eur J Dermatol 1998;8 (7 Suppl):8–12 discussion 20-2.
[46] Steele JC, Gallimore PH. Humoral assays of human sera to disrupted and nondisrupted epitopes of human papillomavirus type 1. Virology 1990;174(2):388–98.
[47] Bonnez W, Da Rin C, Rose RC, Reichman RC. Use of human papillomavirus type 11 virions in an ELISA to detect specific antibodies in humans with condylomata acuminata. J Gen Virol 1991;72 (Pt 6):1343–7.
[48] Salunke D, Caspar D, Garcea R. Self-assembly of purified polyomavirus capsid protein VP1. Cell 1986;46(6):895–904.
[49] Zhou J, Sun XY, Stenzel DJ, Frazer IH. Expression of vaccinia recombinant HPV 16 L1 and L2 ORF proteins in epithelial cells is sufficient for assembly of HPV virion like particles. Virology 1991;185 (1):251–7.
[50] Kirnbauer R, Booy F, Cheng N, Lowy DR, Schiller JT. Papillomavirus L1 major capsid protein self assembles into virus like particles that are highly immunogenic. Proc Natl Acad Sci USA 1992;89 (24 Dec 15):12180–4.
[51] Kirnbauer R, Taub J, Greenstone H, Roden R, Durst M, Gissmann L, et al. Efficient self-assembly of human papillomavirus type 16 L1 and L1-L2 into virus-like particles. J Virol 1993;67(12):6929–36.
[52] Rose RC, Bonnez W, Reichman RC, Garcea RL. Expression of human papillomavirus type 11 L1 protein in insect cells: in vivo and in vitro assembly of viruslike particles. J Virol 1993;67(4):1936–44.
[53] Rose RC, Reichman RC, Bonnez W. Human papillomavirus (HPV) type 11 recombinant virus like particles induce the formation of neutralizing antibodies and detect HPV specific antibodies in human sera. J Gen Virol 1994;75(Pt 8 Aug):2075–9.
[54] Stanley M. HPV vaccines. Best Pract Res Clin Obstet Gynaecol 2006;20(2):279–93.
[55] Suzich JA, Ghim SJ, Palmer Hill FJ, White WI, Tamura JK, Bell JA, et al. Systemic immunization with papillomavirus L1 protein completely prevents the development of viral mucosal papillomas. Proc Natl Acad Sci USA 1995;92(25 Dec 5):11553–7.
[56] Kirnbauer R, Chandrachud LM, O'Neil BW, Wagner ER, Grindlay GJ, Armstrong A, et al. Virus-like particles of bovine papillomavirus type 4 in prophylactic and therapeutic immunization. Virology 1996;219(1):37–44.
[57] Breitburd F, Kirnbauer R, Hubbert NL, Nonnenmacher B, Trin-Dinh-Desmarquet C, Orth G, et al. Immunization with viruslike particles from cottontail rabbit papillomavirus (CRPV) can protect against experimental CRPV infection. J Virol 1995;69(6):3959–63.
[58] Ghim S, Newsome J, Bell J, Sundberg JP, Schlegel R, Jenson AB. Spontaneously regressing oral papillomas induce systemic antibodies that neutralize canine oral papillomavirus. Exp Mol Pathol 2000;68 (3):147–51.
[59] Inglis S, Shaw A, Koenig S. Chapter 11: HPV vaccines: commercial research & development. Vaccine 2006;24(Suppl 3):S3/99–3/9105.

[60] Harro CD, Pang YY, Roden RB, Hildesheim A, Wang Z, Reynolds MJ, et al. Safety and immunogenicity trial in adult volunteers of a human papillomavirus 16 L1 virus-like particle vaccine. J Natl Cancer Inst 2001;93(Feb 21. 4):284−492.

[61] Evans TG, Bonnez W, Rose RC, Koenig S, Demeter L, Suzich JA, et al. A Phase 1 study of a recombinant viruslike particle vaccine against human papillomavirus type 11 in healthy adult volunteers. J Infect Dis 2001;183(May 15. 10):1485−93.

[62] Fife KH, Wheeler CM, Koutsky LA, Barr E, Brown DR, Schiff MA, et al. Dose-ranging studies of the safety and immunogenicity of human papillomavirus Type 11 and Type 16 virus-like particle candidate vaccines in young healthy women. Vaccine 2004;22(21-22):2943−52.

CHAPTER

CLINICAL TRIALS OF HUMAN PAPILLOMAVIRUS VACCINES

19

Jorma Paavonen[1], Suzanne M. Garland[2,3,4] and David Jenkins[5]

[1]University of Helsinki, Helsinki, Finland [2]Department of Obstetrics and Gynaecology, University of Melbourne, VIC, Australia [3]Director Centre Women's Infectious Diseases Research, VIC, Australia [4]Honorary Research Fellow, Infection & Immunity, Murdoch Children's Research Institute, VIC, Australia [5]Emeritus Professor in Pathology, University of Nottingham, Nottingham, United Kingdom

INTRODUCTION

The background evidence for the potential public health value of primary prevention of HPV infection is strong. High-risk human papillomaviruses (hrHPV) are proven oncogenes in the anogenital tract and oropharynx in women and men and account for approximately 5% of all cancers. Other HPV types designated as low-risk for cancer (lrHPV) are important in causing genital warts, as well as low-grade lesions detected by cervical screening programs, laryngeal papillomatosis, and are only occasionally associated with anogenital cancer.

The initial discovery of hrHPV, that is, HPV16/18 led 10 years ago to the Nobel Prize in Physiology or Medicine awarded to Dr. Harald zur Hausen in 2008. There are currently at least 13 HPVs defined as high risk and carcinogenic.

Anogenital HPV infection is highly prevalent from the start of sexual activity, with global rate of cervical HPV in women without cervical lesions being in the range of 12%. Highest rates for age are reported in Latin America, Eastern Europe, and sub-Saharan Africa [1]. Most HPV infections are asymptomatic.

The disease burden caused by HPV is enormous both in high income countries and in low income countries although the latter bear the biggest burden largely as a result of no screening or poor screening and treatment of precancerous lesions. A large part of the drive to develop preventive HPV vaccination came from HPV being recognized as the etiological agent of cervical cancer, the fourth most common cancer globally in women. In Europe, approximately 60,000 new cases are diagnosed annually, and cervical cancer kills approximately 25,000 women. Globally, more than half a million cervical cancer cases occur annually. Most new cervical cancer cases (85%) and deaths (88%) occur in low and middle income countries, in which health care systems are fragmented or fragile, and in which organized national cervical cancer screening programs have not been implemented.

Human Papillomavirus. DOI: https://doi.org/10.1016/B978-0-12-814457-2.00019-2

LIMITATIONS OF SECONDARY PREVENTION

The opportunity of preventing cervical cancer by interrupting the long latency period through which cervical cancer develop from chronic HRHPV infection was put into widespread practice by higher income countries many years before the key role of HPV in carcinogenesis was recognized and established. Secondary prevention was based on cervical cytology (Pap smears) as a screening test and surgical removal or destruction of lesions considered precancerous. Although the principle is simple, in practice there was considerable variation in screening programs between countries.

Secondary prevention by Pap smear testing also has multiple steps. This chain of multiple steps must be systematic and organized, in order to prevent cancer in real life. Thorough quality control is important in any organized cytological screening program. In lower and middle income countries the expertise to provide cytology screening and the expert pathology support may not be available or is limited mainly to women living in metropolitan areas. For instance, in the United States there were significant differences in availability of screening, affected by race and geography.

In general, early detection of cervical cancer precursors by organized screening has shifted the disease burden from cancer to precancer, but cannot fully eradicate the disease burden. Even in the best programs whilst incidence and mortality has markedly reduced, they have plateaued. Despite secondary prevention, cervical cancer incidence rates have recently increased in young women in many countries, including the Nordic countries (ref.). Furthermore, secondary prevention byscreening is not available for other major HPV-related diseases, such as vulvar, anal, or oropharyngeal neoplasia.

PRIMARY PREVENTION

The ultimate goal for primary prevention is to eradicate HPV infection and HPV-related disease burden. The first generation HPV vaccines licensed the quadrivalent vaccine (Gardasil, Merck Co.) and the bivalent vaccine (Cervarix, GSK) both target HPV16/18 with some degree of cross-protection against other high risk HPV types which are phylogenetically related. The most recent second generation vaccine, that is, the nonavalent vaccine (Gardasil9, Merck Co.) also targets the five next most common oncogenic high risk types, HPV31/33/45/52/58.

TIMELINE OF HPV VACCINE SUCCESS STORY

The development of prophylactic HPV vaccines has been a major success story. The discovery that the major capsid antigen L1 self-assembles into empty virus-like-particles (VLPs) was a striking innovation. These VLPs are highly immunogenic when injected and by avoiding the limitations of a natural infection "trick the immune system" into producing type-specific neutralizing antibody responses, opening the possibility of primary prevention of cervical cancer and ultimately other HPV related cancers by vaccination, by preventing the infection and ultimately the disease states preceding neoplasia.

DESIGNING CLINICAL TRIALS OF PREVENTIVE HPV VACCINES

Proving the efficacy, safety, and immunogenicity of primary prevention approach was a major challenge. It required randomized trials of large target populations representative of those who would be vaccinated in real life. Establishing large enough multinational studies and deciding on appropriate endpoints for efficacy (virological and pathological) acceptable to regulatory authorities, as well as ensuring safety was a major task. The only bodies with the resources to do this were large, multinational pharmaceutical corporations, with the support of appropriate academic scientific staff and expertise.

Such an exercise took more than 10 years in order to cover all virological and disease endpoints and of different age groups. The initial phase 1 trials established the immediate safety and immunogenicity of the injected VLP vaccines and the associated adjuvants. Phase 2 studies were randomized controlled trials (RCT) intended to establish the principle that vaccination with VLPs could prevent persistent infection with the HPV types included in the VLP's. Global phase 3 RCTs were aimed at establishing the efficacy, safety, and immunogenicity of vaccines against important clinical endpoints in the development of cervical and other female lower genital tract cancers in a wide variety of settings.

CLINICAL TRIAL ENDPOINTS

Both virological and disease endpoints have been used in clinical trials. Virological endpoints are more practical and easier to use than disease endpoints which require extremely large trials and thorough quality control, that is, reproducibility of disease endpoint definitions. The endpoints for the clinical trials had to be agreed with the regulatory authorities to provide evidence of clinical utility and establish the range of protection against specific virological and disease endpoints. Although it was proven that persistent infection with hrHPV was necessary for the development of moderate to severe squamous intraepithelial lesions (SIL) of the cervix thus providing a surrogate endpoint, there was uncertainty as to the required duration of persistence (6, 12, or 24 months). Similarly, detection of the virus by different HPV DNA technologies has been highly variable and only few tests have been properly validated and universally accepted.

For licensure the regulatory authorities required evidence of efficacy against lesions that in clinical practice were treated as precancerous. In clinical practice high-grade squamous intraepithelial neoplasia was accepted as the disease endpoint although there is evidence of regression of a substantial proportion of CIN2 cases in young women [2]. CIN2 diagnosis made by pathological examination of a biopsy remains highly subjective with strikingly low interobserver reproducibility between experts. CIN3 is a significantly stronger endpoint with less variability in making the diagnosis and less regression.

High grade lesions such as cervical intraepithelial neoplasia grade 3 (CIN3) is certainly the best surrogate marker for cervical cancer. Global Phase III clinical efficacy trials have been large enough for using CIN3 as a definitive disease endpoint. Comparable endpoint data have been collected on equivalent intraepithelial lesions in other areas of the anogenital region.

An important additional part of defining endpoints was linking the lesion to the HPV type producing it. Research to support this led to increased understanding through laser capture microscopy

(LCM) which could attribute the viral type causing the lesion as they are clonal. LCM assisted where multiple cervical HPV infections (up to 20% of HPV infections seen in whole tissue sections) producing complex lesions in the cervix of different grades with different HPV types. This also increased awareness of the performance of different HPV DNA tests applied to cervical swabs and biopsies, and of the relation between persistence of a type-specific infection and lesion development, also called type-assignment algorithms to define the specific HPV type driving the lesion development. There are multiple such algorithms such as proportional, hierarchical, single, or any type.

Since the first licensure in 2006, many high income countries have implemented national HPV immunization programs. More recently international organizations, such as GAVI, the Vaccine Alliance, and the pan-American Health Organization (PAHO), have been instrumental in making HPV vaccines available at low price in low income countries, supporting the goal of global eradication of cervical cancer and other HPV related disease.

HPV VACCINATION PHASE II/III CLINICAL EFFICACY TRIALS IN WOMEN WITH ACTIVE FOLLOW-UP

EARLY CLINICAL TRIALS

The first real proof that HPV vaccines would be successful came from the monovalent double-bind HPV 16 randomized controlled trial published by Koutsky et al. [3]. Just over 2000 young women, 16–23 years of age were randomly assigned to receive three doses of monovalent HPV16 VLP vaccine given at day 0, month 2, and month 6, or placebo. In a primary analysis limited to women who were negative for HPV16 DNA and HPV16 antibodies at enrollment and followed for a median of 17.4 months, a significant reduction in persistent HPV16 infections as detected by HPV DNA (3.8/100 woman-years in the placebo arm and 0 in those vaccinated ($P < 0.001$). All nine histologically proven HPV16 related CIN lesions occurred in those unvaccinated, showing a vaccine efficacy of 100% for this outcome. In addition, the serological response was excellent. The antibody levels among vaccinated individuals were many folds greater than in the placebo arm from natural infection. The seroconversion rate was 99.7%. Tolerability was excellent, with the most frequent adverse experience being pain, redness, and swelling at the injection site.

Harper et al. [4] reported a randomized trial of 1113 15- to 25-year-old women receiving three doses of either a bivalent HPV16/18 L1 VLP vaccine formulated with AS04 adjuvant or placebo (hepatitis A) on month 0, 1, and 6. The vaccine was generally safe and well tolerated, and highly immunogenic. In the according-to-protocol (ATP) analyses, vaccine efficacy was 91.6% (95% CI: 64.5–98.0) against incident infection and 100% against persistent infection (47.0–100) with HPV16/18. In the intention-to-treat (ITT) analyses vaccine efficacy was 95.1% (63.5–99.3) against persistent cervical infection with HPV16/18 and 92.9% (70.0–98.3) against cytological abnormalities associated with HPV16/18 infection.

In a Phase II, randomized, double-blind, placebo-controlled study conducted in Brazil, Nordic countries and the United States, Villa et al. [5,6] randomized 1158 young women aged 16–23 years to receive one of three formulations of a quadrivalent HPV6/11/16/18 VLP vaccine or one of two placebo formulations. All three HPV vaccine formulations were highly immunogenic.

Antibody levels remained stable 3 years through the end of study. Over 5 years of follow-up, the incidence of HPV6/11/16/18 related disease was reduced by 100% among women who received at least one dose, compared with placebo recipients. There were no breakthrough cases of disease through the entire study. No vaccine-related serious adverse effects (SAE) were reported. Vaccine efficacy correlated with a high serological response to each of the genotypes.

PHASE 3 EFFICACY TRIALS

The purpose of the phase 3 trials was to provide exact data on the efficacy of the vaccine tested against important clinical endpoints and to show the safety of vaccination in a wide range of populations globally. These trials were large, involving up to 20,000 participants, extending over many geographical parts of the world.

QUADRIVALENT HPV VACCINE PHASE 3 TRIAL (FUTURE* I/II)

The phase 3 FUTURE studies (females united to unilaterally reduce ecto/endo cervical disease) were multinational (25 countries) clinical trials of a quadrivalent HPV L1 VLP vaccine (HPV6/11/16/18). The FUTURE I and II randomized double-blind placebo-controlled trials enrolled 5455 and 17,622 women aged 16–24 years, 5 × 26 years respectively, between 2001 and 2003. For FUTURE 1 the co-primary composite endpoint was incidence of genital warts, vulvar/vaginal intraepithelial neoplasia, or cancer or CIN or cancer to the four viral types targeted by the vaccine. There was 6 monthly follow-up (more intense to standard of care) of the women for 3 years and there was on 100% effective in reducing the risk of HPV16/18-related high grade cervical, vulvar, and vaginal lesions and the risk of HPV6/11 related genital warts. In the ITT population, vaccination also reduced the risk of any vulval or vaginal regardless of viral type by 34%, and cervical lesions regardless of causal type by 20%. Whilst FUTURE 11 enrolled 12,167 women 15–26 years the primary composite end point was cervical intraepithelial neoplasia grade 2 or 3, adenocarcinoma in situ, or cervical cancer related to HPV-16 or HPV-18. Vaccine efficacy for the prevention of the primary composite end point was 98% (95.89% confidence interval [CI], 86 to 100) in the per-protocol susceptible population and 44% (95% CI, 26 to 58) in the ITT population of all women. The estimated vaccine efficacy against all high-grade cervical lesions, regardless of causal HPV type, in this ITT population was 17% (95% CI, 1 to 31).

In the final analysis with a mean of 3.6 years follow up it was found that Pap abnormalities, and definitive surgical procedures irrespective of causal HPV type were also significantly reduced [7,8]. (Fig. 19.1).

The 4vHPV vaccine was highly immunogenic (seroconversion of at least 99.5% for each of the HPV types), and levels of neutralizing antibody reached were many fold greater than hat achieved by natural infection. Side effects were minimal and those significant related to increased redness, swelling, and pain at the injection site, with no indicators of serious adverse effects.

The clinical implication was that high-coverage HPV vaccination programs among adolescents and young women result in a rapid reduction of genital warts, cervical cytological abnormalities,

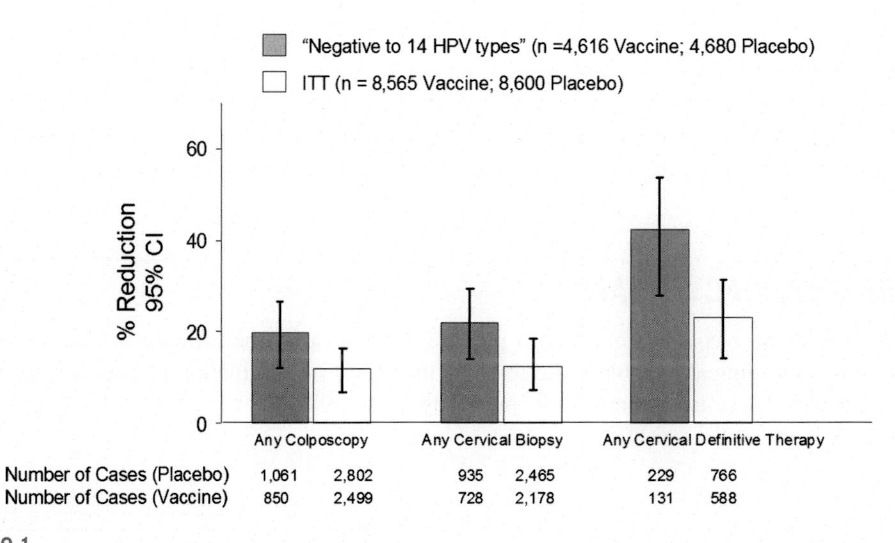

FIGURE 19.1

Reductions in cervical procedures.

Adapted from Muñoz N, Kjaer SK, Sigurdsson K, Iversen OE, Hernandez-Avila M, Wheeler CM, et al. Impact of human papillomavirus (HPV)-6/11/16/18 vaccine on all HPV-associated genital diseases in young women. J Natl Cancer Inst 2010;102(5):325–39.

Table 19.1 Efficacy against external genital lesions irrespective of HPV type.

Endpoint/HPV-native[a]	Vaccine (N = 4616)	Placebo (N = 4680)	VE % (95% CI)
• Genital warts[b]	29	169	82.8 (74.3–88.8)
• VIN2/3;ValN2/3[b]	7	31	77.1 (47.1–91.5)
Endpoint/ITT[b]	**Vaccine (N = 8562)**	**Placebo (N = 8598)**	**VE % (95% CI)**
• Genital warts[b]	134	351	62.0 (53.5–69.1)
• VIN2/3;ValN2/3[b]	30	61	50.7 (22.5–69.3)

[a]*Subjects received at least 1 vaccination, were seroneg and DNA neg to HPV6/11/16/18 and DNA neg to 10 nonvaccine types and had normal Pap test.*
[b]*All subjects who received at least 1 vaccination and had follow-up regardless of the presence of HPV infection or HPV-related disease at enrollment;* VIN, *Vulvar intraepithelial neoplasia;* ValN, *vaginal intraepithelial neoplasia;* ITT, *intention-to-treat.*
Modified from Muñoz N, Kjaer SK, Sigurdsson K, Iversen OE, Hernandez-Avila M, Wheeler CM, et al. Impact of human papillomavirus (HPV)-6/11/16/18 vaccine on all HPV-associated genital diseases in young women. J Natl Cancer Inst 2010;102 (5):325–39 [8].

as well as diagnostic, and therapeutic procedures. The FUTURE 1 vaccine trial not only showed protection against genital warts (by virtue of prevention of infection with HPV 6, 11), but in the placebo arm defined the natural history of genital warts in young women (Table 19.1) [9].

Further follow-up out to 42 months, showed sustained protection against low grade lesions attributable to the vaccine HPV types, as well as a substantial reduction in cervical disease [10].

Moreover, in ad hoc retrospective analysis of the FUTURE I/II studies, it was found that those who were vaccinated and subsequently required surgical treatment for HPV-related disease, there was a 46% reduced risk of subsequent HPV-related disease, including high grade disease, in comparison with those in the placebo arm [11].

PHASE 3 STUDIES OF THE BIVALENT HPV16/18 VACCINE

The goal of a multinational phase 3 study The PApilloma TRIal against Cancer In young Adults (PATRICIA) was to assess the efficacy of the ASO4 adjuvanted bivalent HPV16/18 L1 VLP prophylactic vaccine against infection and disease endpoints, primarily CIN2 and CIN3 [12–14] (Fig. 19.2). A total of 18,644 women aged 15–25 years were randomly assigned to receive either HPV16/18 vaccine or hepatitis A vaccine at 0, 1, and 6 months [12–14]. The mean follow-up was 43.7 months. Vaccine efficacy against CIN3 + associated with HPV16/18 was 100% in the HPV naïve population and 45.7% (95% CI: 22.9–62.2) in the total vaccinated cohort (TVC). Vaccine efficacy against all CIN3 + irrespective of HPV type was 93.2% (78.9–98.7) in the HPV naïve population and 45.6% (28.8–58.7) in the TVC. Vaccine efficacy against high grade CIN was highest in the 15–17 year age group and progressively decreased in the 18–20 year and 21–25 year age groups [15]. Vaccine efficacy against all adenocarcinoma-in-situ (AIS) was 100% (31.0–100) in the HPV naïve population and 76.9% (26.0–95.8) in the TVC.

Just as in the FUTURE 11 trial, a significant reduction in cytologic atypias, colposcopy biopsies, and cervical procedures in comparison to the control arm were observed in the HPV vaccine recipients (Fig. 19.2).

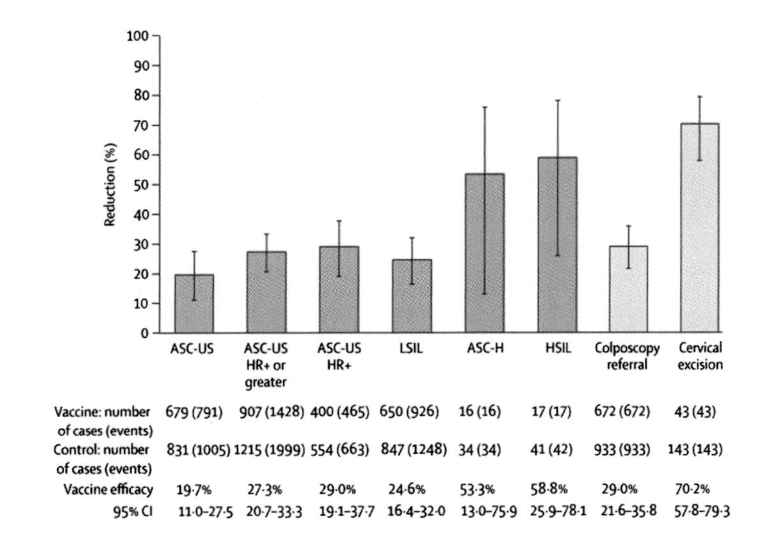

FIGURE 19.2

Reduction in cytological atypias, colposcopy referrals and cervical excision procedures.

COMPARISON OF QUADRIVALENT AND BIVALENT VACCINE EFFICACY AGAINST CERVICAL HSIL BASED ON CLINICAL TRIALS

Tables 19.2 and 19.3 [16] show the phase 3 trial efficacy rates of the quadrivalent and bivalent HPV vaccine against the disease endpoint CIN3 + in the HPV naïve population (Table 19.2) and in the total vaccinated population (Table 19.3). When interpreting the data it is important to keep in mind that there was no face-to-face comparison of the vaccines. Also, the baseline characteristics of the subjects enrolled differed. For instance, the proportion of HPV-naïve subjects was higher in

Table 19.2 Efficacy against HPV16/18 related CIN3 +.

	Vaccine	Control	VE (%)
HPV naive[a]	**No./Total**	**No./Total**	**95%CI**
• Quadrivalent	0/4616	34/4680	97.2 (91.5–99.4)
• Bivalent	0/5466	27/5452	100 (85.5–100)
Intention-to-treat (ITT)[b]			
• Quadrivalent	100/8562	177/8598	43.5 (27.3–56.2)
• Bivalent	51/8694	94/8708	45.7 (22.9–62.2)

VE, *Vaccine efficacy.*
[a]*HPV native population: Subjects received at least 1 vaccination, were seroneg and DNA neg to vaccine types and DNA neg to 12 nonvaccine types and had normal Pap test.*
[b]*All subjects who received at least 1 vaccination and had follow-up regardless of the presence of HPV infection or HPV-related disease at enrollment.*
Modified from Lehtinen M, Dillner J. Clinical trials of human papillomavirus vaccine and beyond. Nat Rev Clin Oncol 2013;10:400–10 [13].

Table 19.3 Efficacy against any CIN3 +.

	Vaccine	Control	VE (%)
HPV naive[a]	**No./Total**	**No./Total**	**95%CI**
• Quadrivalent	36/4,616	64/4,680	43.0 (13.0–63.2)
• Bivalent	3/5,466	44/5,452	93.2 (78.9–98.7)
Intention-to-treat (ITT)[b]			
• Quadrivalent	237/8,562	284/8,598	16.4 (9.4–30.0)
• Bivalent	86/8,694	158/8,708	45.6 (28.8–58.7)

VE, *Vaccine efficacy.*
[a]*HPV native population: Subjects received at least 1 vaccination, were seroneg and DNA neg to vaccine types and DNA neg to 12 nonvaccine types and had normal Pap test.*
[b]*All subjects who received at least 1 vaccination and had follow-up regardless of the presence of HPV infection or HPV-related disease at enrollment.*
Modified from Lehtinen M, Dillner J. Clinical trials of human papillomavirus vaccine and beyond. Nat Rev Clin Oncol 2013;10:400–10 [16].

PATRICIA, and the baseline rates of HPV16/18 DNA positive subjects and subjects with cytologic atypia were similarly lower in PATRICIA. Furthermore, the colposcopic algorithms and the endpoint definitions were not identical, as were the various HPV DNA and antibody assays utilized. Also, the vaccine adjuvants were different.

The data show that both vaccines are highly effective against HPV16/18 related disease in HPV naïve populations. However, the data also show that the bivalent HPV16/18 vaccine was more effective in the ITT population suggesting differences in cross-protection against other non-16/18 high risk HPV types.

THE OVERALL IMPLEMENTATION OF HPV VACCINATION

The initial analyses of the FUTURE trial and the PATRICIA trial rapidly led to the licensure of the both prophylactic HPV vaccines 2006/2007. Subsequently, the vaccines have been included in national vaccination programs in more than 100 countries, with at least 270 million doses of the vaccines administered so far. In addition, GAVI and other international organizations have been highly successful in making vaccines available also in low income countries.

SUBSEQUENT ADDITIONAL PHASE 3 TRIALS

The CVTA, community based randomized phase 3 trial, was conducted in Costa Rica [17], to provide additional independent evaluation of the bivalent HPV16/18 ASO4 adjuvanted vaccine against the disease endpoint CIN2 + . During 2004–2005, a total of 7466 18- to 25-year-old women were randomized to the HPV arm or hepatitis A vaccine (HAV) control arm, and followed for 53.8 months. Vaccine efficacy was 89.8% against HPV16/18 associated CIN2 + , 59.9% against non-HPV16/18 associated CIN2 + , and 61.4% against CIN2 + irrespective of HPV type.

Combined analysis of data from the CVT and PATRICIA trials showed that 4 years after vaccination of women aged 15–25 years, one and two doses of the HPV16/18 vaccine seemed to protect against cervical HPV16/18 infection, similar to the protection provided by the three-dose schedule [18]. Subsequently, many countries have rapidly moved to a two-dose schedule and following endorsement from WHO.

These phase 3 efficacy trials have been the basis upon which public health programs in many countries have introduced HPV vaccination into national immunization programs. As an example, Fig. 19.3 gives a summary of HPV vaccination programs in the Nordic countries (see also Chapter 21).

BROAD SPECTRUM NONAVALENT HPV VACCINE TRIAL

Built on the success of the 4vHPV vaccine and looking at the next most common oncogenic types causing cervical cancer globally, RCTs of a nonavalent (9vHPV) vaccine as a next generation of vaccines and include VLPs of HPV31/33/45/52/58, in addition to HPV6/11/16/18. Between 2007 and 2009

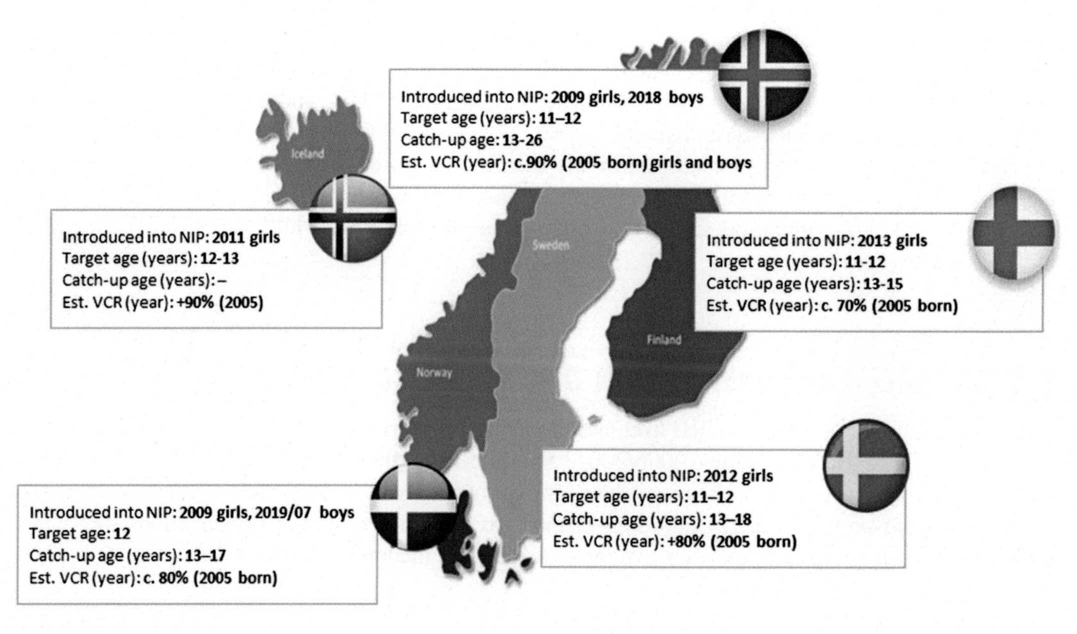

FIGURE 19.3

Summary of the implementation of HPV vaccination programs in the Nordic countries. Courtesy of Annette Gylling, PhD.

over 14,000 16–25 year old women in multiple countries were recruited and randomly assigned to receive 9vHPV vaccine or quadrivalent HPV (qHPV) vaccine [19,20]. These results demonstrated 97% prevention of high grade cervical, vulvar, and vaginal disease associated with HPV31/33/45/52/58 and induced a noninferior HPV 6, 11, 16, 18 antibody responses as compared to the 4vHPV vaccine (Fig. 19.4; Table 19.4). Thus, the 9vHPV vaccine provides the potential to prevent approximately 99% of cervical cancers and HPV-related, vulvar, vaginal and anal cancers, as well as nearly 90% of genital warts worldwide.

Even in those previously vaccinated with the 4vHPV vaccine the 9vHPV vaccine was safe and immunogenic. Injection site pain redness and swelling were higher in the 9vHPV vaccine group than in the 4vHPV group, which was not unexpected. Vaccine related serious adverse events (SAE) were comparable between the groups. Three dose regimen of 9vHPV vaccine to adolescent girls and young women with prior 4vHPV vaccine was highly immunogenic with respect to HPV31/33/45/52/58 and generally well tolerated. Efficacy for the prevention of 31/33/45/52/58 related persistent infection was 90–100% and seroconversion occurred in ≥97% for each of the HPV types. The 9vHPV vaccine efficacy was 97.4% (95% CI: 85.0–99.9) against high grade cervical, vulvar, and vaginal disease related to HPV31/33/45/52/58 and showed that the broad spectrum HPV vaccine prevents infection, cytological abnormalities, high grade lesions, and cervical procedures related to these five additional high risk HPV types (Table 19.4). The 9vHPV vaccine was FDA approved in 2014, and is now the only HPV vaccine available in the United States and Australia.

In a sub-analysis, safety, immunogenicity, and efficacy of the 9vHPV vaccine was reported for Asian populations [21].

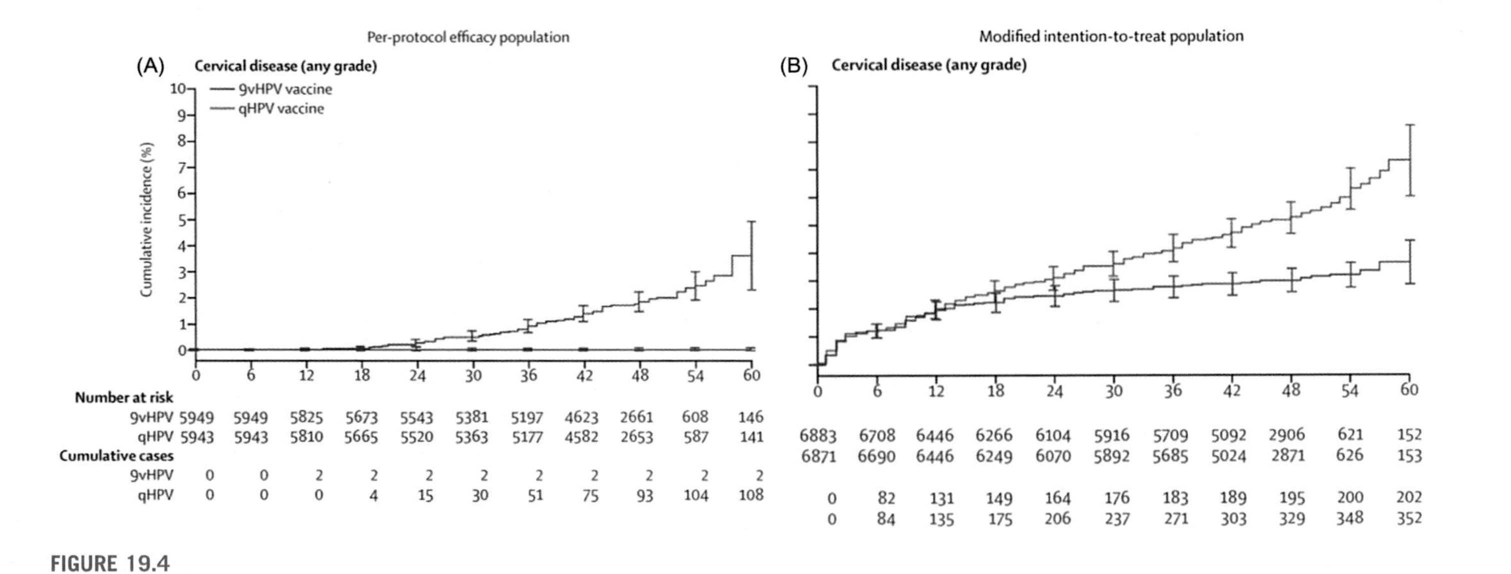

FIGURE 19.4

Time to the development of cervical disease related to HPV31/33/45/52/58 [19] (Sept 5).

Adapted from Huh WK, Joura EA, Giuliano AR, Iversen OE, de Andrade RP, Ault KA, et al. Final efficacy, immunogenicity, and safety analyses of a nine-valent human papillomavirus vaccine in women aged 16-26 years: a randomised, double-blind trial. Lancet 2017; 390(10108):2143–59, Fig. 2A/2B.

Table 19.4 Nine-valent vaccine efficacy against high grade cervical, vulvar, vaginal disease, and persistent infection.

High grade SIL	9vHPV	qHPV	RiskRed
Related to			
• HPV31/33/45/52/58	1/6016	30/6017	96.7%[a]
Persistent infection[c]			
Related to			
• HPV31/33/45/52/58	35/5939	810/5953	96.0%[b]

[a]95% CI 79.5–99.8
[b]94.4–97.2
[c]6 months
Modified from Joura EA, Giuliano AR, Iversen OE, Bouchard C, Mao C, Mehlsen J, et al. Broad spectrum HPV vaccine study. A 9-valent HPV vaccine against infection and intraepithelial neoplasia in women. N Engl J Med 2015;372(8):711–23 [20].

In a combined analysis of seven phase 3 clinical trials, the safety profile of the 9-valent HPV vaccine was comparable to that of the 4vHPV, across age groups and genders with similar AEs, although with more local injection site AEs.

Also in a combined analysis of five phase III clinical trials in girls and boys 9–15 years of age and young women 16–26 years of age, the immunogenicity of the 9-valent HPV vaccine was comparable across subjects of different races and from different regions.

Moreover, in a comparison with an historic placebo population those vaccinated with 9vHPV were found to have a reduction in incidence of high-grade cervical disease and cervical surgery related to the nine HPV types of 98.2% and 97.8%, respectively.

HPV VACCINE TRIALS IN WOMEN OLDER THAN 25 YEARS

Not only was efficacy and safety, and immunogenicity demonstrated in women up to 26 years, but also in older women 24–45 years. The 4vHPV vaccine was shown in an RCT to be efficacious against vaccine-type HPV infection and disease among those not infected with the relevant types at enrollment, and it was also immunogenic and safe. Similarly, bivalent HPV vaccine was also shown to be effective in women older than 25 years (VIVIANE study) [22]. In the Phase III, double blind, randomized controlled trial healthy women older than 25 years (age stratified 26–35 years, 36–45 years and >45 years) were randomly assigned to receive HPV16/18 vaccine or aluminum hydroxide adjuvant control. Of these, 4407 (vaccine 2877, control 2198) were in the ATP cohort for efficacy, and 5747 (vaccine 2877, control 2870) in the TVC. After 7 year follow-up, in all age groups, HPV16/18 vaccine continued to protect against infection, cytological abnormalities, and lesions associated with HPV16/18 and CIN-lesions irrespective of HPV type, and infection with nonvaccine HPV types 31/45. Among women seronegative to the corresponding HPV type, vaccine efficacy against 6-month persistent infection, or CIN associated with HPV16/18 was significant in all age groups combined (90.3%, 96.2% CI: 78.6–96.5). The study showed that adult women also benefit from vaccination although the effectiveness and cost-effectiveness decrease with increasing age.

PHASE 4 TRIALS TO EVALUATE VACCINE PERFORMANCE IN POPULATIONS

A large community randomized HPV16/18 vaccine trial was conducted in Finland [23]. This trial was conducted to answer the simple question of the added value of vaccinating boys as well. A total of 20,513 12- to 15-year-old girls and 11,662 boys from 33 communities in Finland were randomized in three arms to receive gender-neutral HPV16/18 vaccination (11 communities), or women only HPV16/18 vaccination (boys received Hepatitis B control vaccine (HBV) (11 communities), or gender-neutral HBV vaccination (11 communities) in three doses. The vaccination coverage was 51–53% in women and 22–32% in men. Four years later, at age 18, cervico-vaginal samples were tested for HPV DNA. The results of this large community randomized trial on the impact of HPV vaccination on the herd effect and overall protective effectiveness supported gender neutral vaccination when low to moderate vaccination coverage and cross-protection applied. The trial was launched at the time when no national vaccination program had been implemented in Finland. Specifically, vaccine efficacy against HPV16/18 varied between 89% and 95% across birth cohorts in HPV vaccinated study arms. Vaccine efficacy rates against nonvaccine types 31/33/45 consistent with cross-protection were higher by gender-neutral vaccination strategy over women-only vaccination strategy. Not surprisingly, gender neutral vaccination programs have already been implemented in at least 19 countries.

HPV VACCINE EFFICACY AGAINST EXTERNAL GENITAL LESIONS

GENITAL WARTS

Investigation of the placebo arms of the FUTURE 1 and 11 studies allowed for defining the natural history of genital warts. In just under 9000 women seen up to nine visits over 4 years, it was noted that 3.4% of young women developed genital warts. All lesions suggestive of genital warts were biopsied and HPV DNA assays performed. Of those where HPV DNA was detected (91%) 95% were HPV 6/11, confirming the majority of the warts were caused by these two HPV types. Not surprisingly, being positive for 6/11 genital warts corresponded to being positive for HPV DNA 6/11 at baseline, acquisition of a new partner, and high number of sexual partners overall [9].

A systematic review of 10 years real-world experience of the impact and effectiveness of the 4vHPV vaccine (58 studies) showed maximal reductions of approximately 90% for genital warts [24].

HIGH GRADE VULVAR AND VAGINAL INTRAEPITHELIAL NEOPLASIA (VIN2-3, VAIN2-3)

Furthermore, looking at other external genital lesions such as high-grade vulvar and vaginal intraepithelial neoplasia (VIN/VaIN) in the FUTURE trials, the 4vHPV showed excellent efficacy against HPV-related high grade VIN and high grade VaIN among 18,174 16–26 year old women (7811 vaccine recipients; 7785 placebo recipients) [25]. Among women naïve to HPV16/18 the vaccine was 100% effective against VIN2-3 or VaIN2-3 associated with HPV16 or HPV18. In the ITT population vaccine efficacy was 71% (95% CI: 37–88). The vaccine was 49% (95% CI: 18–69) affective against all high grade VIN or VaIN irrespective of the detection of HPV DNA (Fig. 19.5).

More recently, A post hoc analysis of prospectively diagnosed vulvar and vaginal LSILs and HSILs among females 15–26 years of age enrolled in the placebo arms of two phase 3, HPV vaccine trials assessed 14 pre-specified HPV genotypes associated with cervical cancers or anogenital warts using a type-specific multiplex polymerase chain reaction assay. The frequency of lesions associated with specific HPV genotypes was estimated by proportional and other attribution methods. Over 4 years of follow-up, in 8,798 females, 40 vulvar LSILs and 46 vulvar HSILs were diagnosed in 68 females, and 118 vaginal LSILs and 33 vaginal HSILs were diagnosed in 107 females. Of those developing vulvar (41.2%) or vaginal (49.5%) lesions also had cervical lesions, whereas 6.5% of females with cervical lesions had vaginal or vulvar lesions. At least 1 of the 14 HPV genotypes was detected in those with vulvar LSIL (72.5%), vulvar HSIL (91.3%), vaginal LSIL (61.9%), and vaginal HSIL (72.7%). Considering only HPV-positive lesions, the nine most common genotypes causing cervical cancer and anogenital warts (6, 11, 16, 18, 31, 33, 45, 52, and 58) were found in 89.4% of vulvar LSILs, 100% of vulvar HSILs, 56.0% of vaginal LSILs, and 78.3% of vaginal HSILs. This suggests that with time, vaccination will result in reduced rates of HPV-related vulvar and vaginal cancers.

CROSS-PROTECTIVE EFFICACY AGAINST NONVACCINE HIGH RISK HPV TYPES

QUADRIVALENT HPV6/11/16/18 (4VHPV) VACCINE

For the quadrivalent vaccine, infection, and disease end points associated with 10 HPV types (31, 33, 35, 39, 45, 51, 52, 56, 58, and 59) were examined in an end-of-study analysis of FUTURE after 4 years of follow-up. The analyses were restricted to women who were seronegative for HPV-6, -11, -16, and -18 and DNA negative for the 14 HPV types analyzed. In the in generally naïve young women there was cross-protective efficacy for the phylogenetically related oncogenic, but nonvaccine HPV types 31/33/45/ as a composite and against various combinations for persistent infection and CIN1–3/AIS [26] and also for persistent infection in the intention to treat [27] group but not significant protection against CIN2–3/AIS associated with any combination of HPV types. In the detailed analysis of HPV types, these effects were largely driven by HPV 31, the type with the greatest homology to HPV16.

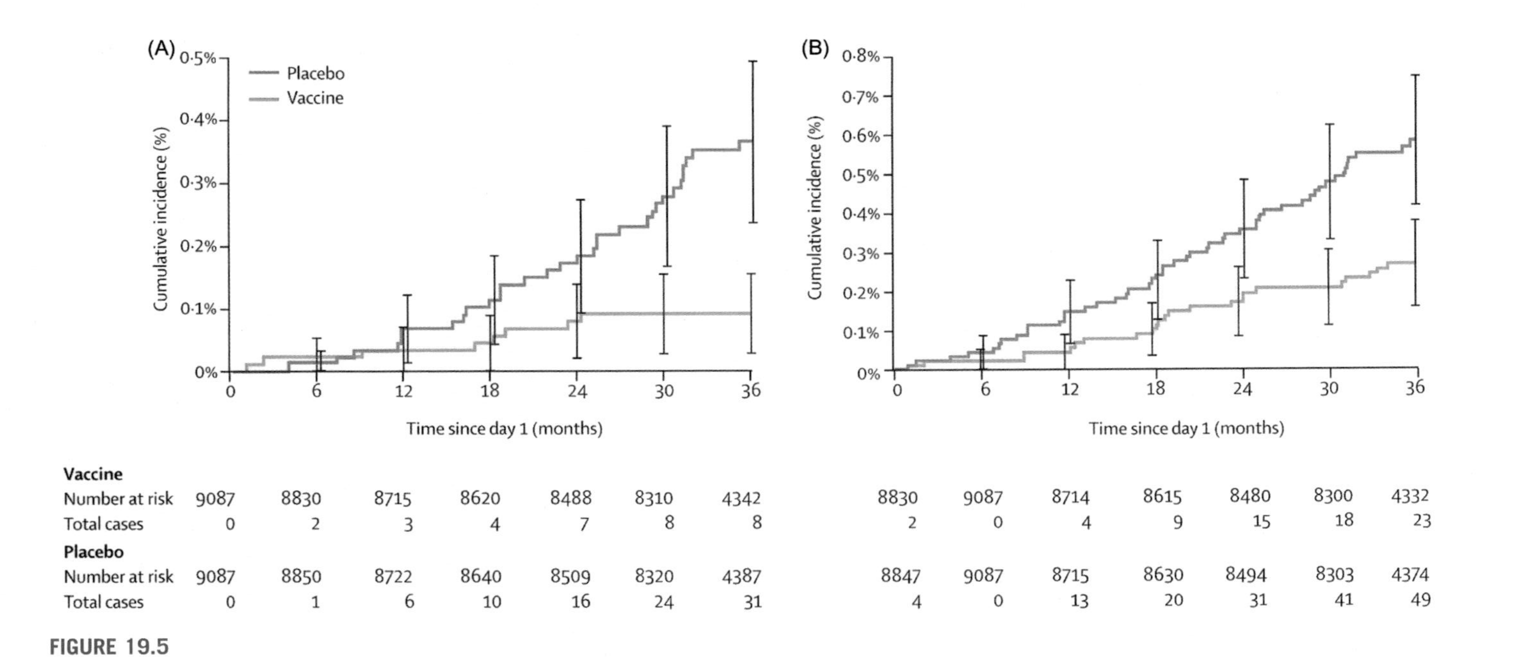

FIGURE 19.5

(A and B) Time to event in the intention-to-treat population. (A) HPV16/18-related VIN2-3 or VaIN2-3. (B) Any VIN2-3 or VaIN2-3 irrespective of causal HPV type.

Adapted Fig. 2 from Joura EA, Leodolter S, Hernandez-Avila M, Wheeler CM, Perez G, Koutsky LA, et al. Efficacy of a quadrivalent prophylactic human papillomavirus (types 6, 11, 16, and 18) L1 virus-like-particle vaccine against high-grade vulval and vaginal lesions: a combined analysis of three randomised clinical trials. Lancet 2007;369(9574):1693–1702 [25].

BIVALENT HPV16/18 (2VHPV) VACCINE

For the bivalent vaccine [28] data from the end-of-study analysis of the PATRICIA trial after 4 years of follow-up showed cross-protective efficacy of the HPV16/18 vaccine against oncogenic nonvaccine HPV types 31/33/45/51 across the cohorts. Consistent vaccine efficacy against persistent infection and CIN2 + (with or without HPV16/18 coinfection) was seen for four oncogenic types. Vaccine efficacy against CIN2 + associated with the composite 12 nonvaccine HPV types was 56.2% (95% CI: 37.2–69.9) in the HPV naïve population and 34.2% (20.4–45.8) in the TVC. Corresponding efficacy rates for CIN3 + were 91.4% (65.0–99.0) and 47.5% (22.8–64.8). Overall, the results showed substantial additional protection over and above that achieved by efficacy against HPV16/18, but long-term follow-up is needed to confirm this and the duration of cross-protective efficacy. The extent and duration of cross-protection are the key elements of overall vaccine efficacy. Analysis of the most comparable subcohorts of FUTURE and PATRICIA trials showed that cross-protective vaccine efficacy estimates against persistent infection and CIN2 + disease associated with HPV31/33/45 were significantly higher for the HPV16/18 vaccine than the HPV6/11/16/18 vaccine [16]. However, cross-protective efficacy of the bivalent vaccine seemed to wane with increasing follow-up. Thus, more data are needed to establish duration of cross-protection. One explanation is that the ASO4 adjuvant in the bivalent vaccine is more immunogenic than the AAHS adjuvant of the quadrivalent vaccine.

Malagon et al. [29] performed a systematic review and meta-analysis of cross-protective efficacy of the bivalent and quadrivalent vaccines. Their interpretation was that the bivalent vaccine seemed more efficacious against nonvaccine HPV types 31/33/45 than the quadrivalent vaccine, but the differences were not all significant and might be attributable to differences in trial designs. Efficacy against persistent infection with types 31/45 seemed to decrease in bivalent vaccine trials with increased follow-up suggesting waning of cross-protection.

Another meta-analysis by Mesher et al. [30] included nine HPV vaccination studies. They studied HPV prevalence rates pre-vaccination and post-vaccination to identify changes in nonvaccine HPV types after introduction of vaccines that confer protection against HPV16/18. Evidence was found of cross-protection for HPV31, but little evidence for reductions of HPV33/45. Slight nonsignificant increases were found for HPV52. The authors concluded that there was no evidence for type-replacement, but that continued monitoring of the vaccinated populations is needed.

HPV VACCINATION PHASE III CLINICAL EFFICACY TRIALS IN MEN WITH ACTIVE FOLLOW-UP

VIRAL AND DISEASE ENDPOINTS USED IN TRIALS AMONG MEN

Similar to the FUTURE trials in young women, the quadrivalent HPV vaccine was shown to prevent infection with HPV-6, 11, 16, and 18 and the development of related external genital lesions in males 16 to 26 years of age. In just over 4000 healthy boys and men 16 to 26 years of age, enrolled from 18 countries in a randomized, placebo-controlled, double-blind trial, in per-protocol population, and in an ITT population the vaccine was shown to be safe and efficacious.

Moreover, in a sub-study of this study whereby 602 healthy men who have sex with men, 16 to 26 years of age, were randomized to receive either qHPV or placebo, with the primary efficacy being to prevent anal intraepithelial neoplasia or anal cancer related to infection with HPV-6, 11, 16, or 18, it was 50.3% (95% confidence interval [CI], 25.7 to 67.2) in the ITT population and 77.5% (95% CI, 39.6 to 93.3) in the per-protocol population [31].

It is to be noted that HPV vaccine related infection and disease burden in men following women only HPV vaccination is seen where high female vaccination coverage is achieved as reported early in Australia [32]. It has resulted from herd protection.

PRIOR HPV VACCINATION PREVENTS RECURRENT HPV RELATED HIGH GRADE DISEASE

Previous vaccination with quadrivalent HPV vaccine among women who had surgical treatment for HPV related disease significantly reduced the incidence of subsequent HPV related disease, including high grade disease [11]. In the FUTURE I/II trials among 17,622 women aged 15–26 years, a total of 587 vaccine recipients and 763 placebo recipients underwent cervical surgical procedure. Vaccination was associated with a significant reduction in risk of any subsequent high grade disease of the cervix by 64.9% (20.1–86.3). A total of 229 vaccine recipients and 475 placebo recipients were diagnosed with genital warts, vulvar intraepithelial neoplasia or vaginal intraepithelial neoplasia, and the incidence of any subsequent HPV related disease was 20.1% in the vaccine recipients and 31% in the placebo recipients, that is, a significant reduction (Fig. 19.6).

In a similar post hoc analysis of the PATRICIA trial, prior vaccination with HPV16/18 vaccine prevented recurrent high grade CIN after definitive surgical therapy [33]. The main outcome was the incidence of subsequent CIN2+ (60 days or more post-surgery). Of the TVC of 18,644 women, 190 HPV vaccine recipients and 264 control vaccine recipients underwent an excisional cervical procedure during the trial. Efficacy against subsequent CIN2+ irrespective of HPV DNA results was 88.2% (14.8–99.7). Thus, women who undergo surgical therapy for cervical lesion after vaccination continue to benefit from vaccination, with a reduced risk of developing subsequent cervical high grade disease.

HPV VACCINATION AFTER TREATMENT FOR CIN PREVENTS RECURRENT HPV RELATED HIGH GRADE DISEASE

Increasing evidence suggests that vaccination after cervical surgery reduces the risk of relapse in women with high grade CIN. This is somewhat unexpected since the current HPV vaccines are prophylactic, not therapeutic. In SPERANZA project [34], 536 women who underwent cervical LEEP surgery for CIN2+ were enrolled and randomized to receive 4vHPV vaccination in three doses (248 women). Control group consisted of 276 women who did not receive HPV vaccination. Recurrence rate was 6.4% in the no-vaccine group compared to 1.2% in the vaccinated group. Thus, HPV vaccination was associated with a significantly reduced risk of subsequent HPV-related high grade CIN after LEEP by 81.2% (95% CI: 34–95), irrespective of causal HPV type. There

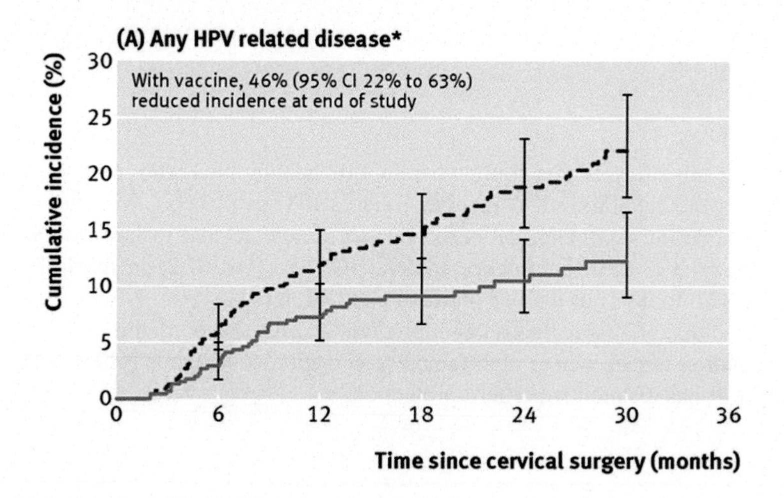

FIGURE 19.6

Impact of quadrivalent vaccine on new disease after definitive therapy.

Adapted Fig. 3A from Joura EA, Garland SM, Paavonen J, Ferris DG, Perez G, Ault KA, et al. Effect of the human papillomavirus (HPV) quadrivalent vaccine in a subgroup of women with cervical and vulvar disease: retrospective pooled analysis of trial data. BMJ 2012;27;344:e1401.

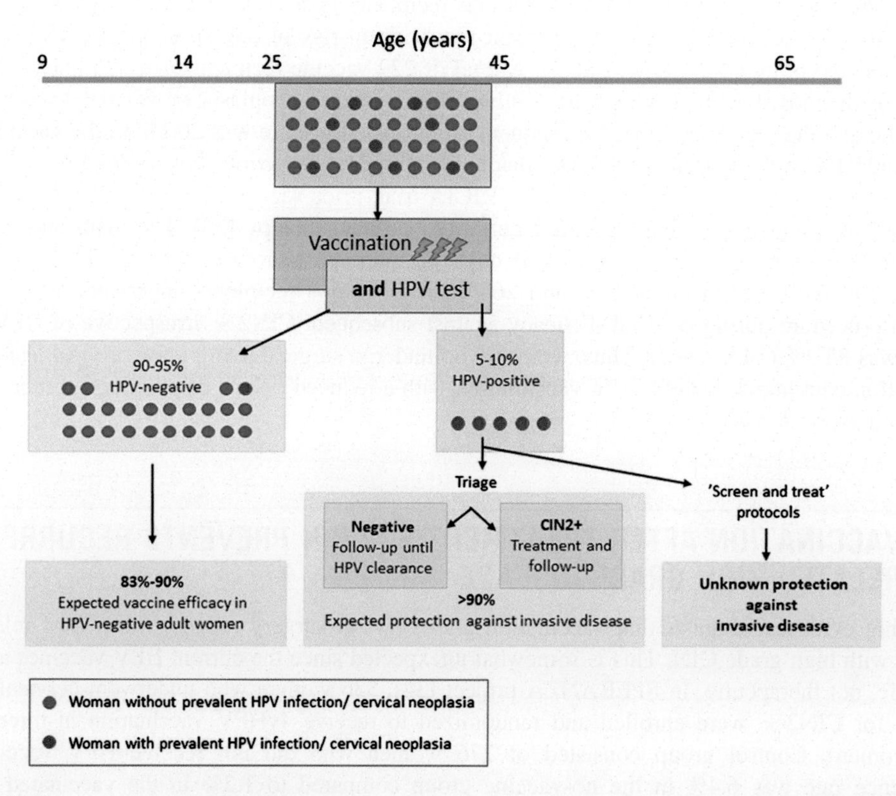

FIGURE 19.7

HPV FASTER: combined screening and vaccination.

Bosch FX, Robles C, D'ıaz M, Arbyn M, Baussano I, Clavel C, et al. HPV-FASTER: broadening the scope for prevention of HPV-related cancer. Nat Rev Clin Oncol 2016;13(2):119–32 [35].

are several explanations; one is primary prevention among women not previously exposed to HPV vaccine types in whom vaccination protects against new HPV infection. Alternatively, HPV vaccination may prevent reactivation of any residual disease by immunological booster effect, or HPV vaccination may simply prevent HPV reinfection.

The clinical implications of these results are obvious. The results may influence the post-treatment management of HPV disease. If these early results are reproduced in randomized trials HPV vaccination post-treatment may play an important role as an adjuvant to surgery.

HPV VACCINE SAFETY

CLINICAL TRIAL SAFETY DATA

Vaccine safety was systematically monitored during the active follow-up period of the vaccine efficacy trials [8,12,13,19,36]. Generally, minor local or systemic vaccine related adverse effects have been more common in the HPV vaccine arms than in the control arms, but these rarely led to discontinuation of the participation in the trials. Serious adverse effects have been rare, and not different in the HPV vaccine arms compared to the control arms. Passive health registry based monitoring of adverse effects continues, and so far no significant signals of serious harm caused by vaccination have been observed.

HEALTH REGISTRY BASED SAFETY DATA

Important safety topics include adverse pregnancy outcome, autoimmune disease (AID), and selected rare syndromes, such as chronic regional pain syndrome (CRPS), postural orthostatic tachycardia syndrome (POTS), chronic fatique syndrome (CFS), and Guillan-Barre syndrome (GBS).

A large nationwide study in Dernmark showed that quadrivalent HPV vaccination during pregnancy was not associated with a significantly higher risk of adverse pregnancy outcomes than no such exposure [37]. The cohort included all women in Denmark who had a pregnancy ending between 2006 and 2013. During the study period, 649,389 records of pregnancies were identified, of which 581,550 were eligible for inclusion in the study cohort. Using nationwide health registers, information on vaccination, adverse pregnancy outcomes, and potential confounders among women in the cohort were linked. Outcomes included spontaneous abortion, stillbirth, major birth defects, small size for gestational age, low birth weight, and preterm birth. In matched analyses, exposure to the quadrivalent HPV vaccine was not associated with significantly higher risks than no exposure for any of the adverse outcomes (Table 19.5). However, many pregnancy outcomes are relatively rare, and even larger studies would be needed to address such outcomes. The data from Denmark are important and reassuring and strongly support the safety of HPV vaccines if inadvertently given in pregnancy. This complements previous safety reports of HPV vaccines in nonpregnant women (Lancet Editorial 2017).

Safety and risk for new onset autoimmune diseases by the HPV16/18 vaccine was assessed during the community randomized trial in Finland in which 14,838 adolescents (females and males)

Table 19.5 qHPV vaccine and adverse pregnancy outcome: Denmark 2006-2013 Joining the dark side: a pathology academic joins a vaccine manufacturer.

Outcome	Vaccine −	Vaccine +	RR (95%CI)
MBD	220/6660	65/1665	1.19 (0.9−1.58)
Spont Ab	131/1852	20/463	0.71 (0.45−1.14)
Preterm	407/7096	116/1774	1.15 (0.93−1.42)
Low BW	277/7072	76/1768	1.10 (0.85−1.43)
SFGA	783/7072	171/1768	0.86 (0.72−1.02)
Stilbirth	4/2004	2/501	2.43 (0.45−13.21)

MBD *Major birth defects;* Low BW, *low birth weight;* SFGA, *small for gestational age.*
Modified from Scheller NM, Pasternak B, Mølgaard-Nielsen D, Svanström H, Hviid A. Quadrivalent HPV vaccination and the risk of adverse pregnancy outcomes. N Engl J Med 2017;376(13):1223−33 [37].

received the HPV vaccine and 17,338 adolescents received HBV control vaccine [38]. In an interim analysis, based on active monitoring during the trial and subsequent passive safety surveillance by health registries, the rates of serious adverse effects did not differ between the study arms. No increase in the incidence rates of a total of 18 new onset autoimmune diseases (including ulcerative colitis, juvenile arthritis, celiac disease, insulin dependent diabetes mellitus [IDDM], Crohn's disease, and others) were observed between the study arms. Lower incidence of IDDM was observed in the HPV vaccine recipients. These safety results are also reassuring. However, safety monitoring continues, and more population-based data will emerge.

The bivalent HPV16/18 vaccine was introduced into the Finnish national vaccination program in November 2013 for girls aged 11−13 with a catch-up for girls aged 14−15. The association between the HPV vaccine and selected AIDs and clinical syndromes was evaluated by conducting a nationwide retrospective register-based cohort study. Vaccination records and national hospital discharge register were linked. Of all girls eligible for vaccination 56% were vaccinated. After adjustments HPV vaccination was not associated with any of the 38 selected outcomes. These results provide valid population based evidence against public skepticism, fear of adverse effects and opposition to HPV vaccination, and can increase HPV vaccination coverage rates in Finland and elsewhere [39].

Similar results have been reported from elsewhere as well [40−42].

Health register based cohort study from Sweden 2006−2012 found that the quadrivalent vaccine was not associated with new onset AIDs (including 49 pre-specified AIDs) in 70,265 women with pre-existing AID [43]. Furthermore, another health register based study from Sweden 2006−2012 of girls with pre-existing AID, showed no increase in new-onset AIDs in this preselected population [44].

European Medicines Agency (EMA) has performed a comprehensive review of chronic regional pain syndrome (CRPS), postural orthostatic tachycardia syndrome (POTS), and chronic fatigue syndrome (CFS) evidence and concluded that the existing evidence does not support any causal association between HPV vaccines and these rare syndromes. The conclusion by EMA was that benefits of HPV vaccination continue to outweigh the risks. Individuals reporting such rare SAEs often had such signs or symptoms already pre-vaccination [45].

HPV vaccination coverage in relation to vaccine safety is an extremely sensitive issue [46]. A striking real life example took place in Denmark where national school-based HPV vaccination program in females was introduced 2009 with high vaccination coverage exceeding 80%. However, after heavy social media reporting of adverse effects by a strong anti-vaccine campaign HPV vaccine confidence dropped dramatically leading to major hesitancy among adolescents. Vaccination coverage rapidly dropped first to 29% and then to 15%. Only after organized systematic efforts and an organized Stop HPV-campaign the HPV vaccination coverage is finally creeping up, currently already exceeding 50%. Unfortunately, this episode reflects lost opportunity of primary prevention of cervical and other cancers in Denmark [47,48].

Overall, HPV vaccines are the best examples of human vaccines most thoroughly evaluated in major exceptionally large global clinical trials. The carefully conducted clinical trials and post-licensure studies have ensured vaccine safety.

CONCLUSION

The development of prophylactic HPV vaccines has been a major success story. HPV vaccines have met and exceeded expectations in preventing infection and disease caused by HPVs. High efficacy against infection and disease endpoints demonstrated in clinical trials have translated into high efficacy in real life. Emerging cancer registry based data demonstrates that HPV vaccination prevents not only high grade precancers, but also invasive HPV-related cancers. Clinical trials have proven that the vaccination program should be gender neutral. This is even more important in situations when the vaccination coverage is low to moderate.

Strategies for best combinations of HPV vaccination and HPV screening need to be evaluated. Vaccination coverage remains still unsatisfactory and too low also in many high income countries, to optimize health benefits. The ultimate goal is the best use of resources to have the maximum impact in preventing HPV-associated cancers. The main problem is the lack of access to HPV vaccination in low income countries where cervical screening is inadequate or absent and where most cancer cases and cancer deaths occur.

Extension of routine vaccination programs to older women aged 25–45 paired with at least one "once-in-a-lifetime" HPV screening test at age 30 years or older has the potential to accelerate the decline in cervical cancer incidence rates in many high income and low income countries. This strategy has been introduced as HPV-FASTER [35] in order to broaden the scope for prevention of HPV-related cancers. Thus, one strategy is to screen and then vaccinate HPV DNA negative individuals (i.e., screen and vaccinate). However, such strategy is not necessarily cost-effective. Alternatively, using vaccinate and screen strategy, vaccinated women who test hrHPV positive would remain as target for diagnostic and therapeutic procedures (Fig. 19.7).

A VIGNETTE: JOINING THE DARK SIDE: A PATHOLOGY ACADEMIC JOINS A VACCINE MANUFACTURER

David Jenkins, emeritus professor of Pathology, University of Nottingham, UK and former Director of HPV 16/18 prophylactic vaccine clinical development, GlaxoSmithKline, Belgium

In 2003 after years in the British National Health Service and University medical schools, I moved as a specialist in the clinical science of HPV to run the clinical trials of the bivalent HPV16/18 vaccine developed by GlaxoSmithKline, later marketed as Cervarix. My pathology is very much about developing diagnosis in clinical practice and the link between pathology and HPV vaccine clinical trials followed naturally from fascination with developing cervical screening and management of cervical neoplasia. I was an established professor and head of department, but did not enjoy removing university status from and decanting into the NHS other senior pathology academics who were distinguished experts in diagnosis and wrote standard pathology textbooks but were deemed by dogmatic university management not to fit their vision of a brave new world of molecular pathology.

"Joining the dark side" was a casual comment from a battle-hardened leader of a rival company's research. This was not unfair, as relations between academics and industry in the UK seemed to be based on mutual suspicion, and clinical research was the loser. But, only the dark side seemed interested and capable of running large scale clinical trials that would determine the value of HPV vaccines. The dark side could be very bright. The phases 2 and 3 trials of the HPV16/18 vaccine was a huge, well-resourced research program with high standards of clinical research, HPV testing, and pathology demanded by regulatory authorities in the USA and the European Union. Certainly, the marketing men and women could oversimplify the science, and the color of packaging could excite management more than issues of serology or molecular HPV testing. Competition between the companies made minor differences in study design and results more important than the huge potential for public health benefits of both vaccines, but also was a challenge to make sure that all the testing was as good as possible.

Most senior academics at the trial centers were very happy to work with an established scientist although marketing always ensured that the role of the company scientist was hidden in the middle of the author list in any publication, however, key the role. Phases 2 and 3 trials of HPV vaccine raised all sorts of technical and scientific issues and problems of study design, with important issues of developing reliable sensitive HPV testing and standardization and reliability of biopsy pathology endpoints. The reproducible definition and clinical meaning of high-grade lesions was and still remains a big public health problem. However, the quality of monitoring of data collection on the huge multi-national phase 3 trial with over a hundred centers exceeded anything an academic epidemiology study usually achieved.

Apart from the trials there were epidemiology studies to work on to provide background data on countries that had escaped previous academic study and where the vaccine might be marketed, and also to support development of virological endpoints of persistent infection and serological studies for future vaccine trials. Teaching was also important as background knowledge of HPV was very variable in different parts of the world involved in the trials. The worlds of "medical affairs" and "political affairs" were mysterious, separate, corporate areas, attempting to influence important "key opinion leaders," and political players, but as long as the science was kept good and straight this was not a real problem.

The science was exciting: the high level of protection against infection and lesions with the sudden realization of cross-protection appeared to the surprise of the pundits. Proving this finding was fascinating, and developed into delving into the complications of multiple HPV infections, commoner than expected, and the complex cervical lesions they could produce. We studied this through several new techniques, including laser-capture microscopy combined with PCR. The HPV 16/18 vaccine clearly worked and seemed potent, even with incomplete dosage.

There is a certain enforced anonymity in working in the pharmaceutical world and in the long term, no way to develop a real disease speciality. If you stay in that world, you would end up doing whatever research the company demands. As an exercise in early retirement it was good science and threw up more interesting questions, many of which still provide me with plenty of excitement. The commercial decision not to promote or develop Cervarix was disappointing. The issues around the limitations of relying heavily on business methods and enterprise aimed at profit by exploiting patents to deliver good global public health benefits equally and fairly remain to my mind a huge, unresolved problem for the world.

REFERENCES

[1] Roden RBS, Stern PL. Opportunities and challenges for human papillomavirus vaccination in cancer. Nat Rev Cancer 2018;18(4):240–54.

[2] Tainio K, Athanasiou A. Tikkinen Kao, Aaltonen R, et al. Clinical course of untreated cervical intraepithelial neoplasia garde 2 under active surveillance; A systmeatic review and meta-analysis. BMJ 2018;360: K409.

[3] Koutsky L, Ault K, Wheeler C, Brown D, Barr E, Alvarez F, et al. A controlled trial of a human papillomavirus type 16 vaccine. N Engl J Med 2002;347:1645–51.

[4] Harper D, Franco E, Wheeler C, Ferris D, Jenkins D, Schuind A, et al. Efficacy of a bivalent L1 virus-like particle vaccine in prevention of infection with human papillomavirus types 16 and 18 in young women: a randomised controlled trial. Lancet 2004;364(9447):1757–65.

[5] Villa LL, Ault KA, Giuliano AR, Costa RL, Petta CA, Andrade RP, et al. Immunologic responses following administration of a vaccine targeting human papillomavirus Types 6, 11, 16, and 18. Vaccine 2006;24 (27-28):5571–83.

[6] Villa LL, Costa RL, Petta CA, Andrade RP, Paavonen J, Iversen OE, et al. High sustained efficacy of a prophylactic quadrivalent human papillomavirus types 6/11/16/18 L1 virus-like particle vaccine through 5 years of follow-up. Br J Cancer 2006;95(11):1459–66.

[7] Kjaer SK, Sigurdsson K, Iversen OE, Hernandez-Avila M, Wheeler CM, Perez G, et al. A pooled analysis of continued prophylactic efficacy of quadrivalent human papillomavirus (Types 6/11/16/18) vaccine against high-grade cervical and external genital lesions. Cancer Prev Res (Phila) 2009;2(10):868–78.

[8] Muñoz N, Kjaer SK, Sigurdsson K, Iversen OE, Hernandez-Avila M, Wheeler CM, et al. Impact of human papillomavirus (HPV)-6/11/16/18 vaccine on all HPV-associated genital diseases in young women. J Natl Cancer Inst. 2010;102(5):325–39.

[9] Garland SM, Steben M, Sings HL, James M, Lu S, Railkar R, et al. Natural history of genital warts: analysis of the placebo arm of 2 randomized phase III trials of a quadrivalent human papillomavirus (types 6, 11, 16, and 18) vaccine. J Infect Dis 2009;199:805–914.

[10] The FUTURE I/II Study Group. Four year efficacy of prophylactic human papillomavirus quadrivalent vaccine against low grade cervical, vulvar, and vaginal intraepithelial neoplasia and anogenital warts: randomised controlled trial. BMJ Cite this as: BMJ 2010;340:c3493. Available from: https://doi.org/10.1136/bmj.c3493.

[11] Joura EA, Garland SM, Paavonen J, Ferris DG, Perez G, Ault KA, et al. Effect of the human papillomavirus (HPV) quadrivalent vaccine in a subgroup of women with cervical and vulvar disease: retrospective pooled analysis of trial data. BMJ 2012;27(344):e1401.

[12] Lehtinen M, Paavonen J, Wheeler CM, Jaisamrarn U, Garland SM, Castellsagué X, et al. Overall efficacy of HPV-16/18 AS04-adjuvanted vaccine against grade 3 or greater cervical intraepithelial neoplasia: 4-year end-of-study analysis of the randomised, double-blind PATRICIA trial. Lancet Oncol 2012;13(1):89–99.

[13] Paavonen J, Naud P, Salmerón J, Wheeler CM, Chow SN, Apter D, et al. Efficacy of human papillomavirus (HPV)-16/18 AS04-adjuvanted vaccine against cervical infection and precancer caused by oncogenic HPV types (PATRICIA): final analysis of a double-blind, randomised study in young women. Lancet 2009;374(9686):301–14.

[14] Paavonen J, Jenkins D, Bosch FX, Naud P, Salmerón J, Wheeler CM, et al. Efficacy of a prophylactic adjuvanted bivalent L1 virus-like-particle vaccine against infection with human papillomavirus types 16 and 18 in young women: an interim analysis of a phase III double-blind, randomised controlled trial. Lancet 2007;369(9580):2161–70.

[15] Apter D, Wheeler CM, Paavonen J, Castellsagué X, Garland SM, Skinner SR, et al. Efficacy of human papillomavirus 16 and 18 (HPV-16/18) AS04-adjuvanted vaccine against cervical infection and precancer in young women: final event-driven analysis of the randomized, double-blind PATRICIA trial.

[16] Lehtinen M, Dillner J. Clinical trials of human papillomavirus vaccine and beyond. Nat Rev Clin Oncol 2013;10:400–10.

[17] Hildesheim A, Wacholder S, Catteau G, Struyf F, Dubin G, Herrero R. CVT Group. Efficacy of the HPV-16/18 vaccine: final according to protocol results from the blinded phase of the randomized Costa Rica HPV-16/18 vaccine trial. Vaccine 2014;32(39):5087–97.

[18] Kreimer AR, Struyf F, Del Rosario-Raymundo MR, Hildesheim A, Skinner SR, Wacholder S, et al. Efficacy of fewer than three doses of an HPV-16/18 AS04-adjuvanted vaccine: combined analysis of data from the Costa Rica Vaccine and PATRICIA Trials. Lancet Oncol 2015;16(7):775–86.

[19] Huh WK, Joura EA, Giuliano AR, Iversen OE, de Andrade RP, Ault KA, et al. Final efficacy, immunogenicity, and safety analyses of a nine-valent human papillomavirus vaccine in women aged 16-26 years: a randomised, double-blind trial. Lancet 2017;390(10108):2143–59.

[20] Joura EA, Giuliano AR, Iversen OE, Bouchard C, Mao C, Mehlsen J, et al. Broad Spectrum HPV Vaccine Study. A 9-valent HPV vaccine against infection and intraepithelial neoplasia in women. N Engl J Med 2015;372(8):711–23.

[21] Toh ZQ, Kosasih J, Russell FM, Garland SM, Mulholland EK, Licciardi PV. Recombinant human papillomavirus nonavalent vaccine in the prevention of cancers caused by human papillomavirus. Infect Drug Resist 2019;12:1951–67.

[22] Wheeler CM, Skinner SR, Del Rosario-Raymundo MR, Garland SM, Chatterjee A, Lazcano-Ponce E, et al. Efficacy, safety, and immunogenicity of the human papillomavirus 16/18 AS04-adjuvanted vaccine in women older than 25 years: 7-year follow-up of the phase 3, double-blind, randomised controlled VIVIANE study. Lancet Infect Dis 2016;16(10):1154–68.

[23] Lehtinen M, Söderlund-Strand A, Vänskä S, Luostarinen T, Eriksson T, Natunen K, et al. Impact of gender-neutral or girls-only vaccination against human papillomavirus-Results of a community-randomized clinical trial (I). Int J Cancer 2018;142(5):949–58.

[24] Garland SM, Kjaer SK, Muñoz N, Block SL, Brown DR, DiNubile MJ, et al. Impact and effectiveness of the quadrivalent human papillomavirus vaccine: a systematic review of 10 years of real-world experience. Clin Infect Dis 2016;63(4):519–27.

[25] Joura EA, Leodolter S, Hernandez-Avila M, Wheeler CM, Perez G, Koutsky LA, et al. Efficacy of a quadrivalent prophylactic human papillomavirus (types 6, 11, 16, and 18) L1 virus-like-particle vaccine against high-grade vulval and vaginal lesions: a combined analysis of three randomised clinical trials. Lancet 2007;369(9574):1693–702.

[26] Brown DR, Kjaer SK, Sigurdson K, et al. The impact of quadrivalent human papillomavirus (HPV; types 6, 11, 16, and 18) L1 virus-like particle vaccine on infection and disease due to oncogenic nonvaccine HPV types in generally HPV-naive women aged 16–26 years. J Infect Dis 2009;199(7):926–35.

[27] Wheeler CM, Kjaer SK, Sigurdson K, et al. The impact of quadrivalent human papillomavirus (HPV; types 6, 11, 16, and 18) L1 virus-like particle vaccine on infection and disease due to oncogenic nonvaccine HPV Types in sexually active women aged 16–26 years. J Infect Dis 2009;199(7):936–44.

[28] Wheeler CM, Castellsagué X, Garland SM, Szarewski A, Paavonen J, Naud P, et al. Cross-protective efficacy of HPV-16/18 AS04-adjuvanted vaccine against cervical infection and precancer caused by non-vaccine oncogenic HPV types: 4-year end-of-study analysis of the randomised, double-blind PATRICIA trial. Lancet Oncol 2012;13(1):100–10.

[29] Malagón T, Drolet M, Boily MC, Franco EL, Jit M, Brisson J, et al. Cross-protective efficacy of two human papillomavirus vaccines: a systematic review and meta-analysis. Lancet Infect Dis 2012;12 (10):781–9.

[30] Mesher D, Soldan K, Lehtinen M, Beddows S, Brisson M, Brotherton JM, et al. Population-level effects of human papillomavirus vaccination programs on infections with nonvaccine genotypes. Emerg Infect Dis 2016;22(10):1732–40.

[31] Palefsky JM, Giuliano AR, Goldstone S, Moreira Jr ED, Aranda C, Jessen H, et al. HPV vaccine against anal HPV infection and anal intraepithelial neoplasia. N Engl J Med 2011;365(17):1576–85.

[32] Chow EPF, Machalek DA, Tabrizi SN, Danielewski JA, Fehler G, Bradshaw CS, et al. Quadrivalent vaccine-targeted human papillomavirus genotypes in heterosexual men after the Australian female human papillomavirus vaccination programme: a retrospective observational study. Lancet Infect Dis 2017;17(1):68–77.

[33] Garland SM, Paavonen J, Jaisamrarn U, Naud P, Salmerón J, Chow SN, et al. Prior human papillomavirus-16/18 AS04-adjuvanted vaccination prevents recurrent high grade cervical intraepithelial neoplasia after definitive surgical therapy: post-hoc analysis from a randomized controlled trial. Int J Cancer 2016;139(12):2812–26.

[34] Ghelardi A, Parazzini F, Martella F, Pierall A, Bay P, Tonetti A, Svelato A, et al. SPERANZA project. Gynecol Oncol 2018;151(2):229–34.

[35] Bosch FX, Robles C, D'ıaz M, Arbyn M, Baussano I, Clavel C, et al. HPV-FASTER: broadening the scope for prevention of HPV-related cancer. Nat Rev Clin Oncol 2016;13(2):119–32.

[36] Moreira Jr ED, Block SL, Ferris D, Giuliano AR, Iversen OE, Joura EA, et al. Safety profile of the 9-valent HPV vaccine: a combined analysis of 7 phase III clinical trials. Pediatrics 2016;138(2): e20154387.

[37] Scheller NM, Pasternak B, Mølgaard-Nielsen D, Svanström H, Hviid A. Quadrivalent HPV vaccination and the risk of adverse pregnancy outcomes. N Engl J Med 2017;376(13):1223–33.

[38] Lehtinen M, Lagheden C, Luostarinen T, Eriksson T, Apter D, Harjula K, et al. Ten-year follow-up of human papillomavirus vaccine efficacy against the most stringent cervical neoplasia end-point-registry-based follow-up of three cohorts from randomized trials. BMJ Open 2017;7(8):e015867.

[39] Skufca J, Ollgren J, Artama M, Ruokokoski E, Nohynek H, Palmu AA. The association of adverse events with bivalent HPV vaccination: A nationwide register-based cohort study in Finland. Vaccine 2018;36(39):5926–33.

[40] Willame C, Rosillon D, Zima J, Angelo MG, Stuurman AL, Vroling H, et al. Risk of new onset autoimmune disease in 9- to 25-year-old women exposed to human papillomavirus-16/18 AS04-adjuvanted vaccine in the United Kingdom. Hum Vaccine Immunother 2016;12(11):2862–71.

[41] Frisch M, Besson A, Clemmensen KKB, Valentiner-Branth P, Mølbak K, Hviid A. Quadrivalent human papillomavirus vaccination in boys and risk of autoimmune diseases, neurological diseases and venous thromboembolism. Int J Epidemiol 2018 Feb 7. Available from: https://doi.org/10.1093/ije/dyx273 [Epub ahead of print].

[42] Grinaldi-Bensouda L, Rossignol M, Koné-Paut I, Krivitzky A, Lebrun-Frenay C, Clet J, et al. Risk of autoimmune diseases and human papilloma virus (HPV) vaccines: Six years of case-referent surveillance. J Autoimmun 2017;79:84–90.

[43] Arnheim-Dahlström L, Pasternak B, Svanström H, Sparén P, Hviid A. Autoimmune, neurological, and venous thromboembolic adverse events after immunisation of adolescent girls with quadrivalent human papillomavirus vaccine in Denmark and Sweden: cohort study. BMJ 2013;9(347):f5906.

[44] Grönlund O, Herweijer E, Sundström K, Arnheim-Dahlström L. Incidence of new-onset autoimmune disease in girls and women with pre-existing autoimmune disease after quadrivalent human papillomavirus vaccination: a cohort study. J Intern Med 2016;280(6):618–26.

[45] Mølbak K, Hansen ND, Valentiner-Branth P. Pre-Vaccination Care-Seeking in Females Reporting Severe Adverse Reactions to HPV Vaccine. A registry based case-control study. PLoS One 2016;11(9): e0162520.

[46] Brisson M, Bénard É, Drolet M, Bogaards JA, Baussano I, Vänskä S, et al. Population-level impact, herd immunity, and elimination after human papillomavirus vaccination: a systematic review and meta-analysis of predictions from transmission-dynamic models. Lancet Public Health 2016;1(1):e8–e17.

[47] Hammer A, Petersen LK, Rolving N, Boxill MF, Kallesøe KH, Becker S, et al. Possible side effects from HPV vaccination in Denmark. Ugeskr Laeger 2016;27:178 (26).

[48] Kjaer SK, Nygård M, Dillner J, Brooke Marshall J, Radley D, Li M, et al. A 12-year follow-up on the long-term effectiveness of the quadrivalent human papillomavirus vaccine in 4 Nordic countries. Clin Infect Dis 2018;66(3):339–45.

FURTHER READING

Bruni L, Diaz M, Barrionuevo-Rosas L, Herrero R, Bray F, Bosch FX, et al. Global estimates of human papillomavirus vaccination coverage by region and income level: a pooled analysis. Lancet Glob Health 2016;4 (7):e453–63.

Garland S.M., Joura E.A., Ault K., Bosch X., Brown D., Castellsagué X., et al. Human papillomavirus genotypes from vaginal and vulvar intraepithelial neoplasia in young women. Obstet Gynecol. 2018;132 (2):261–270. Available from: https://doi.org/10.1097/AOG.0000000000002736.

Garland SM, Cheung T-H, McNeill S, Petersen LK, Romaguera J, Vazquez-Narvaez J, et al. Safety and immunogenicity of a 9-valent HPV vaccine in females 12–26 years of age who previously received the quadrivalent HPV vaccine. Vaccine 2015;33(48):6855–64.

Ginsburg O. Global disparities in HPV vaccination. Lancet Glob Health 2016;4(7):e428–9.

Giuliano AR, Palefsky JM, Goldstone S, Moreira Jr ED, Penny ME, Aranda C, et al. Efficacy of quadrivalent HPV vaccine against HPV Infection and disease in males. N Engl J Med 2011;364(5):401–11.

Joura EA, Kjaer SK, Wheeler CM, Sigurdsson K, Iversen OE, Hernandez-Avila M, et al. HPV antibody levels and clinical efficacy following administration of a prophylactic quadrivalent HPV vaccine. Vaccine 2008;26(52):6844–51.

Koutsky LA, Ault KA, Wheeler CM, Brown DR, Barr E, Alvarez FB, et al. A controlled trial of a human papillomavirus type 16 vaccine. N Eng J Med 2002;347(21):1645–51.

Lehtinen M, Eriksson T, Apter D, Hokkanen M, Natunen K, Paavonen J, et al. Safety of the human papillomavirus (HPV)-16/18 AS04-adjuvanted vaccine in adolescents aged 12-15 years: Interim analysis of a large community-randomized controlled trial. Hum Vaccin Immunother 2016;12(12):3177–85.

Luxembourg A, Moeller E. 9-Valent human papillomavirus vaccine: a review of the clinical development program. Expert Rev Vaccin 2017;16(11):1119–39.

Muñoz N, Manalastas Jr R, Pitisuttithum P, Tresukosol D, Monsonego J, Ault K, et al. Safety, immunogenicity, and efficacy of quadrivalent human papillomavirus (types 6, 11, 16, 18) recombinant vaccine in women aged 24–45 years: a randomised, double-blind trial. The Lancet 2009;373(9679):1949–57.

Schiffman M, Saraiya M. Control of HPV-associated cancers with HPV vaccination. Lancet Infect Dis 2017;17(1):6–8.

SECTION

4

ACCUMULATING LONG-TERM EVIDENCE FOR GLOBAL CONTROL AND ELIMINATION OF HUMAN PAPILLOMAVIRUS INDUCED CANCERS

INTRODUCTION

A COMPLEX ISSUE

From the scientific point of view, the basic and the clinical science to show the importance of Human Papillomavirus (HPV) in cancers and to find ways of prevention through screening and prophylactic vaccination has been a challenge but is, in many ways, the easy bit. Fitting change based on these research works into existing health systems and interesting both the public and political decision-makers to recognize the continuing value and safety of these approaches on a global scale is not simple. Throughout its history (Chapter 1: A Brief History of Cervical Cancer) cervical screening has had important issues of failing to reach women at most risk as well as problems with deciding the best management and treatment of screen-detected abnormalities and some of these issues apply also to HPV vaccination. Both screening and vaccination have been affected by changing science, politics of health, and variable public enthusiasm for participation. No public health action works without widespread participation and good coordination. In the UK the death of the "celebrity" Jane Goody from cervical cancer produced a spike in desire for being screened by conventional cytology but this eventually fell off and there is a reluctance in young women to participate in screening in 2019. This has led to a new national campaign to encourage participation, but one uncoordinated with the contemporaneous reduction in cytology provision as part of the switch in the UK to primary HPV screening. The result has problems with the system and long delays in sending out results.

The example of the UK shows an important general message. It is always an issue that the availability of good scientific evidence has to be combined with acceptability to patients or those being screened, to the finance availability and to political and personal priorities in different settings, and then carefully planned. Introducing developments also makes doctors and pharmaceutical and diagnostic companies prey to dramatic impact of media stories that can be positive or can create scares about the safety or benefits of the intervention involved. The development of the internet has allowed pressure groups a means of publicly promoting their own views, not demanding an evidence-base, including attempts to discredit investigators and their science, as seen with anti-vaccination campaigns.

COMPILING EVIDENCE

The medical and scientific approach to implementing developments is to look at increasing the strength of the evidence for and against a change and then make a balanced assessment. In making this assessment the cost, either increasing or saving, it is also important as for introducing new treatments or preventive measures and the outcome of health economic studies is key. In many countries the final decision is based on the assumption that the cost per quality adjusted life year saved should not exceed the GDP on a per-head basis. Although rather arbitrary, this has become an important practical political and commercial standard. However, this takes no account of the individual and their different attitudes to risk and to particular interventions. Chapter 20, The Key Role of Mathematical Modeling and Health Economics in the Public Health Transitions in Cervical

Cancer Prevention, explains how the use of mathematical modeling has evolved from early Markov chain models exploring the role of possible changes in cervical screening interval, coverage, and testing procedures that provided support for the importance of high population coverage and for primary HPV screening over cytology screening with or without HPV testing in triage. This approach has now developed into the current dynamic models including the processes of acquiring infection and HPV persistence and also detailed costing data to provide more detailed comprehensive economic modeling of different preventive interventions, their interactions, and their costs.

However, the same evidence can still lead to different conclusions: supporting the decision to introduce primary HPV screening in the Netherlands; while in the USA the same evidence has led to compromise proposals on combining HPV testing and cytology in primary screening.

INTRODUCING HUMAN PAPILLOMAVIRUS VACCINATION IN PRACTICE

Chapter 21, Twelve Years of Vaccine Registration and the Consequences, and Chapter 22, Political and Public Responses to Human Papillomavirus Vaccination, deal with the additional evidence and issues that have arisen concerning HPV vaccination since the first registration of the quadrivalent vaccine. Ongoing study in practice provides information about longer-term effectiveness, the strength of the protective response, and the impact of vaccination on screening as the predicted fall in positive predictive value of cytology and colposcopy for precancer in vaccinated women has actually happened [1]. Post-trial experience has also begun to yield the first evidence of an effect of vaccination on reducing the incidence of invasive cervical cancer, on the ability to change from a three-dose to a two-dose vaccination schedule and the possibility of a one-dose schedule, on the very high level of vaccine safety, and the overall benefits of vaccinating boys as well as girls.

THE IMPORTANCE OF VACCINE SAFETY

Nonetheless vaccination in general, including HPV vaccination, despite being probably the most effective medical intervention in diseases ever, remains a controversial issue, with some strong opposition to all vaccination and a widespread sensitivity, sometimes leading to strong public, political, and media reaction to the possibility of side effects. The opposition has not necessarily been dispelled by the extensive evidence of the overall safety of HPV vaccination. The issues around vaccination safety are complicated and individual concerns are not all the same. Problems with vaccine acceptance range from outright opposition based on personal beliefs that do not recognize or trust how science works, through deliberate spreading of misinformation, to reasonable concerns about being presented with adequate, clear evidence on the relative risks of being vaccinated versus the risks of cervical or other precancer and cancer. The issue of personal responsibility is important, particularly that of parents in making a choice to vaccinate their child. There is natural concern that choosing vaccination, could, however unlikely, result in a healthy child suffering a serious side-effect. These issues have surfaced over a number of vaccines. Here freedom of individual choice can be seen in conflict with best public health policy in preventing disease in the whole population [2].

It is important to be aware that inoculation to prevent infection has been controversial since its first introduction as variolation-the administration of serum exuded from the lesions of a mild case of smallpox under controlled conditions to protect against a serious epidemic infection. This was documented in relation to variolation during a smallpox epidemic in 1721 in Boston USA. Between 1770 and 1796 Edward Jenner in England developed vaccination with cowpox as a safer alternative that formed the basis for the eradication of smallpox until it was completed in 1980. Although this vaccine was very effective, there were problems with bacterial contamination leading to deaths from tetanus in the USA in 1901. Subsequent development in the 20th century of polio, typhoid, and other vaccines were generally more successful, although there were cases of polio from the live Sabin vaccine, and in 1982 it became evident that whole-cell pertussis vaccine was associated with serious side effects, particularly infrequent brain damage. This led to failure of vaccination campaigns and also major changes in attitudes to vaccine regulation [3].

The issue of brain-damage related to Pertussis vaccine was succeeded by worries about the relation of vaccines to the increased frequency of autism in the USA and Europe. In the USA the issue of injecting unacceptably high levels of mercury, as thiomersal preservative used in vaccines, became a major cause for parents of autistic children in a psychiatric environment that often blamed them for their child's autism, and was also pursued actively by trial lawyers. In 1998 concerns about the relation of vaccines to autism switched to the MMR vaccine as a result of research by Andrew Wakefield in London. Poor quality research he carried out on his hypothesis that measles virus drove inflammatory bowel disease led eventually to a small, very selective case series of autism and inflammation of the bowel being published in the Lancet. Although scientifically the results were very inadequate and suspect, at most equivocal, it was promoted as proof of a link between the measles virus component of the MMR vaccine and autism. His work was criticized by other researchers from the start and later discredited and retracted by the Lancet. The suggested link with the measles component of MMR however brought a tidal wave of militant "anti-vax" activism that has continued, particularly in North America for 20 years despite extensive epidemiological investigation disproving any link between MMR and autism [4].

Together with other issues, the suspicions about autism provided a very negative background for HPV vaccination in certain parts of the USA. Conservative views about sexual activity before marriage, often based on religion, and political and public health responses to possible side-effects have also been issues in developing vaccine acceptance. Some of the varied and complex issues in different countries including Denmark, Japan, and India are discussed in detail in Chapter 22, Political and Public Responses to Human Papillomavirus Vaccination.

THE MIXED HISTORY OF THE MEDICAL PROFESSION AND THE PHARMACEUTICAL INDUSTRY

The past history of the medical profession and of the pharmaceutical industry has not always helped create an atmosphere of trust. Clinical researchers have not always held to the highest ethical standards of informed consent. The story of HeLa cells, the cultured cell line originally taken in 1951 at Johns Hopkins Hospital Baltimore, USA, from the cancer of the cervix, of a black woman from Virginia, Henrietta Lacks, is an example of the studies conducted without consent in the past. Between 1954 and 1966 the US cancer researcher Chester Southam injected HeLa cells into patients and prisoners without clear consent, producing tumor nodules that generally regressed. This series of "illegal, immoral and deplorable" studies (as some doctors asked to participate labeled them) led in the USA to the introduction of independent research review boards to ensure that ethical standards were met. HeLa cells were later shown to be driven by HPV 18. The position of HeLa cells was further complicated by their successful and profitable commercialization, while her poor family was left wondering about the soul of Henrietta and unable to pay for decent healthcare [5].

Research ethics and safety monitoring have substantially improved and become increasingly rigorous over the 40 years of HPV research. However, issues of drug safety with VIOXX affecting Merck and Seroxat affecting GSK contributed to public suspicion toward these companies. This has led to considerable attention being paid to safety issues by the vaccine manufacturers and regulators.

GOOD EVIDENCE OR BELIEF?

Collection and analysis of good data are important but people are individuals with their own view of life that does not necessarily correspond to the way in which committed scientists and public health experts interpret laboratory and epidemiological data. Perception of the nature of vaccination and the individual assessment of risks of disease versus those of vaccination are important and are complicated by the huge amount of misleading information and emotional pressure that can be imposed on people through the internet by those who see opposition to vaccination as a personal, political crusade or Jihad, against an authoritarian regime of medicine and big pharma. It is important to educate and protect from the actions of these groups, but also to understand that for many more people there are real concerns about how they judge their or their child's risk or benefit that are not simply allayed by presenting statistical evidence. For many concerned parents, any possible harmful effect of a vaccine on the individual is more important than statistical population assessment of epidemiologists and health economists [2]. This becomes particularly important when the key outcome of screening and vaccination is prevention of a relatively infrequent but serious outcome, cancer, of a common infection by HPV has to be judged against the more frequent, unpleasant inconvenience with possible, infrequent serious side effects of precancerous lesions, and the possible risks of screening or vaccination.

CANCER PREVENTION MEETS POLITICS GLOBALLY

Science cannot escape politics, however much scientists try to avoid this through collecting "evidence" and its logical analysis. "In a way there is only political life." writes Adam Phillips as a psychoanalyst. A former freedom fighter, officer of the armed wing of the South African ANC and a close colleague of Nelson Mandela, put it as "All life is politics." The issues of preventing HPV-related cancer are inevitably political because it's about people and their lives and how people with different views about their lives live together. Ideological or party politics may help or hinder the understanding of the interpersonal issues around HPV, cancer prevention, screening, and vaccination.

HPV screening and vaccination are being introduced in increasing numbers of high-income countries, although acceptance, particularly of vaccination is very variable. Major current concerns in the USA and Europe are about integrating these changes and implementing high levels of coverage. High coverage of vaccination has been achieved most effectively through school-based delivery rather than by individual pediatricians or general practitioners.

In Chapter 22, Political and Public Responses to Human Papillomavirus Vaccination, Gregory Zimet and his coauthors discuss what has happened in different situations when HPV vaccination has been introduced and what can be done to limit the public concerns and panic about events that occur after vaccination that might represent possible side effects. An important aim of any program of vaccination must include close monitoring of possible side effects, their rapid investigation, and provision of accurate information about any risks.

The greatest global burden of HPV-related neoplasia, however, falls on low and middle income countries, with up to 14%–15% of cancer in sub-Saharan Africa and India attributable to HPV. Over 85% of cancer attributable to HPV globally is cervical cancer, which remains the third most common female malignancy worldwide [6]. HPV is nearly exclusively sexually transmitted and is not strongly affected by general improvements in medical and living standards. In some parts of the world the experience of women is no different from that reported historically in Europe centuries ago.

Screening and vaccination, however, provide the opportunity for control and even elimination of cervical and other HPV-related cancers. In Chapter 23, Expanding Prevention of Cervical Cancer in Low- and Middle-Income Countries, Lynette Denny and colleagues discuss the different high- and low-tech approaches to prevention that have been explored and started in some low and middle-income countries. Preventing cancer globally is the most important challenge for using the knowledge of HPV and cervical cancer, screening, and vaccination that has been accumulated over the past 40 years. In meeting this challenge, ethical ideas of equity in provision of preventive services globally and the personal and national beliefs, identities, and ideologies become critical. In this forum it is even more important that good science-based judgements and humanist values prevail.

REFERENCES

[1] Palmer T, Wallace L, Pollock KG, et al. Prevalence of cervical disease at age 20 after immunisation with bivalent HPV vaccine at age 12–13 in Scotland: retrospective population study. BMJ 2019;365:l1161.
[2] Bliss E. On immunity: an inoculation. Minneapolis, MN: Graywolf Press; 2014.
[3] Allen A. Vaccine: the controversial story of medicine's greatest lifesaver. New York: WW Norton and Company; 2007.
[4] Mnookin S. The panic virus. New York: Simon and Schuster; 2011.
[5] Skloot R. The immortal life of Henrietta Lacks. London: Pan; 2011.
[6] Forman D, de Martel C, Lacey CJ. Global burden of human papillomavirus and related diseases. Vaccine 2012;30S:F12–23.

CHAPTER

20

THE KEY ROLE OF MATHEMATICAL MODELING AND HEALTH ECONOMICS IN THE PUBLIC HEALTH TRANSITIONS IN CERVICAL CANCER PREVENTION

Karen Canfell[1,2,3] and Johannes Berkhof[4]

[1]Cancer Research Division, Cancer Council NSW, Sydney, NSW, Australia [2]Prince of Wales Clinical School, The University of New South Wales, Sydney, NSW, Australia [3]School of Public Health, University of Sydney, Sydney, NSW, Australia [4]Department of Epidemiology and Biostatistics, Amsterdam UMC, Vrije Universiteit, Amsterdam, The Netherlands

Everything should be made as simple as possible, but not simpler

Albert Einstein

INTRODUCTION

Mathematical modeling has played a key role in the implementation of major public health transitions in the prevention of cervical and other human papillomavirus (HPV)-related cancers. The natural history of HPV infection and cervical cancer is the best understood of any cancer [1], and this has facilitated the development of comprehensive models. In the 1970s, evidence was building on the epidemiological links between sexual behavior exposures and cervical cancer outcomes [2], and in the 1980s and 1990s, HPV was unequivocally causally linked to cervical cancer [3,4]. Through the 1990s and 2000s, a wealth of evidence has emerged on the natural history of HPV and the progression of HPV infection to precancer and cancer at several anogenital and oropharyngeal sites in women and men. The attributable fraction of HPV-related cancer at any site is, however, highest for cervical cancer, and this cancer is one of the most common in women globally with 569,847 new cases and 311,365 deaths estimated from cervical cancer in 2018 [5]. The key insight that almost all cervical cancers are caused by HPV infection [4], has underpinned the development of prophylactic vaccines against HPV infection, and has also underpinned the development of HPV DNA/RNA testing as a primary screening approach. These emergent insights into the natural history of HPV infection, and the policy evaluation needs generated by the availability of these new primary and secondary prevention opportunities, together drove the development of an entirely new field: HPV and cervical cancer modeling.

Human Papillomavirus. DOI: https://doi.org/10.1016/B978-0-12-814457-2.00020-9

Cervical cancer models have brought together and harnessed techniques in infectious disease modeling, but also draw heavily on cancer biology and the epidemiological understanding of cancer at the population level. Their development has thus required major multidisciplinary collaboration. As the natural history of cervical cancer is comparatively very well characterized, success in cervical cancer modeling has also been an exemplary case from the broader perspective of developments in modeling cancer prevention and screening at other cancer sites [6,7]. For cervical cancer prevention, modeling has not only helped characterize health economic outcomes for new vaccination and screening approaches, but impact modeling has been particularly important to the understanding of population-level effects and timing, given the very long timeframe between intervention (e.g., vaccination of preadolescents and young adolescents), and prevention of cancer in mid-adult and older women. Models have played a key role in public-health decision making since trials cannot cover the complexities of all feasible implementation strategies in real-world settings and across all countries. Driven by all these factors, the last 25 years has seen extraordinary development of the field, and major advances in platforms to simulate the natural history of HPV infection, precancer, cancer, and vaccination and screening interventions. Over this period, models have benefited from the emergence of observational evidence on long-term outcomes after HPV infection, major randomized controlled trials for both vaccination and screening, technological improvements including increases in computing power, and advances in data fitting techniques.

In May 2018, the Director-General of the World Health Organization (WHO) announced a call-to-action for cervical cancer elimination as a public health problem, and in January 2019, the Executive Board of the WHO endorsed the development of a strategic plan for achieving elimination, to be presented at the World Health Assembly in 2020. Modeling is central to this endeavor and informing the targets for cervical cancer elimination and the development of an investment case to underpin elimination efforts. In this context, this chapter will provide a historical overview of the development of the field, describe the policy challenges and achievements that modeling has contributed to, will provide an overview of the current state-of-the-art modeling platforms and approaches, and will discuss the coming-of-age of cervical cancer modeling in the context of its central role in the WHO elimination efforts.

GENERAL BACKGROUND TO MATHEMATICAL MODELS IN POPULATION HEALTH

Mathematical models have long been used in fields such as civil and electrical engineering (e.g., in control theory), and have also increasingly been harnessed in medicine over the last few decades. In early applications of modeling to questions in medicine or public health, a *Markov* implementation was often used. Static Markov cohort approaches, first described in the early twentieth century, involve the description of a set of health states, with transition probabilities between states. The main assumption of Markov models is that the probability distribution of future health states only depends on the current distribution and not on previous transitions. This assumption makes it easy to model the health distribution over time for a large population. For example, in relation to cervical cancer, a simple description of a commonly implemented static platform could involve setting up a synthetic cohort of 100,000 women from birth to death. Over time, proportions of the cohort progress (and for some stages, can also regress) through states representing healthy/uninfected, HPV infected, cervical precancer [e.g., cervical intraepithelial neoplasia stages 2 and 3 (CIN2/3)], cervical cancer at each clinical stage, and cervical cancer death and other-cause death [8]. An alternate implementation approach is to retain the idea of health states but to explicitly model outcomes for each simulated individual, and this is known as "microsimulation." In a microsimulation, the probability of an individual progressing through the various health states is sampled from a population-level distribution informed by empirical data. As microsimulation stores individual health trajectories, both current and previous health states are allowed to influence future health. Another important concept for models of infectious disease, and thus applicable to the field, is dynamic modeling, wherein the transmission of an infection within a population is explicitly modeled and the age-dependent rates of new infections among susceptible individuals can change over time—this is known as dynamic modeling. For cervical cancer modeling, microsimulation for the natural history component, coupled with a dynamic approach to modeling infection and vaccination, is now considered as state-of-the-art.

Two major challenges in model development are as follows: (1) choosing an appropriate structure, and (2) choosing appropriate parameters which define the probability of transitioning between states. The model structure is influenced by the policy question being considered and must be based on the best understanding of the biology and epidemiology of the disease. Many of the underlying states defined in a natural history model are, by definition, unobservable, and hence model parameters must be fit (or "calibrated") to observable data. Population models of this type are ideally informed by a large range of datasets. As one example, the natural history state of CIN3 and its prevalence in the real-world population of women is unknown, however, as CIN3 is a screen-detected state, the prevalence of detected CIN3 is an observable outcome which can be used as a calibration "target" for models [8]. Following the calibration of the model, independent validation against new datasets is a key aspect of understanding model performance and validity [9]. The most comprehensive models have now been developed over many years and have undergone multiple cycles of (re)calibration and (re)validation [10–14]. It is important to distinguish between these complex platforms, and simpler health economic models which are often spreadsheet based, developed de novo for specific evaluations, and which may only capture simple assumptions about disease. For specific cost-effectiveness assessments, simpler models may suffice but comprehensive model platforms seem indispensable in areas such as modeling the combined impact of HPV vaccination and cervical screening [15] and modeling the timeline to cervical cancer elimination as a public health problem.

HISTORY OF THE EARLY DEVELOPMENT OF CERVICAL CANCER MODELS: THE 1980S AND 1990S

A timeline of some of the key historical events in HPV and cervical cancer modeling over the last three decades is shown in Fig. 20.1. The earliest models of cervical cancer emerged in the 1970s and 1980s; but these early models did not consider HPV infection as a precursor state in development of cervical cancer. One of the earliest accounts is a statistical model by Barron and Richart in 1968 and the early 1970s [16,17]. A more elaborate model was then developed in the mid-1980s by the MISCAN group, early pioneers in this area, from Erasmus University in Rotterdam, The

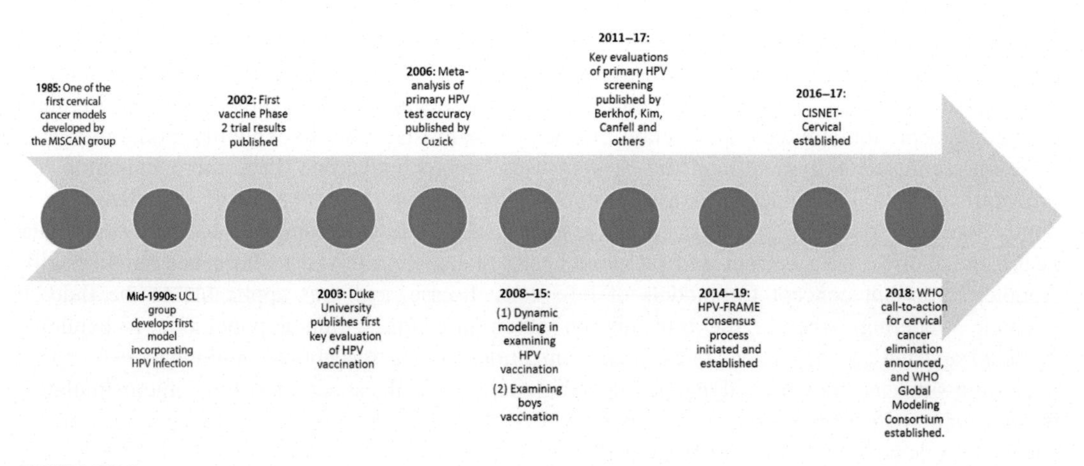

FIGURE 20.1

Timeline of key developments in cervical cancer modeling.

Netherlands [18]. Habbema and colleagues used data from British Columbia and the pilot screening rounds for a national cervical screening program in The Netherlands to inform the development of their models. They constructed a model which simulated states for "no invasive cancer," "preclinical disease", "preclinical invasive disease," and "clinical cervical cancer." They calibrated their model-based assumptions for the mean duration (also known as "*dwell time*") of preclinical disease and for regression of preclinical lesions in the context of limited data sources and only in the context of screening (using the data available at the time), and subsequently reported early illustrative examples of policy evaluations of the health impacts of various screening initiation ages and screening intervals [18–20] as well as the cost-effectiveness of various policies.

In the mid-1990s, a new model platform was developed by a team at the University College London (UCL) [21,22], and the second reported iteration of this platform was the first known model to incorporate HPV infection in pathways for the natural history of disease [22]. This Markov-like platform incorporated states for low- and high-risk (oncogenic) HPV infection, as well as CIN1, CIN2, CIN3, and invasive cervical cancer. Thus its structure formed the basis of many of the models still used today. At the time, the development of the model structure was driven by the need to incorporate HPV infection in the natural history in order to evaluate the potential role of HPV testing in screening, either as a triage for low-grade cytology or as a primary screening test. Thus the UCL group's work represented the first of a large number of evaluations (likely to extend now to several hundred in number) on the use of HPV testing in cervical cancer screening in many countries over the next 25 years. The UCL work was also important because rather than using data from a single screening program, model parameterization was informed from several different data sources from different countries and obtained at different times; this is now a standard approach to cervical cancer modeling (i.e., that the most informative available data worldwide are synthesized in the model). For example, the model used data from early longitudinal studies of the natural history of CIN as well as data from the unethical "unfortunate experiment" in New Zealand to inform the transition rate from CIN3 to invasive cervical cancer; analogous (but updated) data sources and updated analyses of the NZ data are still used by model platforms today [23]. David Jenkins, involved with the UCL team at the time, reports that there was a high level of controversy around the use of early models [24]; their reception was not universally positive and it was "seen as [a] revolutionary idea to use modeling to go beyond the randomized controlled trial" (Personal Communication, D. Jenkins, 2019). In the UCL 1996 evaluation of HPV testing in cervical screening, it was concluded that HPV triage of cytological low-grade could be effective and also that "the potential value of HPV testing as a primary screening method is strongly dependent on the proportion of neoplasias that are HPV-negative." This latter finding prefigures what is still a current discussion about the proportion of cervical cancers that are HPV-negative, and the implications for primary HPV screening; although it should be noted that the current evidence consistently suggests that nearly all cervical cancers are caused by HPV [4], that HPV screening is more sensitive than cytology for the detection of both CIN2+ and CIN3+ [25], and that women who are positive for an oncogenic HPV type are at a much higher risk of developing CIN3+ over time [26–28].

THE 2000S: A NEW FIELD EMERGES

The development of cervical cancer models in the 2000s was greatly facilitated by the fully developed articulation of the role of oncogenic HPV in cervical cancer and the population-level and age-specific patterns of acquisition, progression, regression, and persistence of an infection to CIN3 (precancer) and invasive cervical cancer. This was best expressed in a landmark review by Mark Schiffman, Phil Castle and colleagues from the National Cancer Institute in the United States, which was published in *The Lancet* in 2007 [29]. Their synthesis of the worldwide evidence on the natural history of HPV and cervical cancer, informed by major studies including their large cohort study in Guanacaste, Costa Rica, has been cited on over 2000 occasions to date. Schiffman and colleagues summarized their conceptual model as follows:

> Infection is extremely common in young women in their first decade of sexual activity. Persistent infections and precancer are established, typically within 5–10 years, from less than 10% of new infections. Invasive cancer arises over many years, even decades, in a minority of women with precancer [CIN], with a peak or plateau in risk at about 35–55 years of age. Each genotype of HPV acts as an independent infection, with differing carcinogenic risks linked to evolutionary species.

These concepts, particularly that the risks associated with different oncogenic types differ, with HPV16 associated with the greatest risk of progression [30], and that the HPV infections that eventually lead to most cancers are acquired at a young age, are the basis of the underlying natural history for current simulation platforms of cervical cancer. For example, a recent model-derived estimate of the median age of acquisition of a "causal" HPV infection leading to cervical cancer puts this age at about 21 years [31].

In parallel to this clear articulation of the conceptual model underpinning the development of actual simulations, a wealth of new natural history data became available to inform models of HPV infection and cervical cancer over the decade from 2000 to 2010, as did new trial-based and prospective observational evidence on the effects of primary and secondary prevention. For example, in 2002, first results from vaccine trials showing very high effectiveness of the new prophylactic vaccines in preventing new persistent HPV infections and CIN were reported [32]. This galvanized the field and was closely followed by the first evaluations of cost-effectiveness of HPV vaccination [33,34]. For example, Kulasingam and Myers used a Markov model with states for HPV infection, histological LSIL and HSIL and cervical cancer to evaluate the role potential health and economic effects of an HPV vaccine in a setting with established cervical screening [34]. It should be noted, however, that this early model of vaccination was static and did not consider the additional impact of herd immunity protection conferred by HPV vaccination. Even with this conservative approach, they concluded that HPV vaccination in combination with screening can be cost-effective, but that cost-effectiveness depended on maintaining vaccine effectiveness during the ages of peak HPV incidence. Shalini Kulasingam recalls that the paper was initially rejected by a major scientific journal because the Editor thought at the time that the vaccine was "fictional" (see Box 1). Needless to say, this misunderstanding was rapidly counteracted by events; the decade subsequently saw the publication of key Phase 3 results from prophylactic HPV vaccine trials [35,36]. Over the next 5 years, more than 40 modeled health economic evaluations of vaccination of young females were reported and these underpinned the rapid roll-out of HPV vaccination across high-income countries [37].

In parallel with all these developments, advances in the development of dynamic models of infection were being developed to model infectious disease epidemics, grounded in seminal work from the 1980s and 1990s by Anderson and May [38]. The first publication of a dynamic model of HPV infection, which directly modeled partnership and viral transmission (and could thus take into account herd immunity), was reported in 2006 by Ruanne Barnabas, Geoff Garnett and Colleagues from the University of Oxford and Imperial College London [39]. Barnabas and Garnett, working with Finnish colleagues, developed a single-type HPV16 model of the type known as "*S-I-R-S*" [which considers states for "susceptible to infection," "infected," "recovered from infection/immune" and again, "susceptible" (i.e., the individual has again become susceptible to a new HPV infection)]. They calibrated the dynamic model to HPV16 seroprevalence data from Finland and then used it to derive the first ever estimate of the per-partnership probability of HPV transmission, which they estimated to be 0.6. This estimate and models based on similar dynamic platforms were to inform subsequent policy-decision making in HPV vaccination over the next 15 years [40].

Building on this influential work, several research groups including those at the London School of Hygiene and Tropical Medicine and the UK Health Protection Agency, University Laval, Quebec, Canada, and others, led the development of detailed, calibrated dynamic models of HPV transmission and vaccination. Mark Jit's 2008 analysis of the cost-effectiveness of vaccination in the UK was a landmark, and represented one of the first models to fully take into account herd immunity for each vaccine-included type [41]. This study concluded that universal vaccination would be cost-effective in young females, but not in males, and a series of analyses of male vaccination in the United Kingdom and the United States over the following years reached the same conclusion [37]. Later analyses were able to confirm that the cost-effectiveness of male vaccination in any setting is driven by coverage in females (which provides herd immunity and thus some protection to heterosexual males), the set of HPV-associated tumor sites in females and males included in the analysis and also vaccine price [37,42]. Regarding the latter, strong price reductions as a result of tender-based centralized procurement of the vaccine have led to a more favorable cost-effectiveness profile in some countries [43]. A key follow-up analysis concluded that the incremental benefit of vaccinating boys if vaccine uptake among girls is high is driven by the prevention of anal carcinomas, which the authors felt, underscored the relevance of HPV prevention efforts in men-who-have-sex-with-men (MSM) [44]. Subsequent evaluations considered HPV transmission and targeted vaccination in MSM [45].

Other important milestones in the 2000s included longitudinal cohort data describing outcomes after positive and negative cytology and HPV tests, including cumulative CIN3 + incidence according to HPV type at baseline [26,46]. Newly synthesized data from metaanalysis of HPV and cytology test sensitivity and specificity [47] were published, as were outcomes from a number of RCTs of primary HPV screening versus cytology [48–50]. This emergence of trial data on HPV-based screening greatly facilitated the development of models of the longer term impact of primary HPV screening. For example, in the Netherlands, data from the POBASCAM randomized trial of HPV versus cytology screening [51], initiated in the early 2000s [52] was used to develop and parameterize health economic models that informed the policy decision to introduce HPV as a primary screening test in the Netherlands [53,54]. Their model-based assessment, consistent with an independent assessment by the MISCAN modeling group [53] informed the policy decision to introduce HPV as a primary screening test in the Netherlands, which was implemented in 2017.

In parallel with this work focusing on high-income countries, models were also being reimplemented, adapted, and applied to key issues in cervical cancer control in low- and middle-income country (LMIC) settings. Goldie et al. published an influential initial evaluation of pragmatic strategies for once, twice and thrice-lifetime screening in several LMIC [55]. The first models of HPV vaccination developed by in-country analysts included a Thai model by HITAP [56] and a South African model by Sinanovic et al. [57], both based loosely on the Kulasingam/Myers model [34]. The HITAP model was particularly influential because it led to a delay in the introduction of HPV vaccination in Thailand, as it showed that vaccination was actually not cost-effective at the very high prices that the vaccines were being offered to Thailand at that time, and it would be more cost-effective to introduce organized cytological screening instead.

RECENT DEVELOPMENTS IN CERVICAL CANCER MODELING: 2010–2019

In the decade since 2010, a massive increase in parallel computing power has enabled automated fitting techniques and probabilistic sensitivity analysis, and models have thus been better able to capture the effects of parametric uncertainty on outcomes. Despite these technical advances, valid population model platforms for health outcomes remain grounded in a good understanding of disease and intervention epidemiology, and thus the leading disease modeling groups are led by specialist epidemiologists and/or public health physicians. In addition to provision of essential information for policy making, mathematical models have been increasingly used for studying the dynamics of HPV infections at population, individual and cellular level. Examples include assessments of the impact of a sudden drop in the vaccine uptake, as witnessed in Japan and Denmark on herd effects [58], and the impact of immunity and cellular stochasticity on the clearance of an HPV infection [59].

The clinical and epidemiological evidence base to support cervical cancer modeling has continued to grow, notably with new results from vaccine trials in specific subpopulations, including older women [60], Phase 3 trial results for the next generation nonavalent vaccine [61], and evidence to support the efficacy of two-dose vaccine regimes [61–63]. Models validated against these trial results have now been used for initial evaluation of extended vaccination in adults (sometimes known as "HPV-FASTER" [10]), the cost-effectiveness of next generation vaccines [64–67], and the cost-effectiveness of reduced dose schedules [68–70]. The emergence of real-world impact data from the implementation of vaccination programs in high-income countries worldwide has also enabled model platforms to be subject to "fourth order validation", that is, the model's prior predictions are compared with subsequent real-world health outcomes, including model predictions for herd immunity [71].

From the screening perspective, a key development has been the publication of the pooled analysis of data from four European trials which demonstrated the increased protective effect of primary HPV testing against the development of invasive cervical cancer, compared to cytology-based screening [72]. In addition, a wealth of information on triage testing for HPV-positive women has emerged (as examples, see [73,74]), and within the last few years there has been emerging evidence on the performance of primary HPV screening in HPV-vaccinated populations [75]; these data sources are supporting important modeled evaluations of optimized approaches to screening in the context of HPV vaccination [11,76,77]. A range of important evaluations have also directly underpinned policy decisions to transition to primary HPV screening in national programs, for example, in The Netherlands, Australia, and New Zealand [11,78].

In recent years, an important development has involved the development of consensus standards for reporting HPV models. A key meeting of the active modeling groups at the time was held at the International Papillomavirus Meeting in Malmo, Sweden, in 2009, and this resulted in the publication of a summary of considerations in developing high-quality HPV models [79]. In the subsequent decade, the HPV and cervical cancer modeling community have worked together to produce HPV-FRAME, an initiative to develop a consensus statement and a quality framework specifically for HPV models [8,80]. HPV-FRAME aims to better enable target audiences (e.g., policy-makers) to understand a model's strength and weaknesses in relation to a specific policy question and ultimately improve a model's contribution to informed decision-making. It takes the form of a CONSORT-style itemized checklist encapsulating consensus-based reporting standards for models of interventions in HPV prevention, with the purpose of complementing existing general frameworks. The framework was developed according to an established process for stakeholder involvement and involved a number of workshops held at major international scientific meetings over a period of 5 years. The first version of HPV-FRAME is expected to be published in 2019 and it is designed to be a "living" standard that will need to be adapted and refined according to needs in HPV modeling as they emerge [80].

A related development has been the emergence of key coalitions of leading modeling groups who have come together to perform comparative modeling exercises. The underlying rationale of comparative modeling is that for key policy questions, the robustness of the findings can be increased by bringing together independently developed models which use common data sources for the same policy evaluation. The extent of the variation in outcomes from different models can be very informative in understanding the impact of various structural and parameter assumptions on the final results. One of the first examples from an HPV vaccination modeling perspective was initiated by the WHO [15]. A subsequent large-scale investigator-initiated example led by Marc Brisson from Universite Laval was a comparison of the predicted population-level impact of vaccination on HPV 6, 11, 16, and 18 infections in high-income countries, involving comparison of results from 16 of the 19 HPV transmission dynamic models that had been published at the time [40]. Importantly, this comparative analysis concluded that "*although HPV models differ in structure, data used for calibration, and settings, our population-level predictions were generally concordant.*" This is a key result which speaks to the underlying consistency of modeled analyses of the impact and cost-effectiveness assessment of vaccination in many settings.

In recent years, two global modeling consortia (overlapping with each other in membership) have been facilitated to address the most pressing policy questions in the field. The first of these is the US National Cancer Institute (NCI)-funded CISNET-Cervical collaboration, which was established in 2016, and consists of five established modeling groups (Harvard, Cancer Council NSW, Australia, The University of Minnesota, The University of Washington and Erasmus University, The Netherlands) and a coordinating center at Harvard University, with Jane Kim as the coordinating center Primary Investigator. The primary focus is optimizing vaccination and cervical screening approaches in the United States, and initial collaborative evaluations include modeling a number of strategies for primary HPV screening and cotesting that was recently considered by the US Preventative Services Taskforce. CISNET-Cervical also aims to leveraging opportunities made possible by the collaboration to "*disseminate findings and strengthen transparency, understanding, and confidence in model-based analyses of cervical cancer control strategies.*" [81]

The second major global modeling consortium has been convened by the WHO to support strategic planning for cervical cancer elimination as a public health problem. The WHO has called for global action to scale-up vaccination, screening, and treatment of precancer, early detection and prompt treatment of early invasive cancers, and palliative care, and a strategic plan for elimination will be presented at the World Health Assembly in 2020. A 2019 single-model global elimination modeling study helped inform initial discussion of coverage and elimination targets; this quantified the potential cumulative effect of scaled up global vaccination and screening coverage on the number of cervical cancer cases and to predict the timeline to elimination in each of 181 countries, in categories of countries defined by the Human Development Index (HDI), and globally [10]. Related work currently underway by the WHO Global Cervical Cancer Elimination Modeling Consortium involves comparative modeling using HPV-ADVISE (Universite Laval), the Harvard Model and Policy1-Cervix (Cancer Council NSW Australia). This WHO-sponsored modeling involves simulation of outcomes and cost-effectiveness in 78 low- and lower-middle income countries and is designed to inform scale-up targets for vaccination and screening coverage, an understanding of final impact, timeline, and selection of elimination definition and thresholds, and the global cost-effectiveness analysis and the business case.

CONCLUSIONS

Developing, calibrating, and validating models of cervical cancer has been a successful enterprise because the natural history is uniquely well documented, and there are a wealth of data to inform models, including sexual behavioral studies, studies of HPV infection prevalence, progression and regression of precancer, and progression to CIN3 and invasive cervical cancer. Modeling has been critical to the public health advances achieved in the prevention of cervical and other HPV-related cancers in the last few decades, with many examples of modeling providing an essential underpinning to policy change. If anything, the importance of modeling is increasing over time since the complexities of interaction between HPV vaccination of young cohorts and cervical screening at older ages are substantial and ever-changing as vaccinated cohorts age within screening programs.

Modeling can never be conducted as a purely technical or analytical enterprise. There has been a wealth of data emerging from trials and cohort studies in the field, and models must be continually adjusted to reflect new insights. A deep understanding of disease epidemiology and population dynamics is required for informative evaluation of new interventions. This requires research investment into both people and infrastructure over long periods of time. A major priority for the future is to invest in capacity-building for LMICs to ensure that these countries also develop expertise in predictive modeling and health economic evaluation. Because of the close connection between models and data, HPV and cervical cancer modeling has become a specialist field led by a few groups although nonspecialist groups are also publishing their findings. One challenge is therefore to help the decision-maker appreciate the strengths and limitations of different modeled evaluations, and emergent quality frameworks for reporting modeled evaluations and comparative modeling consortia will play a key role in this respect. In general terms, cervical cancer modeling has come of age over the last 25 years, to the point that modeling sits firmly at center stage as a key tool supporting the strategic planning for the WHO initiative to eliminate cervical cancer as a public health problem over the next century.

ACKNOWLEDGMENTS

The authors would like to acknowledge all their many collaborators and friends in HPV and cervical cancer modeling who have contributed so much to this field over the last several decades. We also thank Ms. Harriet Hui for editorial support.

CONFLICT OF INTEREST

KC receives salary support from the National Health and Medical Research Council (NHMRC) Australia (CDFAPP1082989). She declares that she is co-PI of an investigator-initiated trial of primary HPV screening in Australia ("Compass") which is conducted and funded by the VCS Foundation, a government-funded health promotion charity, which has received a funding contribution from Roche Molecular Systems and Roche Tissue Diagnostics, AZ, USA. JB has received consultancy fees from GSK and Merck and travel support from DDL Diagnostics; consultancy fees were incurred by his employer.

LIST OF ABBREVIATIONS

CIN Cervical Intraepithelial Neoplasia
HDI Human Development Index
HSIL High-Grade Squamous Intraepithelial Lesion
LMIC Low and Middle Income Country
LSIL Low-grade Squamous Intraepithelial Lesion
USPSTF US Preventative Services Task Force
WHO World Health Organization

A VIGNETTE: THE STORY BEHIND THE FIRST COST-EFFECTIVENESS EVALUATIONS OF HUMAN PAPILLOMAVIRUS VACCINATION (TOLD BY SHALINI KULASINGAM)

I had recently joined Duke University as a postdoctoral student for Dr. Evan Myers when it became clear that an human papillomavirus (HPV) vaccine would be a viable prevention option in the future. Evan had developed the Duke Cervical Cancer model and there was interest in determining how an HPV vaccine might affect screening. We examined how screening might need to change, highlighting the need to consider increasing the age of first screening and/or a change in the screening interval to accommodate the significantly lower prevalence of disease in a vaccinated population. The first time the paper was submitted a major journal, it was actually rejected because the Editor was not convinced that there was an actual vaccine. We had to resubmit the paper along with the publication summarizing the results of the first trial of the HPV vaccine; it was only then that the paper was accepted!

REFERENCES

[1] Moscicki AB, Schiffman M, Burchell A, et al. Updating the natural history of human papillomavirus and anogenital cancers. Vaccine 2012;30(Suppl 5):F24–33.
[2] Beral V. Cancer of the cervix: a sexually transmitted infection? Lancet (London, England) 1974;1(7865):1037–40.
[3] zur Hausen H, de Villiers E-M, Gissmann L. Papillomavirus infections and human genital cancer. Gynecol Oncol 1981;12(2):S124–8.
[4] Walboomers JM, Jacobs MV, Manos MM, et al. Human papillomavirus is a necessary cause of invasive cervical cancer worldwide. J Pathol 1999;189(1):12–19.
[5] Bray F, Ferlay J, Soerjomataram I, Siegel RL, Torre LA, Jemal A. Global cancer statistics 2018: GLOBOCAN estimates of incidence and mortality worldwide for 36 cancers in 185 countries. CA Cancer J Clin 2018;68(6):394–424.
[6] Mandelblatt J, Schechter C, Levy D, Zauber A, Chang Y, Etzioni R. Building better models: if we build them, will policy makers use them? Toward integrating modeling into health care decisions. Med Decis Making 2012;32(5):656–9.
[7] Jeon J, Holford TR, Levy DT, et al. Smoking and lung cancer mortality in the United States from 2015 to 2065: a comparative modeling approach. Ann Intern Med 2018;169(10):684–93.
[8] Canfell K, Jit M, Kim J, et al. HPV-FRAME: a quality framework for reporting epidemiologic and economic models of interventions in HPV Prevention. International Papillomavirus Conference IPVC2018, Sydney, Australia; 2018.
[9] Eddy DM, Hollingworth W, Caro JJ, Tsevat J, McDonald KM, Wong JB. Model transparency and validation: a report of the ISPOR-SMDM Modeling Good Research Practices Task Force–7. Med Decis Making 2012;32(5):733–43.

[10] Simms KT, Steinberg J, Caruana M, et al. Impact of scaled up human papillomavirus vaccination and cervical screening and the potential for global elimination of cervical cancer in 181 countries, 2020–99: a modelling study. Lancet Oncol 2019;20(3):394–407.
[11] Lew J-B, Simms KT, Smith MA, et al. Primary HPV testing versus cytology-based cervical screening in women in Australia vaccinated for HPV and unvaccinated: effectiveness and economic assessment for the National Cervical Screening Program. Lancet Public Health 2017;2(2):e96–e107.
[12] Kim JJ, Burger EA, Regan C, Sy S. USPSTF Modeling Study: Screening for Cervical Cancer in Primary Care. J Am Med Assoc 2018;320(7):706–14.
[13] Brisson M, Laprise J-F, Drolet M, Van de Velde N, Boily M-C. Technical Appendix HPV-Advise CDC. 2014.
[14] Drolet M, Bénard É, Jit M, Hutubessy R, Brisson M. Model comparisons of the effectiveness and cost-effectiveness of vaccination: a systematic review of the literature. Value Health 2018;21(10):1250–8.
[15] Jit M, Demarteau N, Elbasha E, et al. Human papillomavirus vaccine introduction in low-income and middle-income countries: guidance on the use of cost-effectiveness models. BMC Med 2011;9(1):54.
[16] Barron BA, Richart RM. Statistical model of the natural history of Cervical Carcinoma. II. Estimates of the transition time from dysplasia to carcinoma in situ. J Natl Cancer Inst 1970;45(5):1025–30.
[17] Barron BA, Richart RM. A statistical model of the natural history of cervical carcinoma based on a prospective study of 557 Cases. J Natl Cancer Inst 1968;41(6):1343–53.
[18] Habbema JD, van Oortmarssen GJ, Lubbe JT, van der Maas PJ. Model building on the basis of Dutch cervical cancer screening data. Maturitas 1985;7(1):11–20.
[19] Koopmanschap MA, van Oortmarssen GJ, van Agt HM, van Ballegooijen M, Habbema JD, Lubbe KT. Cervical-cancer screening: attendance and cost-effectiveness. Int J Cancer 1990;45(3):410–15.
[20] Koopmanschap MA, Lubbe KT, van Oortmarssen GJ, van Agt HM, van Ballegooijen M, Habbema JK. Economic aspects of cervical cancer screening. Soc Sci Med 1990;30(10):1081–7 (1982).
[21] Sherlaw-Johnson C, Gallivan S, Jenkins D, Jones MH. Cytological screening and management of abnormalities in prevention of cervical cancer: an overview with stochastic modelling. J Clin Pathol 1994;47(5):430–5.
[22] Jenkins D, Sherlaw-Johnson C, Gallivan S. Can papilloma virus testing be used to improve cervical cancer screening? Int J Cancer 1996;65(6):768–73.
[23] McCredie MR, Sharples KJ, Paul C, et al. Natural history of cervical neoplasia and risk of invasive cancer in women with cervical intraepithelial neoplasia 3: a retrospective cohort study. Lancet Oncol 2008;9(5):425–34.
[24] Jenkins DA, Sperrin M, Martin GP, Peek N. Dynamic models to predict health outcomes: current status and methodological challenges. Diagnost Prognost Res 2018;2(1):23.
[25] Koliopoulos G, Nyaga VN, Santesso N, et al. Cytology versus HPV testing for cervical cancer screening in the general population. Cochrane Database Syst Rev 2017;(8)).
[26] Dillner J, Rebolj M, Birembaut P, et al. Long term predictive values of cytology and human papillomavirus testing in cervical cancer screening: joint European cohort study. Brit Med J 2008;337:a1754.
[27] Gage JC, Katki HA, Schiffman M, et al. Age-stratified 5-year risks of cervical precancer among women with enrollment and newly detected HPV infection. Int J Cancer 2015;136(7):1665–71.
[28] Gage JC, Schiffman M, Katki HA, et al. Reassurance against future risk of precancer and cancer conferred by a negative human papillomavirus test. J Natl Cancer Inst 2014;106:8.
[29] Schiffman M, Castle PE, Jeronimo J, Rodriguez AC, Wacholder S. Human papillomavirus and cervical cancer. Lancet 2007;370(9590):890–907.
[30] Schiffman M, Glass AG, Wentzensen N, et al. A long-term prospective study of type-specific human papillomavirus infection and risk of cervical neoplasia among 20,000 women in the Portland Kaiser Cohort Study. Cancer Epidemiol Biomarkers Prev 2011;20(7):1398–409.

[31] Burger EA, Kim JJ, Sy S, Castle PE. Age of acquiring causal human papillomavirus (HPV) infections: leveraging simulation models to explore the natural history of HPV-induced cervical cancer. Clin Infect Dis 2017;65(6):893–9.
[32] Koutsky LA, Ault KA, Wheeler CM, et al. A controlled trial of a human papillomavirus type 16 vaccine. New Engl J Med 2002;347(21):1645–51.
[33] Sanders GD, Taira AV. Cost-effectiveness of a potential vaccine for human papillomavirus. Emerging Infect Dis 2003;9(1):37–48.
[34] Kulasingam SL, Myers ER. Potential health and economic impact of adding a human papillomavirus vaccine to screening programs. J Am Med Assoc 2003;290(6):781–9.
[35] Paavonen J, Naud P, Salmerón J, et al. Efficacy of human papillomavirus (HPV)-16/18 AS04-adjuvanted vaccine against cervical infection and precancer caused by oncogenic HPV types (PATRICIA): final analysis of a double-blind, randomised study in young women. Lancet 2009;374 (9686):301–14.
[36] Ault KA. Effect of prophylactic human papillomavirus L1 virus-like-particle vaccine on risk of cervical intraepithelial neoplasia grade 2, grade 3, and adenocarcinoma in situ: a combined analysis of four randomised clinical trials. Lancet (London, England) 2007;369(9576):1861–8.
[37] Canfell K, Chesson H, Kulasingam SL, Berkhof J, Diaz M, Kim JJ. Modeling preventative strategies against human papillomavirus-related disease in developed countries. Vaccine 2012;30(Suppl 5): F157–67.
[38] Anderson RM, May RM. Infectious diseases of humans: dynamics and control. New York: Oxford University Press; 1992.
[39] Barnabas RV, Laukkanen P, Koskela P, Kontula O, Lehtinen M, Garnett GP. Epidemiology of HPV 16 and cervical cancer in Finland and the potential impact of vaccination: mathematical modelling analyses. PLoS Med 2006;3(5):e138.
[40] Brisson M, Bénard É, Drolet M, et al. Population-level impact, herd immunity, and elimination after human papillomavirus vaccination: a systematic review and meta-analysis of predictions from transmission-dynamic models. Lancet Public Health 2016;1(1):e8–e17.
[41] Jit M, Choi YH, Edmunds WJ. Economic evaluation of human papillomavirus vaccination in the United Kingdom. Brit Med J 2008;337:a769.
[42] Suijkerbuijk AW, Donken R, Lugner AK, et al. The whole story: a systematic review of economic evaluations of HPV vaccination including non-cervical HPV-associated diseases. Expert Rev Vaccines 2017;16(4):361–75.
[43] Qendri V, Bogaards JA, Berkhof J. Pricing of HPV vaccines in European tender-based settings. Eur J Health Econ 2019;20(2):271–80.
[44] Bogaards JA, Wallinga J, Brakenhoff RH. Meijer CJLM, Berkhof J. Direct benefit of vaccinating boys along with girls against oncogenic human papillomavirus: bayesian evidence synthesis. BMJ 2015;350: h2016.
[45] Lin A, Ong KJ, Hobbelen P, et al. Impact and cost-effectiveness of selective human papillomavirus vaccination of men who have sex with men. Clin Infect Dis 2017;64(5):580–8.
[46] Katki HA, Schiffman M, Castle PE, et al. Benchmarking CIN 3+ risk as the basis for incorporating HPV and Pap cotesting into cervical screening and management guidelines. J Low Genit Tract Dis 2013;17(5 Suppl 1):S28–35.
[47] Cuzick J, Clavel C, Petry K-U, et al. Overview of the European and North American studies on HPV testing in primary cervical cancer screening. Int J Cancer 2006;119(5):1095–101.
[48] Huijsmans CJ, Geurts-Giele WR, Leeijen C, et al. HPV prevalence in the Dutch cervical cancer screening population (DuSC study): HPV testing using automated HC2, cobas and Aptima workflows. BMC cancer 2016;16(1):922.

[49] Ogilvie GS, van Niekerk D, Krajden M, et al. Effect of screening with primary cervical HPV testing vs cytology testing on high-grade cervical intraepithelial neoplasia at 48 months: the HPV FOCAL randomized clinical trial. J Am Med Assoc 2018;320(1):43–52.
[50] Ronco G, Giorgi-Rossi P, Carozzi F, et al. Efficacy of human papillomavirus testing for the detection of invasive cervical cancers and cervical intraepithelial neoplasia: a randomised controlled trial. Lancet Oncol 2010;11(3):249–57.
[51] Rijkaart DC, Berkhof J, Rozendaal L, et al. Human papillomavirus testing for the detection of high-grade cervical intraepithelial neoplasia and cancer: final results of the POBASCAM randomised controlled trial. Lancet Oncol 2012;13(1):78–88.
[52] Bulkmans NW, Rozendaal L, Snijders PJ, et al. POBASCAM, a population-based randomized controlled trial for implementation of high-risk HPV testing in cervical screening: design, methods and baseline data of 44,102 women. Int J Cancer 2004;110(1):94–101.
[53] van Rosmalen J, de Kok IM, van Ballegooijen M. Cost-effectiveness of cervical cancer screening: cytology versus human papillomavirus DNA testing. BJOG 2012;119(6):699–709.
[54] Berkhof J, Coupé VM, Bogaards JA, et al. The health and economic effects of HPV DNA screening in The Netherlands. Int J Cancer 2010;127(9):2147–58.
[55] Goldie SJ, Gaffikin L, Goldhaber-Fiebert JD, et al. Cost-effectiveness of cervical-cancer screening in five developing countries. New Engl J Med 2005;353(20):2158–68.
[56] Tantivess S, Yothasamut J, Putchong C, Sirisamutr T, Teerawattananon Y. The role of health technology assessment evidence in decision making: the case of human papillomavirus vaccination policy in Thailand. Thailand: Ministry of Public Health; 2009.
[57] Sinanovic E, Moodley J, Barone MA, Mall S, Cleary S, Harries J. The potential cost-effectiveness of adding a human papillomavirus vaccine to the cervical cancer screening programme in South Africa; 2009.
[58] Elfstrom KM, Lazzarato F, Franceschi S, Dillner J, Baussano I. Human papillomavirus vaccination of boys and extended catch-up vaccination: effects on the resilience of programs. J Infect Dis 2016;213 (2):199–205.
[59] Ryser MD, Myers ER, Durrett R. HPV clearance and the neglected role of stochasticity. PLoS Comput Biol 2015;11(3):e1004113.
[60] Skinner SR, Szarewski A, Romanowski B, et al. Efficacy, safety, and immunogenicity of the human papillomavirus 16/18 AS04-adjuvanted vaccine in women older than 25 years: 4-year interim follow-up of the phase 3, double-blind, randomised controlled VIVIANE study. Lancet 2014;384(9961):2213–27.
[61] Joura EA, Giuliano AR, Iversen O-E, et al. A 9-valent HPV vaccine against infection and intraepithelial neoplasia in women. New Engl J Med 2015;372(8):711–23.
[62] Iversen O-E, Miranda MJ, Ulied A, et al. Immunogenicity of 9-valent HPV vaccine using a 2- vs a 3-dose regimen. J Am Med Assoc 2016;316(22):2411–21.
[63] Ogilvie G, Sauvageau C, Dionne M, et al. Immunogenicity of 2 vs 3 doses of the quadrivalent human papillomavirus vaccine in girls aged 9 to 13 years after 60 Months. J Am Med Assoc 2017;317(16):1687–8.
[64] Simms KT, Laprise J-F, Smith MA, et al. Cost-effectiveness of the next generation nonavalent human papillomavirus vaccine in the context of primary human papillomavirus screening in Australia: a comparative modelling analysis. Lancet Public Health 2016;1(2):e66–75.
[65] Drolet M, Laprise J-F, Boily M-C, Franco EL, Brisson M. Potential cost-effectiveness of the nonavalent human papillomavirus (HPV) vaccine. Int J Cancer 2014;134(9):2264–8.
[66] Chesson HW, Meites E, Ekwueme DU, Saraiya M, Markowitz LE. Cost-effectiveness of nonavalent HPV vaccination among males aged 22 through 26 years in the United States. Vaccine 2018;36 (29):4362–8.
[67] Brisson M, Laprise J-F, Chesson HW, et al. Health and economic impact of switching from a 4-valent to a 9-valent HPV vaccination program in the United States. J Natl Cancer Inst 2015;108:1.

[68] Jit M, Brisson M, Laprise J-F, Choi YH. Comparison of two dose and three dose human papillomavirus vaccine schedules: cost effectiveness analysis based on transmission model. BMJ 2015;350:g7584.

[69] Laprise J-F, Drolet M, Boily M-C, et al. Comparing the cost-effectiveness of two- and three-dose schedules of human papillomavirus vaccination: a transmission-dynamic modelling study. Vaccine 2014;32 (44):5845–53.

[70] Laprise J-F, Markowitz LE, Chesson HW, Drolet M, Brisson M. Comparison of 2-dose and 3-dose 9-valent human papillomavirus vaccine schedules in the United States: a cost-effectiveness analysis. J Infect Dis 2016;214(5):685–8.

[71] Drolet M, Laprise J-F, Brotherton JML, et al. The impact of human papillomavirus catch-up vaccination in Australia: implications for introduction of multiple age cohort vaccination and postvaccination data interpretation. J Infect Dis 2017;216(10):1205–9.

[72] Ronco G, Dillner J, Elfström KM, et al. Efficacy of HPV-based screening for prevention of invasive cervical cancer: follow-up of four European randomised controlled trials. Lancet 2014;383 (9916):524–32.

[73] Arbyn M, Depuydt C, Benoy I, et al. VALGENT: a protocol for clinical validation of human papillomavirus assays. J Clin Virol 2016;76(Suppl 1):S14–21.

[74] Oštrbenk A, Xu L, Arbyn M, Poljak M. Clinical and analytical evaluation of the anyplex II HPV HR detection assay within the VALGENT-3 framework. J Clin Microbiol 2018;56(11) e01176-18.

[75] Canfell K, Caruana M, Gebski V, et al. Cervical screening with primary HPV testing or cytology in a population of women in which those aged 33 years or younger had previously been offered HPV vaccination: results of the compass pilot randomised trial. PLoS Med 2017;14(9):e1002388.

[76] Smith MA, Gertig D, Hall M, et al. Transitioning from cytology-based screening to HPV-based screening at longer intervals: implications for resource use. BMC Health Serv Res 2016;16:147.

[77] Hall MT, Simms KT, Lew J-B, et al. The projected timeframe until cervical cancer elimination in Australia: a modelling study. Lancet Public Health 2019;4(1):e19–27.

[78] Lew J-B, Simms K, Smith M, Lewis H, Neal H, Canfell K. Effectiveness modelling and economic evaluation of primary HPV screening for cervical cancer prevention in New Zealand. PLoS One 2016;11(5): e0151619.

[79] Craig BM, Brisson M, Chesson H, Giuliano AR, Jit M. Proceedings of the modeling evidence in HPV pre-conference workshop in Malmo, Sweden, May 9–10, 2009. Clin Therap 2010;32(8):1546–64.

[80] HPV-frame: framework for modeling HPV prevention, <http://www.hpv-frame.org/2019>, Last Accessed September 2019.

[81] National Cancer Institute. Cancer intervention and surveillance modeling network (CISNET), <https://cisnet.cancer.gov/cervical/2019>, Last Accessed September 2019.

CHAPTER

21

TWELVE YEARS OF VACCINE REGISTRATION AND THE CONSEQUENCES

Suzanne M. Garland[1,2,3], Lauri E. Markowitz[4], Heather Cubie[5] and Kevin G. Pollock[6]

[1]Department of Obstetrics and Gynaecology, University of Melbourne, VIC, Australia [2]Director Centre Women's Infectious Diseases Research, VIC, Australia [3]Honorary Research Fellow, Infection & Immunity, Murdoch Children's Research Institute, VIC, Australia [4]Associate Director for Science, HPV National Center for Immunizations and Respiratory Diseases, Centers for Disease Control and Prevention, Atlanta, GA, United States [5]Global Health Academy, University of Edinburgh, Edinburgh, United Kingdom [6]School of Health and Life Sciences, Glasgow Caledonian University, Glasgow, Scotland

In this chapter, we describe specific details on the three regions of the world from where we come, to highlight the differences in approaches to HPV vaccination, delivery, challenges and outcomes in high income countries where vaccination programs were first initiated. In addition, we describe a global viewpoint within the context of space allowed. This chapter deals with a rapidly changing area, particularly since the call to action by the Director General of WHO, Dr. Tedros, for elimination of cervical cancer as a public health matter [1]. This is a massive goal to accomplish, but achievable given political will and that we have the tools to do so.

HPV VACCINES AND THEIR INTRODUCTION

REGISTERED HPV VACCINES

Three prophylactic vaccines have been registered to date. The bivalent (2vHPV) vaccine Cervarix which targets 16/18, the quadrivalent (4vHPV) vaccine GARDASIL which addition to 16/18 targets 6/11, and the nonavalent (9vHPV) vaccine GARDASIL®9 which targets the 4vHPV types and the next five most common cervical cancer causing types (HPV31/33/45/52/58). All have been shown in phase 3 and 4 trials to have excellent safety, immunogenicity, and efficacy against the specific vaccine genotype infections and HPV-related diseases [2].

INTRODUCTION INTO NATIONAL IMMUNIZATION PROGRAMS

Through 2018, over 80 countries have introduced HPV vaccination into their national immunization program [3]. While recommendations and strategies vary, most countries with HPV vaccination included in their national immunization programs have recommended HPV vaccines for girls, and

Human Papillomavirus. DOI: https://doi.org/10.1016/B978-0-12-814457-2.00021-0

adopted a target age in early adolescence, with catch-up practice ranging from none to HPV vaccine available up to age 26 years.

National immunization programs were endorsed firstly in 2006 in the United States for 4vHPV and shortly thereafter in Australia, followed by many other, largely well-resourced, countries. In the United States, the program targeted 11- to 12-year-old girls and the vaccine was administered primarily by pediatricians and/or family medicine practitioners, with funding through federal programs or private health insurance. In Australia, the program was school-based targeting 12- to 13-year-old girls, free for all, with no reimbursement step and was delivered by school nurses qualified to administer vaccines. Both countries adopted a catch-up program for women through age 26 years (as an ongoing program in the USA and for 2 years only in Australia, to the end of 2009). In the United Kingdom, the 2vHPV vaccine was used for the school-based program starting 2008, where 12- to 13-year old girls were eligible, with catch-up targeting 14- to 17-year-old girls. In 2012, the UK changed to the 4vHPV as a result of their tendering process. Subsequent to introduction in these early adopting countries, most countries in Western Europe introduced HPV vaccination [3].

Some middle-income countries (MIC) that have successfully introduced vaccination include Malaysia, where in 2010 the 2vHPV vaccine was introduced in a school-based program targeting 13-year-old girls. They then made a switch to 4vHPV with a catch- up program initiated in 2012 to 18 years of age and in 2018 free vaccine for unmarried women up to 27 years of age. With multi-sectoral collaboration Malaysia has achieved near universal coverage, with an excess of 90% in 13 year olds [4]. Another example is South Africa which introduced a two-dose school based delivery to 9-year-old girls in 2014 and achieved coverage exceeding 80% for the first dose [5].

Through the Pan American Health Organization's Revolving Fund, starting in 2011, HPV vaccines were available for a substantially lower price than in high income countries, allowing almost all countries in Central and South America to introduce vaccination.

In resource-poor countries, where there is the greatest burden of disease, vaccination initially commenced as specific sub-national demonstration initiatives [6]. For example, in Bhutan a low-income country (LIC), HPV vaccination commenced as a school-based pilot program in 2009, as an initiative of the Government in partnership with the Australian Cervical Cancer Foundation and Merck's Gardasil Access program. This was implemented and expanded by 2012 to a national HPV vaccination program. By 2010, more than 90% of girls aged $12-18$ years had been vaccinated with three doses of the HPV vaccine. In 2011, Rwanda was the first country in Africa to introduce HPV vaccine, also through a donation program from the manufacturer and national school-based delivery; high coverage was achieved (98%) [7]

Starting in 2012, eligible countries have been able to apply for support through Gavi, the Vaccine Alliance. Between 2013 and 2016, Gavi provided support to over 20 eligible countries (those with GNI per capita $\leq$\$1580 US) for 2-year HPV vaccine demonstration projects (e.g., Malawi and Tanzania 2014; Uganda in 2015), with the anticipation that these would lead to national introduction. Between 2014 and 2018, over 43 LIC and lower-middle-income countries (LMICs) gained much experience with HPV vaccine delivery either through demonstration or pilot programs, with some introducing national HPV vaccination programs [3].

Fig. 21.1 shows global HPV vaccine introduction through to January 2019 by type of program (national, demonstration, where national immunization programs have been adopted, demonstration status, projected National, sub-National and demonstration completed. Additional countries in Africa introduced HPV vaccination in 2018 and 2019 including Ethiopia and Senegal.

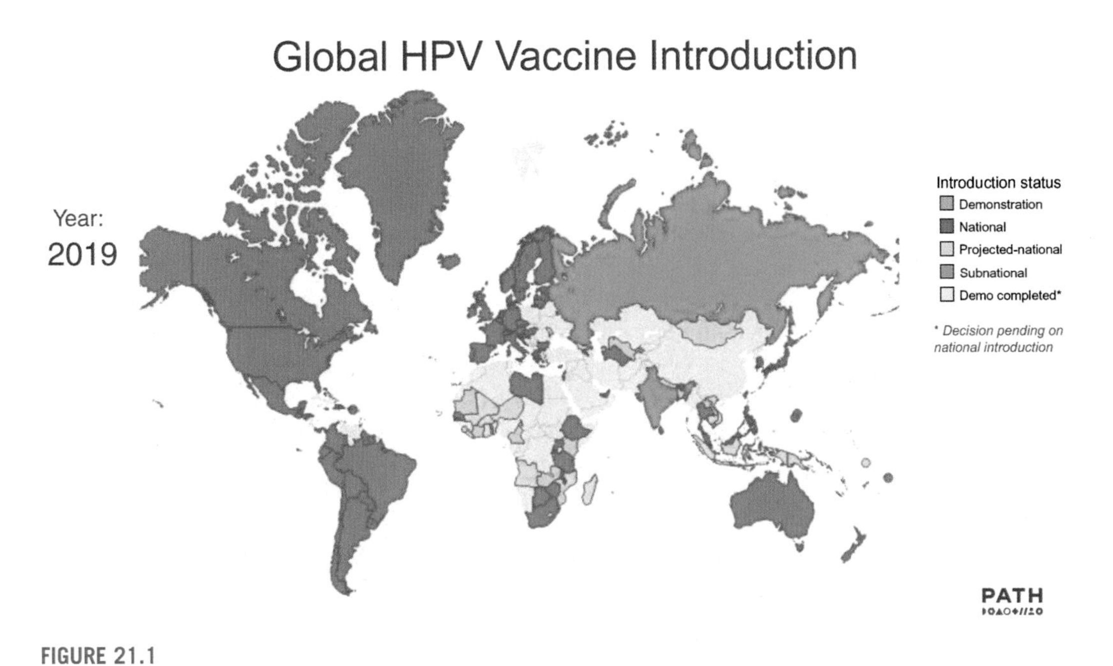

FIGURE 21.1

Global HPV vaccine introduction. Published with permission.

MEASURES OF VACCINATION PROGRAMS

Vaccine Coverage

Variable coverage has been achieved in different countries, with generally higher coverage achieved with school- based vaccination. A few countries have faced challenges from anti-vaccine groups and have experienced declines in coverage after initially successful introductions (see below).

Good communication to the general public and vaccine providers is key to achieving high coverage of any vaccine. In Scotland, an extensive communication campaign was run over the school holiday period immediately before the program started in August 2008, with eye-catching, age-appropriate TV, radio, and news coverage. As in Australia, communication to the lay public and medical/nursing professions focused on a vaccine that would prevent the majority of cervical cancers and this may have contributed to high initial uptake.

When HPV vaccination was introduced into the national immunization program in Australia, the childhood vaccination registry did not have the capacity to include the adolescent HPV vaccination data. Therefore, a separate registry was established, the National HPV Vaccination Program Register (NHVPR). Data from HPV vaccines being administered at schools is obligatory, whereas for the catch-up program it was voluntary. The NHVPR has played an essential role in monitoring and evaluation of the program by recording information about HPV vaccine doses administered in the country. NHVPR shows high coverage at 84% for two doses for girls turning 15 years of age in 2016 (79% for three doses). For boys, the comparable coverage achieved for two doses was 78% (73% for three doses), with some variations by State and Territory. Whilst coverage within the catch-up cohorts show greater variation, they did have an unexpectedly high uptake United States rate for those 20 to 26 years of age at the commencement of the program (NHVPR, 2018). The reported two-dose coverage within a three-dose regimen for females aged 14–26 years has varied between 25–81%, 39–81% and 43–80% for 2007, 2008, and 2009 respectively, although there is under-reporting due to the voluntary nature of catch-up component of data registration. Two-dose coverage for the male catch-up program for the cohort 14–15 years within a three-dose regimen was 52% in 2013 and 73% in 2014. All HPV vaccine data have recently been incorporated into the Australian Immunization Register (AIR), a whole of life register with the ability to record all vaccinations for people of all ages given by a registered vaccination provider.

In Scotland, school-based uptake of both the bivalent and quadrivalent HPV vaccines in girls aged 12–13 has been impressive, with vaccine uptake sustained at levels approximating 90% since 2008 [8]. In addition, a 3-year (from September 2008 to 2011) catch-up campaign offered vaccine to all girls aged 14–17, with uptake recorded between 80% and 30% in younger and older girls at age of vaccination, respectively [9]. The differential uptake in girls offered vaccine through primary care or in a school-based setting exemplifies that the latter approach results in a ''captive'' cohort, which can be well-coordinated by the school nurses and head teachers. Despite recent communications to head teachers and school nurses from anti-HPV vaccine groups in the UK, there is strong commitment in the program to ensure that high uptake of the vaccine continues.

In contrast with Australia and UK, almost all vaccinations in the United States are delivered by primary care providers in clinic settings. Vaccine coverage is measured by the National Immunization Survey (NIS-Teen), annually among 13 – 17 year olds in the United States. At the beginning of the HPV vaccination program, uptake of HPV vaccine paralleled other recently recommended vaccines for adolescents. However, HPV vaccine coverage did not increase comparably within 3 years after introduction. HPV vaccine in coverage among females aged 13-17 years was 69% for at least one dose and 53% for all doses depending on age recommendation [10]. Coverage in males began to increase after the routine recommendation in 2011; in 2016, at least one-dose and up-to-date coverage among males aged 13–17 years reached 63% and 44%, respectively. There are large differences by state, with at least one-dose coverage in 2017 ranging from 47% to 92% [10].

Multiple studies have examined reasons for the low HPV vaccine coverage as well as practices that could improve coverage [11]. Many studies have identified lack of strong provider recommendation for vaccination at the recommended age as a major reason. Efforts to increase coverage in the United States have included providing education, tools and communication messages for vaccine providers. However, reasons for low coverage are multi-factorial and include factors at the parent and health system levels, in addition to providers. While coverage remains lower than target goals, coverage continues to increase yearly with larger increases in males than in females [10].

Vaccine Impact and Effectiveness

Vaccine effectiveness can be estimated by comparing prevalence or incidence in vaccinated versus unvaccinated individuals within similar populations. Impact studies compare infection or disease in pre- and post-vaccination periods. Both types have been conducted to evaluate the real-world effects of vaccination.

HPV DNA Genoprevalence

In a systematic review and meta-analysis conducted in 2018 to assess the population-level consequences and herd effects after female HPV vaccination programs, where coverage was greater than 50%, to there were significant reductions in HPV 16/18 between the pre- and post-vaccine periods. This was 80% in girls aged 15–19 years and 65% for those aged 20–24 years. There were also significant reductions (50%) in HPV types 31, 33 and 45 in this age group suggesting cross-protection [12].

In another systematic review of outcomes in real-world settings at 10 years for the 4vHPV vaccine, maximal reductions of 90% were reported for the four genotypes covered in the vaccine [13]. The greatest impact was seen where the vaccine is routinely administered before HPV exposure.

Due to prior genoprevalence data, Australia was able to show the impact of high coverage vaccination. There was a 77% reduction for HPV 6/11/16/18 infection within 5 years of implementing the national program. More recently, findings are consistent with a 92% decline in vaccine-related types from 22.7% in the pre-vaccine era to only 1.5%, observed in young women presenting to Australian Family Planning Clinics for Pap cytology cervical screening (Machalek et al., 2018), with an adjusted vaccine effectiveness against vaccine-targeted HPV types for fully vaccinated women compared to unvaccinated women of 86%. There was evidence of herd protection in females, as those unvaccinated post-vaccination implementation had lower rates of the vaccine-targeted types, compared to those prior to the program. Moreover, in the state of Victoria in a more generalizable population of young women age 18−25 years recruited through social media (Facebook) who self-collected vaginal swabs, there was consistent and very low prevalence of vaccine-targeted HPV types at only 1.7%, with a significant difference in prevalence of any of HPV 31/33/45 collectively of 2% for those fully vaccinated to 6.8 % for those unvaccinated $P = .01$ [14]. Furthermore, as a result of the female only program, there is evidence of considerable herd protection in males [15].

Similarly, in Scotland, the effect of bivalent vaccine on the prevalence of high-risk HPV types in women is remarkable. When the national vaccine program commenced, Scotland also established population-based surveillance of young women attending their first cervical screen at age 20 years, linked to individual vaccination status. Therefore, it was one of the few countries in the world able to detect early impact of vaccine through HPV DNA genoprevalence. In the youngest birth cohorts (those born in 1995 associated with 90% uptake of vaccine at age 13), there was a virtual abolition of HPV 16 and 18 infection, in addition to a profound reduction in cross-protective types (HPV 31/33, 45). No evidence of HPV type replacement was evident and significant herd protection in non-vaccinated females was observed [16].

Although lower coverage has been achieved in the United States, impact of the vaccination program on several outcomes has been demonstrated. Within 4 years of vaccine introduction, data from national surveys showed declines in HPV vaccine-type prevalence among females— a 56% decrease in HPV vaccine types in cervical-vaginal samples from 14- to 19-year olds [17]. Further decreases have been demonstrated, including declines among 20- to 24-year olds observed within 6 years of vaccination introduction; by 8 years after introduction, there was a 71% decrease in HPV vaccine type prevalence among 14- to 19-year olds and a 61% reduction among 20- to 24-year olds [18,19]. Other examples of reduction in vaccine-related HPV genotypes since introduction of HPV vaccination include studies from Denmark, Germany, New Zealand, Sweden, and Canada [12].

Genital Warts

Declines in genital warts were seen very rapidly in Australia at 3 years postvaccination program commencing, reflecting the short incubation period for this disease. There was evidence of herd protection in young males, as well as females [20–22]. In the United States declines have also been documented in genital warts [23] in addition to reports from other countries that introduced quadriavlent HPV vaccine such as Denmark, Sweden, Germany, Canada, and New Zealand.

In the systematic review of Drolet et al. [12] after 5–8 years of vaccination, countries with multi-cohort vaccination and high coverage (≥50%) had greater reductions in genital warts, 44 and 85 percentage points among girls and boys aged 15–19 years, respectively, than countries with single-cohort vaccination and/or low vaccination coverage.

Cervical Disease

When HPV vaccination commenced in Australia, the precancer screening program targeted women aged 18–69 years for 3 year cervical cytology. Within 3 years, there was evidence of a decline in the youngest women for histologically diagnosed high-grade lesions (also termed cervical intraepithelial neoplasia [CIN] grade 2+). Through 2016, declines in histologically proven high grade abnormalities of 65% in women aged <20 years since 2007, 40% in women 20–24 years since 2010, and 13% in women aged 25–29 years since 2013 have been documented [24].

When the impact of vaccine on CIN was assessed in catch-up women in Scotland, vaccine effectiveness was greater against CIN 2 and 3 lesions, in those women from the most deprived backgrounds [25–27]. Compared with the most deprived, unvaccinated women, the relative risk of CIN 3 in fully vaccinated women in the same deprivation group was 0.29 (95% CI: 0.2–0.43) compared with 0.62 (95% CI: 0.4–0.97) in vaccinated women in the least-deprived group [27]. These data are welcoming and allay the concern that inequalities in cervical cancer may persist or increase following the introduction of the vaccines in the UK.

In the United States, declines in CIN2+ have been challenging to monitor because cervical cancer screening recommendations and managment of lesions have changed since vaccine was introduced; therefore, cervical lesions should ideally be informed by changes in screening rates. Data from several sources, including a single state-wide monitoring, a project monitoring CIN2+ in sites in five states and health claims data, show declines in cervical lesions among screened women under age 25 years [28–30].

A recent systematic review and meta-analysis, including data on CIN2 + from nine studies in six high income countries, evaluated CIN2 + among screened women, by age group. CIN2 + decreased significantly *by* similarly 50% among girls aged 15–19 years, and 30% among women aged 20–24 years [12].

Juvenile Onset Recurrent Respiratory Papillomatosis

In a prospective study of the incidence juvenile onset recurrent respiratory papillomatosis (JoRRP) after implementation of the Australian National HPV Vaccination Program, for the first-time a reduction in JoRRP has been documented reflecting the impact of immunization and reduced transmission from mother to baby [31]. Canada has also reported a decline in JoRRP [32].

Cervical Cancer

Vaccine impact on cervical cancer will take longer to observe. Some data now available from the Finnish cancer-registry based on passive follow-up of vaccinated young women from the phase 3, 2vHPV, and 4vHPV clinical trials, and a community-randomized phase 4 trial with age-aligned non-vaccinated cohorts might suggest evidence of effectiveness against invasive HPV-related cancers [33]. There is also some evidence from evaluations after national introductions: in Scotland, national statistics demonstrate a decline in cervical cancer in 20–24-year olds (13 in 2012 vs 4 cases in 2017), the majority of those being vaccine eligible[34]; in recent U.S. data, the 4-year average annual incidence rates for cervical cancer in young women <25 years of age in 2011–2014 were 29% lower than those in the pre-vaccine era of 2003–2006 (6.0 vs 8.4 per 1,000,000 people); while these might suggest evidence for an effect on early-onset cervical cancers, declines could also be due to changes in screening [35].

EVOLVING POLICY ISSUES

Vaccine policy has evolved since the 2vHPV and 4vHPV vaccines were first licensed as a result of data on efficacy and immunogenicity for males, for different number of doses, licensure of an additional multivalent vaccine, and health economic considerations.

Male Vaccination

Both 2vHPV and 4vHPV vaccines were first licensed for use in females. In 2009, after data from trials in males became available, HPV vaccines were licensed for use in males as well [36,37]. Considerations for including males in national programs include the burden of disease in males, which varies by country and vaccine coverage achieved in females, as both affect cost-effectiveness of male vaccination [38]. In 2011, vaccination for boys was included in the national immunization program in the United States [39] and in 2013 in Australia [40]. Australia had a catch-up program for boys up to 15 years of age for a period of 2 years. In the United States, catch-up is ongoing and recommended through age 21 years and through age 26 for young men who have sex with men (MSM). Other countries in the Americas and Europe have also included males in their programs. Targeted free 4vHPV vaccination programs for MSM were introduced in the Australia in State of Victoria and in Scotland in mid-2017 followed by England in April 2018. The UK extensions also include gay and bisexual men up to age 45 through sexual health clinics and HIV clinics. The program was initiated based on advice from the UK Joint Committee on Vaccination and Immunization, which recognized that MSM receive little benefit from the current national female only HPV program. The vaccination program, designed to be opportunistic, has been accommodated into pre-existing sexual health services and is popular with MSM attending the service. Furthermore, the first dose uptake of 64% correlates well with estimates used in cost-effectiveness models [41].

Finally, in July 2018, the UK stated that they would introduce gender-neutral vaccination, with boys being offered the vaccine from September 2019 (UK Government, 2018). This decision was based solely on equality and not cost-effectiveness, a remarkable precedent for UK vaccine policy.

Two Dose Schedules

While all HPV vaccines were first studied, licensed, and recommended as three-dose schedules, interest in two-dose schedules arose from a post hoc analysis of a trial of 2vHPV vaccine [42]. Subsequently, non-inferiority was observed in antibody responses after a two-dose vaccine schedule (given at an interval of 6–12 months) in those under 15 years of age [43]; WHO endorsed such a schedule [44] (WHO, 2014). Many countries that had started with three doses have changed recommendations for this age group. A two-dose schedule should facilitate delivery of HPV vaccine and reduce costs. Three vaccine doses are still recommended for those starting the vaccination series after age 15 and for persons who are immunocompromised [45].

Nonavalent Vaccine [9vHPV]

Some countries starting with 4vHPV vaccine have now adopted 9vHPV vaccine, based on the wider coverage, from 70% of cervical cancers are caused by genotypes HPV 16/18 to 93% for the genotypes targeted by 9vHPV vaccine (HPV16/18/31/33/45/52/58). In the United States, 9vHPV was introduced in 2015 and since the end of 2016 has been the only HPV vaccine available. In Australia, a switch to the 9vHPV commenced in 2018 for 12–13-year-old boys and girls, as a two-dose program, and with a catch-up to 19 years of age (3 doses for those >15 years of age).

Other Vaccines on the Horizon

China and India are making progress on developing [46] HPV vaccines that are being trialed for efficacy and immunogenicity and should provide alternatives to the currently licensed vaccines [47]. Two are bivalent vaccines and one is a quadrivalent vaccine. These products will contribute to the global vaccine supply which is currently not sufficient to meet the growing demand [48].

Challenges

There have been numerous challenges during the first decade of HPV vaccination. In addition to program challenges of delivering vaccine to an adolescent age group in some countries and vaccine cost, in recent years, the phenomena of anti-vaccine campaigns and vaccine hesitancy have emerged as threats to uptake and impact of HPV vaccine programs. The situation in Japan is one example. HPV vaccine was included in the national immunization program from December 2010, for girls aged 12–16 years; coverage reached ~70% for dose one. However, in June, 2013, the Japanese Ministry of Health, Labor, and Welfare suspended proactive recommendations for the HPV vaccine after unconfirmed reports of adverse events following vaccination appeared in the media. In January, 2014, the Vaccine Adverse Reactions Review Committee investigating these adverse events concluded that there was no evidence to suggest a causal association between the HPV vaccine and the reported adverse events after vaccination, but a proactive recommendation for HPV vaccination was not reinstated. In Denmark, uptake of vaccine decreased from 90% to 20% after clinical reports of a condition called postural orthostatic tachycardia syndrome (POTS), a vague disease for which no association with vaccine has been found. Studies from other countries suggest that vaccine confidence can be partly restored, but the time required can be long in some settings. In Ireland, a recent crisis of confidence in HPV vaccine was reversed through cross sector collaboration with a wide range of stakeholders within a short time frame [49].

When vaccine programs are introduced into a country, it is important that a mechanism to deal with reports of potential adverse events can be implemented rapidly and the events be assessed professionally. Many countries have systems in place, such as the Adverse Events Following Immunization—Clinical Assessment Network (AEFI-CAN) program in Australia which aims to mitigate against loss of public confidence for any vaccine in the national program; there are long standing vaccine safety infrastructures in the United States, the UK and other developed countries. The WHO Global Advisory Committee on Vaccine Safety has played a major role in reviewing safety concerns about HPV vaccine and assisting countries [50].

Impact on Current Cervical Screening Programmes

While increasing coverage of HPV vaccine, some countries have considered the impact on the cervical cytology screening programs because the positive predictive value of cytology for predicting underlying high-grade dysplasia decreases dramatically with vaccination. Therefore, the use of more sensitive and objective assays for primary cervical screening such as HPV nucleic acid testing (NAT) is being considered. The Netherlands were the first to accomplish this in 2017 and using one HPV DNA platform (Roche Cobas 4800 system) screening across the country, utilizing five large pathology laboratories and integrating molecular biology in combination with a completely harmonized laboratory workflow [51]. As of December 1, 2017, Australia changed from cervical cytology from 18 to 69 years of age, with three yearly screening, to the HPV NAT tests utilizing those with limited genotyping (HPV16/18) capability, commencing at 25 years of age with immediate triage to cytology and colposcopy for those 16/18 positive, and five yearly screening for those HPV DNA negative. This transition has offered a ready and passive mechanism for monitoring the impact of vaccination on infection prevalence among women attending such screening [52]. In the USA, while cervical cancer screening recommendations have been evolving, with HPV DNA testing incorporated into recommendations that also include cytology, no changes have been made directly related to the HPV vaccination program.

The massive reductions in the most carcinogenic types clearly have implications for associated disease and the way it is managed clinically. There has already been a reduction in referrals for colposcopy and less disease detected at colposcopy [53]. Fewer women have therefore needed ablative treatment reducing the burden of complications from the procedure. Low levels of disease following routine immunization clearly have ramifications for screening and management of vaccinated women. Whilst significant disease is reduced, it has not been eradicated and continued screening is therefore necessary. The performance of cytology-based and indeed HPV-based screening is likely to deteriorate in immunized women. This can be attributed to reduced disease levels leading to lower specificity and positive predictive value for significant disease. The reduction in colposcopy performance in women who have been vaccinated is already at the lower acceptable level within UK national cervical screening program guidelines. Thus, the effectiveness of HPV primary screening in highly immunized populations will require close monitoring and novel methods for both improving uptake of screening and the effectiveness of the screening test will be needed in order to maintain the performance of cervical screening as a process [54].

CONCLUSION

While challenges remain for HPV vaccination programs, there has been interest in an initiative to eliminate cervical and other HPV-driven cancers as a public health problem [1]. Although tools are available to reach this goal (safe effective prophylactic vaccines, HPV NAT testing, and treatment of cervical cancer particularly if detected early), substantial work will be needed to reach high coverage of HPV vaccination for adolescents and high coverage of cervical screening, with appropriate treatment of all women. The remarkable reductions in vaccine-targeted HPV genoprevalence, with subsequent reductions in associated short incubation outcomes in countries with high coverage vaccination underscores the potential to achieve this goal globally.

Health economic models can help to determine the optimal combination of HPV vaccination and screening in public health programs and to guide policy and program decisions in different population and geographic areas to achieve further reductions in HPV related disease.

REFERENCES

[1] WHO Director-General calls for all countries to take action to help end the suffering caused by cervical cancer. 2018 [cited 2018]; Available from: https://www.who.int/reproductivehealth/call-to-action-elimination-cervical-cancer/en/.

[2] Paavonen J.J., Garland S.M. Chapter 16, Clinical trials of HPV vaccines. HPV Story. 2018.

[3] Gallagher KE, LaMontagne DS, Watson-Jones D. Status of HPV vaccine introduction and barriers to country uptake. Vaccine 2018;36(32, Part A):4761–7.

[4] Buang SN, Ja'afar S, Pathmanathan I, Saint V. Human papillomavirus immunisation of adolescent girls: improving coverage through multisectoral collaboration in Malaysia. BMJ 2018;363.

[5] Delany-Moretlwe S, Kelley KF, James S, Scorgie F, Subedar H, Dlamini NR, et al. Human papillomavirus vaccine introduction in South Africa: implementation lessons from an evaluation of the National School-based Vaccination Campaign. Global Health: Sci Pract 2018;6(3):425–38.

[6] LaMontagne DS, Bloem PJN, Brotherton JML, Gallagher KE, Badiane O, Ndiaye C. Progress in HPV vaccination in low- and lower-middle-income countries. Int J Gynecol Obstet 2017;138 (S1):7–14.

[7] Gatera M, Bhatt S, Ngabo F, Utamuliza M, Sibomana H, Karema C, et al. Successive introduction of four new vaccines in Rwanda: high coverage and rapid scale up of Rwanda's expanded immunization program from 2009 to 2013. Vaccine 2016;34(29):3420–6.

[8] Sinka K, Kavanagh K, Gordon R, Love J, Potts A, Donaghy M, et al. Achieving high and equitable coverage of adolescent HPV vaccine in Scotland. J Epidemiol Community Health 2014;68 (1):57–63.

[9] Potts A, Sinka K, Love J, Gordon R, McLean S, Malcolm W, et al. High uptake of HPV immunisation in Scotland—perspectives on maximising uptake. Eurosurveillance 2013;18(39):20593.

[10] Walker TY, Elam-Evans LD, Singleton JA, Yankey D, Markowitz LE, Fredua B, et al. National, regional, state, and selected local area vaccination coverage among adolescents aged 13-17 Years - United States, 2016. MMWR Morb Mortal Wkly Rep 2017;66(33):874–82.

[11] Smulian EA, Mitchell KR, Stokley S. Interventions to increase HPV vaccination coverage: a systematic review. Hum Vaccin Immunother 2016;12(6):1566–88.

[12] Drolet M, Bénard E, Pérez N, Brisson M, For the HPV Vaccination Impact Study Group. Population-level impact and herd effects following the introduction of Human Papillomavirus vaccination programs: updated systematic review and meta-analyis. Lancet 2019;294(10197):497–509.

[13] Garland SM, Kjaer SK, Muñoz N, Block SL, Brown DR, DiNubile MJ, et al. Impact and effectiveness of the quadrivalent human papillomavirus vaccine: a systematic review of ten years of real-world experience. Clin Infect Dis 2016;63(4):519–27.

[14] Garland SM, Cornall AM, Brotherton JML, Wark JD, Malloy MJ, Tabrizi SN. Final analysis of a study assessing genital human papillomavirus genoprevalence in young Australian women, following eight years of a national vaccination program. Vaccine 2018;36(23):3221–30.

[15] Chow EP, Machalek DA, Tabrizi SN, Danielewski JA, Fehler G, Bradshaw CS, et al. Quadrivalent vaccine-targeted human papillomavirus genotypes in heterosexual men after the Australian female human papillomavirus vaccination programme: a retrospective observational study. Lancet Infect Dis 2017;17(1):68–77.

[16] Kavanagh K, Pollock KG, Cuschieri K, Palmer T, Cameron RL, Watt C, et al. Changes in the prevalence of human papillomavirus following a national bivalent human papillomavirus vaccination programme in Scotland: a 7-year cross-sectional study. Lancet Infect Dis 2017;17(12):1293–302.

[17] Markowitz LE, Hariri S, Lin C, Dunne EF, Steinau M, McQuillan G, et al. Reduction in human papillomavirus (HPV) prevalence among young women following HPV vaccine introduction in the United States, National Health and Nutrition Examination Surveys, 2003–2010. J Infect Dis 2013; jit192.

[18] Markowitz LE, Meites E, Unger ER. Two vs three doses of human papillomavirus vaccine: new policy for the second decade of the vaccination program. JAMA 2016;. Available from: https://doi.org/10.1001/jama.2016.16393.

[19] Oliver SE, Unger ER, Lewis R, McDaniel D, Gargano JW, Steinau M, et al. Prevalence of human papillomavirus among females after vaccine introduction—National Health and Nutrition Examination Survey, United States, 2003–2014. J Infect Dis 2017;216(5):594–603.

[20] Donovan B, Franklin N, Guy R, Grulich AE, Regan DG, Ali H, et al. Quadrivalent human papillomavirus vaccination and trends in genital warts in Australia: analysis of national sentinel surveillance data. Lancet Infect Dis 2011;11(1):39–44.

[21] Ali H, Guy RJ, Wand H, Read TR, Regan DG, Grulich AE, et al. Decline in in-patient treatments of genital warts among young Australians following the national HPV vaccination program. BMC Infect Dis 2013;13(1):1–6.

[22] Harrison CBH, Garland SM, Conway L, Stein A, Pirotta M. Fairley decreased management of genital warts in young women in Australian General Practice Post Introduction of National HPV Vaccination Program: results from a Nationally Representative Cross-Sectional General Practice Study. PLoS One 2014;9(9):e105967.

[23] Flagg EW, Schwartz R, Weinstock H. Prevalence of Anogenital Warts Among Participants in Private Health Plans in the United States, 2003–2010: potential impact of human papillomavirus vaccination. Am J Public Health 2013;103(8):1428–35.

[24] Australian Institute of Health and Welfare 2018. Cat no CAN 111 Cervical screening in Australia 2018. Canberra: AIHW; 2018.

[25] Pollock KGJ, Kavanagh K, Potts A, Love J, Cuschieri K, Cubie H, et al. Reduction of low- and high-grade cervical abnormalities associated with high uptake of the HPV bivalent vaccine in Scotland. Br J Cancer [Epidemiol] 2014;111:1824.

[26] Kavanagh K, Pollock KGJ, Potts A, Love J, Cuschieri K, Cubie H, et al. Introduction and sustained high coverage of the HPV bivalent vaccine leads to a reduction in prevalence of HPV 16/18 and closely related HPV types. Br J Cancer [Epidemiol] 2014;110:2804.

[27] Cameron RL, Kavanagh K, Cameron Watt D, Robertson C, Cuschieri K, Ahmed S, et al. The impact of bivalent HPV vaccine on cervical intraepithelial neoplasia by deprivation in Scotland: reducing the gap. J Epidemiol Community Health 2017;71(10):954–60.
[28] Gargano JW, Park IU, Griffin MR, Niccolai LM, Powell M, Bennett NM, et al. Trends in high-grade cervical lesions and cervical cancer screening in 5 States, 2008–2015. Clin Infect Dis 2018; ciy707-ciy.
[29] Benard VB, Castle PE, Jenison SA, et al. Population-based incidence rates of cervical intraepithelial neoplasia in the human papillomavirus vaccine era. JAMA Oncol 2017;3(6):833–7.
[30] Flagg EW, Torrone EA, Weinstock H. Ecological association of human papillomavirus vaccination with cervical dysplasia prevalence in the United States, 2007–2014. Am J Public Health 2016;106 (12):2211–18 [Article].
[31] Novakovic D, Cheng ATL, Zurynski Y, Booy R, Walker PJ, Berkowitz R, et al. A Prospective study of the incidence of juvenile-onset recurrent respiratory papillomatosis after implementation of a National HPV Vaccination Program. J Infect Dis 2018;217(2):208–12.
[32] Campisi P. Should we be screening for oral HPV-related diseases? 32nd IPVC congress 2018; Sydney 2018.
[33] Luostarinen T, Apter D, Dillner J, Eriksson T, Harjula K, Natunen K, et al. Vaccination protects against invasive HPV-associated cancers. Int J Cancer 2018;142(10):2186–7.
[34] ISD Scotland. Cancer Statistics - Female Genital Organ Cancer. Available from: https://www.isdscotland.org/Health-Topics/Cancer/Cancer-Statistics/Female-Genital-Organ/#carcinoma; 2018.
[35] Guo F, Cofie LE, Berenson AB. Cervical cancer incidence in young U.S. females after human papillomavirus vaccine introduction. Am J Prev Med 2018;55(2):197–204.
[36] Giuliano AR, Palefsky JM, Goldstone S, Moreira Jr ED, Penny ME, Aranda C, et al. Efficacy of quadrivalent HPV vaccine against HPV infection and disease in males. N Engl J Med 2011;364 (5):401–11.
[37] Palefsky JM, Giuliano AR, Goldstone S, Moreira ED, Aranda C, Jessen H, et al. HPV vaccine against anal hpv infection and anal intraepithelial neoplasia. N Engl J Med 2011;365(17):1576–85.
[38] Harder T, Wichmann O, Klug SJ, van der Sande MAB, Wiese-Posselt M. Efficacy, effectiveness and safety of vaccination against human papillomavirus in males: a systematic review. BMC Med 2018;16 (1):110.
[39] Markowitz LE, Gee J, Chesson H, Stokley S. Ten years of human papillomavirus vaccination in the United States. Acad Pediatr 2018;18(2, Supplement):S3–10.
[40] Brotherton JM, Batchelor MR, Bradley MO, Brown SA, Duncombe SM, Meijer D, et al. Interim estimates of male human papillomavirus vaccination coverage in the school-based program in Australia. Commun Dis Intell Q Rep 2015;39(2):E197–200.
[41] Pollock KG, Wallace LA, Wrigglesworth S, McMaster D, Steedman N. HPV vaccine uptake in men who have sex with men in Scotland. Vaccine 2018;.
[42] Kreimer AR, Hildesheim A, Rodriguez AC, Porras C, Solomon D, Lowy DR, et al. Proof-of-principle evaluation of the efficacy of fewer than three doses of a bivalent HPV16/18 vaccine. J Natl Cancer Inst 2011;103(19):1444–51.
[43] Dobson SM, McNeil S, Dionne M, et al. Immunogenicity of 2 doses of hpv vaccine in younger adolescents vs 3 doses in young women: a randomized clinical trial. JAMA 2013;309(17):1793–802.
[44] Human papillomavirus vaccines: WHO position paper, October 2014. Wkly Epidemiol Rec 2014;89 (43):456–92.
[45] Garland SM, Brotherton JML, Moscicki AB, Kaufmann AM, Stanley M, Bhatla N, et al. Vaccination of immunocompromised hosts. Papillomavirus Res 2017;4:35–8.

[46] WHO. Global market Study HPV. Available from: https://www.who.int/immunization/programmes_systems/procurement/v3p/platform/module2/WHO_HPV_market_study_public_summary.pdf?ua = 1; 2018.
[47] Yin F, Wang Y, Chen N, Jiang D, Qiu Y, Wang Y, et al. A novel trivalent HPV 16/18/58 vaccine with anti-HPV 16 and 18 neutralizing antibody responses comparable to those induced by the Gardasil quadrivalent vaccine in rhesus macaque model. Papillomavirus Res 2017;3:85–90.
[48] WHO. Meeting of the Strategic Advisory Group of Experts on immunization, October 2018 - Conclusions and recommendations. Wkly Epidemiol Rec 2018;49(93):661–80.
[49] Corcoran BCA, Barrett T. Rapid response to HPV vaccination crisis in Ireland. The Lancet 2018;391 (10135):2103.
[50] WHO. The Global Advisory Committee on Vaccine Safety. Available from: https://www.who.int/vaccine_safety/committee/en/; 2019.
[51] Framework for the Execution of Cervical Cancer Population Screening National Institute for Public Health and the Environment (RIVM); 2017. Available from: www.rivm.nl/en/Documents_and_publications/Common_and_Present/Publications/Disease_prevention_and_healthcare/cervical_cancer_screening/Framework_for_the_Execution_of_Cervical_Cancer_Population_Screening; 2019.
[52] Brotherton JML, Hawkes D, Sultana F, Malloy MJ, Machalek DA, Smith MA, et al. Age-specific HPV prevalence among 116,052 women in Australia's renewed cervical screening program: a new tool for monitoring vaccine impact. Vaccine 2019;37(3):412–6.
[53] Cruickshank M, Pan J, Cotton S, Kavanagh K, Robertson C, Cuschieri K, et al. Reduction in colposcopy workload and associated clinical activity following human papillomavirus (HPV) catch-up vaccination programme in Scotland: an ecological study. BJOG: Int J Obstet Gynaecol 2017;124(9):1386–93.
[54] Munro A, Gillespie C, Cotton S, Busby-Earle C, Kavanagh K, Cuschieri K, et al. The impact of human papillomavirus type on colposcopy performance in women offered HPV immunisation in a catch-up vaccine programme: a two-centre observational study. BJOG: Int J Obstet Gynaecol 2017;124 (9):1394–401.

CHAPTER

POLITICAL AND PUBLIC RESPONSES TO HUMAN PAPILLOMAVIRUS VACCINATION

22

Gregory D. Zimet[1], Beth E. Meyerson[2], Tapati Dutta[3], Alice Forster[4], Brenda Corcoran[5] and Sharon Hanley[6]

[1]Department of Pediatrics, Indiana University School of Medicine, Indianapolis, IN, United States [2]Department of Applied Health Science, Rural Center for AIDS/STD Prevention, Indiana University Bloomington School of Public Health, Bloomington, IN, United States [3]Department of Applied Health Science, Prevention Insights, Rural Center for AIDS/STD Prevention, Indiana University Bloomington School of Public Health, Bloomington, IN, United States [4]Research Department of Behavioural Science and Health, University College London Institute of Epidemiology and Health Care, London, United Kingdom [5]National Immunisation Office (formerly), Dublin, Ireland [6]Department of Reproductive and Developmental Medicine, Hokkaido University Graduate School of Medicine, Sapporo, Japan

INTRODUCTION

In this chapter, we will cover some of the political and policy issues that have been encountered with implementing and sustaining human papillomavirus (HPV) immunization programs. Through case presentations focused on challenges and successes, we hope to identify strategies that can prospectively minimize obstacles as well as address and overcome challenges when they arise. The successful implementation of vaccination programs requires consistent multilevel messaging (i.e., at the national, state, local, and clinical practice levels) directed at the population in general, healthcare providers, and patients. Such messaging carried out in advance of program implementation can help to ensure buy-in from political partners and from the general public, and help planners and others to anticipate potential obstacles and concerns about HPV vaccine initiatives, thereby serving to provide some inoculation from backlash.

Prior collaboration on HPV-related issues by coalitions might also make a difference, as was seen in a 2016 case from the US state of Indiana. Here, an effort by antivaccine activists and the governor to remove reminder recall letters to parents encouraging HPV vaccination was partially thwarted by a broad-based, advocacy effort [1]. Advocacy work on state HPV vaccination legislation the year prior facilitated the cross-sector working partnerships that eventually engaged the media. This coalition was not established in the planning process for the state immunization program, yet it facilitated the successful response to protect the reminder recall program from further interruption. In a similar situation, a number of publicly funded Catholic Schools in Alberta, Canada decided in 2008 to ban HPV vaccine, despite national approval and public funding for vaccination. A similar broad-based coalition of existing and new partners formed to overturn the ban and were ultimately successful after a sustained, multiyear effort [2].

Human Papillomavirus. DOI: https://doi.org/10.1016/B978-0-12-814457-2.00022-2

Having prior relationships among government and nongovernmental organizations (NGOs) and across communities certainly helps, yet several national and state vaccination programs have not benefitted from prior community-engaged planning. A recent study of African country cervical cancer prevention plans found that while the commonly engaged partners were around the planning table, such as the ministries of health, technical writing groups, and topic experts; communities themselves were not engaged beyond top-down implementation planning [3]. Early engagement of partners should occur at the planning stages. While this seems straightforward, it is clear that government entities need to reach further into the community to identify partners for this process. Usually, the planning for vaccination programs is tasked to a government Ministry of Health at the national or state level. This is part of the challenge, because government partners do not necessarily recognize the broad expanse of potential partners in public health. Unanticipated opposition to HPV vaccination and attendant drops in vaccination rates observed in Japan [4–6], Ireland [7], Denmark [8,9], and Columbia [10] (among other countries) point to the continued need for vaccine planners and their partners to engage with a broad coalition and directly in the community. Other countries, such as the United States, have encountered very slow, and geographically variable increases in HPV vaccination rates over time [11,12]. Further, in India, there was a strong backlash against an HPV vaccine demonstration project. The reaction was fueled largely by false claims of harm and the incorrect characterization of the project as a clinical trial, and this impeded the broader implementation of HPV vaccination programs in India [13–15].

Anticipation of, and preparation for, reactions and misunderstandings about any type of public health program are essential to optimizing success and are key aspects of communication and planning [16,17]. Ideally, communication strategies should be developed for before, during, and after the roll-out of new or revised initiatives. Related to the need for broad engagement in planning is the critical element of including viewpoints that are not in agreement with the program at hand. Engaging across conceptual divides during and even after the planning process can inform issues identification, partner strategies, and messaging [18,19].

CASE STUDIES

As of June, 2017, 91 countries had implemented national HPV vaccination programs and 38 countries had introduced pilot programs [20]. As shown in Fig. 22.1, provided by PATH, by the end of January, 2019, this number had risen to over 100.

In countries that have introduced HPV vaccination the mean coverage has been estimated to be 60%, but this figure varies widely both between and within countries [21]. In addition, many countries do not have national programs; and relatively few national programs target both males and females. Thus, the vast majority of adolescents worldwide remain unvaccinated and, therefore vulnerable to HPV-related diseases.

Moreover, despite the overwhelming evidence of HPV vaccine efficacy and safety [22], barriers to achieving high rates of HPV vaccination continue to exist many countries. Failure to achieve high coverage with HPV vaccine is clearly due to multiple factors that may differ from country to country, and likely include financial barriers (particularly in low- and middle-income countries),

FIGURE 22.1

Global HPV Vaccine Introduction as of January 28, 2019 (care of PATH).

public health policy issues, health communication difficulties, and parental hesitancy. Changes in HPV vaccination recommendations (e.g., change in the number of recommended doses, moving from female only to gender neutral vaccination, and changes in targeted ages) can create uncertainty and communication challenges; and yet, over time, may simplify collective understanding of the requirements. That said, a recent study found widespread confusion among HCPs across five countries about age guidelines for HPV vaccination [23].

Vaccine cost is an issue raised by Dutta and Meyerson's review of African country cervical prevention plans [3], and the issue is likely shared among several developing countries [24]. The financial cost burden is recognized, as many low and lower-middle income countries receive financial assistance for HPV vaccine implementation from GAVI, PAHO, and WHO [21,25].

In several countries, including Japan, Ireland, Denmark, Columbia, and Austria, initially successful HPV vaccination programs encountered crises that led to precipitous drops in vaccination rates. In Japan, there has been little to no recovery of vaccination rates after the steep drop from over 70% series completion to about 1% [4]. In some cases (e.g., Austria, Denmark, and Ireland) the programs have recovered to some extent (Ireland and Denmark) or completely (Austria) [7,26–28]. India, which encountered extreme opposition to HPV vaccination demonstration projects that led to the suspension of those programs, has also re-initiated its HPV vaccine programming on a pilot basis [29]. It may be instructive to examine the strategies used in these cases to work to overcome the threats and reestablish success.

CASE STUDY: SLOW AND UNEVEN PROGRESS; HUMAN PAPILLOMAVIRUS VACCINATION IN THE UNITED STATES

Although HPV vaccine was licensed and made available in the United States in 2006, coverage rates have increased only very gradually and nationally remain well-below the Healthy People 2020 target of 80% series completion for males and females [12]. Furthermore, rates of series initiation vary dramatically across the United States, ranging, in 2017, from 46.9% in Wyoming to 91.9% in the District of Columbia (see Fig. 22.2). To achieve high rates of childhood and adolescent vaccination, the primary public health tools used in the United States are school-entry requirements, which vary from state to state. Only one state in the United States, Rhode Island, currently has a true gender-neutral HPV vaccine school-entry requirement, and early evidence indicates that it has been successful in increasing vaccination coverage, particularly among boys [30].

Some of the difficulties encountered with HPV vaccination would be true of any vaccine for which there is not a school-entry requirement. However, there have been unique barriers in the United States that have interfered with broader adoption of HPV vaccination, including widespread distrust provoked by efforts to legislate HPV vaccine school-entry requirements in 2006–07, immediately after licensure [31], poor and inconsistent communication practices by HCPs related to HPV vaccine recommendation [32,33], false or misleading information on social media and the internet in general [34], parental concerns about safety and several other issues [35], and a

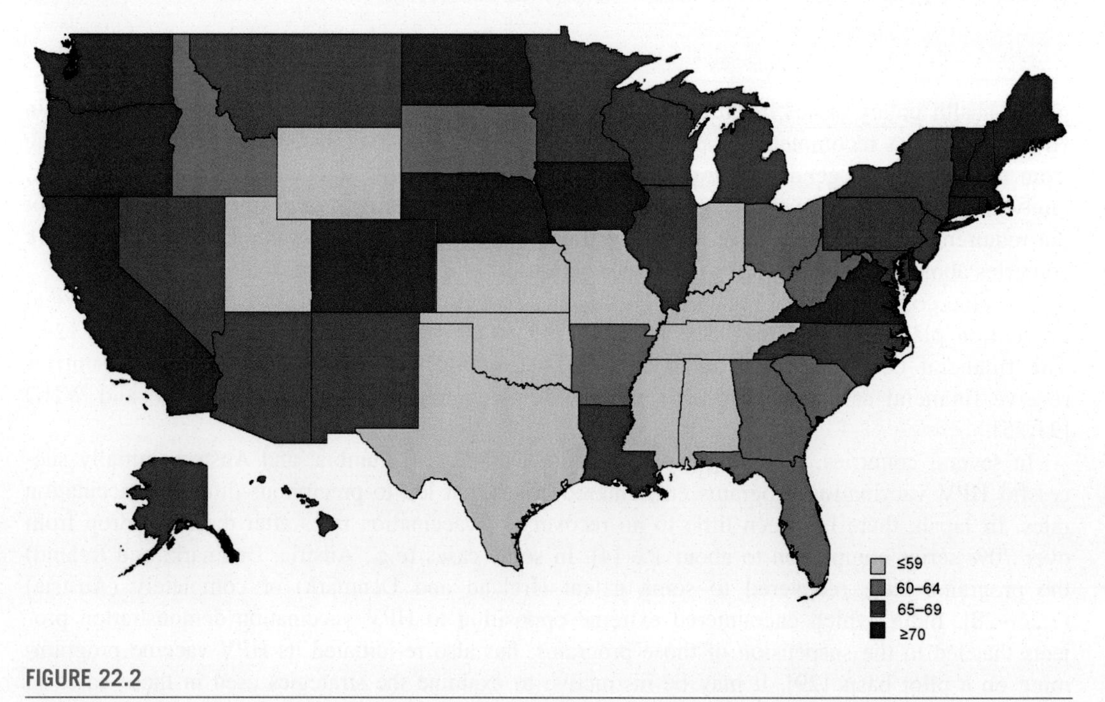

FIGURE 22.2

Estimated coverage with ≥ 1 doses of HPV vaccine among adolescents aged 13–17[†] years—National Immunization Survey-Teen, United States, 2017 (https://stacks.cdc.gov/view/cdc/58072).

confusing patchwork of state laws that include simply requiring education about HPV and HPV vaccination, unenforceable school-entry requirements (e.g., Virginia), and enforceable school-entry requirements (e.g., Rhode Island) [36,37].

Although vaccine recommendations, including those for HPV vaccine, are issued nationally by the Advisory Committee on Immunization Practice (ACIP), how these recommendations are implemented depends on state immunization policies, which vary from state to state [38], and on the practice approaches of healthcare systems and individual healthcare providers. Top-down policy approaches to improving HPV vaccination rates include instituting state-level school-entry requirements and institution by health systems and/or insurance companies of benchmarks for vaccine coverage. However, implementing school-entry requirements for HPV vaccine is politically challenging and may not be feasible in many states in the United States [36]. The lack of a unified national policy along with a clinic/HCP-based delivery system, in part, explain both the slow increase over time in HPV vaccine coverage and the disparities from state to state (as well as within states) [12,39]. Much of the effort in the United States has focused on developing and testing interventions to improve vaccination rates at the clinic level, which is an important focus, but results of most intervention studies have modest to moderate effects sizes [40]. Furthermore, dissemination and implementation of some of the interventions can be very challenging and costly. Because it is unlikely that the overall approach to health and public health in the United States will change appreciably over the next several years, arguably the best approach to improving HPV vaccine coverage is a multilevel, multimodal approach. This approach should include broad-based messaging directed at parents and adolescents (including via social media), efforts to counter misinformation disseminated on the internet, interventions implemented at the HCP and practice levels, changes in health system policies, efforts to move states toward implementing school-entry requirements for HPV vaccination, and development of alternative sites for vaccine delivery (e.g., pharmacies) [41]. Importantly, there are several coalitions in the United States composed of governmental, nonprofit, academic, and for-profit organizations that help to promote accurate information about HPV vaccination and to disseminate practical tools for improving vaccination rates. Two examples are the National HPV Roundtable (http://hpvroundtable.org/) and the UNITY Consortium (http://www.unity4teenvax.org/).

CASE STUDY: RECOVERY; HUMAN PAPILLOMAVIRUS VACCINATION IN AUSTRIA, DENMARK, AND IRELAND

In Austria, Denmark, and Ireland, initial successes with implementation of national HPV vaccination programs encountered unanticipated obstacles that led to sharp drops in vaccination rates. In Austria, HPV vaccine was first introduced in 2007. Shortly after introduction, there was a report that a young woman had died three weeks after vaccination, with her parents and news media attributing her death to HPV vaccination [42]. Shortly thereafter, the initial, promising start to the HPV vaccination program came to an abrupt halt, with attendant criticism that the National Immunization Program lacked transparency in decision-making and communication and widespread questions about vaccine safety and effectiveness [26]. Subsequently, in 2013, HPV vaccine was successfully reintroduced in Austria, after opposition had diminished and in the context of overwhelming evidence of the safety and effectiveness of HPV vaccination. HPV vaccine was included in Austria's National Immunization Program and administration was targeted to both boys and girls at age 9 years, which helped to convey the message that HPV vaccination was a regular part of the childhood immunization program [43].

In Denmark and Ireland, there was a somewhat different set of trajectories. In both countries, HPV vaccination programs had been established and had achieved high sustained vaccination coverage. In Denmark, after having established very high coverage for the vaccine, HPV vaccination rates dropped sharply [8]. The sharp decline in vaccination was attributed to negative coverage of HPV vaccine by the media and the airing of an inaccurate, negative documentary, suggesting that monitoring of media reporting on HPV vaccination and, particularly, their dissemination of misinformation, may be an important part of counteracting the effects of such coverage [9]. In response to these negative media reports, the Danish Health Authority surveyed parents to better understand their concerns, then formed a partnership with nongovernmental organizations, such as the Danish Cancer Society and the Danish Medical Association to develop an informational campaign to educate parents and counter the misinformation [27]. Initial results have been encouraging [44].

Ireland encountered a similar drop in HPV vaccination rates. The school-based program in Ireland began in 2010 for girls aged 12–13 years. Initially over 80% of eligible girls completed the HPV vaccine series and this increased every year to a peak of 87% during 2014–15 [45]. However, by the next academic year, that figure had dropped to 72%, then to 51% during 2016–17 [45], largely due to the propagation of misinformation about vaccine safety and efficacy by groups of concerned parents, with support from local and national media. This publicity resulted in a widely viewed television documentary "Cervical cancer vaccine—is it safe?" which contained extracts from a similar Danish documentary [7]. To reverse this decline in vaccination and counter the misinformation that had been propagated, the Irish health authority established cross-sectoral alliances with governmental and nongovernmental organizations, implemented a comprehensive training program for a wide range of health professionals, and provided enhanced information for parents after focus group discussions and analysis of issues raised on social media. These actions along with vaccine promotion by a very powerful patient advocate (a 25-year-old woman with terminal cervical cancer) led to a rapid rise with uptake of the first vaccine dose at 65% in 2017–18 increasing to 70% in 2018–19 [28].

CASE STUDY: NO RECOVERY; HUMAN PAPILLOMAVIRUS VACCINATION IN JAPAN

In contrast to the experiences of Austria, Denmark,and Ireland, Japan experienced a dramatic drop in HPV vaccination coverage from which it has not recovered. Free vaccination against HPV began in December 2010 with a special fund for girls aged 12–16 years [4]. Owing to unexpectedly high demand, vaccine stock decreased dramatically between March and August 2011, which meant many girls could not initiate vaccination until after September 2011. During this period the media fiercely criticized the government for not stocking enough of the cancer preventing vaccine [46]. However, around the time it became clear the HPV vaccine was to be included in the National Immunisation Programme (NIP) in April 2013, the media stance began to change. Increasing unconfirmed reports of Adverse Events Following Immunization (AEFI) began to appear in the media and a "Victims" support group was established [47].

Two months later, after including the HPV vaccine in the NIP, on June 14, 2013, the Ministry of Health, Labour and Welfare (MHLW) suspended proactive recommendations for the vaccine and uptake very quickly dropped from >70% to <1% [4]. Almost 6 years later, the suspension and low-to-no HPV vaccine coverage remains, despite no evidence to suggest the vaccine was responsible for any of the reported symptoms [48,49].

The reasons for this sustained drop in coverage are complex and varied. The first reason is poor risk management by the MHLW. Japan has inadequate disease registries and no HPV vaccine registry, which makes it difficult to ascertain background incidence rates on neurological and autoimmune conditions claimed by the Victims' support group to be caused by the HPV vaccine. This makes it is almost impossible to carry out observed versus expected analyses, and consequently, for health officials to show no significant increase in these specific conditions after HPV vaccine introduction.

The second reason is poor risk communication by the MHLW. Upon suspension of the proactive recommendations for the HPV vaccine, the MHLW told all prefectural governors not to actively recommend the vaccine and to cease all vaccine promotion. However, at the same time, the vaccine remained in the NIP and health facilities were told to continue to offer full support to parents seeking the vaccine and facilitate vaccine access. Therefore the decision to suspend proactive recommendations was not based on scientific evidence and caused enormous confusions for public health officials, doctors, and the general public. This, combined with the fact that many cervical cancer advocacy groups also stopped promoting the HPV vaccine, made it difficult for parents to access high-quality educational information about cervical cancer and HPV.

Finally, a persistently negative, unchallenged mainstream media bias against the HPV vaccine, where media rhetoric quickly became scientific fact, and an anti-vaccine group more organized than the government, especially on social media, has made it difficult for the government to restart proactive recommendations for HPV vaccination.

CASE STUDY: SUCCESSFUL AND SUSTAINED IMPLEMENTATION; HUMAN PAPILLOMAVIRUS VACCINATION IN AUSTRALIA AND THE UNITED KINGDOM

Australia and the United Kingdom are excellent examples of successful, sustained HPV vaccination strategies. In both countries, HPV vaccines are routinely administered in schools and the vaccine is covered by the national health insurance and/or immunization programs. Australia began its national HPV vaccination program in 2007. In addition to routine vaccination of girls ages 12–13 years, Australia also implemented an extensive 2-year catch-up program for women ages 14–26 years, which took place both in schools and in general primary care practices. In 2013, the program was extended to include boys ages 12–13 years with a one year catch-up program for boys 14–15 [50]. In 2017, a second catch-up program up to the age of 19 years commenced for both sexes and in 2018, a routine 2-dose schedule began for the target cohort of adolescents aged 12–13 years [50]. Coverage is high in both males and females at around 80% for series completion. With implementation of primary HPV testing for cervical cancer screening, a large female catch-up cohort, and a gender-neutral HPV vaccination program, researchers estimate that cervical cancer mortality will be virtually eliminated in Australia (i.e., < 1 death per 100,000 women) in about 20 years [51].

In the United Kingdom, universal school-based bivalent HPV vaccination of girls ages 12–13 years was introduced in 2008. Three-dose completion rates of over 80% were achieved shortly thereafter, facilitated by delivery of the vaccine in schools, an approach already used for delivering vaccines to school-aged children [52]. Subsequently, in 2012, the UK Departments of Health switched to the quadrivalent HPV vaccine and in April, 2018, the recommendation for HPV vaccination was extended to men who have sex with men up through age 45 delivered via sexual health services [53]. During pilot implementation of this program, around 46% of eligible men received the first-dose of the vaccine [53]. Other groups for whom HPV vaccination was newly recommended in 2018 included transgender persons, if they met certain behavioral risk criteria. Risk-based vaccination strategies are often problematic for a variety of reasons, including difficulties assessing risk on the part of health care providers and determining denominators to monitor uptake. An impending rollout of a two-dose gender-neutral HPV vaccination policy that will include adolescent boys [54], will mean that there will eventually need to be a re-evaluation of the cost-effectiveness of the risk-based strategy [55]. The UK HPV vaccine program has not been without its own controversy. In 2009, a girl died shortly after receiving the vaccine, which was widely reported in the mainstream media. Her death was later attributed to an unrelated health condition, and swift intervention by health authorities prevented her death from having any impact on uptake of the program [56].

CASE STUDY: FALSE START; HUMAN PAPILLOMAVIRUS VACCINATION IN INDIA

The Indian Academy of Pediatrics Committee on Immunization (IAPCOI) recommends HPV vaccination for all children, with a two-dose schedule for children 9–14 years of age and a trhree-dose schedule for those who are 15–18 years of age [57]. In addition, in 2016, GAVI approved 500 million USD for India's vaccination programs over 2016–21, including costs to introduce HPV vaccine [58]. In 2016, the states of Delhi and Punjab announced their plans to introduce HPV vaccine [59,60].

However, the introduction of the vaccine continues to face vehement opposition [61]. In response to these objections, the Deputy Director General of the World Health Organization and former Director General of the Indian Council of Medical Research (ICMR), urged the Indian Government to introduce the vaccine since it could help avert a large number of cervical cancer deaths in the country [62].

There is a complicated history behind the objections to HPV vaccination in India [63]. In 2009, PATH, an international nonprofit organization collaborated with the Ministry of Health and Family Welfare (MoHFW) and other partners in India to implement and evaluate a relatively small-scale demonstration project for HPV vaccine implementation [64]. Under this post-licensure demonstration project, HPV vaccines were introduced among 14,000, 10–14-year-old girls in three *mandals* (i.e., a sub-district, an administrative division of some states in India) of Andhra Pradesh and among 16,000 girls in three *Blocks* (i.e., rural areas administratively earmarked for development) of Vadodara District in Gujarat. The vaccine was administered through a camp approach in hostels and school campuses. However, several months into the trial, the government halted the project after news that seven girls had died after being vaccinated. Not surprisingly, investigations confirmed that the deaths were unrelated to vaccination and absolved the program's managers—PATH and the ICMR—of any responsibility in the deaths. Nonetheless, ICMR, under some pressure, later reported that ethical guidelines were not adequately followed in the course of this project [65]. In contrast, PATH representatives and others continued to defend the demonstration projects, stating that they adhered to ethical practices [14,15]. Regardless of the merits or lack of merits associated with the criticism of the demonstration project, opponents of HPV vaccination were able to reach and influence the highest levels of government in India and their efforts received widespread coverage in the news [65,66]. As a result, wider implementation of HPV vaccination programs in India have been delayed and HPV vaccination continues to generate controversy [61,63].

In looking back at the HPV vaccine demonstration projects it is impossible to know if the problems that arose could have been anticipated and adequately prepared for. However, the misunderstandings related to the projects (i.e., false attributions of harm and mischaracterization of the projects as clinical trials) point to the importance of community engagement, community preparation, and effective communication strategies from the community level to the governmental level.

CASE STUDY: INTRODUCTION OF HUMAN PAPILLOMAVIRUS VACCINATION PROGRAMS IN LOW-/MIDDLE-INCOME COUNTRIES

Several other low-/middle-income countries have initiated HPV vaccination programs since 2015 [67,68]. Today 14% of low/middle income countries have a national HPV vaccine program, as compared with 55% of high/upper-middle income countries. But for those LMIC countries, HPV vaccine program scale up to the national level remains challenging due to the lack of sustainable funding [24]. The first low-/middle-income country and the first Asian country to implement a national cervical cancer vaccination program was Bhutan. With support from the Australian Cervical Cancer Foundation (ACCF) and vaccine donations through Merck's Global Access Plan (GAP), a school-based pilot program with national-scale up vaccinated around 50,000 girls aged 12–18 years between 2009 and 2010; around 96% of the eligible cohorts. A national program for girls aged 12 years was established in 2011. In 2012 and 2013, the program shifted from the school to the community and as a result uptake dropped to below 70%. From 2014 onwards vaccination returned to the school and coverage has been >90% thereafter [69]. In the national program, ACCF provides financial support to the Royal Government of Bhutan to secure doses of GARDASIL at the access price offered by Merck.

In Rwanda, the first African country to have an HPV vaccination program, high uptake was achieved by school-based vaccination [70]. As Torres-Rueda and colleagues report, the campaign allowed for the leveraging of additional services for vaccine recipients such as health promotion focused on other topics [70]. Sustainability will be a challenge, as the commitment to this program was from 2014–17 by GAVI after the government of Rwanda financed the initial program in 2011.

Using schools as the vehicle for vaccine delivery was also the decision in Malaysia; which implemented their HPV vaccination program in 2010 [71]. The focus, like with Bhutan and Rwanda, was girls, with an impressive 1st dose uptake of 99% and an estimated population coverage for series completion of 83% to 91%.

DISCUSSION

As illustrated by the case studies presented in this chapter, there have been a wide range of experiences associated with implementation of HPV immunization programs. There have been notable, sustained successes, set-backs followed by recoveries, slow and uneven progress with ongoing challenges, and sustained failures with no recovery. These various trajectories are a function of many factors, including HPV vaccination policies, which vary considerable from country to country in terms of national versus regional authority, coverage of vaccine costs, age and gender targets of vaccination, and vaccination venue. The nature of HPV vaccine hesitancy also can vary from country to country, as reported by Karafillakis et al. in a recent paper on determinants of HPV vaccine hesitancy in Europe, further suggesting that strategies to improve HPV vaccination rates will need to take into account country- and region-specific policies and attitudes [72]. At the same time, there are commonalities that are largely consistent across successes and failures. For instance, it is clear that strong support for HPV vaccination programs from relevant government agencies is an absolutely necessary, though not sufficient, element for success. Conversely, lack of government support leaves countries vulnerable to anti-HPV-vaccine initiatives. Formations of coalitions that include governmental, nongovernmental, and grass-roots organizations can help to inoculate vaccination programs from unwarranted, but sometimes effective attacks. Furthermore, the global, rapid dissemination of false information on vaccines, including HPV vaccine, via social media, such as Facebook, Youtube, Twitter, and Instagram means that we have to focus our attention and efforts at all levels: global, national, and regional. For instance, a recent analysis of a coordinated antivaccine social media attack on a single primary care practice located in the US state Pennsylvania found that only 5 of 136 posters were from Pennsylvania, with 24 from California, and several from other countries, such as Australia and Canada [73]. Similarly, a 2014 study found that Japan's MHLW 2013 suspension of the recommendation for HPV vaccination resulted in a rapid, global spread of mostly negative HPV vaccine posts on various online media platforms [5]. There are ongoing efforts to describe and evaluate social media content related to HPV vaccination, as reflected in a recently published systematic review [74]. Monitoring media information on HPV vaccination as well as developing and disseminating effective social media messaging campaigns to anticipate, and respond to, inaccurate information about HPV vaccination will be an important focus of research efforts. Moreover, it is notable and important that social media platforms have begun to take action against the misinformation on vaccines that proliferate throughout social media (For example, see: https://edition.cnn.com/2019/09/04/health/facebook-vaccine-education-bn/index.html and https://newsroom.pinterest.com/en/post/bringing-authoritative-vaccine-results-to-pinterest-search).

Some broad policy approaches that could serve to maximize sustained HPV vaccination coverage involve disassociating the vaccine from sexual behavior, which can be accomplished in several ways. Included among these are gender neutral vaccination, targeting vaccination to children who are 9 years of age, rather than ages 11–12 years, and emphasizing cancer prevention as the primary vaccination message. A gender-neutral approach may not be financially viable for some countries and would depend on an adequate supply of vaccine to immunize all boys and girls. However, this approach has the added benefit of moving countries more quickly towards the ultimate goal of HPV-cancer elimination.

In general, as noted in this chapter, the countries with the greatest successes with HPV vaccination have used school-based approaches, which help to maximize efficiency of delivery and coverage, minimize disparities in vaccination, and tend to create a pro-HPV-vaccine social norm. Though school-based vaccination has great potential for success [75], not all countries have the political will or infrastructure to implement wide-spread school-based HPV immunization programs. When this is the case, the WHO suggests that high rates of vaccination may be achieved through a combination of strategies that include clinic-based approaches, outreach, and use of alternative venues (e.g., pharmacies) [76]. When multiple strategies are deployed simultaneously, however, it becomes particularly important to have a functioning vaccine registry in order to keep track of HPV vaccine doses delivered.

There is enough global experience with HPV vaccination to engage in scenario building with partners at the planning stages. What, for example, might be the community response to an HPV vaccination campaign in school-aged children in selected communities? These scenarios should be thoroughly discussed, and plans made for the use of risk communications strategies, timing of response, messengers, and messages. Timing is critical, as governments cannot and should not "hide" from public scrutiny. We have already seen situation where ministries of health or NGO subcontractors have been isolated during crises, ultimately forcing higher level government response to pause or even cancel vaccine programming. For example, Japan's Ministry of Health Labour and Welfare missed critical opportunities following false attributions of harm to clarify the science of HPV vaccination and reduce confusion among the medical community and public [4]. For any vaccine campaign to be successful requires a well-functioning immunization program, high-level political commitment to vaccination, and coordination among immunization, child/adolescent health, cancer prevention, and community groups.

REFERENCES

[1] Meyerson BE, Zimet GD, Adams K. Try this at home: Rapid response coalition building and evidence-based advocacy. Case from Indiana, USA. EUROGIN 2017; 10/09/2017, 2017; Amsterdam, The Netherlands.

[2] Guichon JR, Mitchell I, Buffler P, Caplan A. Citizen intervention in a religious ban on in-school HPV vaccine administration in Calgary, Canada. Prev Med 2013;57(5):409–13.

[3] Dutta T, Meyerson B, Agley J. African cervical cancer prevention and control plans: A scoping review. J Cancer Policy 2018;16:73–81.

[4] Hanley SJ, Yoshioka E, Ito Y, Kishi R. HPV vaccination crisis in Japan. Lancet 2015;385(9987):2571.

[5] Larson HJ, Wilson R, Hanley S, Parys A, Paterson P. Tracking the global spread of vaccine sentiments: the global response to Japan's suspension of its HPV vaccine recommendation. Hum Vaccin Immunother 2014;10(9):2543–50.

[6] Ueda Y, Enomoto T, Sekine M, Egawa-Takata T, Morimoto A, Kimura T. Japan's failure to vaccinate girls against human papillomavirus. Am J Obstet Gynecol 2015;212(3):405–6.

[7] Corcoran B, Clarke A, Barrett T. Rapid response to HPV vaccination crisis in Ireland. Lancet 2018;391 (10135):2103.

[8] Lubker CL, Lynge E. Stronger responders-uptake and decline of HPV-vaccination in Denmark. Eur J Public Health 2018;29(3):500–5.

[9] Suppli CH, Hansen ND, Rasmussen M, Valentiner-Branth P, Krause TG, Molbak K. Decline in HPV-vaccination uptake in Denmark - the association between HPV-related media coverage and HPV-vaccination. BMC Public Health 2018;18(1):1360.

[10] Simas C, Munoz N, Arregoces L, Larson HJ. HPV vaccine confidence and cases of mass psychogenic illness following immunization in Carmen de Bolivar, Colombia. Hum Vaccin Immunother 2019;15 (1):163–6.

[11] Stokley S, Jeyarajah J, Yankey D, et al. Human papillomavirus vaccination coverage among adolescents, 2007-2013, and postlicensure vaccine safety monitoring, 2006-2014 - United States. MMWR 2014;63 (29):620–4.

[12] Walker TY, Elam-Evans LD, Yankey D, et al. National, regional, state, and selected local area vaccination coverage among adolescents aged 13-17 years - United States, 2017. MMWR. 2018;67 (33):909–17.

[13] Larson HJ, Brocard P, Garnett G. The India HPV-vaccine suspension. Lancet 2010;376(9741):572–3.

[14] LaMontagne DS, Sherris JD. Addressing questions about the HPV vaccine project in India. Lancet Oncol 2013;14(12):e492.

[15] Tsu VD. Indian vaccine study clarified. Nature 2011;475(7356):296.

[16] O'Mara-Eves A, Brunton G, Oliver S, Kavanagh J, Jamal F, Thomas J. The effectiveness of community engagement in public health interventions for disadvantaged groups: a meta-analysis. BMC Public Health 2015;15:129.

[17] Schoch-Spana M, Franco C, Nuzzo JB, Usenza C. Working Group on community engagement in health emergency planning. community engagement: leadership tool for catastrophic health events. Biosecur Bioterror 2007;5(1):8–25.

[18] Ward JK, Cafiero F, Fretigny R, Colgrove J, Seror V. France's citizen consultation on vaccination and the challenges of participatory democracy in health. Soc Sci Med 2019;220:73–80.

[19] Marshall HS, Proeve C, Collins J, et al. Eliciting youth and adult recommendations through citizens' juries to improve school based adolescent immunisation programs. Vaccine 2014;32 (21):2434–40.

[20] Cervical Cancer Action. Global Maps: Global Progress in HPV Vaccination: Status: June 2017, <http://www.cervicalcanceraction.org/comments/comments3.php>; 2017 [accessed 03.01.19].

[21] Brotherton JM, Zuber PL, Bloem PJ. Primary prevention of HPV through vaccination: update on the current global status. Curr Obstet Gynecol Rep 2016;5(3):210–24.

[22] Arbyn M, Xu L, Simoens C, Martin-Hirsch PPL. Prophylactic vaccination against human papillomaviruses to prevent cervical cancer and its precursors. Cochrane Database Syst Rev 2018;(5).

[23] Topazian HM, Dizon AM, Di Bona VL, et al. Adolescent providers' knowledge of human papillomavirus vaccination age guidelines in five countries. Hum Vaccin Immunother 2019;15(7–8):1672–7.

[24] Gallagher KE, LaMontagne DS, Watson-Jones D. Status of HPV vaccine introduction and barriers to country uptake. Vaccine 2018;36:4761–7.

[25] Sabeena S, Bhat PV, Kamath V, Arunkumar G. Global human papilloma virus vaccine implementation: an update. J Obstet Gynaecol Res 2018;44(6):989–97.

[26] Paul KT. "Saving lives": adapting and adopting human papilloma virus (HPV) vaccination in Austria. Soc Sci Med 2016;153:193–200.

[27] WHO Europe. Danish health literacy campaign restores confidence in HPV vaccination, <http://www.euro.who.int/en/countries/denmark/news/news/2019/01/danish-health-literacy-campaign-restores-confidence-in-hpv-vaccination>; 2019 [accessed 30.01.19].

[28] Health Service Executive. HSE confirms HPV vaccine uptake is now 70%, <https://www.hse.ie/eng/services/news/media/pressrel/hse-confirms-hpv-vaccine-uptake-is-now-70-.html>; 2019 [accessed 14.03.19].

[29] World Health Organization. Punjab launches HPV vaccine with WHO support, <http://www.searo.who.int/india/mediacentre/events/2016/Punjab_HPV_vaccine/en/>; 2016 [accessed 01.02.19].

[30] Thompson EL, Livingston 3rd MD, Daley EM, Zimet GD. Human papillomavirus vaccine initiation for adolescents following Rhode Island's school-entry requirement, 2010–2016. Am J Public Health 2018;108(10):1421–3.

[31] Haber G, Malow RM, Zimet GD. The HPV vaccine mandate controversy. J Pediatr Adolesc Gynecol 2007;20:325–31.

[32] Gilkey MB, Malo TL, Shah PD, Hall ME, Brewer NT. Quality of physician communication about human papillomavirus vaccine: findings from a national survey. Cancer Epidemiol Biomarkers Prev 2015;24(11):1673–9.
[33] Sturm L, Donahue K, Kasting M, Kulkarni A, Brewer NT, Zimet GD. Pediatrician-parent conversations about human papillomavirus vaccination: an analysis of audio recordings. J Adolesc Health 2017;61:246–51.
[34] Margolis MA, Brewer NT, Shah PD, Calo WA, Gilkey MB. Stories about HPV vaccine in social media, traditional media, and conversations. Prev Med 2019;118:251–6.
[35] Hirth JM, Fuchs EL, Chang M, Fernandez ME, Berenson AB. Variations in reason for intention not to vaccinate across time, region, and by race/ethnicity, NIS-Teen (2008-2016). Vaccine 2019;37:595–601.
[36] Keim-Malpass J, Mitchell EM, DeGuzman PB, Stoler MH, Kennedy C. Legislative activity related to the human papillomavirus (HPV) vaccine in the United States (2006-2015): a need for evidence-based policy. Risk Manag Healthc Policy 2017;10:29–32.
[37] Hoss A., Meyerson B.E., Zimet G.D. State statutes and regulations related to human papillomavirus vaccination, Hum Vaccin Immunother 2019;15(7–8):1672–1677.
[38] Immunization Action Coalition. State mandates on immunization and vaccine-preventable diseases, <http://www.immunize.org/laws/>; 2018 [accessed 18.01.19].
[39] Kepka D, Spigarelli MG, Warner EL, Yoneoka Y, MCConnel N, Balch AH. Statewide analysis of missed opportunities for human papillomavirus vaccination using vaccine registry data. Papillomaivrus Res 2016;2:128–32.
[40] Smulian EA, Mitchell KR, Stokley S. Interventions to increase HPV vaccination coverage: A systematic review. Hum Vaccin Immunother 2016;12(6):1566–88.
[41] Daley E, Thompson E, Zimet G. Human papillomavirus vaccination and school entry requirements: Politically challenging, but not impossible. JAMA Pediatr 2019;173(1):6–7.
[42] oe24. Student died after a cancer vaccine, <https://www.oe24.at/oesterreich/chronik/oberoesterreich/Studentin-starb-nach-einer-Krebs-Impfung/228130>; 2008 [accessed 17.01.19].
[43] Paul KT, Wallenburg I, Bal R. Putting public health infrastructures to the test: introducing HPV vaccination in Austria and the Netherlands. Sociol Health Illn 2018;40(1):67–81.
[44] Statens Serum Institut. Twice as many received HPV vaccination in 2017 as in 2016, <https://en.ssi.dk/news/news/2018/2018---01---hpv>; 2018 [accessed 15.03.19].
[45] Health Protection Surveillance Centre. HPV Immunisation Uptake Statistics. <http://www.hpsc.ie/a-z/vaccinepreventable/vaccination/immunisationuptakestatistics/hpvimmunisationuptakestatistics/>; 2018 [accessed 15.03.19].
[46] 子宮頸がんワクチン供給不足　厚労省、助成条件緩和, <https://www.nikkei.com/article/DGXNASDG07035_X00C11A3CR8000/>; 2011 [accessed 14.03.19].
[47] Tsuda K, Yamamoto K, Leppold C, et al. Trends of media coverage on human papillomavirus vaccination in Japanese newspapers. Clin Infect Dis 2016;63(12):1634–8.
[48] Suzuki S, Hosono A. No association between HPV vaccine and reported post-vaccination symptoms in Japanese young women: results of the Nagoya study. Papillomavirus Res 2018;5:96–103.
[49] Saitoh A, Okabe N. Progress and challenges for the Japanese immunization program: beyond the "vaccine gap". Vaccine 2018;36(30):4582–8.
[50] Brotherton JM. Human papillomavirus vaccination update: nonavalent vaccine and the two-dose schedule. Aust J Gen Pract 2018;47(7):417–21.
[51] Hall MT, Simms KT, Lew JB, et al. The projected timeframe until cervical cancer elimination in Australia: a modelling study. Lancet Public Health 2019;4(1):e19–27.
[52] Public Health England. Human papillomavirus (HPV) vaccine coverage in England, 2008/09 to 2013/14: a review of the full six years of the three-dose schedule. London, England: Public Health England; 2015.

[53] Public Health England. Human papillomavirus (HPV) vaccination for men who have sex with men (MSM): 2016/17 pilot evaluation. London, England: Public health England; 2018.
[54] Department of Health and Social Care. HPV vaccine to be given to boys in England. <https://www.gov.uk/government/news/hpv-vaccine-to-be-given-to-boys-in-england/>; 2018 [accessed 22.02.19].
[55] Forster AS, Gilson R. Challenges to optimising uptake and delivery of a HPV vaccination programme for men who have sex with men. Hum Vaccin Immunother 2019;15(7–8):1541–3.
[56] Adetunji J. Cervical cancer vaccination 'most unlikely' to have caused girl's death, <https://www.theguardian.com/lifeandstyle/2009/sep/29/cervical-cancer-hpv-girl-death>; 2009 [accessed 15.03.19].
[57] Balasubramanian S, Shah A, Pemde HK, et al. Indian Academy of Pediatrics (IAP) Advisory Committee on Vaccines and Immunization Practices (ACVIP) Recommended Immunization Schedule (2018-19) and Update on Immunization for Children Aged 0 Through 18 Years. Indian Pediatr 2018;55(12):1066–74.
[58] GAVI. Historic partnership between GAVI and India to save millions of lives, <https://www.gavi.org/library/news/press-releases/2016/historic-partnership-between-gavi-and-india-to-save-millions-of-lives/>; 2016 [accessed 03.01.19].
[59] The Times of India. Vaccination for cervical cancer: Delhi does it first, <https://timesofindia.indiatimes.com/city/delhi/Vaccination-for-cervical-cancer-Delhi-does-it-first/articleshow/51200809.cms>; 2016 [accessed 03.01.19].
[60] The Tribune. Cervical cancer vaccine for Class VI girls in state, <https://www.tribuneindia.com/news/punjab/cervical-cancer-vaccine-for-class-vi-girls-in-state/267475.html>; 2016 [accessed 03.01.19].
[61] Narayanan N. Efficacy, safety, cost: India's decade-old debate on the cervical cancer vaccine erupts again, <https://scroll.in/pulse/865284/efficacy-safety-cost-indias-decade-old-debate-on-the-cervical-cancer-vaccine-erupts-again>; 2018 [accessed 03.01.19].
[62] Ghosh A. India fit case for cervical cancer vaccine, says WHO official, <https://indianexpress.com/article/india/india-fit-case-for-cervical-cancer-vaccine-says-who-official-5023708/>; 2018 [accessed 03.01.19].
[63] Jain A. The head of ICMR speaks about the health challenges facing the country, and the importance of data-driven policy making, <https://fountainink.in/qna/public-health-imperatives>; 2016 [accessed 03.01.19].
[64] LaMontagne DS, Barge S, Le NT, et al. Human papillomavirus vaccine delivery strategies that achieved high coverage in low- and middle-income countries. Bull World Health Organ 2011;89(11):821–830b.
[65] Bagla P. Indian parliament comes down hard on cervical cancer trial, <https://www.sciencemag.org/news/2013/09/indian-parliament-comes-down-hard-cervical-cancer-trial>; 2013 [accessed 03.01.19].
[66] Sinha K. Brinda to Azad: Take action against violators of HPV project, <https://timesofindia.indiatimes.com/india/Brinda-to-Azad-Take-action-against-violators-of-HPV-project/articleshow/10514461.cms>; 2011 [accessed 14.01.19].
[67] LaMontagne DS, Bloem PJN, Brotherton JML, Gallagher KE, Badiane O, Ndiaye C. Progress in HPV vaccination in low- and lower-middle-income countries. Int J Gynaecol Obstet 2017;138(Suppl 1):7–14.
[68] Gallagher KE, Howard N, Kabakama S, et al. Lessons learnt from human papillomavirus (HPV) vaccination in 45 low- and middle-income countries. PLoS One 2017;12(6):e0177773.
[69] Dorji T, Tshomo U, Phuntsho S, Tamang TD, et al. Introduction of a national HPV vaccination program into Bhutan. Vaccine 2015;33:3726–30.
[70] Torres-Rueda S, Rulisa S, Burchett HE, Mivumbi NV, Mounier-Jack S. HPV vaccine introduction in Rwanda: impacts on the broader health system. Sex Reprod Healthc 2016;7:46–51.
[71] Muhamad NA, Buang SN, Jaafar S, et al. Achieving high uptake of human papillomavirus vaccination in Malaysia through school-based vaccination programme. BMC Public Health 2018;18(1):1402.
[72] Karafillakis E, Simas C, Jarrett C, et al. HPV vaccination in a context of public mistrust and uncertainty: a systematic literature review of determinants of HPV vaccine hesitancy in Europe. Hum Vaccin Immunother 2019;15(7–8):1615–27.

[73] Hoffman BL, Felter EM, Chu K-H, et al. 266. The emerging landscape of anti-vaccination sentiment on Facebook. J Adolesc Health 2019;64(2):S136.
[74] Ortiz RR, Smith A, Coyne-Beasley T. A systematic literature review to examine the potential for social media to impact HPV vaccine uptake and awareness, knowledge, and attitudes about HPV and HPV vaccination. Hum Vaccin Immunother 2019;15(7–8):1465–77.
[75] Perman S, Turner S, Ramsay AI, Baim-Lance A, Utley M, Fulop NJ. School-based vaccination programmes: A systematic review of the evidence on organisation and delivery in high income countries. BMC Public Health 2017;17(1):252.
[76] World Health Organization. Human papillomavirus vaccines: WHO position paper, May 2017-Recommendations. Vaccine 2017;35(43):5753–5.

CHAPTER

EXPANDING PREVENTION OF CERVICAL CANCER IN LOW- AND MIDDLE-INCOME COUNTRIES

23

Lynette Denny[1,2], Heather Cubie[3] and Neerja Bhatla[4]

[1]Department of Obstetrics and Gynaecology, University of Cape Town/Groote Schuur Hospital, Cape Town, South Africa [2]South African Medical Research Council Gynaecology Cancer Research Centre, University of Cape Town, Cape Town, South Africa [3]Global Health Academy, University of Edinburgh, Edinburgh, United Kingdom [4]Department of Obstetrics and Gynaecology, All India Institute of Medical Sciences, New Delhi, India

Prevention of cervical cancer using mass cytology-based national programs dramatically reduced the mortality and morbidity of cervical cancer in the last century where programs worked and were correctly implemented [1]. The successes in many countries, particularly the Nordic countries, were not however transported to low- and middle-income countries (LMICs) where the infrastructure, human and financial resources and competing health needs prohibited the establishment and sustainability of cytology-based programs. In the past more than 20 years, efforts to use alternative screening algorithms have emerged and shown considerable potential, albeit large-scale roll-out of alternative programs has not been done in most LMICs.

The two alternative tests to cytology that have been most studied are Visual Inspection with Acetic Acid (known as VIA) and testing for HPV DNA of high-risk types known to have a causal relationship with cervical cancer. VIA involves exposing the cervix using a speculum and covering the cervix with 3%–5% acetic acid for a period of 1–2 minutes. Thereafter, the cervix is examined for what are known as aceto-white lesions. Precancerous lesions interact with dysplastic or abnormal epithelium by turning white.

Many studies using VIA as a primary screening test have been performed involving thousands of women in research and demonstration projects. This heterogenous group of studies has a wide range of sensitivity and specificity for the detection of high-grade cervical cancer precursors, many reflect verification bias and most are cross-sectional [2]. WHO published on a demonstration project they performed in six African countries which included Malawi, Madagascar, Nigeria, Uganda, Tanzania, and Zambia [3]. Between September 2005 and May 2009, 19,579 clients were screened from the six countries. Eleven percent ($n = 1980$) of women were VIA positive and 88% ($n = 1737$) were eligible for cryotherapy. Of those eligible for cryotherapy, 61% ($n = 1058$) received cryotherapy, 35% ($n = 601$) were lost to follow up, and 5% ($n = 78$) of women did not receive treatment for various reasons.

Of note, 1.7% ($n = 326$) women were found to have a cervix suspicious for cancer. Only 30% ($n = 96$) of these women were investigated, of whom 79 had cancer confirmed, 9 did not have cancer and in 8 cases there was no information on the outcome. There was also no information on the

Human Papillomavirus. DOI: https://doi.org/10.1016/B978-0-12-814457-2.00023-4

230 women who were not investigated. One of the obstacles to being investigated was the fact that women had to pay for their biopsies and it is assumed most could not afford to. There is an interesting anomaly in medical care in a number of SSA (sub-Saharan African) sites. For instance, HIV care and treatment provided by PEPFAR (the United State's President Emergency Plan for AIDS relief) funded treatment is free, but cancer services are not.

One of the most important features of successful screening programs is the coverage of the target population. In this study, the target population was women aged 30–50 years within the catchment area of the clinical site. After 3 years of running the project, the coverage was very low, ranging from 0.4 % in Zambia to 7.1% in Madagascar.

This project illustrates the many pitfalls in establishing screening programs, regardless of the screening test used. Clearly laboratory-based testing adds complexity to the system, but even point-of-care tests do not eliminate the difficulties of recruiting women, keeping them in the system for treatment and follow-up. Without these factors working, impact on cervical cancer reduction is likely to be minimal. For cervical cancer screening programs to be effective, a strong infrastructure needs to be developed.

However, Shastri et al. [4] reported on a randomized controlled trial in which 75,360 women from 10 clusters in the screening group and 76,178 women from 10 comparable clusters in the control group were recruited in Mumbai. Women aged 35–64 years were recruited and screened by primary healthcare workers. The screening group received four rounds of cancer education and VIA at 24-month intervals, whereas the control group received one round of cancer education at the time of recruitment. Both groups were actively monitored at 24-month intervals for cervical cancer incidence and mortality. Women who were positive on VIA were referred to the local tertiary hospital where they received repeat VIA, colposcopy, and Pap smears. Those diagnosed with cervical cancer precursors or invasive cancers were treated according to standard protocols.

In the screening group, 89% of women participated and 79.4% complied with the protocol. After 12 years of follow-up, the incidence of invasive cervical cancer was 26.74/100,000 in the screening group and 27.49/100,000 in the control group. There were 67 and 98 cervical cancer deaths in the screening and control groups, respectively. This translated into a 31% reduction in cervical cancer mortality in the screening group compared with the control group (Mortality RR = 0.69; 95% CI: 0.54–0.88; $P = .003$).

Parham and colleagues [5] from 2006 to 2013, up-scaled screening services (using VIA) for cervical cancer from 2 to 12 clinics in Lusaka, Zambia, through which 102,942 women were screened. The majority (72%) was in the target age range of 25–49 years and 28% were HIV positive. In all, 20% of women were VIA positive of whom 56.4% were treated with cryotherapy and 44% were referred for histopathological evaluation. Most women received same day services including the 82% who were VIA negative but only 5% underwent same day treatment. Among those referred for histopathological examination, 44% had CIN 2 or greater. Detection rates for CIN 2 + and cancer were 17 and 7 per 1000 women screened, respectively. This study shows that screening can be performed and up-scaled in remarkably low-resourced settings. This project included many partners and leveraged funding from HIV resources that were provided by a number of organizations such as PEPFAR. The project is yet to report on the ongoing impact on cervical cancer reduction and/or prevention.

The discovery of the causal relationship of infection of the cervix with oncogenic types of human papillomavirus (HPV) with cervical cancer by Zur Hausen [6], led the way for the use of molecular testing for HPV infection of the cervix. Walboomers et al. [7] showed that 99.7% of cervical cancers worldwide were causally associated with infection of the cervix by high-risk or oncogenic types of HPV.

Munoz et al. [8] pooled data from 11 case-control studies involving 1918 women with histologically confirmed squamous cell carcinoma of the cervix and 1928 control women. The pooled odds ratio for cervical cancer associated with the presence of any HPV infection was 158.2 (95% CI: 113.4–220.6). Based on the pooled data, 15 HPV types were classified as high-risk (types 16, 18, 31, 33, 39, 45, 51, 52, 56, 58, 59, 68, 73 and 82) and are considered carcinogenic.

In a metaanalysis of HPV types found in 10,058 invasive cervical cancers worldwide [9] (which included squamous cell carcinomas, adenocarcinomas, and adeno-squamous carcinomas), a high prevalence of HPV was confirmed with HPV 16 (51%) and 18 (16.2%) being the commonest types in different regions of the world. However, more than several other types of HPV were also associated with cervical cancer, of which types 45, 31, 33, 58, and 52 were the most prevalent. Further, HPV type 16 was more prevalent in squamous carcinomas and HPV type 18 more prevalent in adenocarcinomas of the cervix. Overall, HPV prevalence differed little between geographical regions (83%–89%) but was low compared to the almost 100% prevalence in studies that have used the most sensitive methods of detection for HPV. There is good evidence that HPV infection precedes the development of cervical cancer and that persistent infection with high-risk types of HPV is necessary for the development and progression of precancerous lesions of the cervix, either to higher grades of precancerous disease or to cancer [10], a process that can take 10–30 years. Further, infection of the cervix with high-risk types of HPV was associated with abnormal cytology in women previously cytological negative [11]. Rijkaart et al. [12] performed a randomized trial of women aged 29–56 years allocated to cotesting with HPV and cytology versus cytology alone, at two screening visits 5 years apart. They showed that women allocated to the intervention group (cotesting) had significantly less CIN 3 at round 2 and lower rate of cervical cancer compared to the control group (cytology alone). Ronco et al. [13] reported similar results with marked reduction in disease in women screened with HPV DNA testing in high-resource settings. It seems to me that one problem we have found in cervical adenocarcinomas is one of accuracy of diagnosis and quality of information and the difficulty of applying trials and epic studies in the developed countries relying on high quality analyses and accurate data to the less well resourced parts of the world without being aware of the basic problems like accuracy of diagnosis.

Of importance, HPV infection is associated with other cancers besides cervical cancer: about 25% of head and neck, 79% of vaginal, 88% of anal, 43% of vulvar, and 50% of penile cancers.

Critical to successful screening in LMICs is to ensure that screening is affordable, accessible, feasible, and effective. Point-of-care testing has received attention recently. A study by Cubie et al. [14] is presented below as a case study of what can be achieved with new and modern technology.

CASE STUDY

Malawi is an example of a very low-income country. Indeed, it was given the lowest global country ranking from the World Bank in 2016 [15]. It is a land-locked country of ~18 million people, with 84% living in rural areas and 45% under 15 years of age. Malawi also has the world's second highest rate of cervical cancer and highest mortality rate [16]. Population-based cervical screening and treatment is desperately needed if Malawi is to reduce the increasing burden of cervical cancer, fueled by HIV positivity and poverty.

The challenges in delivering a cervical cancer prevention program in Malawi have been well documented [17] and primarily relate to lack of workforce capacity with appropriate skills, a lack of essential medical equipment and consumables, nonavailability of treatment in local health facilities, and disparity of access between urban and rural areas, together with poor awareness of cervical cancer prevention, long distances to health facilities, and misconceptions about cervical screening, lack of supportive supervision, and use of male service providers [18]. An early demonstration pilot found VIA positivity rates of 12.4% in Malawi, but reported limited ability to provide same day treatment [3] due to inconsistent gas supplies, fragility of cryo-guns, problems with transportation and high cost.

The Malawian Ministry of Health has had a clear commitment to reducing deaths from cervical cancer as outlined in their 2011–16 Strategic Health Plan [HSSP I] [19]. National policy included treatment of early lesions with cryotherapy, in line with the WHOs current guidelines [20]. The updated strategic plan [HSSP II] [21] allows for thermo-ablation (previously known as cold or thermo-coagulation) as an alternative to cryotherapy, largely following the success of the Nkhoma Cervical Cancer Screening Programme (CCSP) in Central Malawi.

NKHOMA "SCREEN AND TREAT" PROGRAM

The Nkhoma CCSP has been running since 2013, based on Nkhoma CCAP Hospital, a regional referral hospital and 10 surrounding Health Centers (some Government and some church-funded). From the start, hospital screening services were integrated within an existing Reproductive Health Unit, allowing broadening of staff skills and ensuring consistency of messages to women. Early results on the first 7500 women using VIA and thermo-ablation demonstrated feasibility, acceptability and increased coverage, with >70% of those with early lesions receiving same day treatment [22]. The hospital offers daily screening clinics, most health centers deliver weekly clinics and two very rural centers have had week-long campaign services, which will be repeated at annual intervals, weather permitting (Beatrice Kabota, personal communication). Both regular clinics and campaign models of delivery were effective. By December 2017, nearly 26,000 women had been screened across all facilities and for >85% it was a first ever screening visit. Almost 65% of those who could benefit from early treatment received it the same day, and 90% within one month. With an estimated female population of child-bearing age of 92,000 in the region, cumulative coverage since 2013 is >28% (unpublished data; manuscript in preparation). Thermo-ablation is considered "old technology" by many, but the compact nature of the equipment, ease of use and low cost led to rapid acceptance in Nkhoma and its associated health centers and also in the Malawi Ministry of Health where it has been accepted as an alternative to cryotherapy. Not only is the equipment small, portable, and durable, it uses minimal electricity and can be run from a generator or car battery. Recent developments include hand-held devices powered by rechargeable battery or solar power. An additional benefit is the ability to take punch biopsies in the screening clinic, using the equipment for hemostasis. A WHO Guideline Review Committee and Guideline Development Group recently approved guidelines for treatment with thermal ablation [May 2019, Linda Eckert, personal communication].

Evidence has accumulated over three decades demonstrating the effectiveness of thermo-ablation since Gordon and Duncan [1991] [23] reported on experience in Dundee, Scotland. Cure rates for CIN 1-2 of 95% at 1 year and 92% at 5 years in 1628 patients treated over 14 years were reported, with no increase in rates of miscarriage, preterm or operative delivery in 266 posttreatment pregnancies. A retrospective review of 577 patients treated between 2001 and 2011 in England reported a cure rate of 95.7% at 1 year [24]. Within the Nkhoma CCSP, 421 women had returned for 1-year follow-up by October 2016 and 95% were VIA negative, giving a treatment failure rate of 5% [25].

The Nkhoma model is now being expanded to "hub and spokes" centers in each Region of Malawi, through a Scottish Government funded initiative (Campbell and Cubie, personal communication). This will not only impact significantly on population coverage, but also on the number of women receiving timely treatment for precancerous lesions in LMICs.

Cervical Screening Using Human Papillomavirus Testing

Given the move to HPV primary screening in high-resourced countries and the relative lack of trained VIA providers, consideration was given to the use of objective HPV testing, either as an alternative to VIA or as a primary screen to be followed by VIA of positives, as recommended from studies in other low-income settings [26,27]. Few readily available HPV tests are currently suited to LMICs, but the Xpert platform (Cepheid, USA), which uses a system based on real-time polymerase chain reaction (PCR), with all reagents and processes contained within a single-use cartridge, has been widely introduced in Africa for tuberculosis detection. Xpert HPV had been trialed in Edinburgh [28] prior to the commencement of the NKhoma program and the platform was known in Nkhoma Hospital laboratory. Xpert HPV is a qualitative test for detection of high risk (Hr)-HPV and can identify separately HPV 16; HPV 18 and 45 and an aggregate result for "other" Hr-HPV types. Review of individual channel data allows further breakdown into HPV31 and related [31, 33, 35, 52, 58]; HPV51/59 and HPV39 and related [39, 56, 66, 68].

Samples were obtained from 763 women attending VIA for the first time. Of the 98% with valid HPV test results, almost 20% were Hr-HPV positive. HPV 16 and HPV 18/45 were detected in 4.8% each of the total, but Hr-HPV "other types," particularly HPV 31-related, occurred much more frequently (12.7% of total). Both HPV prevalence and multiple infections were twice as frequent in women infected with HIV [14]. By the end of 2016, over 2000 HPV tests had been completed. Hr-HPV prevalence remained at 20%, and VIA positivity around 6%.

Xpert HPV testing proved straightforward and reproducible and a turnaround time from clinic to laboratory to clinic of 2 hours was achieved. Running the test in the clinic was also possible, thus making this a potential candidate for near patient testing. However, the manufacturer requires cytology-based medium (Preservcyt, Hologic Inc.) to be used, which is very wasteful if cytology is not being performed, involves transport of large volumes with high alcohol content and creates problems for waste disposal. We have shown that the test gives reproducible results with several alternative, cheaper media [29], but a validated comparison needs to be done.

We have also shown that it works well with self-collected vaginal samples taken in the clinic setting, with a slightly higher overall Hr-HPV prevalence rate [30]. Given the choice, more women opted for self-sampling than VIA and staff found that women obtained the samples without problems, including those who were illiterate. Women were prepared to wait 2 hours for the result, at least in this rural setting, knowing they would also receive same day treatment if needed. Self-sampling using a rapid HPV test and triage by VIA is therefore amenable to "same day 'screen and treat'" programs. High-throughput HPV testing in central labs might be more economical in urban settings, but would require two visits, with potential loss to follow up. Nkhoma is currently investigating whether limited HPV testing, for reassurance of clearance of disease after treatment ("test of cure") might be useful, as introduced in Scotland in 2012 [31].

Suspect cancers showed good agreement (90%) between HPV and VIA. As expected from a more sensitive primary test, there were far more HPV + than VIA + samples. However, agreement between HPV and VIA results was poor, with >70% of VIA + being Hr-HPV negative (50% in HIV + women), despite quality assurance programs in place both for HPV testing and for the clinical VIA service. It is well known that VIA skills are difficult to maintain, but much had been done to mitigate this in Nkhoma, including keeping records of case-load, group discussion of difficult cases, use of image banks, continuous professional development, and annual objective assessment of competence. All screening modalities have limitations and there is a recognized challenge of over-diagnosis with VIA in screen and treat settings. However, it enables detection and treatment of early lesions in many women where no alternative is available. Given the current prohibitive HPV test costs for low-resourced countries like Malawi, same day screen and treat based on VIA is likely to continue as the cervical screening modality of choice.

PERSPECTIVE FROM ASIA

The Asian continent, which has half the global burden of cervical cancer, probably has the greatest heterogeneity in sociocultural milieu, economic development, and healthcare delivery systems. This has precluded implementation of uniform systems of cervical screening. Despite ample evidence on the superiority of HPV testing as a screening tool, in the LMICs, the emphasis remains on VIA as the screening method for reasons of affordability and logistic simplicity. Bangladesh, India, and parts of Thailand and China have adopted the VIA strategy.

Bangladesh launched a VIA-based program in 2004. It was first introduced in 16 districts in 2005 and gradually scaled up to the rest of the country [32]. VIA screening is offered at 15 district hospitals and nearly 300 sub-district centers, urban centers, family welfare centers, and by some NGOs. About 350 doctors and nearly 2000 paramedical workers and nurses are part of this program. Screening is recommended at 3-yearly intervals in women aged >30 years. Women who test positive are referred to colposcopy centers at the medical colleges. About 20% of the referred women have been detected with CIN 2 + lesions. Key challenges identified have been lack of an organized screening program and low coverage—about 2 million women, that is, about 10% of the eligible population have been screened since the inception of the program over a decade ago. Some women are being screened repeatedly, others not at all.

With the changing trends in women's cancers, breast cancer has now overtaken cervical cancer in many countries. Hence, the programs now focus on cervical and breast cancer screening, the latter through clinical breast examination (CBE) and also teaching breast self-examination (BSE) to the women. In Bangladesh, CBE, and BSE were added to the cervical screening program since 2005.

In India, the southern state of Tamil Nadu was the first to implement VIA-based opportunistic screening in a phased manner from 2007, starting with two districts and expanding to the rest of the state in 2010 [33]. The program is coordinated with the help of NGOs, village level volunteers and a large number of government personnel. In this program, the focus is on four main noncommunicable diseases (NCDs). NCD nurses are employed to offer blood pressure and blood sugar checks along with breast examination and VIA. Coverage was impressive—about 14 million women were screened in a relatively short span of time. However, there were significant dropouts at each stage—lack of compliance with colposcopy, biopsy, and treatment. As a result, the diagnosis of disease was far less compared to what had been seen previously in an IARC study done in the same state. [34]

The single visit approach is being advocated by the WHO SEARO for this region, to deal with the problem of loss to follow up. In a study in New Delhi, it was seen that offering same day colposcopy to screen positive women and treatment to women with a Reid score >3 minimized the over-treatment rate to $<12.5\%$ with an efficacy of 81.3%, while reducing the loss to follow-up to one-third compared with the multi-step approach [35]. In India, a national program for screening for breast, cervical, and oral cancer has been rolled out since 2016 based on the Tamil Nadu experience. It is being piloted in the various states. The challenges of training the workforce, quality control, and establishing referral linkages still need to be addressed, although the program does envisage following the screen-and-treat model for eligible cases, namely, where the lesion is fully visualized, the lesion size is no more than two quadrants and there is no suspicion of invasive cancer, nor any symptoms like postcoital or postmenopausal bleeding. In Bangladesh, the single visit approach has been tried since 2010. The introduction of portable colposcopes has allowed nurses to refine the detection of disease at the periphery and minimize referral. Portable colposcopes can also be used by health workers to capture and transmit images through laptops, tablets or smartphones for expert evaluation. In this way, women can receive services at the last mile facility, with only very few needing to be referred for specialist care.

Another problem faced in LMICs has been the lack of appropriate, locally relevant guidelines. In 2018, the Federation of Obstetrics and Gynaecological Societies of India (FOGSI) brought out resource-based recommendations on screening and treatment of preinvasive cervical lesions [36]. These recommendations provide simple algorithm-based management charts for good and low-resource situations using cytology, VIA or HPV testing. The screen-and-treat approach is advocated in these recommendations, in line with the recommendations from WHO and the Ministry of Health.

Eventually, an affordable point-of-care HPV test will be required for effective screening, especially in the postvaccination era. A major advantage of HPV testing is the feasibility of self-sampling. The vaginal self-sample has been seen to be acceptable, feasible and reliable with a high level of concordance with a physician-collected cervical sample [37]. Even in situations where the woman is unable to provide the sample, it can be collected by a health worker, obviating the need for a sterile speculum, light source, etc.

GLOBAL PERSPECTIVE

In May 2018, the WHO announced a call for elimination of cervical cancer as a public health problem by 2030. As part of this effort it is proposed that 70% of women should receive screening by an effective method (preferably HPV testing) at 35 and 45 years of age. Also, that 90% of women detected with a cervical lesion should receive treatment. HPV vaccination of 90% of girls aged <15 years with two doses is also recommended. It is expected that a combined strategy of screening and vaccination can bring about a significant decline in the incidence rates of this cancer at an accelerated pace. Countries now need to develop national strategic plans to adapt the global mission to their local requirements. Cervical cancer has enormous social impact as it impacts women in their prime as well as their families. It is time now to find the ways and means to end the scourge.

REFERENCES

[1] Hakama M, Chamberlain J, Day NE, Miller AB, Prorok PC. Evaluation of screening programmes for gynaecological cancer. Br J Cancer 1985;52(4):669–73.

[2] Sauvaget C, Favette JM, Muwonge R, Wesley R, Sankaranarayanan R. Accuracy of visual inspection with acetic acid for cervical cancer screening. Int J Gynaecol Obstet 2011;113(1):14–24.

[3] Prevention of cervical cancer through screening using visual inspection with acetic acid (VIA) and treatment with cryotherapy. World Health Organisation, Geneva, Switzerland; 2012.

[4] Shastri SS, Mittra I, Gauravi AM, GuptaS, Dikshit R, Badwe RA, et al. Effect of VIA screening by primary health workers: randomized controlled study in Mumbai, India. J Natl Cancer Inst 2014;106(3): dkju009. Available from: https://doi.org/10;1093/jnci/dju009.

[5] Parham GP, Mwanahamuntu MH, Kapambwe S, Muwonge R, Bateman AC, Blevins M, et al. Population-level scale-up of cervical cancer prevention services in a low-resource setting: development, implementation, and evaluation of the cervical cancer prevention program in Zambia. PLoS One 2015;10(4):e0122169. Available from: https://doi.org/10.1371/journal.pone.0122169.

[6] zur Hausen H. Papillomaviruses and cancer: from basic studies to clinical application. Nat Rev Cancer 2001;2:342–50.

[7] Walboomers JM, Jacobs MV, Manos MM, Bosch FX, Kummer JA, Shah KV, et al. Human papillomavirus is a necessary cause of invasive cervical cancer worldwide. J Pathol 1999;189(1):12–19.

[8] Munoz N, Bosch FX, de Sanjose S, Herrero R, Castellsague X, Keertie, et al. Epidemiologic classification of human papillomavirus types associated with cervical cancer. N Eng J Med 2003;348:518–27.

[9] Clifford GM, Smith JS, Plummer M, Munoz N, Franceschi S. Human papillomavirus types in invasive cervical cancer worldwide: a meta-analysis. Br J Cancer 2003;88:63–73.

[10] Wright TC, Kurman RJ, Ferenczy A. Precancerous lesions of the cervix. In: Kurman RJ, editor. Blaustein's Pathology of the Female Genital Tract. New York: Springer-Verlag; 2002.

[11] Kjaer S, Hogdall E, Frederiksen K, et al. The absolute risk of cervical abnormalities in high-risk human papillomavirus-positive cytologically normal women over a 10 year period. Cancer Res 2006;66 (21):10360–636.

[12] Rijkaart DC, Berkhof J, Rozendal L, van Kemenade FJ, Bulkmans NWJ, Heideman DAM, et al. Human papillomavirus testing for the detection of high-grade cervical epithelial neoplasia and cancer: final results of the POBASCAM randomised controlled trial. Lancet Oncol 2012;13:78–88.

[13] Ronco G, Dilner J, Elfstrom KM, Tunesi S, Snijders PJ, Arbyn M, et al. Efficacy of HPV based screening for prevention of invasive cervical cancer: follow up of four European randomised controlled trials. Lancet 2014;383(9916):524–32.

[14] Cubie HA, Morton D, Kawonga E, Mautanga M, Mwenitele I, Teakle N, et al. HPV prevalence in women attending cervical screening in rural Malawi using the cartridge-based Xpert HPV Assay. J Clin Virol 2017;87:1–4. Available from: https://doi.org/101016/j.

[15] World Bank. <http://data.worldbank.org/country/malawi> [accessed 13.12.16].

[16] Bray F, Ferlay J, Soerjomataram I, Siegel RL, Torre LA, Jemal A. Global cancer statistics 2018: GLOBOCAN estimates of incidence and mortality worldwide for 36 cancers in 185 countries. <https://onlinelibrary.wiley.com/doi/full/10.3322/caac.21492>.

[17] Maseko FC, Chirwa ML, Muula AS. Health systems challenges in cervical cancer prevention program in Malawi. Glob Health Action 2015;8:26282. Available from: https://doi.org/10.3402/gha.v8.26282 eCollection 2015.

[18] Munthali AC, Ngwira BM, Taulo F. Exploring barriers to the delivery of cervical cancer screening and early treatment services in Malawi: some views from service providers. Patient Prefer Adherence 2015;9:501–8. Available from: https://doi.org/10.2147/PPA.S69286 eCollection 2015.

[19] The Malawi Health Sector Strategic Plan (HSSP II) 2017–2022. Towards Universal Health Coverage. <www.health.gov.mw/index.php/policies-strategies?download = 47:hssp-ii-final> [accessed 19.02.18].

[20] The Malawi Health Sector Strategic Plan (HSSP II) 2017-2022. Towards Universal Health Coverage. <www.health.gov.mw/index.php/policies-strategies?download = 47:hssp-ii-final> [accessed 19.02.18].

[21] Comprehensive cervical cancer control: a guide to essential practice. 2nd Edition. World Health Organization 2014. <http://apps.who.int/iris/bitstream/10665/144785/1/9789241548953_eng.pdf> [accessed 19.02.18].

[22] Campbell C, Kafwafwa S, Brown H, Walker G, Madetsa B, Deeny M, et al. Use of thermo-coagulation as an alternative treatment modality in a 'screen and treat' programme of cervical screening in rural Malawi. Int J Cancer 2016;. Available from: https://doi.org/10.1002/ijc.30101.

[23] Gordon HK, Duncan ID. Effective destruction of cervical intraepithelial neoplasia (CIN) 3 at 100°C using the Semm cold coagulator: 14 years' experience. Br J Obst Gynae 1991;98:14–20.

[24] Parry-Smith W, Underwood M, De Bellis-Ayres S, Bangs L, Redman CW, Panikkar J. Success rate of thermo-coagulation for the treatment of cervical intraepithelial neoplasia: a retrospective analysis of a series of cases. J Low Genit Tract Dis 2015;.

[25] Campbell C, Kabota B, Kafwafwa S, Madetsa B, Morton D, Walker H, et al., Implementation of a 'hub and spokes' model of delivery of cervical screening in rural Malawi. IPV 2017, Cape Town, February 2017; Poster HPV.

[26] Sankaranarayanan R. Screening for cancer in low- and middle-income countries. Ann Glob Health 2014;80(5):412–17. Available from: https://doi.org/10.1016/j.aogh.2014.09.014.

[27] Joshi S, Kulkarni V, Darak T, Mahajan U, Srivastava Y, Gupta S, et al. Cervical cancer screening and treatment of cervical intraepithelial neoplasia in female sex workers using "screen and treat" approach. Int J Womens Health 2015;7:477–83. Available from: https://doi.org/10.2147/IJWH.S80624 eCollection 2015.

[28] Cuzick J, Cuschieri K, Denton K, Hopkins M, Thorat MA, Wright C, et al. Performance of the Xpert HPV assay in women attending for cervical screening. Papillomavirus Res 2015;. Available from: https://doi.org/10.1016/j.pvr.2015.05.002 online June16. ISSN 2405-8521.

[29] Cubie H, Kawonga E, Mwenitete I, Mautanga M, Kabota B, Morton D, et al. Low cost collection systems for Xpert® HPV in low income countries. IPV 2017, Cape Town, February 2017; Poster HPV-143.

[30] Cubie HA, Stanczuk G, Kawonga E, Mautanga M, Mwenitete I, Kabota B, et al. Cervical screening in rural Malawi using Xpert® HPV and self-taken vaginal samples. EUROGIN2016, Salzburg; Abstract00556.

[31] Cubie HA, Canham M, Moore C, Pedraza J, Graham C, Cuschieri K. Evaluation of commercial HPV assays in the context of post-treatment follow-up: Scottish Test of Cure Study (STOCS-H). J Clin Pathol 2014;67(6):458–63.

[32] Nessa A. Progress in implementing VIA screening in Bangladesh. Orlando, FL: IFCPC World Congress; 2017.

[33] Cervical cancer preventions screening and treatment—a pilot initiative of Tamil Nadu Health Systems Project. <https://mail.google.com/mail/u/0/?tab = rm#inbox/FMfcgxwCgVbLmjVRsWrFtgxcskDJHxnK?projector = 1&messagePartId = 0.2>; 2019 [accessed 25.04.19].

[34] Sankaranarayanan R, Esmy PO, Rajkumar R, Muwonge R, Swaminathan R, Shanthakumari S, et al. Effect of visual screening on cervical cancer incidence and mortality in Tamil Nadu, India: a cluster-randomised trial. Lancet 2007;370(9585):398–406.

[35] Singla S, Mathur S, Kriplani A, Agarwal N, Garg P, Bhatla N. Single visit approach for management of cervical intraepithelial neoplasia by visual inspection & loop electrosurgical excision procedure. Indian J Med Res 2012;135(5):614–20.

[36] FOGSI GCPR on screening and management of preinvasive lesions of cervix and HPV vaccination. <http://www.fogsi.org/wp-content/uplods/2018/03/FOGSI-GCPR-Final-March-2018.pdf> [accessed 25.04.19].

[37] Bhatla N, Dar L, Patro AR, Kumar P, Kriplani A, Gulati A, et al. Can human papillomavirus DNA testing of self-collected vaginal samples compare with physician-collected cervical samples and cytology for cervical cancer screening in developing countries? Cancer Epidemiol 2009;33(6):446–50.

FURTHER READING

Castle PE, Wacholder S, Sherman MF, et al. Absolute risk of a subsequent abnormal pap among human papillomavirus DNA-positive, cytologically negative women. Cancer 2000;95(10):2145–51.

The Malawi Health Sector Strategic Plan (HSSP). Moving towards equity and quality. <http://www.nationalplanningcycles.org/sites/default/files/country_docs/Malawi/2_malawi_hssp_2011_-2016_final_document_1.pdf>; 2011–2016 [accessed 19.02.18].

Epilogue: Looking forward to cervical cancer elimination

David Jenkins[1,2] and F. Xavier Bosch[3,4,5]

[1]Emeritus Professor of pathology, University of Nottingham, United Kingdom [2]Consultant in Pathology to DDL Diagnostic laboratories, Rijswijk, The Netherlands [3]Catalan institute of Oncology (ICO), Barcelona, Spain [4]Bellvitge Research Institute (IDIBELL), Barcelona, Spain [5]Open University of Catalonia (UOC), Barcelona, Spain

In the field of human papillomavirus (HPV) and cancer, the beginning of the 21st century is characterized by the bold proposal to eliminate cervical cancer globally as a public health problem. The field seems to be ripe to attempt an effort at the level of the one deployed in the past to eradicate small pox or to dramatically reduce (to the level of elimination) polio, measles, and several other communicable diseases caused by infectious agents.

As shown in this book a wealth of scientific documentation on HPV is available to substantiate the objectives, the monitoring as well as the technological and financial aspects of a cervical cancer elimination campaign which in turn should achieve a significant preventive impact on other HPV-induced cancers. The backbone of the project is generalized (gender neutral) vaccination with broad spectrum vaccines and at least one/two HPV-based screening events followed by adequate treatment of invasive cancer, including advanced cancer and the provision of palliative care whenever other options fail. The development and validation of the required tools are the results that the scientific community has provided and evaluated in the last three decades.

In 2018, WHO launched the cervical cancer elimination campaign and is currently generating the guidelines and basic documents to provide the theoretical framework for the elements of the strategy and the intermediate monitoring endpoints. Other aspects of the campaign will be developed such as the provision of funds, guidance on vaccine access and procurement, appropriate use of the HPV screening technologies and the educational efforts that need to be put in place and coordinated to cement the operation [1].

With this call and recommendations, the countries will have to match these objectives to their local situation and formulate their national prevention plans. The intermediate objectives are being defined in such a way as to achieve by 2030 a reduction of 30% of the cervical cancer mortality. Model-based predictions for short-term evaluation include:

1. Achieving sustained HPV vaccination of at least 90% of all cohorts of girls under 15 years,
2. Providing two HPV screening episodes to at least 70% of women at ages 35 and 45 years, and
3. Treating adequately 90% of all "cases"—whether screen positive preneoplastic lesions or invasive cancer at any stage including access to diagnosis and palliative care.

In most circumstances, intensive vaccine coverage would achieve the elimination level towards the end of the 21st century. The time to achieving this level of control solely with vaccination is strongly dependent on the baseline levels of cervical cancer incidence. In the higher-risk countries, vaccination only is predicted not to achieve elimination rates within the century. However, the addition of the HPV screening component of the campaign at a high level of coverage and follow-up is predicted to accelerate the time to elimination by two or three decades [2].

The coming years will witness the application of different versions of this general strategy, the deployment of the relevant indicators and the monitoring of progress towards the general goal of having populations largely free of cervical cancer. Developed countries will certainly build upon these objectives and in many instances vaccination and screening plans include the extension of the vaccine indications to boys (23 countries in 2018), a broader age range for vaccination among women, specific ancillary programs to vaccinate high-risk groups, and repeated HPV screening on a 5-year interval.

This book has reviewed in considerable detail the historical perspective the landmarks and the state of the art in HPV research and its clinical applications. The review has also identified gaps in knowledge in which more research is still needed and critical issues that need to be scientifically resolved. Some of the salient points are hereby summarized.

INTEGRATING CERVICAL CANCER PREVENTION IN THE FRAME OF OTHER NATIONAL HEALTH PRIORITIES

Achieving elimination of cervical cancer as a public health problem will require in many instances making choices within the health systems of the countries, when competing for resources with other major health problems. Fig. A.1 shows the relative impact (mortality in thousands of cases in 2017) of all major causes including three major infectious diseases TB/HIV/MALARIA as well as the number of deaths at the major cancer sites including cervical cancer. In the graph, cervical cancer appears as relatively minor condition (1% of the total mortality among women worldwide). However, in the decade 2007–17, the number of cases of cervical cancer has increased by close to 19%, whereas the number of deaths to some selected infectious diseases has significantly decreased by 15% for TBC; 50% for HIV/AIDS and 30% for malaria [3]. Interpretation of time trends for cervical cancer strongly suggests the dominant effect of population growth and longevity. Predictions under current preventive practices are that the number of cases of cervical cancer will double to triple in developing countries by the turn of the century. This somber prognosis is counterbalanced by the options offered by the arrival of HPV vaccines and HPV screening technologies and by the definite commitment of the major public health institutions in the world to focus on the control of HPV and its related diseases.

Other than the crude number of cases or the incidence and mortality rates, social and political issues may be relevant at the time of making decisions to deploy the cervical cancer elimination campaign and on the intensity with which the campaigns will be implemented and sustained over time. Some of these issues (by no means an exhaustive account) could include the following:

1. Cervical cancer mortality clusters dramatically in countries without established screening opportunities and in poor populations of the world including low and marginalized social groups in developed countries.
2. Cervical cancer is a long-term consequence of a sexually transmitted infection and thus is a cancer with a stigma in many populations.

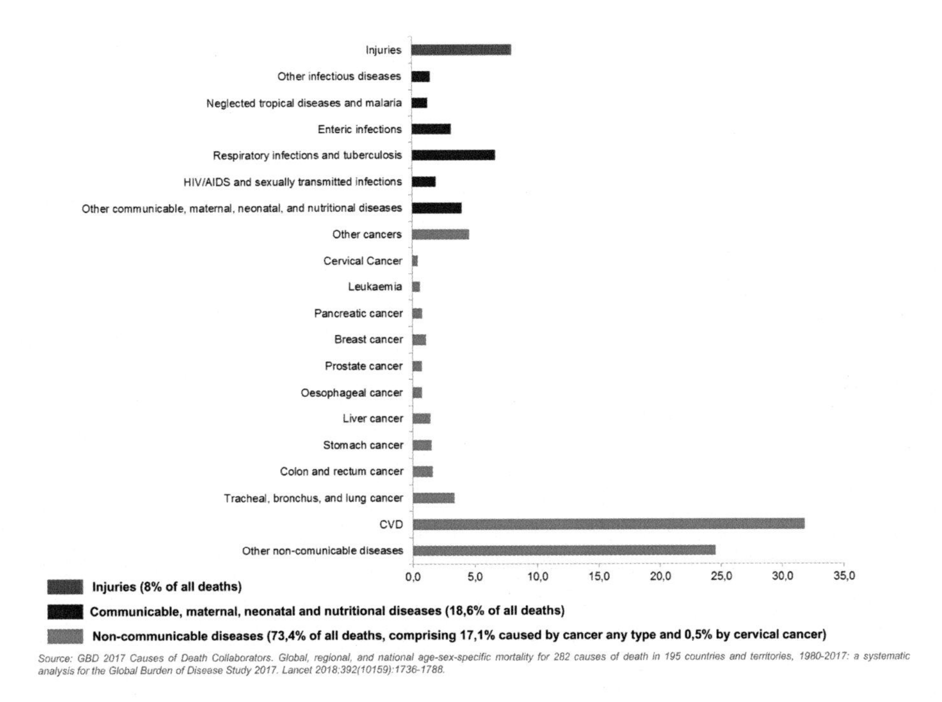

FIGURE A.1

Global causes of death in 2017 (both sexes, all ages, $n = 55{,}945{,}700$).

3. Cervical cancer is a women's disease and the priority of the intervention reflects also the social role of women in the decision making process and their access to preventive care in every country and culture.
4. Children's sexual abuse, adolescent sexual behavior, and sexual patterns along adult lifetime are all relevant to calibrate the amplitude of the HPV preventive programs. However, many of these parameters and social trends are still perceived as taboo and in practice their characteristics and impact remain largely unseen by the decision makers, the health professionals, and the population at large.
5. In developed countries, HPV vaccination and screening practices are implemented and highly supported by the population and the public and private health programs. Modification of the current status of the preventive campaigns may create some transition periods of confusion and will require careful local organization.
6. The early milestones on cancer reduction will occur and be measured in the countries in the developed world that pioneered HPV-based preventive interventions (i.e., cervical cancer reduction in Finland/Australia/Scotland/and others). In all these populations, the information systems and cancer registries are operational and record linkage is feasible. Extrapolation of results and protocols to developing countries may require careful attention and in many instances calibrated algorithms for diagnostics and management will have to recognize and adapt to the socioeconomic variability. It is important that accurate diagnosis of invasive cancers is widespread globally as wrong diagnoses particularly of adenocarcinomas arising in

vaccinated individuals can cause issues in understanding the role of HPV in cervical cancer and apparent failures of HPV screening and vaccination.

NATURAL HISTORY AND SPECTRUM OF DISEASE RELATED TO HUMAN PAPILLOMAVIRUS

While cervical cancer is the most frequent cancer linked to HPV infection and dominated the clinical and epidemiological research in the 1980s and 1990s, progress in the understanding of the etiological role of HPV in other conditions has provided estimates of the other cancer sites affected by the oncogenic capacity of HPV. As a group, these malignancies amount to an expected number of new cases per year worldwide of 630,000 of which 29,000 correspond to oropharyngeal cancer; 556,000 to female genital cancers (cervical, vulvar, and vaginal) and 48,000 to penile and anal cancer [4].

Fig. A.2 provides a numerical estimate of the HPV-linked pathology for 30 countries in the European Union. The graph also displays the relevance of the preneoplastic lesions linked to HPV in countries where cervical screening is in place. Notice that related to an estimate of cervical cancer incidence generated by reliable cancer registries (estimated at some 35,000 cases) the equivalent

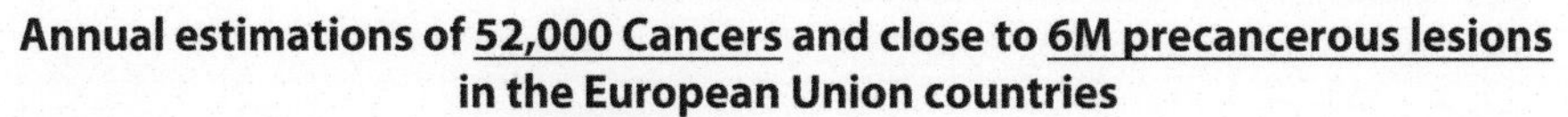

Annual estimations of 52,000 Cancers and close to 6M precancerous lesions
in the European Union countries

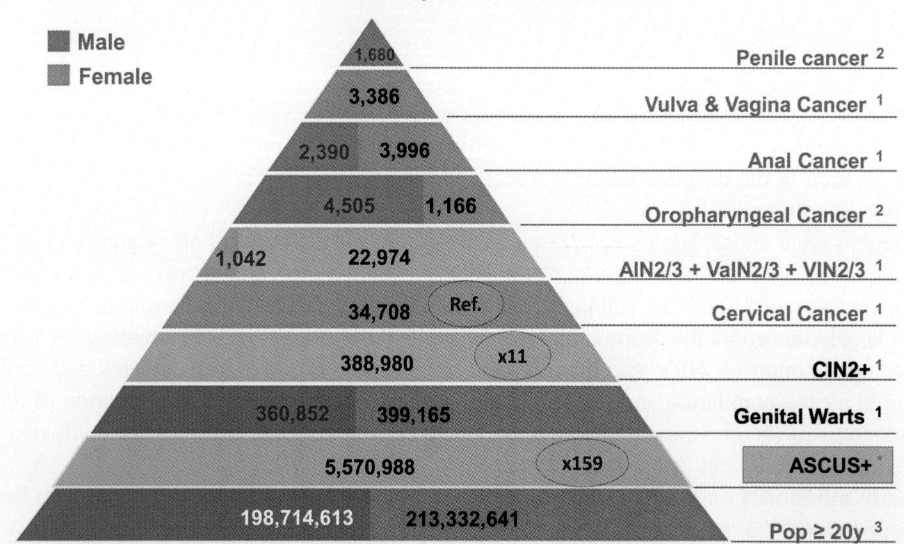

(Austria, Belgium, Bulgaria, Croatia, Cyprus, the Czech Republic, Denmark, Estonia, Finland, France, Germany, Greece, Hungary, Iceland, Ireland, Italy, Latvia, Lithuania, Luxemburg, Malta, Netherlands, Norway, Poland, Portugal, Romania, Slovenia, Slovakia, Spain, Sweden, UK) + Switzerland. * Estimations assuming 3.5% of ASCUS+ among women aged 25-65 years. 1. Hartwig et al. 2017 Infect Agent Cancer; 2. Forman et al .2012 Vaccine; 3. World Population Prospects 2012 Revision

FIGURE A.2

European Union (30 countries) HPV-related disease burden.

estimate for preneoplastic lesions is close to 160-fold higher for atypical or borderline abnormal cytology or worse (ASCUS +) (5.6M) and 11-fold higher for high grade lesions (389,000 cases of CIN2 +). These are conditions requiring further diagnostic procedures and eventually treatment. One of the consequences of using a screening test with low sensitivity and poor reproducibility (i.e. cytology) followed by decision to treat based on colposcopy and biopsies also limited by low sensitivity and repeatability is the undesired overtreatment of lesions that may otherwise regress spontaneously with high frequency (i.e., a significant fraction of the CIN 2 cases).

In many developed countries, the costs and health services requirements are strongly determined by the burden of preneoplastic conditions. This is likely to be an important driver in the decisions to transition screening programs from the cytology-based current situation to the HPV-based programs that are being recommended by most updated guidelines and national programs. The burden created in the health system by the preneoplastic diagnoses is a hidden cost, rarely perceived and evaluated as a side effect of screening programs. The unwanted side effects include not only in the worst case scenario, some obstetric consequences but also the psychological stress related to the diagnosis and treatment as well as the costs to the system and the expense to the women involved. Better triage tests to determine more accurately and reproducibly the need for treatment of HPV-positive women are an important approach as detailed in this book.

All preneoplastic lesions of the other HPV-related cancer sites (anogenital and oropharyngeal) do not follow the same patterns as in the cervix and have not been fully described. Neither has any screening methodology been validated. Limited information from Phase III clinical trials indicates that protection afforded by the HPV prophylactic vaccines can include the HPV infections and preneoplastic lesions of the anal canal, the vulva, and the vagina. It can thus be speculated that vaccinated cohorts will be protected against the relevant fraction of oncogenic HPV-driven cancer cases in all locations.

Understanding the burden of disease linked to HPV infections has greatly benefited from the existence of an IARC-lead international network of cancer registries and a long-term sustained research program that standardizes methodology, assists novel cancer registries in developing countries and reports regularly on cancer incidence, mortality, and time trends [4,5]. Moreover, this program has also developed methodology to produce estimates of cancer incidence for regions where cancer registration is limited or nonexisting [4]. The HPV field has in addition supported the organization and maintenance of a specific information resource named ICO/IARC HPV Information Center that compiles and offers in user-friendly and free-access format summaries of published documentation on HPV, HPV-related cancers, and status of preventive programs. The information is retrievable for each one of the 192 countries in the world or any combination of countries in regions or continents [6]. These resources and the coordination of the health statistical unit at WHO are currently devising the appropriate tools to monitor the deployment of the cervical cancer elimination campaign.

RESEARCH AREAS IN THE SCREENING AND TREATMENT FIELDS

Some of the relevant issues that may be addressed by research projects in the coming years include the development and validation of low cost and point-of-care screening technologies. These alternatives may alleviate the difficulties to adequately follow up women with screen-detected abnormalities, a proven major barrier in most developing countries. Research on screen-and-treat protocols is of importance to ensure that the participants at any screening event are properly managed on time with a significant reduction of the numbers of individuals lost to follow-up [7]. It is important to reach international agreements to accelerate the acceptance of promising new technologies for screening. Currently, academic researchers have generated a relatively rapid validation procedure comparing HPV test performance on predefined sample sets (i.e., the Valgent project) [8] and generating sound summaries of the testing and vaccine literature (i.e., the Cochrane reviews, the IARC monographs or the ICO/IARC HPV information Center) [6,9,10]. The prequalification procedures organized by WHO as well as other major reviews need to join forces with the academic community to accelerate the arrival of critically important technology.

Self-sampling for HPV testing has been extensively proposed and evaluated in many different settings. The efficacy, acceptance and safety of these approaches are important in improving screening coverage in both developing and developed countries. Self-sampling appears to be useful not only among the nonparticipants in current screening programs but is progressively proposed as a generalized alternative for all participants in screening programs. Remarkable examples of high rates of self-sampling coverage and participation have been achieved in environments such as Malaysia and Turkey as well as in developed countries like The Netherlands and Sweden [6,11–13].

In the understanding of the natural history from HPV infection to cancer, biomarkers are useful intermediate diagnostic resources that can serve the purpose of screening and early diagnosis or be part of the triage strategies for HPV-positive individuals to decide on management. The closer to invasive cancer and the higher the consistency of the molecular change the better the predictive value. P16 (p16INK4a) is the longest established biomarker as a surrogate of HR HPV E7 gene activity but is limited on its own as it is also expressed physiologically in cervical metaplasia and is expressed in productive HPV lesions as well as high-grade preneoplasia and cancer (cross-reference to fb-non-chapter 2). The immunohistochemical stain for Ki 67 is a useful addition to cytology and additional biomarkers under study including E4 to identify productive HPV lesions, mRNA tests, and methylation of viral and host genes to define progression of HPV infection to neoplasia. Panels of molecular markers (over 100 studied) are being investigated to achieve discrimination between HPV infections with high risk of progression and transient infections with limited clinical consequences.

It is important to consider any increase in complexity of management and cost resulting from the introduction of new biomarkers. While the adoption of protocols including these markers is likely to be feasible and acceptable in developed countries, they may be more difficult to use in developing countries raising additional barriers in situations where simplicity and reduced follow-up visits seems to be of critical importance.

Imaging of the surface of the cervix using mobile technology, artificial intelligence, and large image banking with rapid access in the internet may provide an alternative for screening and triage in some settings [14].

Nonsurgical treatments for HPV infections/early cervical lesions are the long-term missing link in the chain of options for cervical cancer control. The introduction of HPV testing in primary screening will generate a significant fraction of the screening population (expected average of 10%–12%) with normal cytology and HPV test positive. For these women, the current recommendations are condom use (to prevent transmission) and monitoring (6–12 months repeated visits) without any effective treatment. This is clearly unsatisfactory for both the patient and the healthcare provider. Moreover, HPV positive women remain the most important high-risk group for HPV transmission, contributing to the incidence and high prevalence of HPV infections in the

Table 1 Areas of Research Where Relevant Changes are Likely to Occur in the Coming Years.

HPV and Cancer. Natural History	HPV Screening/Treatment	HPV Vaccination
Natural history of HPV-related head and neck cancers	Self-sampling as the routine practice in screening programs	Impact of one dose protocols in adolescents
HPV in the etiology of cancers of the skin/anal canal/penis/vulva/vagina, and scrotum	Screen-and-treat protocols in low/middle resource settings	Vaccination of adult women (with one/two doses)
HPV/HIV interactions	Point-of-care/low-cost HPV screening technologies	Generalized male vaccination
HPV transmission, sexual dynamics of the populations, and social groups (i.e., MSM, commercial sex workers)	Imaging technologies for screening in low/middle resource settings	Improving herd protection by vaccination HPV + and prevention of transmission
International surveys on trends in sexual behavior and in sub populations	Triage of screen-positives in low/middle resource settings	Therapeutic/mixed vaccines. Natural boosting of vaccinated individuals by HPV infections
Spontaneous HPV exposures as natural boosters	Screening other HPV-sensible organ sites (i.e., anal canal)	Include HPV vaccines into the EPI vaccination schedule/multiple antigen combined vaccines
Development of the methodology to monitor intervention impact under suboptimal conditions of data collection	Treatments of HPV + individuals	Selective vaccination of high-risk groups
Maintenance of the health statistics systems to monitor time trends in vaccination and screening coverage and cancer incidence	Modeling screening requirements in vaccinated cohorts	Low-cost HPV vaccines
	Monitoring screening programs in low/middle income countries	Compliance and COI regulations
	Educational programs on the reshaping of screening protocols in developed countries	Educational efforts on HPV vaccine safety
		Antivax positioning

populations, on average close to 12% in women with normal cytology [15]. Finding effective treatments to resolve the infection/early lesions and to interrupt/decrease transmission is an urgent need to accelerate control of the infection. Several trials using immunological treatments are underway as well as trials using small molecules [16]. In this respect, the opportunity to reduce transmission by vaccinating adult women is an important alternative to explore following the observations of a reduced risk of recurrence of HPV infections at the time of treatment for CIN 2 + cases [17] or the anecdotal observation of reduced clinical symptoms of recurrent respiratory papillomatosis following immunization with a vaccine including antigens of HPV 6/11 [18] (Table 1).

RESEARCH IN THE VACCINATION FIELD

After first licensing in 2006, HPV vaccines are now entering the second decade of follow-up of large populations. Likewise, participants in the early Phase I/II trials already have accumulated 15 + years of observation. The record of vaccine effectiveness and safety is extraordinary and reductions of cervical cancer incidence have already been suggested in some early analysis of the participants in the vaccine trials in the Nordic countries in Europe [19]. Ongoing systematic time-trend analyses of cervical cancer incidence are being conducted and preliminary results are strongly suggestive of a consistent reduction of cervical cancer incidence in the cohorts that have been offered vaccination in their adolescent years.

It is worth remembering the progress made from the early HPV vaccine indications in 2006/07 (girls only, single cohorts, age below 15 years/before sexual initiation, three doses required, etc.) to the gradually expanding indications based on either results of formal trials (male vaccination, widening of the age range, HIV cohorts, MSM, other high-risk groups, etc.) or based on clinical observations and educated interpretation of the evidence outside formal RCT (vaccination as adjuvant to treatments for CIN2 + or as part of the management of recurrent respiratory papillomatosis) [20]. These novel indications translate the underlying observations that:

1. HPV vaccines are highly effective also in adult individuals.
2. HPV-vaccinated women are predicted to reduce dramatically their lifetime screening requirements increasing the cost effectiveness of the programs.
3. HPV vaccines are safe and do not have any formal contraindications. Further, their safety record now includes all special populations examined including, for example, groups of HIV +, HPV +, pregnant and lactating women, individuals with immunosuppression or with autoimmune diseases, and patients with RRP or CIN2 + lesions.
4. Reaching vaccination coverage levels above 50% of the female population generates a very powerful herd protection effect and significant reductions in HPV prevalence and of genital warts has been documented among nonvaccinated females as well as in nonvaccinated males in the community. The nature of the herd protection needs to be further investigated and notably the opportunities to boosting it by reducing the transmission rates of the HPV carriers to their partners by vaccination.
5. Getting prepared for an effective and rapid investigation of any apparent breakthrough cases in vaccinated individulas and possible vaccine side effects can minimize the opportunity for "false-news" supporting antivaccine beliefs.
6. Finally, HPV vaccines are not therapeutic for prevalent infections. Therefore HPV + women, even if they receive some benefit from vaccination (protection against HPV types not prevalent at the time of vaccination) will remain at high risk and should be offered a closer follow-up algorithm.

This continuously expanding panel of plausible indications has been socially encouraged by at least three components:

1. The reduction in the number of doses required (i.e., recommended two doses for girls below age 15 years) with some studies suggesting that two doses may be equally effective at least to age 18 years [21].

2. The important reduction in vaccine prices notably for GAVI countries and large procurements afforded by public national programs of procurement consortiums such as the revolving fund lead by OPS/PAHO.
3. The overwhelming evidence of the efficacy and effectiveness of HPV vaccines that prompted decisions at all levels to generalize promotion of the vaccination programs.

The use of two-dose regimes with 6–12 months interval between doses as opposed to the original indication of three-dose regimes was first suggested in Quebec, Canada and rapidly adopted in Mexico and other countries in public programs with a research component to ensure that third doses would be administered if the interim analyses showed the need to booster the immune response. These exploratory population-based programs were initially inspired by previous examples of dosage reduction in other vaccines and the early findings of the immune response assessment of the HPV vaccines. Further, Phase III and IV trials and unscheduled findings from population vaccination programs provided an ancillary result from groups of women that received one or two doses for a variety of reasons instead of the three doses scheduled. Upon completion of 7 + years of observations of these spontaneously formed subgroups it seems clear than the antibody levels plateau and that no cases of persistent infections/CIN lesions have occurred after receiving one single dose of vaccine. The results attracted enough interest to launch formal trials to compare the value of one dose in terms of protection from infection and disease as well as to allow modeling the potential duration of protection [22].

Increasing indications by expert advisory groups in many countries reflect the recognized value of the HPV vaccines and is creating at present some shortage in the production and supply, a circumstance that may delay introduction and generalization. While efforts to increase the offer are underway, novel vaccines produced in cheaper and more efficient systems are being developed and tested. The expected arrival to the market of these new products from China and India will alleviate the shortage and allow faster deployment of the programs in low- and middle-income countries [23].

COMPREHENSIVE RESEARCH ON COMBINATIONS OF HUMAN PAPILLOMAVIRUS SCREENING AND HUMAN PAPILLOMAVIRUS VACCINATION

Another area of great interest refers to the age range of vaccine indication and particularly how to articulate the extended vaccination indications with the rapidly expanding HPV-based screening protocols. One of such proposals is the HPV FASTER project which has recently been modeled using extensive data sets from population-based programs [2,24].

The proposal of the HPV-FASTER protocol is to offer HPV vaccination to women in a broad age range of 9–45 years (exact limits to be further refined and locally calibrated) irrespective of HPV-infection status. Women of any age above 25/30 years would, in addition to the vaccination, be screened using a validated HPV test as part of their initial visit; women who test HPV-positive would be offered triage and follow-up diagnostic tests and treatment in accordance with recommended guidelines. Screen-and-treat protocols would be useful in many environments working towards the concept of a single visit for the prevention of cervical cancer.

The rationale of this project is that women properly diagnosed as HPV-negative at screening ages and receiving a broad spectrum vaccine should have a subsequent lifetime risk of cervical cancer very low, tending to zero. Likewise HPV-positive women at the time of vaccination adequately followed-up (including treatment of CIN2 + cases) will also have a significantly reduced risk of disease. In both instances, the subsequent needs for screening may be dramatically reduced to one or two in a lifetime, thus increasing sustainability and compliance as well as alleviation of the burden and workload at the health centers, typically overloaded already with patient care.

Vaccinated cohorts will increasingly enjoy a significant reduction in viral circulation, in abnormal pap smears including high-grade lesions and within a few decades, significant reductions in cervical cancer and all other HPV-caused cancers. These changing estimations of disease burden will also reduce the frequency of screening events and drive the replacement of the pap smear method by HPV-based screening technologies. However, for several cohorts that could not benefit from early vaccination, screening will remain their only alternative for prevention. As a consequence and for a number of years the two interventions will need to be combined and the challenge is to accomplish it in the most cost-effective way.

ANTIVACCINES AND ANTIVACCINATION CAMPAIGNS

Vaccines have generated over the years an intriguing effect currently embraced by the concept of "anti-vaxers" and "vaccine skeptics." Among a list of mechanically repeated and unproven arguments (i.e., natural alternatives to vaccination, low number of cancer cases when screening is in place, etc.) vaccine safety concerns are consistently presented, typically claiming that all adverse events occurring soon after a vaccination episode are causally related to the vaccine or the vaccine components such as the adjuvants.

Antivaccine movements are being identified as a significant barrier to achieve universal coverage. Examples are available where cases of diseases that were considered eliminated in given populations reemerged either as isolated cases (i.e., one case of diphtheria in Spain) or full outbreaks (i.e., measles in Europe or the United States). WHO has identified vaccine hesitancy as one of ten most relevant public health problems, deserving understanding and attention [25].

These attitudes include occasionally health professionals that are hesitant or reluctant to recommend HPV vaccines to their patients or display openly defiance to the scientific international consensus on vaccine recommendations. These positions ignore the wealth of scientific evidence accumulated over decades and the many examples in which unvaccinated individuals develop the disease in environments that have been disease-free for many years.

Social scientists investigating the phenomena describe a hysteresis-like effect in which a fake narrative (i.e., measles vaccine causes autism) installs itself in the collective mindset after a clinical anecdotal episode, an alarming publication or simply a reiterated front page in the media. The fake-story strongly influences the time of persistency of the doubts about vaccine, generates hesitancy in a fraction of the population and has the potential to dramatically reduce the vaccine coverage rates for some time [26]. Examples of vaccine crisis for HPV have been experienced in Japan, Colombia, Denmark, and Ireland, countries in different parts of the world with very different cultural and socioeconomic status and with different access to information and to publically supported vaccination programs.

Furthermore, the confidence and credibility crisis has hit some of the most prestigious review bodies, such as the Cochrane collaboration, and remains virulent even if a full audit including scientific reanalysis of the data and inclusion of additional trials are conclusive on the safety of the vaccines. Any initial scientific skepticism concerning evidence of effectiveness and safety are nowadays resolved by the overwhelming evidence generated by public programs including literally tens of millions individuals who have been vaccinated in countries that benefit from careful scientific scrutiny and independent evaluation of any suspicion of a deleterious secondary effects of any clinical significance [9].

COMPLIANCE REGULATIONS AND PUBLIC AND PRIVATE PARTNERSHIPS

Other than safety, an increasingly important issue in HPV research and prevention interventions will be the management of the collaboration between academic scientists, clinical practitioners and the relevant industry either in the diagnostic or vaccine fields or eventually in the promotion of treatments whenever available. Issues of potential conflicts of interest and increasing compliance restrictions to participate in public forums are often aired as arguments to restrict vaccination programs and deserve an in depth discussion in the academic and public health fields.

Close collaboration of academic research and industry support in the HPV field has had a critical impact on the progress towards the call for elimination of HPV-related cancers in at least the following aspects:

1. Providing the basic etiological evidence and HPV type specific attributable fractions worldwide to support the development and evaluation of HPV screening and vaccination alternatives.
2. Helping define the HPV types to be included as probes in diagnostic tests and as antigens in vaccine composition.
3. Describe the international consistency and the limited variability in the HPV types in related cancers. One part of this is ensuring access to high-quality pathology providing consistent accurate diagnosis of cancers studied.
4. Validating HPV diagnostic technologies in large ASCUS triage trials and in very large primary screening trials literally involving millions of women.
5. Recruiting over 100,000 participants in RCT of HPV vaccines and driving a significant number of ancillary studies on HPV natural history.
6. Providing cost reductions and tiered pricing in diagnostic tests and HPV vaccines for GAVI countries and national public programs.
7. Supporting educational and communication programs and events worldwide.
8. Coordinate with the public health communities and institutions the production and distribution of the hundreds of millions doses of HPV vaccines that will be required over time to sustain the cervical cancer elimination campaign.

Despite the importance of close collaboration between academic research and industry in the development of preventive approaches to HPV-associated cancers described in this book, it is no secret that the interactions between industry, health institutions, and health professionals have experienced significant storms and ethical and scientific challenges worldwide.

However, there are substantial differences in the nature of public and private collaborations in relation to the product being promoted by the specific aspects of the product under consideration.

As an example in a recognized oversimplified discussion, one can consider that the ethical/corruption issues are obvious if the product promoted is toxic and health damaging (i.e., tobacco, alcohol, added sugar, etc.). More difficult interpretations of the results and judgements about new interventions occur when considering clinical studies with uncertain and/or borderline results. This has happened most often with recommendations of expensive chemotherapies with very limited therapeutic benefit. Here the health economic impact is often used to make a judgement, but this uses arbitrary levels that are not necessarily matched to the values or desires of individuals.

These situations are very different from the issues surrounding HPV vaccination where there is unanimous agreement that the vaccines are of great benefit to many different populations at large. While, at the time of completion of the phase 3 trials, it was possible to argue that even the large scale trials could not exclude the possibility of a very rare, but serious side effect in a very small number (< 1/100,000) of vaccines. The accumulated data from the 12 years of follow-up and the large numbers vaccinated make this no longer scientifically tenable. The benefit of HPV vaccines should be promoted and openly defended.

Global medicine, all biomedical science, and public health have to face the ethical and scientific challenges of ensuring that in its relations with relevant industries and within its own work, the quality and the analysis of the data and the interpretation lead to improved transparency of evidence and how it was obtained, analyzed and judgements made. It is important to progress beyond the current process of conflict of interest disclosures to facilitate and promote full, open discussion of results and their interpretation. If we fail to expand our understanding and address the consequences of conflicts of interest and develop a full ethical and scientific code of action we will continue to suffer from encountering very damaging situations in which relevant institutions and individuals are challenged and damaged in their scientific credibility by rumors, fake news or convoluted personal histories. It is important that researchers act ethically and are allowed to act responsibly and authoritatively, whether academic, public health, or industry based, in collaboration with regulatory agencies and recommending bodies to ensure that the highest standards of clinical research are followed and that no effort or contribution is bureaucratically dismissed.

We are living in exciting times where the initial hypothesis of a sexually transmitted agent causing cervical cancer has evolved to a massive international campaign to eliminate several important cancers in both genders. The initiative will require multidisciplinary coordinated efforts across very different cultural and economic environments and face significant challenges.

ACKNOWLEDGEMENTS

The work was partially supported by

1. The European Union Seventh Framework Programme (grant agreement #603019; CoheaHr);
2. Grants from the Instituto de Salud Carlos III-ISCIII (Spanish Government) cofunded by FEDER funds/ European Regional Development Fund (ERDF)—a way to build Europe (RD12/0036/0056, CIBERESP and CIBERONC) and
3. The Government of Catalonia via the Agència de Gestió d'Ajuts Universitaris i de Recerca (Agency for Management of University and Research Grant 2014SGR1077 and 2014SGR2016)

The authors acknowledge the support of the Cancer Epidemiology and Research Program (PREC) at ICO and in particular of Dr. Marisa Mena in providing comments and literature reviews during the process of the preparation of the manuscript.

COMPETING INTERESTS

F.X.B., received research funding via their institution from GlaxoSmithKline, Merck, Qiagen, Roche, and Sanofi Pasteur MSD. Personal grants for travel expenses, meetings and/or giving conferences from GlaxoSmithKline, Merck, Qiagen, Hologic, and Sanofi Pasteur MSD.

D.J. has received research funding from GlaxoSmithKline and Merck and was Director of Clinical Research on the Cervarix (HPV16/18) research program. He is a consultant to DDL Diagnostic Laboratories, Rijswijk, The Netherlands.

REFERENCES

[1] World Health Organization (WHO). Cervical cancer: an NCD we can overcome. Geneva: WHO; 2018. Available from: https://www.who.int/reproductivehealth/DG_Call-to-Action.pdf. accessed date: 4 June 2019 https://www.who.int/cancer/cervical-cancer/cervical-cancer-elimination-strategy.

[2] Simms KT, Steinberg J, Caruana M, Smith MA, Lew JB, Soerjomataram I, et al. Impact of scaled up human papillomavirus vaccination and cervical screening and the potential for global elimination of cervical cancer in 181 countries, 2020–99: a modelling study. Lancet Oncol 2019;20(3):394–407.

[3] GBD 2017 Causes of Death Collaborators. Global, regional, and national age-sex-specific mortality for 282 causes of death in 195 countries and territories, 1980-2017: a systematic analysis for the Global Burden of Disease Study 2017. Lancet 2018;392(10159):1736–88.

[4] Ferlay J, Colombet M, Soerjomataram I, Mathers C, Parkin DM, Pineros M, et al. Estimating the global cancer incidence and mortality in 2018: GLOBOCAN sources and methods. Int J Cancer 2019;144 (8):1941–53.

[5] Bray F, Colombet M, Mery L, Piñeros M, Znaor A, Zanetti R, et al., editors. Cancer incidence in five continents, vol. XI (electronic version). Lyon: International Agency for Research on Cancer; 2017. Available from: http://ci5.iarc.fr, accessed date: 4 June 2019.

[6] Bruni L, Albero G, Serrano B, Mena M, Gómez D, Muñoz J, et al. ICO/IARC Information Centre on HPV and Cancer (HPV Information Centre). Human papillomavirus and related diseases in the world. Summary Report 22 January 2019. Available from: http://www.hpvcentre.net [accessed 04.06.2019].

[7] Woo LY. Mega The feasibility and acceptability of self-sampling and HPV testing using Cepheid Xpert® HPV in a busy primary care facility. J Virus Erad 2019;5(Suppl 1):10–11.

[8] Bonde J, Ejegod DM, Cuschieri K, Dillner J, Heideman DAM, Quint W, et al. The Valgent4 protocol: Robust analytical and clinical validation of 11 HPV assays with genotyping on cervical samples collected in SurePath medium. J Clin Virol 2018;108:64–71.

[9] Arbyn M, Xu L. Efficacy and safety of prophylactic HPV vaccines. A Cochrane review of randomized trials. Expert Rev Vaccines 2018;17(12):1085–91.

[10] IARC Working Group on the Evaluation of Carcinogenic Risks to Humans. IARC Monographs on the evaluation of Carcinogenic risks to humans. Hum Papillomaviruses 2007;90:1–636 Monograph.

[11] Polman NJ, de Haan Y, Veldhuijzen NJ, Heideman DAM, de Vet HCW, Meijer CJLM, et al. Experience with HPV self-sampling and clinician-based sampling in women attending routine cervical screening in the Netherlands. Prev Med 2019;125:5–11.

[12] Gustavsson I, Aarnio R, Berggrund M, Hedlund-Lindberg J, Strand AS, Sanner K, et al. Randomised study shows that repeated self sampling and HPV test has more than two fold higher detection rate of women with CIN2 + histology than Pap smear cytology. Br J Cancer 2018;118(6):896–904.

[13] Gultekin M, Zayifoglu Karaca M, Kucukyildiz I, Dundar S, Boztas G, et al. Initial results of population based cervical cancer screening program using HPV testing in one million Turkish women. Int J Cancer 2018;142(9):1952–8.

[14] Hu W, Bell D, Antani S, Xue Z, Yu K, Horning MP, et al. An observational study of deep learning and automated evaluation of cervical images for cancer screening. J Natl Cancer Inst 2019;. Available from: https://doi.org/10.1093/jnci/djy225.

[15] Bruni L, Diaz M, Castellsagué X, Ferrer E, Bosch FX, de Sanjosé S. Cervical human papillomavirus prevalence in 5 continents: meta-analysis of 1 million women with normal cytological findings. J Infect Dis 2010;202(12):1789–99.

[16] Stern PL, van der Burg SH, Hampson IN, Broker TR, Fiander A, Lacey CJ, et al. Therapy of human papillomavirus-related disease. Vaccine 2012;30(S%):F71–82.

[17] Ghelardi A, Parazzini F, Martella F, Pieralli A, Bay P, Tonetti A, et al. SPERANZA project: HPV vaccination after treatment for CIN2. Gynecol Oncol 2018;151(2):229–34.

[18] Ivancic R, Iqbal H, deSilva B, Pan Q, Matrka L. Current and future management of recurrent respiratory papillomatosis. Laryngoscope Investig Otolaryngol 2018;3(1):22–34.

[19] Luostarinen T, Apter D, Dillner J, Eriksson T, Harjula K, Natunen K, et al. Vaccination protects against invasive HPV-associated cancers. Int J Cancer 2018;142(10):2186–7.

[20] Drolet M, Bénard É, Pérez N, Brisson M, HPV Vaccination Impact Study Group. Population-level impact and herd effects following the introduction of human papillomavirus vaccination programmes: updated systematic review and meta-analysis. Lancet 2019; pii: S0140-6736(19)30298-3.

[21] Basu P, Muwonge R, Bhatla N, Nene BM, Joshi S, Esmy PO, et al. Two-dose recommendation for Human Papillomavirus vaccine can be extended up to 18 years - updated evidence from Indian follow-up cohort study. Papillomavirus Res 2019;7:75–81.

[22] Kreimer AR, Herrero R, Sampson JN, Porras C, Lowy DR, Schiller JT, et al. Evidence for single-dose protection by the bivalent HPV vaccine-review of the Costa Rica HPV vaccine trial and future research studies. Vaccine 2018;36(32 Pt A):4774–82.

[23] Roden RBS, Stern PL. Opportunities and challenges for human papillomavirus vaccination in cancer. Nat Rev Cancer 2018;18(4):240–54.

[24] Bosch FX, Robles C, Díaz M, Arbyn M, Baussano I, Clavel C, et al. HPV-FASTER: broadening the scope for prevention of HPV-related cancer. Nat Rev Clin Oncol 2016;13(2):119–32.

[25] World Health Organization. Ten threats to global health in 2019, 2019. Available at https://www.who.int/emergencies/ten-threats-to-global-health-in-2019 [accessed 04.06.2019].

[26] The Vaccine confidence project. 2016. Available from: http://www.vaccineconfidence.org/research/the-state-of-vaccine-confidence-2016 [accessed 04.06.2019].

Index

Note: Page numbers followed by "*f*" and "*t*" refer to figures and tables, respectively.

D

I

J

K

L

P

Q

R

S

T

U

V

Printed in the United States
By Bookmasters